Advances in 21st Century Human Settlements

Indexed by SCOPUS

This Series focuses on the entire spectrum of human settlements—from rural to urban, in different regions of the world, with questions such as: What factors cause and guide the process of change in human settlements from rural to urban in character, from hamlets and villages to towns, cities and megacities? Is this process different across time and space, how and why? Is there a future for rural life? Is it possible or not to have industrial development in rural settlements, and how? Why does 'urban shrinkage' occur? Are the rural areas urbanizing or is that urban areas are undergoing 'ruralisation' (in form of underserviced slums)? What are the challenges faced by 'mega urban regions', and how they can be/are being addressed? What drives economic dynamism in human settlements? Is the urban-based economic growth paradigm the only answer to the quest for sustainable development, or is there an urgent need to balance between economic growth on one hand and ecosystem restoration and conservation on the other—for the future sustainability of human habitats? How and what new technology is helping to achieve sustainable development in human settlements? What sort of changes in the current planning, management and governance of human settlements are needed to face the changing environment including the climate and increasing disaster risks? What is the uniqueness of the new 'socio-cultural spaces' that emerge in human settlements, and how they change over time? As rural settlements become urban, are the new 'urban spaces' resulting in the loss of rural life and 'socio-cultural spaces'? What is leading the preservation of rural 'socio-cultural spaces' within the urbanizing world, and how? What is the emerging nature of the rural-urban interface, and what factors influence it? What are the emerging perspectives that help understand the human-environment-culture complex through the study of human settlements and the related ecosystems, and how do they transform our understanding of cultural landscapes and 'waterscapes' in the 21st Century? What else is and/or likely to be new vis-à-vis human settlements—now and in the future? The Series, therefore, welcomes contributions with fresh cognitive perspectives to understand the new and emerging realities of the 21st Century human settlements. Such perspectives will include a multidisciplinary analysis, constituting of the demographic, spatio-economic, environmental, technological, and planning, management and governance lenses.

If you are interested in submitting a proposal for this series, please contact the Series Editor, or the Publishing Editor:

Bharat Dahiya (bharatdahiya@gmail.com) or
Loyola DSilva (loyola.dsilva@springer.com)

More information about this series at http://www.springer.com/series/13196

T. M. Vinod Kumar
Editor

Smart Environment for Smart Cities

 Springer

Editor
T. M. Vinod Kumar
School of Planning and Architecture
New Delhi (SPA-D)
New Delhi, India

ISSN 2198-2546   ISSN 2198-2554 (electronic)
Advances in 21st Century Human Settlements
ISBN 978-981-13-6821-9   ISBN 978-981-13-6822-6 (eBook)
https://doi.org/10.1007/978-981-13-6822-6

Library of Congress Control Number: 2019932690

This Springer imprint is published by the registered company Springer Nature Singapore Pte Ltd.
The registered company address is: 152 Beach Road, #21-01/04 Gateway East, Singapore 189721, Singapore

Contents

Editor and Contributors

About the Editor

 **Prof. T. M. Vinod Kumar** had 48 years of experience in Urban Planning, as a teacher, researcher, and adviser/consultant and worked in India, China, Bhutan, Nepal, Malaysia, Indonesia and Hawaii USA. He was Dean of Studies, Head of the Department of Urban Planning, Head Centre for Systems Studies and Analysis, Centre for GIS and Remote Sensing, and Centre for Urban Studies of School of Planning and Architecture, New Delhi; Visiting Professor National Institute of Technology, Calicut, and institute of Technology Bandung, Indonesia and Professional Associates, East West Resources Systems Institute Honolulu, Hawaii, Fellow Centre for the Study of Developing Societies, Delhi, Project Manager in Council for Social Development, New Delhi, Regional Program Coordinator at the International Centre for Integrated Mountain Development (ICIMOD) and Planner-Engineer at the Ford Foundation. He is the author of many books and journal articles. He coordinated and edited "Geographic Information System for Smart Cities" (Copal: 2014), "E Governance for Smart Cities" (Springer: 2015) "Smart Economy in Smart Cities" (Springer: 2016), "E-Democracy for Smart Cities" (Springer: 2017) and "Smart Metropolitan Regional Development: Economic and Spatial Design Strategies (Springer-Nature: 2018)", He is the coordinator and editor international book projects "Smart Environment for Smart Cities" to be published by Springer Nature in 2019 and "Smart Living for Smart Cities Vol 1 and 2" to be published by Springer Nature in 2020.

Contributors

Dr. Prabh Bedi is a Geographer, Regional Planner and Geospatial Technology expert. She completed her Post Baccalaureate Certificate in Geographic Information Systems from Pennsylvania State University, USA. A graduate from School of Planning and Architecture, New Delhi (India) with Master of Planning (Regional Planning), she has over 20 years of experience in planning and geospatial technologies, both in industry and academics. She is presently working as Associate Professor with Sushant School of Art and Architecture, Ansal University. She has published articles on NUIS, Smart Cities, urban greens and hydrology in the context of sustainability. Her research interests include information systems, geospatial technologies, NUIS, Big Data, smart cities, demography, hydrology, sustainability and climate change.

Prof. Bhasker Vijaykumar Bhatt received B.E. (Civil Engineering), Master of Town and Regional Planning from NIT-Surat, India with merit medals. He also obtained MBA (Project Management) from Sikkim Manipal University (DLM-2013). He acquired urban planning experience on various assignments across the Gujarat state during 2007–2011. Post field experience on town planner positions, since 2011, he has been serving academic institutions and teaches urban planning. He is serving as an Associate Professor at BCHS-APIED, Vallabh Vidyanagar, Anand for Master of Planning. He guided 26 PG Dissertation research, appointed as Journal Reviewer-Editorial Board Member. He was a mater trainer for faculty members development programs on the theme of Design Thinking in Engineering at the Gujarat Technological university in the duration of the year 2014–2015. He has so far published about 60 research articles in journals and conferences. He has been awarded for best paper in five events of national and international conferences. He authored a chapter in an international collaborative based book publication by Springer-Nature that was published in 2018. He is a member and office bearer to several professional organizations having national importance. He has been awarded with honours for his meritorious and honorary services to the society at several occasions. He is pursuing his doctoral research and exploring the relationship among various climate change parameters and urbanization with an approach of GIS and RS. He has been contributing Open Educational Resources through his web—www.bvbhatt.com.

Prof. (Dr.) Chetan B. Bhatt is a Principal and Professor in Instrumentation and Control Engineering, at Government MCA College, Ahmedabad affiliated to Commissionerate of Technical Education, Government of Gujarat, India. He has received his postgraduate and Ph.D. from Indian Institute of Science, Bangalore, India. He is actively involved in many multidisciplinary research projects for Governments and Industries. His areas of research include cyber physical systems, expert systems, smart sensors for food analysis, industrial data communications, adaptive e-learning, and engineering education. He is a recipient of 'Distinguished Instrumentation Professional' award (2016) by International Society of Automation (ISA) Asia—Pacific Division–14, and 'Excellence in Education' award (2017) by International Society of Automation (ISA).

Prof. Jignesh G. Bhatt received B.E. (Instrumentation and Control Engineering), Gujarat University, India, 1997 and M.Tech. (Electrical Engineering, Specialization: Measurement and Instrumentation) from IIT Roorkee, India, 2010. He acquired industrial experience on Managerial Positions during 1997–99. Post industrial experience on managerial positions during 1997–1999, since 2000, he has been currently serving Dharmsinh Desai University (DDU), India as Sr. Assistant Professor (Instrumentation and Control Engineering Department). He served as Co-ordinator and Principal Investigator for Govt. Sponsored Research and Community Development Projects, PG Thesis Supervisor, Invited Expert-Session Chair in Conferences and Journal Reviewer-Editorial Board Member. His research interests include Automation, e-Governance, e-Learning, Instrumentation, Smart City, Smart Grid, Solar City, Solar Energy and Wireless Sensor Networks.

Dr. P. Bimal is a city planner and an architect. He has graduated B.Arch. from Kerala University and acquired post-graduation in City Planning from Indian Institute of Technology Kharagpur (IIT Kgp). He was awarded Ph.D. for his thesis titled 'An Algorithmic Framework for Surface Modelling of Population Concentration and Distribution' by NIT Calicut. He has been working as assistant professor at NIT Calicut since 2009 and was offering courses for B.Arch, MPlan and Ph.D. students. He had been working with JUSCO, a Tata Steel subsidiary before joining NIT Calicut. His research interest includes Geospatial Techniques (GIS) and its applications in planning, Smart cities, Transportation, Simulation and Modelling, Parametric modelling, Internet of Things, building automation and architectural visualisation.

Leon Cheng is a Research Associate at the Institute for Sustainable Urbanisation, and Assistant Planner at UDP International. He holds a Master's Degree in Urban and Regional Planning from the University of Sydney and a B. Sc. Major in Statistics and Minor in Economics from the University of British Columbia. He is currently working a planning project in Hong Kong, urban design and architectural guidelines project for Amaravati, AP, India and large master planning projects in the Philippines. His research experience includes affordable housing in Sydney, as well as public open space in Hong Kong. Prior to UDP International he was involved with Walk DVRC in Hong Kong.

Hillary Chung is a Research Assistant at the Institute for Sustainable Urbanisation, and Planning Assistant at UDP International. She holds a Master's Degree in International Planning from the University College London, United Kingdom and another Master's Degree in Urban Studies and Planning from University of Sheffield, United Kingdom. She is working on a place-making project in Hong Kong, urban design and architectural guidelines project for Amaravati, AP, India and large master planning projects in the Philippines. Her research experience includes open space in Sheffield, urban regeneration and Smart City in Hong Kong.

Prof. (Dr.) V. Devadas has been working as Professor in the Department of Architecture and Planning, Indian Institute of Technology-Roorkee, India. He had developed specialization in Urban Dynamics, Rural Dynamics, Regional Dynamics, Urban and Rural Development planning, Infrastructure Development Planning, Sustainable Development, Integrated Development, Smart City Planning, Energy Management, Financial Management, Resource Management, Synchronized Development Planning, Modelling Urban Physics, Climate Change and Glacier Melting, etc. He qualified Doctoral Degree in Development Planning through the IIT-Kharagpur, India; M.Phil Degree in Micro Level Planning with Distinction and Postgraduate Degree in Rural Development with university First rank (Gold Medal) through the Gandhigram Rural Institute; Postgraduate Degree in Economics through the Annamalai University. He had won the Khosla Research Prize, in 1998. He had been working in various capacities, which include Project Manager; Project Officer; Research Associate; Chief, State Planning Board, Government of Kerala; etc. He has more than thirty five years of research experience; guided 18 Ph. D. Degree theses and 60 Postgraduate Degree theses successfully. He co-authored a book published through Lambert Academic Publishers, Germany; and authored more than hundred research papers, which are published in International/national Journals. He chaired and delivered keynote addresses in few International/national conferences.

Dr. Sujata S. Govada is an award winning qualified urban designer, certified town planner and a registered architect in India, with over 35 years of diverse international experience on design and planning projects. She is the Founding Director of the Institute of Sustainable Urbanisation and the Founding and Managing Director of UDP International, a boutique global practice. She has extensive experience working on a diverse range award winning planning and design projects in Hong Kong, China, Philippines, US and India. She is currently leading planning and design projects in Hong Kong, urban design and architectural guidelines project for Amaravati, AP, India and large master planning projects in the Philippines. Her expertise and research interests include sustainable urbanization, smart city development, transit and people oriented development, harbour front planning, new town development, urban renewal, affordable housing and community engagement. She was involved as lead researcher in the Hong Kong sections of the collaborative international research publications "Smart Economy in Smart Cities", "E-Democracy for Smart Cities" and "Smart Metropolitan Regional Development" published by Springer. She is an Adjunct Associate Professor at the School of Architecture, Chinese University of Hong Kong, as well as a Global Trustee of the Urban Land Institute, Founding Vice President of the Hong Kong Institute of Urban Design and Past President of the American Institute of Architects Hong Kong Chapter.

Mrs. Kshama Gupta is currently working as Scientist in Indian Institute of Remote Sensing (Indian Space Research Organization), Dehradun. She had completed her M.Tech. in Urban Planning from School of Planning and Architecture, New Delhi, India after the completion of Bachelors in Architecture from Malviya National Institute of Technology, Jaipur. Since then she is working as researcher in the field of remote sensing and GIS applications for urban management and made contributions in many national level projects and research areas. She has more than 60 publications to her name as research papers in international/national journals and conferences and ISRO Technical reports. Her research interest includes smart planning, urban climate and micro climate, urban green spaces and 3D modeling of urban areas.

Dr. P. S. Harikumar is Senior Principal Scientist and Head of Water Quality Division in Centre for Water Resources Development and Management (CWRDM) Calicut. He received the Ph.D. from Cochin University of Science and Technology (CUSAT) and did the post-doctoral studies in KTH Royal Institute of Technology, Stockholm, Sweden. He worked in Indian Petrochemicals Corporation, Vadodara before joining CWRDM. His field of specialization include: Environmental Management, Water Quality Modelling, Water Treatment, Nanotechnology, Wetland Studies, and Urban Water Management. He has 45 journal papers to his credit and published 5 books. Dr. Harikumar submitted 50 research reports to various International/National/State Level agencies. He is a member of State Environmental Expert Appraisal Committee and Food Safety Standards Authority of India.

Dr. Omkar Jani received his Bachelor's degree in Electrical Engineering with Honours from the University of South Carolina, Ph.D. in Electrical Engineering from Georgia Institute of Technology with specialization in Solar Photovoltaic Science and Engineering, and completed his post-doctoral fellowship from the Institute of Energy Conversion, University of Delaware.

Omkar is currently the Director, Research & Culture at Kanoda Energy Systems Pvt. Ltd., a renewable and smart energy solution provider company. Kanoda is among the most quality and performance-conscious solar companies of India, and is driven by a technology-driven approach. Kanoda also recently received an award as one of the best solar companies to work with in India.

Dixit Joshi is presently working in Town Planning and Valuation Department, Government of Gujarat as a Junior Town Planner since August 2018. He worked on various Smart City projects of Karnataka state under Smart Cities Mission India from March 2017 to August 2018. He also worked with Feedback Infra Pvt. Ltd. And DDF Consultants Pvt. Ltd. As Urban Planner and GIS expert. He has done his B.E. Civil from L.D. College of Engineering, Ahmedabad in 2013, M.E. In Town and Country Planning from Sarvajanik College of Engineering and Technology, Surat in 2015 and PG Diploma with specialization in Urban and Regional Studies from Indian Institute of Remote Sensing (IIRS), ISRO, Dehradun in 2016. His area of expertise is in Urban and Regional studies using GIS applications.

Pramod Kumar is Group Head, Urban and Regional Studies Department, IIRS, Dehradun, India. He is an alumnus of IIT, Kharagpur, India and joined Indian Space Research Organisation in 1991. Earlier, he has worked as Assistant Engineer at CES, New Delhi. He has been involved in more than 50 national level/technology demonstration and research projects using geospatial data and techniques to evolve solutions for natural resources management and brought out technical reports and research publications. He has published more than 45 papers in journals and conference proceedings and many technical reports. He is the recipient of ISRO Team Excellence Awards for two projects. Presently, he has research interests in urban water utilities, urban hydrology and regional planning.

Prof. (Dr.) Shashikant Kumar accomplished Ph.D. (Geography, MSU, 2014), M.A. (Geography, MSU, 1992) and PG Diploma in Planning (CEPT, 1996). He has over 21 years of consulting and research experience in Urban and Regional Planning and Geoinformatics (GIS/RS). He is Professor and Principal, Master of Planning at BCHM (APIED), Vallabh Vidyanagar, Anand. He is also Director of Green Eminent, Vadodara, a firm providing consulting services and research in URP and Geoinformatics. He has experience in the field data coordination, evaluation and assessment of about 32 development and technology projects. He has 22 articles published related to planning and GIS technologies. He performed researches for displacement and rehabilitation, evaluation of watershed projects, development plans. He authored 5 chapters in international books. He has developed training literature for the short-term courses in GIS. He has presented research papers at national and international conferences regarding the urban and regional issues, climate change and GIS.

Prof. Dr. Mahavir has been engaged in teaching and practice of Urban and Regional Planning over last 35 years. He holds degrees in Architecture, and Urban and amp; Regional Planning. He also holds a Ph.D., jointly from the University of Utrecht and ITC, The Netherlands. For over 15 years, he has been holding the position of Professor at School of Planning and Architecture: New Delhi (An Institution of National Importance under an Act of Parliament), where he is currently the Dean (Academics). Prior to this, he has headed the Departments of Environmental Planning, Physical Planning and Regional Planning; and Centre for Remote Sensing and GIS. His areas of interest include application of Geo-informatics in Planning, Regional Planning, Planning Techniques, etc. He has several professional projects, research and publications to his credit. He is on several committees of the Government, notably related to the NCR Planning Board and the GIS based Master Planning for AMRUT Cities.

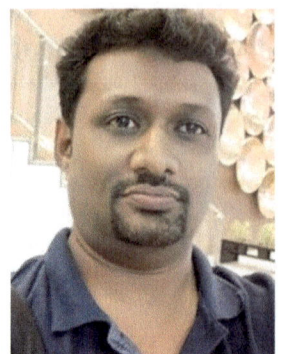

Dr. C. Mohammed Firoz is an architect and urban planner by profession. He holds a Ph.D. degree from Indian Institute of Technology Kharagpur, Post-graduation in Urban and Regional Panning from CEPT University, India and B.Arch. degree from NIT Calicut, India for which he was as a university rank holder. He has been involved in teaching, researching and consulting at NIT Calicut since July 2004. He was also engaged as a visiting teacher at the Architectural Association London (AA London) for the term May–June 2015. His field of interest includes Rural Urban interface studies, Sustainable urbanism, Regional development and Planning etc.

Dr. Silas M. Muketha is a lecturer at the Department of Urban and Regional Planning (DURP), University of Nairobi (UoN) and Director at Alliance Land Surveys Limited. He is a holder of Ph.D. and Master's Degrees in Urban and Regional Planning (UoN) and a BSc (Hons) In Surveying and Photogrammetry (UoN). Dr. Muketha has over 30 years of experience in land surveying, and planning. He worked in the water sector from 1988 to 2014. He served as the chief land surveyor and team leader strategic planning unit at the National Water Conservation and Pipeline Corporation (NWCPC). He also worked as the vice-chairman of the Research Committee at the Corporation. He has been involved in several land surveying, environmental management and physical planning projects in Kenya, Rwanda and Southern Sudan.

His research experience and interests are on land surveying and mapping, conservation of riparian zones, sustainable land use and environmental management. Dr. Muketha is a licensed and registered land surveyor, a registered physical planner and a registered Lead Expert with NEMA. He is a Fellow member-Institution of Surveyors of Kenya, a corporate member of Kenya Institute of Planners and Architectural Association of Kenya. He is also a corporate member of the Environment Institute of Kenya.

Mr. Dennis Mwaniki is the Associate Director for Urban Planning & GIS in GORA Corp. He has more than seven year's hands-on experience in the fields of urban planning, Geographic Information Systems and environmental management. Since the inception of GORA Corp in July 2014, Mr. Mwaniki has been responsible for providing technical assistance to partner governments and local authorities to contextualize and set up Observatories linking Research to Action (ORAs), which are the institutional homes for development of urban indicators; and for training on and administering the UrbanInfo open data platform. He is a member of the Global Human Settlements Working Group of the Group on Earth Observations (GHS WG) and is actively involved in the ongoing development of the Open Cities Institute.

Prior to joining GORA Corp, Mr. Mwaniki worked for 3.5 years as a GIS and research consultant in the Global Urban Observatory (GUO) and the Best Practices Section of UN-Habitat. In GUO, he developed and implemented the methodology for calculating the amount of land allocated to streets for over 100 cities, administrated the UrbanInfo open data software, and contributed to the State of the World's Cities report: Prosperity of Cities in 2012–13. He also worked as GIS and urban planning consultant with the Slum Dwellers International—Kenya; the Center for Urban Research and Innovations at the University of Nairobi; and various private companies based in Nairobi. Mr. Mwaniki's research interests are on the integration of GISs and other ICTs in urban planning, smart city growth, housing and urban environmental management. He holds a Master of Arts degree in Environmental Planning and Management and a Bachelor of Arts degree in Urban and Regional Planning.

Wilfred O. Omollo hold a B.A. degree in Geography (GIS) from Moi University, M.A. (Planning) from the University of Nairobi, and a Ph.D. (Planning) degree from Jaramogi Oginga Odinga University of Science and Technology. He is registered to practice as a Physical Planner by the Physical Planners Registration Board of Kenya, in addition to being a corporate member of the Architectural Association of Kenya (Planning Chapter) (AAK-TP), Town and County Planning Association of Kenya and the National Quality Institute (NQI) of Kenya. He is currently employed at Kisii University as Planning Officer where he is oversees infrastructure planning and development. He has previously worked for the Government of Kenya as a District Planning Officer in Embu, Isiolo, Meru Central, Meru South, Maara and Tharaka Districts. During this period, he was the Chief Government advisor on Urban/Regional Planning within these Districts and prepared several long-term physical development plans for towns such as Embu, Isiolo, Chuka, Chogoria, and Meru. To date, Wilfred has gained over twelve years' experience in professional planning practice, especially in the preparation of urban and regional physical development plans and Environmental Impact Assessment (EIA). He has moreover worked as a part-time lecturer in urban land management for two years (2004–2006) at the University of Nairobi. As a registered and practicing physical planner, he has widely conducted consultancies in diverse fields such as settlement planning, urban transport planning, environmental management and participatory urban planning. He specifically consulted for Practical Action on integrated ward development planning; Ministry of Local Government on the preparation of Nairobi Metropolitan Master Plan; and the National Housing Corporation of Kenya on Non-motorized transport, design for design for storm water drainage (SWD) and road safety initiatives (RSI). His research interests are in land use development planning and control; and applied remote sensing and GIS.

Romanus O. Opiyo is a lecturer at the Department of Urban and Regional Planning (DURP) and coordinator of research with Centre for Urban Research and Innovations (CURI), University of Nairobi (UoN), Kenya. He is a holder of Ph.D. and Masters in Urban and Regional Planning (UoN) and B.A. degree in Social Sciences from Catholic University of Eastern Africa (CUEA). Romanus has Thirteen (13) years University teaching and research supervision experience and more than fifteen (15) years in research work, report writing and proof reading. His research experience and interests are on urban security, transportation, Smart land use planning initiatives, climate change, institutional governance and livelihoods. He has published widely, presented in local (Kenya) and international conferences and has contributed to urban and transport related policy development in Kenya. He has previously contributed in the publication of three editions of Smart cities namely; Smart Economy in Smart Cities, E-Democracy for Smart Cities and recently Smart Metropolitan Regional Development: Economic and Spatial Design Strategies.

Dr. Opiyo is a member of the Environmental Institute of Kenya (EIK). He is also a member of the Kenya Civil Society Network for NMT and Road Safety (CIVNET) and a Member of Kenya Transport Research Network (KTRN).

Mr. Kamal Pandey is working as a scientist at Indian Institute of Remote Sensing, ISRO Dehradun. He is having more than 15 years of working experience in the field of remote sensing and GIS for natural resource management. His area of specialisation includes software application development for RS&GIS using open source and COTS solutions. He is currently involved in the training and capacity building programs of Indian Institute of Remote Sensing and delivers lectures and conduct practical classes on programming concepts, crowdsourcing, GNSS and Geoportals. His current research interests include automation of geospatial workflow using service-oriented architecture. He has guided numerous students in their research work for computer programming in geospatial domain.

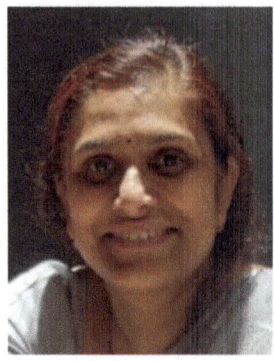

Kshama Puntambekar is currently working as Assistant Professor and Coordinator, Master of Urban & Regional Planning Programme in Department of Planning at school of Planning and Architecture (SPA), Bhopal. She is also centre head for Centre for Geoinformatics at SPA Bhopal. She holds her Bachelor's Degree in Architecture and Master's degree in Urban Planning. Her Ph.D. work has research contributions in the field of 'Land use and Travel Behavior' understanding how they are interrelated to govern the form and flow of the city using GIS for the spatial analysis.

She has contributed in Local Biodiversity Strategies and Action Plan as a Global Studio conducted simultaneously in 8 different countries with UNU-IAS and presented in COP 11 and CBD in 2012. Dr. Puntambekar has contributed to several projects of national and international collaborative research in the institution. She has contributed as co-author in Low Carbon Society, Bhopal-2035.

Sami Rehman is presently working in University of Petroleum and Energy Studies Dehradun as an Assistant Professor in the Department of Electrical and Electronics Engineering since 2015. He has done his B.Tech from Rajasthan University in 2008 and M.Tech from University of Petroleum and Energy Studies, Dehradun. Presently he is pursuing his Ph. D in Energy Domain from University of Petroleum and Energy Studies, Dehradun. He is also an Energy Consultant who has undertaken government projects for Energy Auditing (World Bank Project) and Renewable Energy Feasibility Studies in state of Uttarakhand. His area of expertise is in Solar Thermal Systems, Energy Auditing and Project Management.

Timothy Rodgers is a Research Associate at the Institute for Sustainable Urbanisation, and Project Coordinator and Research Associate at UDP International. He holds a Master's Degree in Transport Policy and Planning from the University of Hong Kong, and a Bachelor's Degree in Economics and Geography (Urban Systems) from McGill University, Canada. His research experience at UDP International includes current issues such as Smart Cities and Affordable Housing, assisting in the development of a Smart City Framework with a focus on People, Place and Planet, as well as conducting research on public and private project proposals and reports. He was the team lead for the Hong Kong chapter of the collaborative international research publications "Smart Economy in Smart Cities" and "Smart Metropolitan Regional Development", and contributed to the Hong Kong chapter of the collaborative international research publication "E-Democracy for Smart Cities".

Dr. Arijit Roy is a Scientist in Indian Institute of Remote Sensing (IIRS), Indian Space Research Organization, Dehradun. He is presently Head, Disaster Management Studies Department at IIRS. Dr. Roy had completed his Ph.D. in Botany with specialization in Restoration Ecology from Banaras Hindu University, Varanasi, India. His research interest is in geospatial modelling of the impact on the terrestrial ecosystems mainly the structure and functioning (biodiversity, nutrient dynamics) because of the climate forcing, anthropogenic influences and natural disasters. He is also interested in exploring the use of temporal hyperspectral satellite data to assess the spatial variations in the ecosystem level processes and flows especially of nitrogen. He has guided more than 15 M.Tech./M.Sc. Dissertations, 1 Ph.D. and 5 Ph.D. students are presently under his supervision. Dr. Roy has more than 80 publications, which include 33 peer reviewed journal publication, 19 peer reviewed international conference papers and has authored one book, numerous book chapters and technical reports.

Dr. Shovan K. Saha obtained his B. Arch degree and Diploma in Town and Country Planning from School of Planning and Architecture, New Delhi and later earned a Dr. Engg. from Kyoto University, Japan. His career represents a mixture of professional work, research dominated by academic pursuits. While based in SPA New Delhi Professor Saha actively participated in research projects and international courses in Urban and Regional Planning, Metropolitan Housing and Urban Conservation during his tenure at the UN the Centre for Regional Development, Nagoya, Japan. At SPA New Delhi he taught Urban Planning, Environmental Planning and successfully guided several doctoral studies. Having completed his tenures as founder Director, SPA Vijayawada and founder Dean, School of Architecture and Planning, Sharda University, Greater Noida, Professor Saha is currently Professor Emeritus, Sharda University.

Mr. Ummer Sahib is a *City Planner and Technologist* with 30 years experience in geospatial technology and development of innovative solutions for City Planning and Management. He has been responsible for process automation in executing large GIS, LIS and Urban mapping projects in Asia, Middle East, Europe and Africa. Since 2005, Mr. Sahib is the Executive Director of Informap Technology Center in Sharjah UAE, heading Future Technologies and application development in Location Services.

Starting his career in GIS software development project for Department of Electronics, Govt. of India, Mr. Sahib's expertise lies in GIS, Open source web mapping applications, navigation data production, Geo-analytics and perfecting a virtual addressing system called GRL (Geo Reference Locator) for the UAE. Mr. Sahib is the driving force behind his company's creative innovation on GPS and NFC technology solutions. He is passionate on integrating IoT, Big Data and cloud technologies for establishing Urban Operating System to realize the true meaning of a SMART CITY.

Mr. Sahib holds Post Graduate Diploma (M.Tech) in City Planning (1985) from School of Planning and Architecture, CEPT University, Ahmedabad, India. He is also the Associate member of the Institute of Town Planners, ITPI India, since 1986 and Fellow in the Institute of Directors (IOD).

He has co-authored book on **E-Governance for Smart Cities** published by Springer in January 2015 and **E-Democracy for Smart Cities in 2017**. He was a panel member in the 4th GCC Municipalities & smart cities conference and speaker at the GITEX Smart Cities conference held in Dubai. He has published several papers on GIS/LIS and accident management and is currently leading a team on designing and developing Mobile Apps for Smart Cities. He is also the co-author in the international book project "**Smart Environment for Smart Cities**" to be published by Springer Nature in 2019.

Informap disrupted conventional need for audio tour guide equipment by launching izi.TRAVEL, Worlds No. 1 GPS aided FREE audio guide APP for tourists and open platform for Tourist Departments, Agencies, Museums and Tour guides to publish audio tours FREE.

Contact Details
Ummer Sahib, Executive Director,
Informap Technology Center, LLC
Office HC2, Tiger Tower 1,
Al Tawun St, Al Tawun Area.,
PO Box 38098, Sharjah, United Arab Emirates.

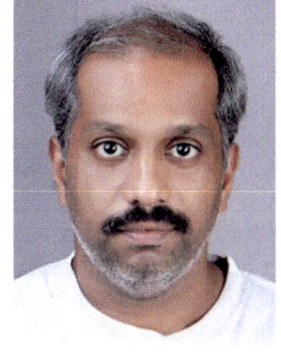

Dr. Praveen Sankaran is an Assistant Professor in the Department of Electronics and Communication Engineering at NIT Calicut. Dr. Sankaran received his Bachelor of Technology in Applied Electronics and Instrumentation Engineering from University of Calicut in 2002, and Master of Science in Electrical Engineering from Old Dominion University, Norfolk, Virginia in 2005. He earned his Ph.D. in Electrical and Computer Engineering from Old Dominion University, Norfolk, Virginia in 2009. His current research focuses on nonlinear feature extraction methods for data classification.

Achintya Kumar Sen Gupta is presently working as Director (Technical) with Institution for Hygiene and Environmental Sanitation as well as independent consultant with Asian Development Bank and World Health Organization. He had done his Bachelor in Civil Engineering (1969) and Masters in Public Health Engineering (1978). He has more than 48 years of experience in handling Water and Sanitation issues, Drinking Water Quality Management, Solid Waste Management including handling of Hazardous Waste, Air and Water Pollution, Climate Change and its impact on health. He had served Government of India and Government of Rajasthan for than 25 years handling water, sanitation and solid and liquid waste management programmes. He had also served various UN Agencies like WHO, The World Bank, UNDP, UNIDO and UNICEF for more than 16 years handling environmental sanitation issues linked with health. He also worked with various NGOs handling community based programmes for more than 5 years. He is presently focusing on Risk Assessment of Wastewater System, Water Safety and Security Plan for Urban and Rural areas, Planning, development and marketing of Mini Water Supply scheme in a selected rural areas, Developing and coordinating School Sanitation Projects, guiding development of the training modules and coordinating training programmes for Water Safety Plan including Water Quality management, Sanitation Safety Plan for sanitation, health and hygiene.

Dr. Mahendra Sethi is an urban environment expert, and his research explores the role of developing cities at the interface of global change and local governance. He is a fellow with Land Use, Infrastructure and Transport group of MCC and TU-Berlin; an early-career researcher of Earth Systems Governance 2014, University of East Anglia, UK; and a recipient of United Nations University —Institute for the Advanced Study of Sustainability (UNU-IAS) Ph.D. Fellowship in Tokyo, Japan. In India, Mahendra is Associate Professor in the Faculty of Architecture at Dr. A. P. J. Abdul Kalam Technical University, Lucknow. In the past, he was associated with the National Institute of Urban Affairs as Editor of the research journal *Urban India*. Mahendra started his career with development consultancy, appraisals and formulation of urban, regional and environmental plans for national/state governments, statutory bodies, private sector and international organizations, including the World Bank and World Wide Fund for Nature. He has published extensively in scientific journals, peer-reviewed books, etc. including *Climate Change and Cities: A Spatial Perspective of Carbon Footprint and Beyond*, a seminal research into economic development, urbanization and GHG patterns of over 200 countries (Routledge–Taylor & Francis) and *Climate Co-benefits in Indian Cities* (Springer Nature), a volume synthesizing findings of researchers from India, Japan, France, UK and USA. Mahendra regularly disseminates research through tutorials, special lectures and trainings on urban environmental, social and policy issues in national and international conferences, workshops, etc. He is a recipient of the Government of India's Ministry of Human Resource Development Scholarship (2005–2007) and was also offered the Liverpool University Scholarship (2005).

Prof. (Dr.) Neerajkumar D. Sharma is currently the Principal at the GIDC Degree Engineering College, Abrama, Navsari (Gujarat-India). He meritoriously accomplished B.E. Civil (2002), M.E. (Environmental Engineering) and Ph.D. at SVNIT, Surat. He has also obtained diploma in Software Engineering and PG Diploma in Environmental and Sustainable Development. He has more than 15+ years of experience academics of UG and PG Civil Engineering programs. He also served for Environmental engineering applications. He is a Ph.D. Supervisor and performs coordination among institutes for the Gujarat Technological University. He has guided 28 PG dissertation and 4 UG civil engineering projects. He is also a life member to professional organizations of Engineering Education, Civil and Environmental Engineering. He has organized a national and an international conference. He has authored 65 research articles in journals and conferences. His area of interest includes environmental science-based applications, construction management and urban-climate studies.

Asfa Siddiqui is currently working as a Scientist at Indian Institute of Remote Sensing (Indian Space Research Organization) since 2014. She did her Bachelors in Architecture from Govt. College of Architecture, Lucknow in 2011 and Masters in Planning with specialization in Urban Planning from School of Planning and Architecture, New Delhi in 2013. She worked at NIT Kozhikode (Calicut) prior to joining ISRO. She is a double Gold medalist and have received government scholarships and awards at college level. She has more than 20 publications in journals, conferences and book chapters to her credit. Her work focuses on urban and regional areas with emphasis on Energy (Solar) and Environment (Urban Heat Island effect and Air Pollution).

Dr. Aki Suwa took a professor position at Kyoto Women's University in 2014, after being a research fellow at the United Nations University Institute of Advanced Studies. Her interests include renewable energy and the local governance to promote the renewables. She is a Japanese national, holding a M.Sc. degree in Environmental Technology and Policy from Imperial College, and Ph.D. from University College London, University of London.

Dr. Neha G. Tripathi is an environmental planner with over 15 years of experience in Environmental Management and Architecture. She is presently working as Assistant Professor in School of Planning and Architecture, New Delhi. She did her post-graduation from School of Planning and Architecture, New Delhi in Master of Planning, specialization in Environmental Planning (2003) and was awarded the Gold medal for overall performance. She is Associate member, Institute of Town Planners, New Delhi, Member, Council of Architecture, New Delhi and Member, Indian Building Council, New Delhi. National Environmental Science Academy. Experience of more than 15 years in the field of environmental planning and architecture and worked on projects like Zonal Development Plan for Eco sensitive Area, Mount Abu, Indian Institute of Management, Indore, etc. She has published articles on field of Ecological foort print, Solar Zoning, Climate change etc. Awarded the first prize for paper presentation, Climate Change and Indian Cities Perspectives presented for World Habitat Day 2011 organized by HUDCO, New Delhi.

Smart Environment for Smart Cities

T. M. Vinod Kumar

Abstract The contours of Smart Environment Governance and Community Management for Smart City are presented in the first chapter. The management of environmental resources which is largely common property call for strengthening smart community capabilities for environment resources management by designing a system and protocol for environment resources management as appropriate and consistent with environmental governance of a smart city. The environment is made itself smart to be self-aware by using IOTs and ICTs and E-Governance tools based on existing environmental legislation and E-Democratic management practices are applied for environment resource management by the smart community to intervene 24 h and 7 days a week. The required directive principles of Environmental Democracy are presented side by side with faith-based approach. This approach requires a well-trained smart community with continuing environmental education and competence to meet environmental challenges which is one of six components of the smart city system. Since humans are one component of Environment and integrated, the smart living that conserves environment is the answer to the smart environment. An integrated global cooperation which considers the environment as shared common resource shall complement this effort. Religious practice is a way of life which in turn leads to a smart environment through smart living. The community shall be guided by their own Faith which is a way of life that considers environment having a distinct interface of deep ecology and religion (whether it is Hinduism, Buddhism, Judaism, Christianity or Islam and so on). This chapter describes an integrated approach to smart environment for smart city, and scientific, and religious faith based, as well as the deep ecology-based interface for environment governance and management. The cultural system underlying the Vedic and Buddhist religion which is in practice in several Asian countries as examples are presented which can further strengthen the Smart Environment for Smart City approach through faith based smart living.

Keywords Smart city and smart environment · Concept · Definition · Management · Protocol · E-Environment Governance and E-Environment

T. M. Vinod Kumar (✉)
School of Planning and Architecture, New Delhi, India
e-mail: tmvinod@gmail.com

Besant Nivas, Jayanthi Road, P.O. Kolathara, Kozhikode 673655, Kerala, India

© Springer Nature Singapore Pte Ltd. 2020
T. M. Vinod Kumar (ed.), *Smart Environment for Smart Cities*, Advances in 21st Century Human Settlements, https://doi.org/10.1007/978-981-13-6822-6_1

Democracy · Ecosystem and Religion for smart community · Vedic, Deep ecology, and smart environment resources management · Vedic and Buddhist cultural system and smart community · Directive principles of Environment Democracy · Practice of E-Environment Governance

1 Smart Environment for Smart Cites Is the Rediscovery of Environment and Cultural System of Cities

This introductory chapter on "Smart Environment for Smart Cities" unveils the design and protocol of ecosystem and cultural system of cities that progress every city to smart city with smart environment. It is abstract since it is not oriented to any specific components of the environment or any city with distinct ecology and unique cultural system. Like Buddhism liberated to common householder, some 2500 plus years ago the well-developed Vedic environmental knowledge base or intellectual property of nature from the monopoly of *Brahmin* upper caste to the benefit of *Kshatriya* the warrior caste, the *Vaisyas* the business class and the *Sudras* the down trodden labour caste and made it universal, the smart city attempt to do the same using ICT and IOT technologies to make it universal, and E-Environment Democracy and E-Environment Governance make it as easy to access environment knowledge base and use it by all as if the entire beneficiary population are specialist environment and urban study academicians. These beneficiaries include children below school going age with definite environment management role, illiterates, and all the remaining in a city irrespective of the caste, class and religion they belong to. The multiple Religion and faith in a secular city democracy, they practice whether by birth or adoption have a great influence in design and management practice of environment resources for the smart environment. The author of this chapter hailing from Asia and has practiced the cultural system given by Hinduism by birth and Buddhism by conviction; confirms the importance of the cultural system for the smart environment. So, the Environmental World view of Vedas and Buddhism practiced for the daily life of a householder is also presented along with smart city model to establish the potential cultural resources of smart communities for conserving, using and developing a smart environment for smart cities. This chapter brings out the design and protocol for the smart environment for smart cities to all and uses it in every second of life by all, the ecosystem-based knowledge base of cities and cultural system. The cultural system discussed in this chapter is what is familiar to the author namely arising out of Hinduism the Vedas and Buddhism of different schools [1–6]. A preliminary exploration, the author undertook but not recorded in this chapter shows that the cultural system practiced in Islam, Judaism and Christianity is also equally valid with unified approach to environment and so also probably other religious practices as smart environment cultural system.

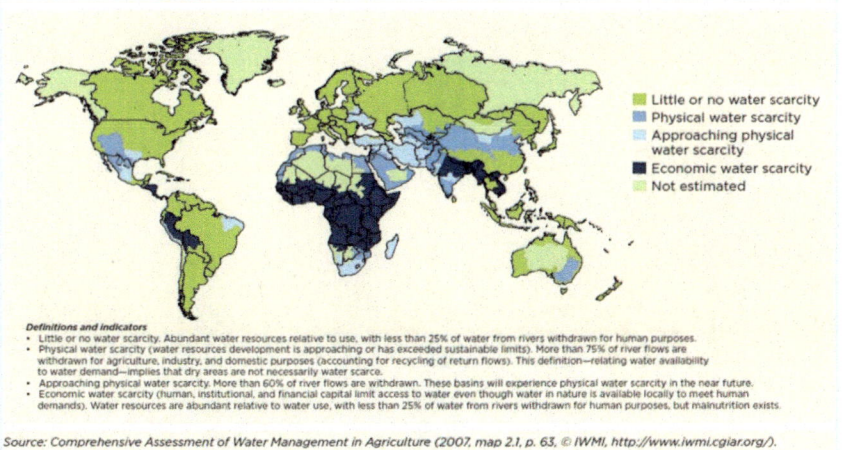

Source: Comprehensive Assessment of Water Management in Agriculture (2007, map 2.1, p. 63, © IWMI, http://www.iwmi.cgiar.org/).

Fig. 1 Water Scarcity Map of the World

1.1 Environment Design, Concept, Management and Protocol

There have been remarkable developments in bringing out environment issues quantitatively with measurements and mapping. The Moors law of computer chip made ICT to aid common people to analyse, model and use environment data 24 h and 7 days a week effortlessly For example, the future global issues as related to Water is mapped in Figs. 1 and 2 based on such research and fully capable of updating.

The following set of maps are the product of empirical research and quantification and discuss in detail how the climate change impact on the universe as a long-term trend. Here how the change in temperature, rain fall will affect the agricultural production and how it will impact on the low-income vulnerable countries and how it affects the rise of sea level and which are the cities that get affected in the long and intermediate range. This prepares all of us to face the environmental future of the world to agree on a global agenda. Micro level interventions are also equally important which can be undertaken by smart communities who feel it and face it daily (Figs. 3, 4, 5, 6 and 7).

Many chapters of this book prove this statement with specific empirical city studies from many countries and cities. These presentations emanating from many sources have generated considerable concerns and attentions in various international organisations that generated a series of international agenda for collective actions. These agenda have got a commitment from member countries of the United Nations as for example UN Sustainable Development Goals 2030.

Also, there has been development in well calibrated, predictive urban environment modelling to describe the future environment issues, the rate at which it emerges and how it can impact the urban dweller's lives. We now know the per capita carbon

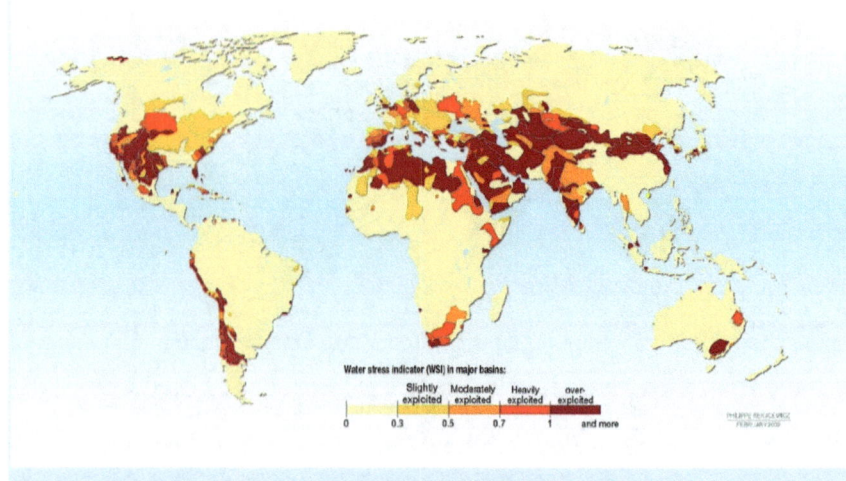

Source: http://www.grida.no/graphicslib/detail/water-scarcity-index_14f3

Fig. 2 Water stressed areas of the World

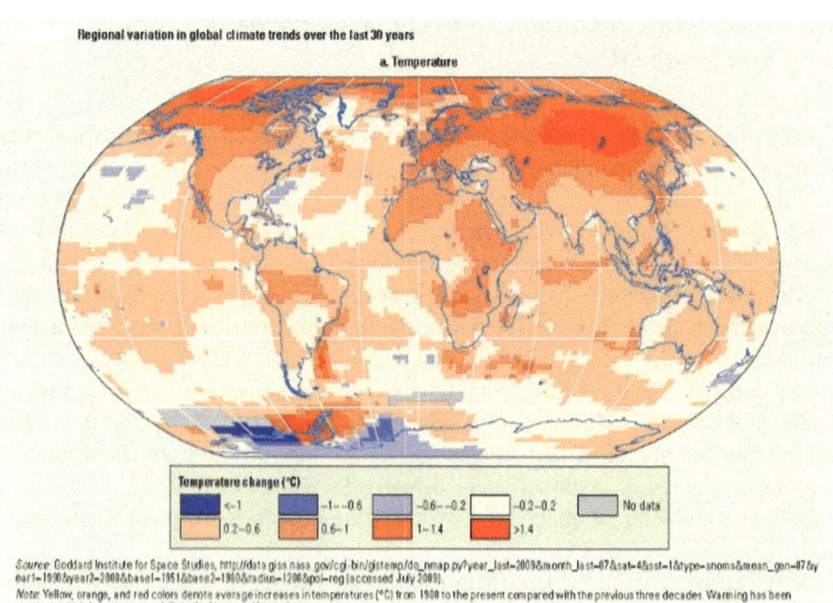

Fig. 3 Global climate trend. *Source* https://siteresources.worldbank.org/INTWDR2010/
Resources/5287678-1226014527953/WDR10-Full-Text.pdf

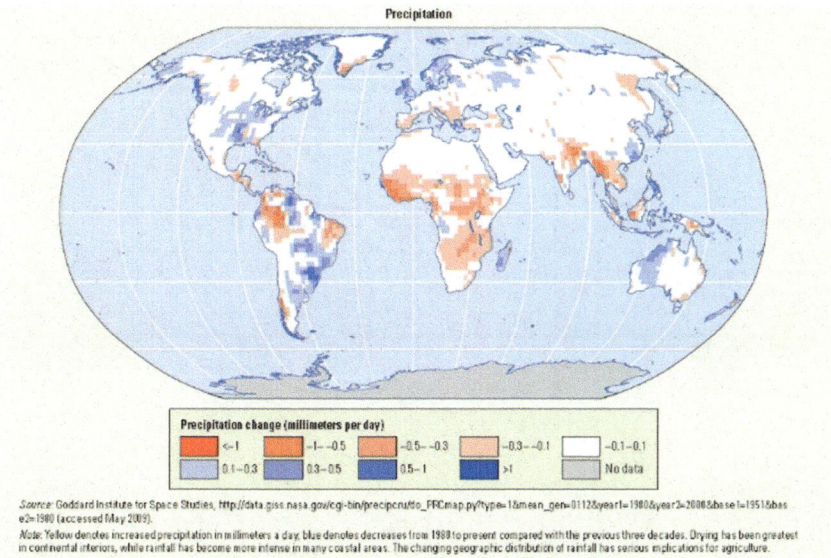

Fig. 4 Change in precipitation. *Source* https://siteresources.worldbank.org/INTWDR2010/Resources/5287678-1226014527953/WDR10-Full-Text.pdf

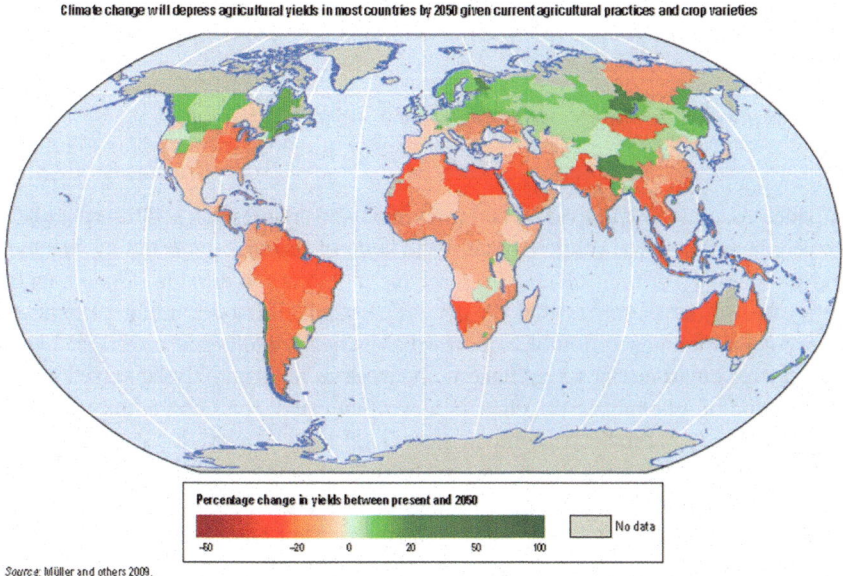

Fig. 5 Climate Change and agricultural yield 2050. *Source* https://siteresources.worldbank.org/INTWDR2010/Resources/5287678-1226014527953/WDR10-Full-Text.pdf

Small and poor countries are financially vulnerable to extreme weather events

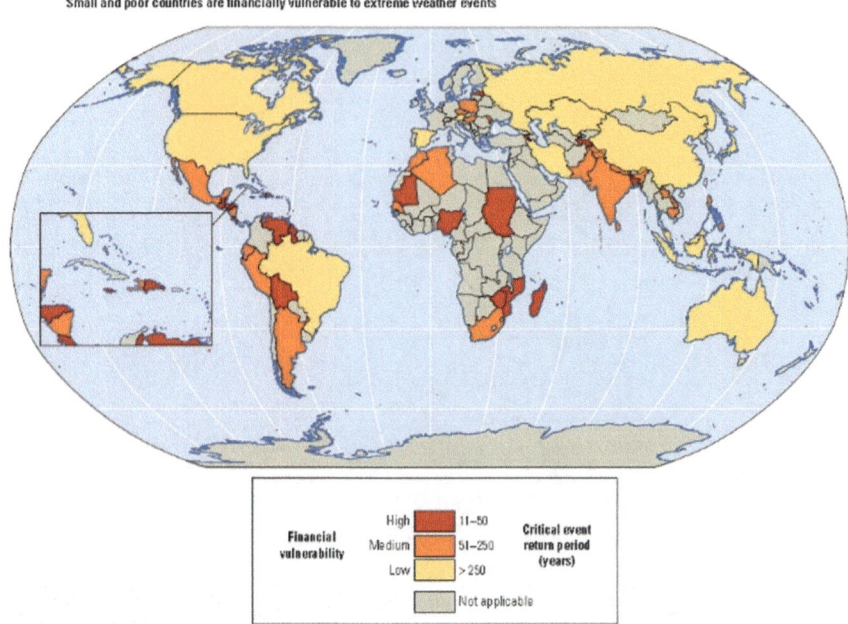

Source: Mechler and others 2009.
Note: The map shows degree to which countries are financially vulnerable to floods and storms. For example, in countries shaded dark red a severe weather event that would exceed the public sector's financial ability to restore damaged infrastructure and continue with development as planned is expected about once every 11 to 50 years (an annual probability of 2–10 percent). The high financial vulnerability of small economies underscores the need for financial contingency planning to increase governments' resilience against future disasters. Only the 74 most disaster-prone countries that experienced direct losses of at least 1 percent of GDP due to floods, storms, and droughts during the past 30 years were included in the analysis.

Fig. 6 Vulnerable counties for climate change. *Source* https://siteresources.worldbank.org/INTWDR2010/Resources/5287678-1226014527953/WDR10-Full-Text.pdf

dioxide production by individual countries and its many presents and future impacts on people and nations. While a pattern and style of resources consuming life style and morphology of their cities that generates more carbon dioxide in one country even affect other countries that do not follow such life style, urban pattern and whose per capita generation of carbon which is negligible because of right living. This proves environment issues have no boundaries and can only be solved by the cooperation of all countries together in macro scale and smart community in micro scale. Now we can predict, how much rise will take place in sea water in future with generated global warming due to carbon dioxide emission and how many countries will disappear submerged in water. This leads to actions.

The two graphs below present the carbon dioxide emission of few selected countries. This shows that urbanisation does not automatically increase carbon dioxide emission, but it is the life style of people with non-renewable fossil fuel-based energy consumer society of the country which determines the load of carbon dioxide (Fig. 8).

Transportation sector of large cities is the major carbon dioxide polluter. We also get to know how our cities shall be reshaped for minimising carbon dioxide emission. Lower the gross urban density higher is the production of carbon dioxide as per Fig. 9.

At risk: Population and megacities concentrate in low-elevation coastal zones threatened by sea level rise and storm surges

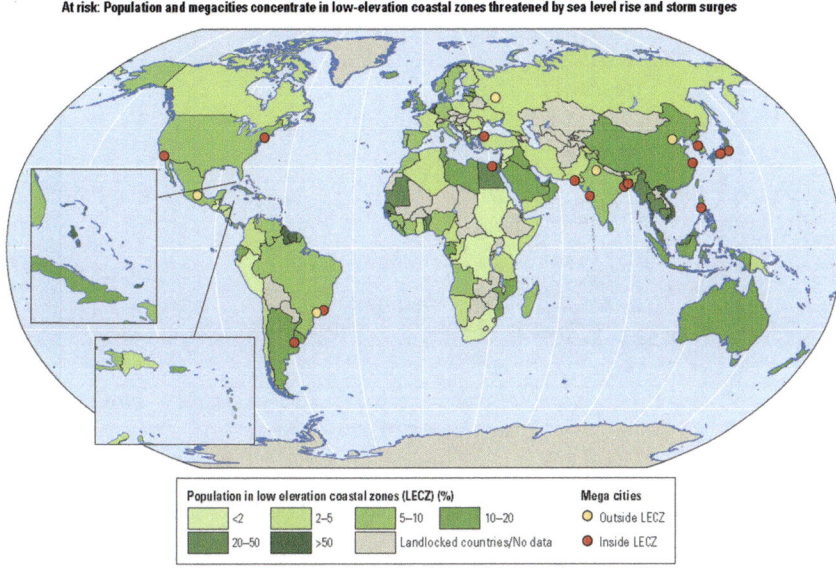

Source: United Nations 2008a.
Note: Megacities in 2007 included Beijing, Bombay, Buenos Aires, Cairo, Calcutta, Dhaka, Istanbul, Karachi, Los Angeles, Manila, Mexico City, Moscow, New Delhi, New York, Osaka, Rio de Janeiro, São Paulo, Seoul, Shanghai, and Tokyo. Megacities are defined as urban areas with more than 10 million inhabitants.

Fig. 7 Risk of submergence of cities by climate change. *Source* https://siteresources.worldbank. org/INTWDR2010/Resources/5287678-1226014527953/WDR10-Full-Text.pdf

Development and CO$_2$ Emissions

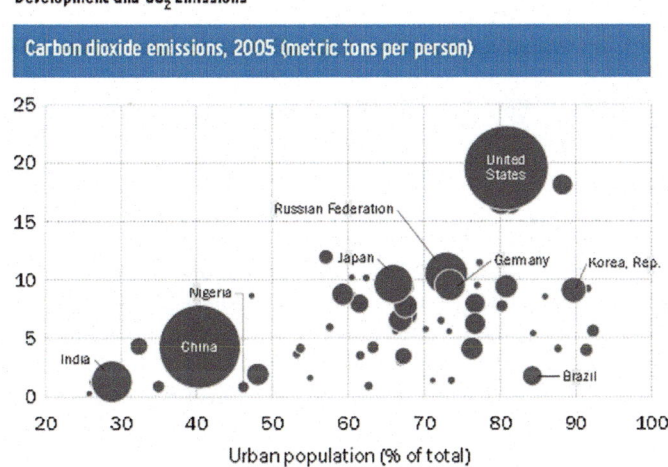

Source: World Bank, 2009a.

Fig. 8 Carbon Dioxide emission of selected countries against the level of urbanisation. *Source* World Bank 2009. World Development Indicators 2009. Washington, DC: World Bank

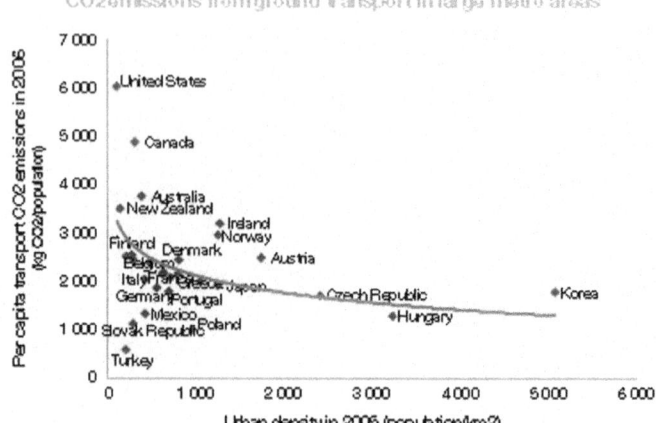

Fig. 9 Carbon Dioxide generation against city density of selected Metropolises. *Source* OECD 2013

Lower urban density generates more commuting distance and hence more carbon dioxide from automobiles. This gives some perspective, how one can re design cities for mitigating Carbon dioxide issues. Accepting urban compaction and higher density which reduces average commuting distance and replacing motorised vehicles that use petroleum vehicles to electric cars, cycle or walking and redesign the cities for environment friendly mobility however large it may be may bring out lesser carbon dioxide emission.

The future of environment has created enough alarms in the International fora to enact global collective actions. The agreed principle that governs is polluter pay for pollution can be circumvented if one country opts out of G20 countries decide to pull out of a global agreement. None can compel any country in this global effort where G19 of G20 agreed. Since all countries have an interconnected environmental future, one country cannot simply ignore and stop acting when the rise in temperature, submergence of cities under sea or disappearance of islands and excess rain fall, fire and floods become a reality due to poor international cooperation for the global environment. The recent floods and fires in 2018 with great loss in USA seems not sufficient to make the USA wake up and be more environmentally responsible for the world.

Many of the collective environmental intervention end up in global investment plans of Governments, and inter Government money transfers from polluters to polluted, from a developed country to developing and specifying global avenues of expenditure to mitigate environment issues. Many a time the financial flow from developed country to developing countries may not take place. Stricter enforcement mechanism works well when a country decides to achieve from their own effort. If some country does not act all other countries who may not be largely responsible for the current state of environment also suffer. Better still be the local nation and local community can perceive this problem before hand and jointly decide how to mitigate these issues by their collective effort.

However, in addition to money flow to tackle the problem, there are distinct roles for everyone in environmental resources management irrespective you are young or old or rich or poor person in sorting out the emerging environment issues. This calls for conceptualising environment in a different way that can generate positive actions to create a desirable urban environment outcome and a protocol how to deal with the emerging environment issues.

1.2 Two Approaches: Scientific and Deep Ecology

If we extend our discussion of Sect. 1 in ecological perspective with all connected elements of living and non-living and abiotic and biotic system of the universe we reach the study of global at the macro level or a city based, or watershed based scientific ecology at the micro level as per the area of study which may focus one aspect of environment. In system theory, we see the city or the world that is more than of its parts, but also itself a part of the larger systems. A family, community, watershed, and metropolitan region be systems where the parts are people and their environment [7].

In fact, there exist two ecologies, the Scientific ecology and Deep ecology, [8–14] both are equally important. The scientific ecology involves the systematic study of the interrelationships between species and their environment, biotic and abiotic relationship giving equal importance to all elements. The focus will be on measurable data and its analysis in an objective manner to find out its interrelationship. The second approach is the Deep Ecology approach and differs from scientific ecology. Here, person experiences as part of the living planet in a household, community, city or region are compelled to find his/her role in protecting the earth and its life. In this approach, the relationship is more of an involved participation which is far from scientific as one who feels connected with as well as part of the world surrounding the self and not detached and objective as we do in scientific ecology. It recognises and involves the role of cultural and religious values and biases in the overall picture of society and the environment. This approach is inclusive and for everybody, not just for scientifically trained ecologists with several university degrees as his continuing academic work, but with each person whether he is a child or adult moved by their own values, experiences and feelings to do their bit for the world or city around them as if 'I am immersed in and involved with the pond of life and must therefore protect

it' which may sound unscientific and even un-*vedantic* where you may be asked
who asked you to do so; but more useful for the cause of ecology. Deep ecology
tends to emphasize the relationships between biotic and abiotic system holistically
and extend it further to bring together personal and social changes by applying a
new world view of the relationship. This emphasizes our belonging in life and our
tendency to act for life, wanting to work for environment resources management
that protects and conserve all forms of materials and life as if it is their own using
all scientific approaches and ICT enabled capabilities. This is absent in scientific
ecology even if it is highly sophisticated and rigorous, but the smart environment for
smart city advocate deep ecology since it is oriented towards environmental action
based on scientific ecology as against knowledge accumulation alone.

2 Towards Deep Ecology Based Smart Environment Management

The deep ecology consists of eight basic principles, or guidelines for a reformed way
of thinking about our environment (not being entirely exclusive to the living plants
and animals, or the paradigmatic thought of the word 'environment', but basically
the world around us, the place we live).

1. Inherent value
   The well-being and flourishing of human and nonhuman life on earth have value
   in them (this is commonly referred to as inherent worth, or intrinsic value). These
   values are independent of the usefulness of the nonhuman world for human
   purposes.
2. Diversity
   Richness and diversity of life forms contribute to the realization of these values
   and are also values in themselves.
3. Vital Needs
   Humans have no right to reduce this richness and diversity except to satisfy vital
   needs.
4. Population
   The flourishing of human life and cultures is compatible with a substantial
   decrease of the human population. The flourishing of nonhuman life requires
   such a decrease.
5. Human Interference
   The present human interference with the nonhuman world is excessive, and the
   situation is rapidly worsening.
6. Policy Change
   Policies must therefore be changed. These policies affect basic economic, technological, and ideological structures. The resulting will be deeply different from
   the present.
7. Quality of Life

The ideological change is mainly that of appreciating life quality (dwelling in situations of inherent value) rather than adhering to an increasingly higher standard of living with the expenditure of money. There will be a profound awareness of the difference between big and great.

8. Obligation of Action

   Those who subscribe to the foregoing points have an obligation directly or indirectly to try to implement the necessary changes.

This chapter focus on Deep ecology based Smart Urban Environment Resources Management to determine how one can conduct system design and postulate protocol of action. To achieve this objective, smart environment management is postulated where smart community, is expected to practice deep ecology. Research studies show that there is a strong relationship of Religion whether it is Hinduism, Islam, Buddhism, Judaism or Christianity and so on with Ecology. Religion can be a bridge that connects Scientific ecology with Deep ecology. Therefore, Vedic and Buddhist approaches are presented in this chapter as examples which can guide the smart community in their perception of environment and action, they need to perform depending upon local reality for environment resources management issues derived from scientific measurement, analysis and modelling. While environmental science is helpful scientifically understand the environmental phenomenon including change, it is movements like deep ecology and those based on faith which makes you act. Undoubtedly, the environmental legislations of the secular or theocratic country is largely influenced by religions and it cannot be ignored in designing smart environment for smart cities. However, it may be kept in mind the religious environmental practice is influenced by the geography where the religion is born but certain religions like Buddhism, Christianity and early Islam in different countries is an adaptation of religion in a geography.

The Vedic insight on environment postulate is the earliest known writing in India preserved even today and being practiced by Indians. The Buddhist Model emanated from many countries in many forms from the south and south-east Asia and later in North America and Europe in many Mahayana, Theravada or Vajrayana schools of philosophy with many sub schools of thoughts geographically spread as shown in Fig. 10. They have a different view of conceptualising the environment and how humans can interact with the environment bringing out distinct conceptualisation and protocol of action for environment management.

First, the smart environment model of smart cities is presented which leads to conceptualisation and system design and protocol. Although these three approaches namely, smart city, Vedic and Buddhist are developed several thousand years apart, there is commonality which can be borrowed from one model to another which is the reason for presenting these models side by side while accepting Smart Environment Model as an operational model.

Where is Buddhism thriving today

Originated in India, 500 B.C.

Spread through the East,
first to Sri Lanka and SE Asia,
then north to China, Tibet

Three major streams:

1) Theravada – SE Asia
2) Mahayana – China, Japan, Korea
3) Vajrayana -- Tibet

All forms have migrated west in
the 20th century.

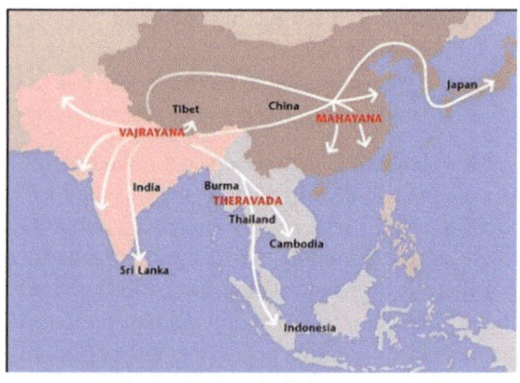

Fig. 10 Buddhist Model of Environment Practiced through local adaptation. *Source* [15]

2.1 Smart City Model

To understand the "Smart Environment" to conduct smart environment governance and management, the Smart city needs to be defined first. Then, we must understand what the smart city system is, consisting of six integrated elements. This will be discussed in Sect. 2.4 after discussing the various definition of smart cities in Sect. 2.2 and smart community in Sect. 2.3 which is the driving force of smart cities and uses E-Democracy and help implement and E-Governance to achieve various ever-changing objectives of smart cities. In Sect. 2.4 smart environment will be detailed out and defined as an integral and not independent part of the six-component system.

Smart cities were also called first, wired cities but the term wired lost meaning in an era of wireless communication technology. Other names are Digital City, Intelligent City, Ubiquitous city, Hybrid city, and Information city. People oriented smart cities are called the Creative city, Learning City, Humane city and Knowledge city. The difficulty of explaining these terms created a situation that these terms became unpopular. All these terms which may be considered as an earlier concept were subsequently replaced by the "Smart city".

Urban Planners consider "smart" as strategic development of a city emerging out of SWOT (Strength, Weakness, Opportunities and Threats) analysis which also forms part of smart city approach. For technologists, Smart implies automatic computing principles and artificial intelligence, like self-configuration, self-healing, self-protection, and self-optimisation which is extensively used in all six smart city components. Smart homes with several systems of automation, smart public buildings, smart airports, smart hospitals are also equipped with sensors and actuators and so also smart cities. The smart ecosystem is an emerging concept which incorporates integration in the development of smart technology in a smart city. There can also be smart individuals, smart community and smart society all which extend this concept. The self-awareness of smart cities is derived out of successfully modelling through ICT and IOT technologies, (such as for example geospatial and sensors technologies and related WIFI and Bluetooth communication and cloud technologies) in a real-world smart city at every instance to a cyberspace, the Smart city model that can be used to intervene in the real-world smart city using for example SCADA. Simplistically, it can be a web site which can be used to intervene automatically in the real-world cities using big data cloud and apps by intelligent entities like the Internet of Things. In summary, using a ubiquitous network of sensors, the smart city can be intervened at cyberspace. A cloud computing platform handles the massive data storage required for such intervention, computation, analysis, and decision-making process using spatial decision support system and artificial intelligence and conducts automated control based on the results of those analyses and decisions. At the level of the smart city, the digital model of the city and physical cities can be linked by the Internet of Things and ICT, thereby forming an integrated cyber-physical space. Smart water supply or intelligent water is a good example of this modelling. Therefore, using a simple formula, we can denote the smart city as follows: Smart City = digital city in cyberspace + Internet of Things + cloud computing. Using radio frequency identification (RFID), infrared sensors, global positioning systems (GPS), laser scanners, and other information-sensing devices, the Internet of Things connects real objects in the city based on protocols that conduct information exchange and communication to achieve intelligent identification, positioning, tracking, monitoring, and management. Cloud computing is an Internet-based computing model for public participation. Its computing resources (including computing power, storage capacity, interactivity, etc.) are dynamic, scalable, virtualized, and provided as a service. Again, cloud computing is a model of Internet-based computing, an extension of distributed computing and grid computing. This cloud computing can rapidly process the huge amounts of data produced by the smart city and simultaneously service millions of users using apps at every instant below micro second. Some time it can decide on their own with no human intervention using artificial intelligence.

2.2 Definition of Smart Cities

A literature survey has shown 10 definitions of smart cities.

1. The use of smart computing technologies to make the critical infrastructure components and service of a city-which include city administration, education, health care, public safety, real estate, transportation, and utilities-more intelligent, interconnected and efficient [16].
2. A city well performing in a forward-looking way in economy, people governance, mobility, environment and living built on the smart combination of endowments and activities self-decisive independent and aware citizens [17].

3. A city striving to make itself smarter more efficient, sustainable, equitable and liveable [18].
4. A city that monitors and integrates, conditions of all its critical infrastructure including roads, bridges, tunnels, rails, subways, airports, seaports, communications, water, power, even major buildings can better optimise its resources, plan its preventive maintenance activities, and monitor security aspects while maximising services to its citizens [19].
5. An instrumented, interconnected and intelligent city instrumentation enables the capture and integration of live real-world-data through the use of sensors, kiosks, meters, personal devices, the web, appliances, camera, smart phones, implanted medical devices, the web, and another similar data-acquisition system including social networks as networks of human sensors interconnected means the integration of those data into an enterprise computing platform and the communication of such information among the various city services. Intelligent refers to the inclusion of complex analytics, modelling, optimisation, and visualisation in the operational business processes to make a better operational decision [20].
6. A city that gives inspiration, shares culture, knowledge, and life; a city that motivates, its inhabitant to create and flourish in their own lives [21].
7. A city where ICT strengthen the freedom of speech and the accessibility to public information and services [22].
8. A city that monitors and integrates conditions of all its critical infrastructure [19].
9. A city connecting the physical infrastructure. The IT infrastructure and social infrastructure and the business infrastructure to leverage the collective intelligence of the city [23].
10. A city combining ICT and web 2.0 technology with other organisational design and planning efforts to be de materialise and speed up the bureaucratic process and help identify new innovative solutions to city management complexity, to improve sustainability and liveability [24].
11. The book "Geographic Information System for Smart Cities" [25] define Smart City as a knowledge based city that develops extra ordinary capabilities to be self-aware, how it functions 24 h and 7 days a week and communicate, selectively, in real time knowledge to citizen end users for satisfactory way of life with easy public delivery of services, comfortable mobility, conserve energy, environment and other natural resources, and create an energetic face to face communities and a vibrant urban economy even at a time there is National economic downturns.

2.3 Smart Communities

Integrated with the smart cities concept is the smart communities which are one component of smart city system namely "smart people". One can form many proactive and well trained and subjected to continuing education, smart environment communities to be a watch dog of well-defined and location specific environment issues at selected geographic locations every moment and act to intervene using digital means in a scientific way using E-Democracy and E-Governance. This constitutes smart environment management which can only be executed by the smart community and none else. Therefore, Smart Communities assume considerable importance in this book since they are fully capable of smart environment management. In most cases they will be managing the common property the environment in many manifestations, and they are legitimate owners of the common property. They can easily help Environment Governance at the local level which cannot be effectively implemented by existing bureaucracy and its way of procedure driven legal working.

City planners divide spatially a city into many communities. This concept of a face to face community is generally implemented by Traditional Neighbourhood Concept. Many a time urban planners and urban designers must reinvent neighbourhood using appropriate urban design or urban land management adjustments spatially since it is often lost in uncontrolled fringe area development, organic urban spread or sprawl that creates urban agglomerations. They use New Urbanism principles to achieve this objective.

The concept of Smart Community was first used in 1993 in Silicon Valley, California when it was under a bad spell of economic recession and down turn which was predicted to last long [26]. Silicon Valley business leaders, community members, government officials and educators decided together to help jump start the region. In Smart Communities Guidebook developed by California Institute for Smart Communities in 1997 at San Diego State University defines Smart Community as follows [26]. A "smart community" is simply that: a community in which government, business, and residents understand the potential of information technology, and make a conscious decision to use that potential of that technology, to transform life and work in their region in significant and positive ways. The implementation guide developed by the same institute elaborate it further. A smart community is a community in which members of local government, business, education, healthcare institutions and the public understand the potential of information technology and form a successful alliance to work together to use technology to transform their community in significant and positive ways. Because of these unified efforts, the community can leverage resources and projects to develop and benefit from telecommunication infrastructure and services much earlier than if otherwise would. Instead of incremental changes, a transformation occurs which increases choice, convenience and control for people in the community as they live, work, travel, govern, shop, and entertain themselves. Smart communities are economically competitive in the new global economy and attract and promote commerce because of an advanced telecommunications infrastructure. Smart community international network (SCIN) defines [27] Smart Community as follows. A smart community is a community with a vision of future that involves the application of information and communication technologies in a new and innovative way to empower its resident, institutions and regions. As such they make the most of the opportunities that new applications afford, and broadband-based services can deliver-such as better health care delivery, better education and training and new business opportunities. In a similar way Australian, Smart Community is defined. Australian Smart Communities are communities with a vision of future that involves harnessing the power of the internet and other ICT technologies in new and innovative ways to empower their residents, institutions, and their citizens. Smart city concept had an international following in the developed world [28–33].

A smart community uses broadband networks to enable a series of applications that the community can leverage for innovative economic development and commerce, top-notch education, first-rate health care, cutting-edge government services, enhanced security and more efficient utility use [34]. Broadband facilitates greater interconnection for intra- and inter community resources. Moreover, broadband enables intelligent networks, making communities smarter, more efficient and better able to prepare their citizens to participate in the global economy. Schools can engage in distance learning to offer courses that would be otherwise unavailable. University Grants Commission of India and All India Council for Technical Education gives the grant to have web-based classes on many topics. These smart communities bring about the smart environment must have an E-Democracy set up given in Sect. 2.4 constitutionally valid Environmental Democracy Directive Principles which is given below in Sect. 2.5.

2.4   E-Democracy for Smart Cities

E-democracy in a smart city is nothing else but a virtual or cyberspace face-to-face democracy practised in ancient Licchavi or Athens discussed in earlier para using multimedia and ICT. This is possible in a smart city with a high-level endowment of broadband and ICT infrastructure even if the city is a meta city with 20 million or above population or megacities with 10 million and above or metro city with one million and above unlike ancient Athens or Licchavi with less than 10,000 population. E-democracy is the technological adaptation of ancient face-to-face democratic tradition in an ICT-enriched smart city. It is the use of information and communication technologies and strategies by 'democratic sectors' within the political processes of local communities, states/regions, and nations. Democratic actors and sectors in this context include, in order of importance, citizens/voters, political organisations, the media, and elected officials and governments. E-democracy, like democracy in its proactive form, is a direct democracy which is the primary requirement of smart cities. E-democracy often refers to technological adjuncts to a smart city republic, i.e. the use of information and communication technologies and strategies in political and governance processes.

In a few cases, the word 'e-democracy' is used to refer to anything political that involves the Internet. It may be also called the Internet of Democracy. E-democracy is concerned with the use of information and communication technologies to engage citizens in supporting the democratic decision-making processes and strengthen representative democracy. E-democracy is said to aim for more active citizen participation by using the Internet, mobile communications, and other technologies in today's representative democracy, as well as through more participatory or direct forms of citizen involvement in addressing public challenges. Constitution of a country shall direct the E-Democracy. Environment related Directive principles for Environment Democracy is enumerated in Sect. 2.5.

2.5 Environmental Democracy Directive Principles

This section is largely adopted from Vandana Shiva's water wars [35]

1. Environment Resources are Natures Gift: Since it is a gift, we must use it as per our sustenance needs and not greed's. This is the basis of ecological democracy which accepts deep ecology principle.
2. Environment Resources are essential to life: All species have their rightful share of environmental resources. This principle also follows deep ecology.
3. Life is interconnected through Environmental Resources: Environmental resources connect all species. We must ensure that our actions do not harm other species and other humans rich or poor. This follows some principles laid out in Buddhist religion and Rig Veda.
4. Environmental Resources must be free for sustenance needs: Since nature gives Environmental Resources to us free of cost, buying and selling it for profit violates our inherent right to nature's gift and denies the poor of their human rights. The first part again follows deep ecology.
5. Environmental Resources is limited and can be exhausted: Environmental Resources are limited and exhaustible if used non-sustainably. Non-sustainable use includes extracting an Environmental Resources from ecosystems than nature can recharge (ecological non-sustainability) and consume. More than one's legitimate share, given the rights of others to a fair share (social non-sustainable). The hydrologic concept of water balancing of ground water resources is a good example of intervention.
6. Environmental Resources must be conserved: Everyone has a duty to conserve Environmental Resources and use Environmental Resources sustainably, within the ecological and just limits. This helps for Environmental Resources use free for sustenance.
7. Environmental Resources is common: Environmental Resources is not a human invention. It cannot be bound and has no boundaries. It is by nature a common. It cannot be owned as private property and sold as a commodity.
8. No one holds a right to destroy: No one has a right to overuse, abuse, waste, or pollute Environmental Resources systems. Governance methods of Tradable-pollution permits violate the very principle of sustainability and just use.
9. Environmental Resources cannot be substituted: Environmental Resources are intrinsically different from other resources and products. It cannot be treated as a commodity.

If we practice Veda or Buddhism in daily life which will be detailed out in the subsequent sections, and for that matter any religion such as Christianity, Judaism, Islam, as a way of life, we are practicing Deep Ecology and E-Environment Democracy directives discussed above that relate man with the environment for sustainable environment resources management. This aspect is discussed in Sect. 3 onwards. Since the smart community is a part of an integrated smart city system, it is discussed in Sect. 2.6.

2.6 Smart City System

A smart city is characterised by six components such as Smart People, Smart Economy, Smart Mobility, Smart Environment, Smart Living and Smart Government [36–38]. The ability of any city to transform and progress towards these six components of smart cities from the existing conditions in an integrated and comprehensive manner is the ability to transform the city into a smart city. Prime movers for this capacity to transform cities, are Smart People and Smart Economy. This proposition contradicts the highly simplified approach of many Governments in developing countries, the world over who considers Smart Cities are there because of investment decisions made for deployment of smart ICT enabled infrastructure technologies such as the Internet of Things and other smart city infrastructure which can be purchased from Multi-National Companies. This is far from the truth. The smart cities shall be considered as centres of emerging third industrial revolution [39, 40] where sharing economy develops and move towards marginal cost of production nearing to zero, triggered by a smart economy which is sustainable and resilient. A Smart city has the capacity to grow more and more smart people to multiply their opportunities for engaging in the smart economy. The emergent needs of the smart economy and smart people fix the configuration of smart mobility, smart environment and resulting in smart living. There can be many variations of these components as per the ecology and local culture. Here the technologies used is evolving over time and what is used some three years before will be totally obsolete afterwards and technologies are developed in these cities by smart people and not imported from multinationals. Without these two primes component of this smart city system, it is impossible to transform a city to smart city and smart city creation cannot take place by simply investing one time, in smart infrastructure and smart ICT enabled technologies in a city.

As per Fig. 11, Smart City System comprises of six key building blocks: (i) smart governance, (ii) smart environment, (iii) smart mobility, (iv) smart people, (v) smart living, and (vi) smart city economy. These six building blocks are closely interlinked and contribute to the 'Smart City System'. Some authors treat the six elements of a Smart City System equally [41]. However, following Vinod Kumar [42], we give prominence to 'smart people' because without their active participation and involvement a Smart City System would not function in the first place. A Smart City System will risk its efficient functioning without Smart People.

Fig. 11 Smart city system. *Source* [36–38]

I. **Smart Governance**

'Smart Governance', has the following attributes.

(1) A smart city practices accountability, responsiveness, and transparency (ART) in its governance.

(2) A smart city uses big data, spatial decision support systems and related geospatial technologies in urban and city regional governance.

(3) A smart city constantly innovates e-governance for the benefit of all its residents.

(4) A smart city constantly improves its ability to deliver public services efficiently and effectively.

(5) A smart city practices participatory policy-making, planning, budgeting, implementation, and monitoring.

(6) A smart city has a clear sustainable urban development strategy and perspectives known to all.

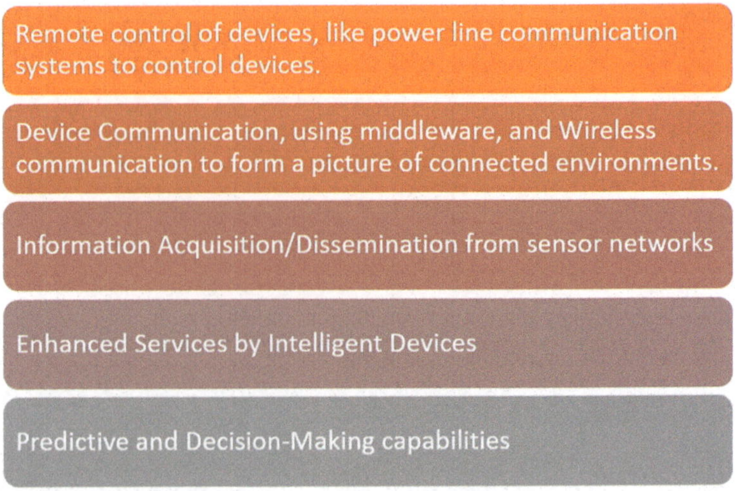

Remote control of devices, like power line communication systems to control devices.

Device Communication, using middleware, and Wireless communication to form a picture of connected environments.

Information Acquisition/Dissemination from sensor networks

Enhanced Services by Intelligent Devices

Predictive and Decision-Making capabilities

Fig. 12 Converting environment to smart environment *Source* Author

(7) A smart city utilizes creative urban and regional planning with a focus on the integration of economic, social, and environmental dimensions of urban development.

(8) A smart city features an effective, efficient, and people-friendly urban management.

(9) A smart city practices E-Democracy to achieve better development outcomes for all.

(10) A smart city embraces a Triple Helix Model in which government, Academia and Business/Industry practice changing roles in Governance.

II. Smart Environment

Smart Environment can be defined in the same way as Smart Cities. Smart Environment is a knowledge based environment that develops extra ordinary capabilities to be self-aware, how it functions 24 h and 7 days a week and communicate, selectively, in real time knowledge to citizen end users for a satisfactory way of life with easy public delivery of services, comfortable mobility, conserve energy, environment and other natural resources, and create energetic face to face communities and a vibrant urban economy even at a time there is National economic downturns. It can also be made smart using ICT and IOT as given in Figs. 12, 13 and 14.

The collection of data for environment management is achieved through hardwires related to smart metering. A diagrammatic presentation is given below.

The smart environment generates data for environment management. Only a well-trained smart community can use the data generated to make appropriate decisions as per the current issue faced by the user population. The following diagrammatically shows the data environment architecture of environmental resources.

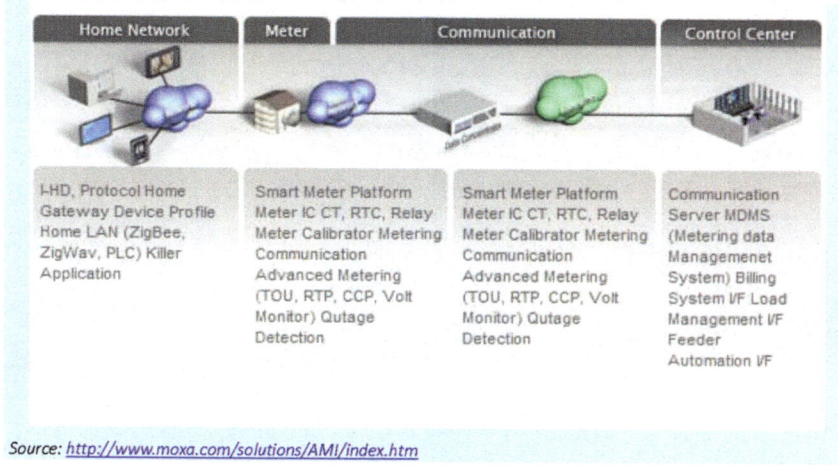

Source: http://www.moxa.com/solutions/AMI/index.htm

Fig. 13 Smart metering setup for smart environment.

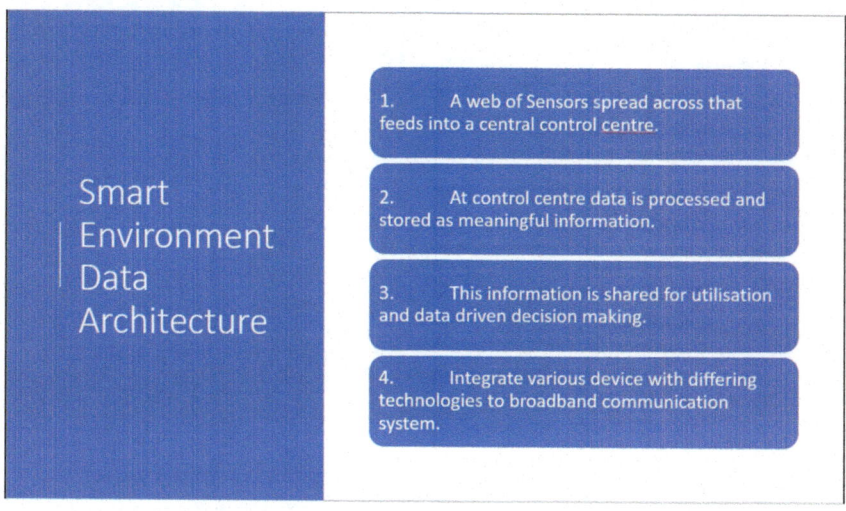

Fig. 14 Smart environment data architecture. *Source* Author

Smart Environment Management need to be taken one component of environment after another. The classification of Vedic Model of Environment given in Sect. 3 shall help us to decipher what to be taken up. The environment issue which threatens a location shall be taken up first. The same six system model of smart cities can be applied to the environment if you can specify one environment resources one by one at a time to avoid abstraction. Figure 15 achieve this taking any environmental resource that needs environment resource governance and management. The same

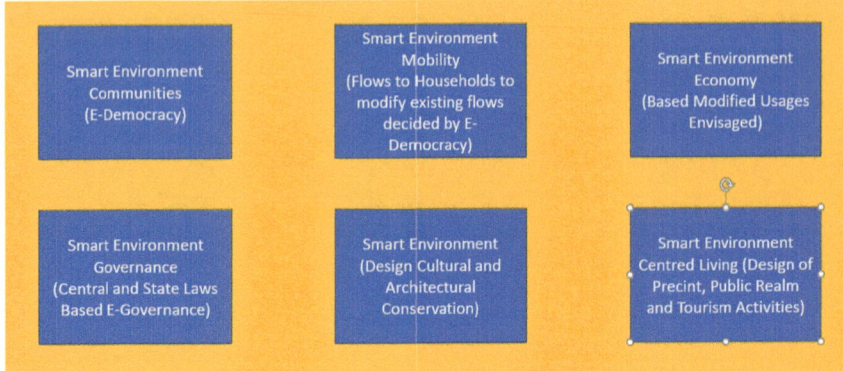

Fig. 15 Smart environment system. *Source* Author

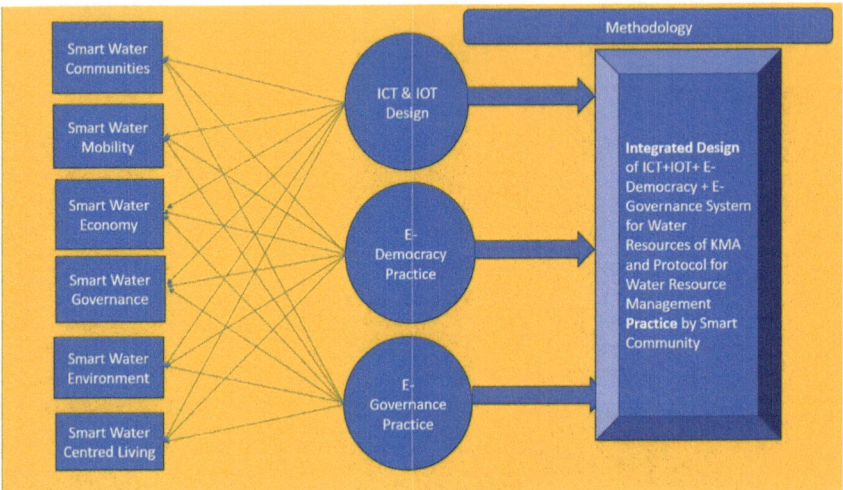

Fig. 16 Methodology of smart Environment Management using IoT and ICT system, E-Democracy and E-Governance (example water resources). *Source* Author

six systems can be applied to say water environments if generic environment is substituted with water resource or urban flood.

The design and protocol for six of the elements of the system need to be developed from three points of views namely ICT and IOT system, E Democracy System and E-Governance System and each of the elopement shall investigate design and protocol for all six elements.

It is represented in the linkage in Fig. 16.

'Smart Environment', the second building block, has the following attributes.

(1) A smart city lives with and protects nature.

(2) A smart city is attractive and has a strong sense of place that is rooted in its natural setting.

(3) A smart city values its natural heritage, unique natural resources, biodiversity, and environment.

(4) A smart city conserves and preserves the ecological system in the city region.

(5) A smart city embraces and sustains biodiversity in the city region.

(6) A smart city efficiently and effectively manages its natural resource base.

(7) A smart city has recreational opportunities for people of all ages.

(8) A smart city is a green city.

(9) A smart city is a clean city.

(10) A smart city has adequate and accessible public green spaces.

(11) A smart city has an outdoor living room. Unlike the indoor living room in houses where we meet others, outdoor living rooms are aesthetically designed intimate, active, and dynamic urban realms where people meet face to face for a culturally and recreationally rich and enjoyable contact as part of living and work.

(12) A smart city has distinctive and vibrant neighbourhoods that encourage neighbourliness and a spirit of community.

(13) Smart city values and capitalizes on scenic resources without harming the ecological system, natural resources, and biodiversity.

(14) A smart city has an integrated system to manage its water resources, water supply system, wastewater, natural drainage, floods and inundation, especially in the watersheds where it is located, especially in view of the (impending) climate change.

(15) A smart city focuses on water conservation and minimizes the unnecessary consumption of water for residential, institutional, commercial, and industrial use, especially in the arid and semi-arid areas.

(16) A smart city has an efficient management system for the treatment and disposal of wastewater, and reuse of treated wastewater, particularly in the arid and semi-arid areas.

(17) A smart city has an efficient management system for the collection, treatment, and disposal of industrial wastewater.

(18) A smart city has an integrated and efficient management system for the collection, transfer, transportation, treatment, recycling, reuse, and disposal of municipal, hospital, industrial, and hazardous solid waste.

(19) A smart city has an efficient system to control air pollution and maintain clear air, especially in the air sheds where it is located.

(20) A smart city has an efficient and effective system for disaster risk reduction, response, recovery, and management.

(21) A smart city has and continually upgrades its urban resilience to the impacts of climate change.

(22) A smart city can create a low-carbon environment with a focus on energy efficiency, renewable energy, and the like.

III. **Smart Mobility**

'Smart Mobility', the third building block of a Smart City System, includes the following features.

(1) A smart city focuses on the mobility of people, and not only that of vehicles [43, 44].
(2) A smart city will advocate walkability and cycling.
(3) A smart city has vibrant streets (at no additional cost).
(4) A smart city effectively manages vehicular and pedestrian traffic, and traffic congestion.
(5) A smart city has pleasurable (say for bicycle and walks) routes.
(6) A smart city has balanced transportation options.
(7) A smart city will have a mass rapid transit system, such as metro rail, light metro, monorail, or 'sky train' for high-speed mobility.
(8) A smart city will have integrated high-mobility system linking residential areas, work places, recreational areas, and transport notes (e.g. bus/railway station/s and airport).
(9) A smart city will practice high-density living, such that benefit of high-speed mobility is uniformly available.
(10) A smart city has seamless mobility for differently-abled (often incorrectly called, disabled) people.

IV. Smart People

'Smart People', the fourth building block of a Smart City System, require many crucial attributes as given.

(1) Smart people excel in what they do professionally and smart community is converted functionally professional using existing population resources by continuing education and training and not awarding degrees and certificates.
(2) Smart people have a high Human Development Index [42].
(3) A smart city integrates its universities and colleges into all aspects of city life.
(4) It attracts high human capital, for example knowledge workers.
(5) A smart city maintains high Graduate Enrolment Ratio and has people with a high level of qualifications and expertise.
(6) Its inhabitants opt for lifelong learning and use e-learning models.
(7) People in a smart city are highly flexible and resilient to the changing circumstances.

V. Smart Living

'Smart living', the fifth building block of a Smart City System, includes the following features.

(1) A smart city has strong and shared values.
(2) A smart city records and celebrates local history, culture, and nature.
(3) A smart city has a vibrant downtown, 24 h and 7 days a week.
(4) A smart city can provide the necessary safety and security to women, children, and senior citizens.

(5) A smart city improves the urban way of life.
(6) A smart city builds natural and cultural assets to build a good quality of life.
(7) A smart city not only understands the big picture of urban liveability, but also pays attention to small details.
(8) A smart city has high-quality open and accessible public spaces.
(9) A smart city has high-quality public services and amenities.
(10) A smart city is an ideal place of living, especially for women, children, and senior citizens.
(11) A smart city organizes festivals that celebrate people, life, and nature in the city.
(12) A smart city has a ritual event (or more) that symbolizes the values and aspirations of the community.
(13) A smart city celebrates and promotes art, cultural, and natural heritage in the city.
(14) A smart city engages artists to improve and enrich the aesthetics of daily life of the city.

VI. Smart City Economy

'Smart City Economy', the sixth building block, requires the following attributes.

(1) A smart city understands its economic DNA.
(2) A smart city is driven by innovation and supported by universities that focus on cutting-edge research, not only for science, industry, and business but also for cultural heritage, architecture, planning, development, and the like.
(3) A smart city highly values creativity and welcomes new ideas.
(4) A smart city has enlightened entrepreneurial leadership.
(5) A smart city offers its citizens diverse economic opportunities.
(6) A smart city knows that all economics works at the local level.
(7) A smart city is prepared for the challenges posed by and opportunities of economic globalization.
(8) A smart city experiment, supports, and promotes sharing economy.
(9) A smart city thinks locally, acts regionally, and competes globally.
(10) A smart city makes strategic investments on its strategic assets.
(11) A smart city develops and supports compelling national brand/s.
(12) A smart city insists on balanced and sustainable economic development (growth).
(13) A smart city is a destination that people want to visit (tourism).
(14) A smart city is nationally competitive on selected and significant factors.
(15) A smart city is resourceful, making the most of its assets while finding solutions to problems.
(16) A smart city excels in productivity.
(17) A smart city has the high flexibility of the labour market.
(18) A smart city welcomes human resources that enhance its wealth.
(19) A smart city's inhabitants strive for sustainable natural resource management and understand that without this its economy will not function indefinitely.

For smart environment management of any common property resources such as for example water, aggregate first detailed scientific evidence of all research studies conducted in the study area. It should be reviewed for gaps and for the case of water resources, use watershed as a unit of GIS analysis and study shall be considered and analysed in detail of all environment issues at present and in future in these watersheds. The cause and effect of pollutions, water scarcity and possibility of water resources augmentation to the actual need by constructing smart water related infrastructure such as check dams, storage systems, piping system and micro hydro and or wells and ponds rehabilitation etc. fitted with sensors and ICT network to manage it; and water pollution abatement measures need to be worked out. This will show the additional information required to detail out the issue and find solutions based on six component of smart city system for smart water resources discussed. Then the spatial decision support system is developed for E-Governance which considers state and central Government legislation and E-Democracy applications as per requirement of each watershed needed is designed along with protocol. This leads to deployment of ICT and IOT network, by location and configuration, tasks to be performed, applications to be developed and gives all necessary decision support system for water Democracy and Water Governance. All these integrated design and protocol based on Spatial Decision Support System is for the smart water community to implement. This is diagrammatically resented in Fig. 17.

When we focus on smart environment resources management in a smart city, all these six components of smart cities get modified and therefore cannot be taken up in isolation. Hence, we require a comprehensive six component view of the city. Again, ICT system designed for environment resources management can also be used and shared for other urban activities not directly related to the environment which leads us to one sharing system for all purposes.

The first industrial revolution made people resource hungry, became global explorers for all types of resources and lead to colonialization, deliberate retarding of development of colony and slavery. It gave rise to economic distribution taught that maximise income and that concentrate on class conflicts like that of Marxism Communism, Socialism on one side and Capitalism and neoliberal capitalism on the other side. These were not a philosophy for the sustainable environment but promoted human greed for wealth at any cost and not based on actual minimal sustainable needs of all humanity and other living and non-living entities. The first industrial revolution is dead and gone and replaced by subsequent industrial revolutions, but this greed still exists as manifested by excess consumption and corruption which has become part and parcel of democracy in all countries and even communist countries where democracy is not practiced. We can see that Communist China and Russia are as greedy and self-centred as democratic USA and UK. How do we liberate our self from greed to save the environment? How do we change our life style to save the environment? These are an important and specific requirement of environmental resources management of any smart cities. It is the change in the attitude of people, a new way of life, a new world view and a value system based on scientific ecology that can save the urban environment. Some initial discussion of this aspect is already covered in Sect. 2.5 Environmental Democracy Directive Principles and on Deep

Fig. 17 Integrating six components of the smart environment by three groups of integrated design to postulate protocol (example water resource) for smart community. *Source* Author

ecology with a promise to explore further. To extend it further, I have taken two sample religious practice still followed to see whether this can take humanity away from ecosystem destruction through greed. This is given in Sects. 3 and 4.

3 The Environment Management as Per Vedic Model [45]

In Sanskrit and in many Indian languages, the word *Paryavarana* is used for the environment, meaning which encircles us, or which is all around in our surroundings that include me. *Paryavaran Bhavan* in New Delhi houses Ministry of Environment and Forest of Government of India. Environment manifests itself as environmental resources are meant for the sustained use of all including vegetation, animals, birds and human in the environment. Undoubtedly environment shall be protected and sustained for the future generation of vegetation, animals, birds, human and others. The Environment (Protection) Act, 1986 of India defines the environment as follows:

'Environment includes water, air and land and the inter-relationship which exists among and between water, air and land and human beings, other living creatures, plants, microorganisms and property [46]. From the above definition, it can be briefly said that we are surrounded by an environment which consists of two components namely biotic (living organisms) and abiotic (non-living materials like land, water and air). The living organisms can be grouped into three types—those living mainly on land, in water and in the air. The non-living materials of the environment are land, air, water, and property. The environment existed several thousands of years and there has been disappearing acts of many biotic environmental resources including many species of animals, birds, fishes, trees and plants. Sadly, they do not exist anymore. When environment become non-liveable in future as discussed in Sect. 1 with climatic change, the rise of water levels, or acute water scarcity, high polluted air, atomic bomb radiations or natural calamities, the civilisation partially or fully vanishes along with all species of the biotic environment in one go. While this possibility stares us all and frightens us, it is important to know how we survived all through these centuries. This exploration leads us to the study the wisdom of environment sustainability contained in many ancient texts such as Vedas the earliest written document in India, and popular Buddhist texts some 2600 years old, for a world religion practiced by common persons. Most of these texts are written for a way of life in a suitable environment and have given due importance to the environment and its relationship with humans which resulted in a protocol to deal with nature as a daily practice.

3.1   Vedic View of Environment

The Vedic view of the environment is well-defined in one verse of the Atharvaveda where three coverings of our surroundings are referred to as Chandamsi: 'Wise utilizes three elements variously which are varied, visible and full of qualities. They are called as Chandamsi meaning coverings available everywhere.' These are water, air and plants or herbs. They exist in the world from the very beginning. It proves the knowledge of Vedic seers about the basic elements of the environment. According to one indigenous theory established in the Upanishads, the universe consists of five basic elements viz., (1) Earth or land, (2) water, (3) light or lustre, (4) air, and (5) ether. Nature has maintained a status of balance between and among these constituents or elements and living creatures. A disturbance in the percentage of any constituent of the environment beyond certain limits disturbs the natural balance and any change in the natural balance cause lots of problems to the living creatures in the universe. Different constituents of the environment exist with set relationships with one another. The relation of a human being with the environment is very natural as he cannot live without it. From the very beginning of creation, he wants to know about the environment and its interrelationship for self-protection and benefit. This guided the life of people who knows Vedas and hence the environmental wisdom.

The Vedic era people were awed by nature. Sand-storm and cyclone, intense lightening, terrific thunderclaps, the heavy rush of rain in monsoon, the swift flood in the stream that comes down from the hills, the scorching heat of the sun, the cracking red flames of the forest fire, all witness to power beyond man's power and the Vedic sages who wrote these texts felt the greatness of these forces. They adored these activities. They appreciated these forces. They worshiped and prayed them due to regard, surprise and fear. They realized instinctively that action, movement, creation, change and destruction in nature are the results of forces beyond men's control. Thus, they attributed divinity to nature and worshipped nature and never tried to exploit nature for greed. If you worship nature, you cannot abuse nature or pollute the environment but only try to conserve, preserve, respect, sustain and live with nature. This gave Vedic population irrespective of their education level, an environmental Vedic code of conduct or protocol to follow in their day to day lives. This Vedic view of environment got lost in modern India, and we seem to work against what is given in Vedic culture transforming our environment beyond repair or redemption by deforestation, soil erosion and degradation, water and air pollution and excretions in the open area with no toilets.

3.2 Bestowing Divinity to Nature

The divinity of nature is acknowledged in Vedic hymns. Rigvedic hymns could be divided into many parts, but their main part belongs to Natural hymns, the hymns related to natural forces. The Vedic hymns addressed environmental forces as deities (Devata) which are the most impressive phenomenon of nature and its varied aspects. The word Devata means divine, and dignified which is bright, strong, donor, and powerful. In these hymns, we find prayers for certain natural elements such as air, water, earth, sun, rain, dawn etc. The glorious brightness of the sun, the blaze of the sacrificial fire, the sweep of the rain-storm across the skies, the recurrence of the dawn, the steady currents of the winds, the violence of the tropical storm and other such natural energies, fundamental activities or aspects are glorified and personified as divinities (Devata). The interaction with nature resulted in appreciation and prayer but, indeed, after a good deal of observation. Attributes assigned to deities fit in their natural forms and activities, as Soma is green, fire is bright, air is fast moving, and the sun is the dispenser of darkness. The characteristics of these forces described in the verses prove that Vedic seers were masters of natural science. In Vedic view, this world consists of Agni i.e., fire or heat and Soma i.e. water. Sun (Surya) is the soul of all which is moving, and of which is not moving. Indra is a most powerful god who kills Vritra, the symbol of the cloud to free waters. Vritra means one who covers and is derived from the root vri, to cover. The main force of expansion in the Vedic cosmology is Indra, and his chief adversary, the main force of contraction, is Vritra [47]. Maruts are Indra's associates. Vedic seers pray boldly to these natural forces and aspects for bestowing plenty and prosperity on them. Aditi is praised as Devamata, the mother of all-natural energies and she symbolizes the Nature.

3.3 Division of Universe

Vedic seers have a great vision about the universe. The universe is made on scientific principles, and that's why it is well measured. Vedic scientists divided the tripartite division of the universe into three regions Prithvi, the earth, Antariksha, the aerial or intermediate region which is between heaven and earth, and Dyau, the heaven or sky is very well established in the Vedic literature. Prithvi can be given a scientific name 'observer space.' It is our space, the space in which we live and die, whatever we can see and observe. From one end of the universe to the other end is the expanse of Prithvi, and that is what the name Prithvi means: the broad and extended one. Dyau can be termed 'Light space' because light propagates in this space. Antariksha can be termed as 'Intermediate space' as this space exists in between observer space and light space. A verse from the Yajurveda states that the division of the universe was done on a subtle level, and not on the gross level [48]. Here; about the environmental study, we regard the division of the universe as the most important concept of the Vedas. Though many gods are described in the hymns, and it is very difficult to arrange them in different classes, but there exists three main Gods: Agni in the earth, Vayu in atmosphere and Sun in heaven. Each one of them is known by various names depending on the different actions performed. These three gods are three major forms of energy, fire on the earth, the air in intermediate space and light in the upper region. Other energies of those regions are related to or under them. So generally, gods are classified into three groups called upper, middle and lower, and, therefore, provide a system to study the atmosphere and it's all aspects. Regarding global harmony, Vedic seers always pray for the welfare of all creatures and all regions.

3.4 The Concept Earth (Prithvi)

The concept of the form of the earth in the Rig-Veda is most fascinating. It is mostly addressed along with heaven into a dual conception (Rodasi, Dyavaprithivi). There is one small hymn addressed to Prithvi, while there are six hymns addressed to Dyavaprithivi. Prithvi is considered the mother and Dyau are considered the father in the Vedas, and they form a pair together. One of the most beautiful verses of the Rig-veda says, 'Heaven is my father, brother atmosphere is my navel, and the great earth is my mother.' Heaven and earth are parents: Matara, Pitara, Janitara in the union while separately called as father and mother. They sustain all creatures. They are parents of all gods. They are great (Mahi) and widespread. Earth is described as a goddess in Rig-veda. In the Atharvaveda, the earth is described in one hymn of 63 verses. This famous hymn called Bhumisukta or Prithivisukta indicates the environmental consciousness of Vedic seers. The seers appear to have an advanced understanding of the earth through this hymn. She is called Vasudha for containing all wealth, Hiranyavaksha for having gold bosom and Jagato Niveshani for being the abode of the whole world. She is not for the different races of men alone but for other creatures also. She is called Visvambhara because she is representative of the universe. She is the only planet directly available for the study of the universe and to realize the underlying truth. This is a wide earth which supports varieties of herbs, oceans, rivers, mountains, hills etc. She has at places different colours as dark, tawny, and white. She is raised at some place and lowered at some places. The earth is fully responsible for our food and prosperity. She is praised for her strength. She is served day and night by rivers and protected by sky. The immortal heart of the earth is in the highest firmament (Vyoma). Her heart is the sun. 'She is one enveloped by the sky or space and causing the force of gravitation. She is described as holding Agni. It means she is described as the geothermal field. She is also described as holding Indra i.e., the geomagnetic field. The earth is described then as being present in the middle of the oceans (sedimentary rocks) and as one having magical movements.' [49]. The hymn talks about different energies which are generated from the form of the earth. 'O Prithvi! thy centre, thy navel, all forces that have issued from thy body- Set us amid those forces; breathe upon us.' [50]. Thus, the earth holds almost all the secrets of nature, which will help us understand the universe. She is invested with divinity and respected as mother 'The earth is my mother and I am Her son.' [51]. The geographical demarcations on this earth have been made by men and not by nature.

3.5   The Concept of Water (Apah)

Water is essential to all forms of life. According to Rig-Veda the water as a part of the human environment occurs in five forms:

1. Rain water (Divyah)
2. Natural spring (Sravanti)
3. Wells and canals (Khanitrimah)
4. Lakes (Svayamjah)
5. Rivers (Samudrarthah) [52].

There are some other classifications also in the Taittiriya Aranyaka, [53] Yajurveda [54] and Atharvaveda [55] as drinking water, medicinal water, stable water etc. Chandogya Upanishad describes qualities of water 'The water is the source of joy and for living a healthy life. It is the immediate cause of all organic beings such as worms, birds, animals, men etc. Even the mountains, the earth, the atmosphere and heavenly bodies are water concretized.' [56]. The cycle of water is described. From ocean waters reach to sky and from sky come back to earth [57]. Rainwaters are glorified. The rain-cloud is depicted as Parjanya god. The fight between Indra and Vritra is a celebrated story from the Rig-veda. It is explained in many ways. According to one view it is a fight for waters. Indra is called Apsu-jit or conquering the waters, while Vritra is encompassing them. Vritra holds the rain and covers waters and thus being faulty is killed by Indra through his weapon called Vajra i.e., thunderbolt. The Indra Vritra fight represents natural phenomenon going on in the aerial space. By the efforts of Indra all the seven rivers flow. The flow of water should not be stopped and that is desired by humanity. The significance of water for life was well known to Vedic seers. They mention—Waters are nectars [58]. Waters are a source of all plants and giver of good health. Waters destroy diseases of all sorts. These waters are for purification. It seems that later developed a cultural tradition of pilgrimage on the river-banks is based on the theory of purification from the water. The ancient Indians knowing water as a vital element for life, were very particular to maintain it pure and free from any kind of pollution. The Manusmriti (Manu is law giver and Smriti is his doctrine) stresses on many instances to keep water clean. The Padma Purana condemns water pollution forcefully saying, 'the person who pollutes waters of ponds, wells or lakes goes to hell.' [59] which in modern days is polluter pays for his sin by environmental laws and governance.

3.6   The Concept of Air (Vayu)

The observer space is the abode of matter particles, light space is the abode of energy and the intermediate space 'Antariksha' is the abode of the field. The principal deity of Antariksha is Vayu (air). Jaiminiya Brahmana quotes, 'Vayu brightens in Antariksha.' The field is another form of energy and, therefore, Yajurveda says, 'Vayu has penetrating brightness.' The meaning of Vayu is made clear in Shatapatha Brahmana in the following Mantra, 'Sun and rest of the universe are woven in the string. What is that string, that is Vayu.' This verse clearly shows that here Vayu cannot mean air alone. The apparent meaning of Vayu is air. The Vedic seers knew the importance of air for life. They understood all about air in the atmosphere and about the air inside the body. The Taittiriya Upanishad throws light on five types of the wind inside the body: Prana, Vrana, Apana, Udana and Samano Air resides in the body as life [60]. Concept and significance of air are highlighted in Vedic verses. Rigveda mentions 'O Air! You are our father, the protector. Air has medicinal values [61] 'Let the wind blow in the form of medicine and brings me welfare and happiness.' [62]. Medicated air is the international physician that annihilates pollution and imparts health and hilarity, life and liveliness to people of the world. Hilly areas are full of medicated air consisted of herbal elements. Another verse describes characteristics of air 'The air is the soul of all deities. It exists in all as life-breath. It can move everywhere. We cannot see it. Only one can hear its sound. We pray to air God. Ancient Indians, therefore, emphasized that the unpolluted, pure air is a source of good health, happiness and long life. Vayu god is prayed to blow with its medicinal qualities.

3.7   The Concept of Ether (Akasha)

Modern environmentalists discuss sound or noise pollution. There is a relation between ether and sound. The sound waves move in the sky at various frequencies. A scientist could see the sky which exists only near earth, but Taittirya Upanishad throws light on two types of ether i.e.: one inside the body and the other outside the body [63]. The ether inside the body is regarded as the seat of the mind. An interesting advice to mankind is found in the Yajurveda 'Do not destroy anything of the sky and do not pollute the sky. Do not destroy anything of Antariksha [64]. Sun shines in Dyuloka and we get light from the sky. The sun rays strengthen our inner power and are essential for our life. Thus, importance and care for ether are openly mentioned in the Vedic verses.

3.8   The Concept of Mind (Manas)

Many prayers are found in Vedas requesting God to keep the mind free from bad thoughts, and bad thinking. In this regard, the Shivasankalpa Sukta of Yajurveda is worth mentioning [65]. Considering the havoc that the polluted minds may create, our ancient sages prayed for a noble mind free from bad ideas. The logicians recognize Manas as one of the nine basic substances in the universe. The mind is most powerful and unsteady. Although the study of mind does not appear directly under the contents of modern environmental science about the cultural environmental consciousness of Vedic seers, we find many ideas discussed in the Vedic literature on the pollution of mind and its precautions meaningful for its impact on the environment. However Buddhism centred itself on mind.

3.9   Animals and Birds

Animals and birds are part of nature and environment. It is natural, therefore, that Vedic seers have mentioned their characteristics and activities and have desired their welfare. Rig-Veda classifies them in three groups—sky animals like birds, forest animals and animals in human habitation [66]. All the three types of living creatures found in the universe have distance environment and every living creature has an environment (Territory) of its own. But when we look from man's perspective all of them constitute his environment. There is a general feeling in the Vedic texts that animals should be safe, protected and healthy [67]. Domestic animals, as well as wild animals along with human beings should live in peace under the control of certain deities like Rudra, Pushan etc. Vedic people have shown anxious solicitude for the welfare of their cattle, cows, horses etc. The cow as the symbol of wealth and prosperity, occupied a very prominent place in the life of the people in Vedic times.

3.10   Plant and Herbs (Oshadhi)

The knowledge about the origin and significance of plants can be traced out from Vedic literature in detail. In Rigveda one Aranyani sukta is addressed to the deity of the forest [68]. Aranyani, queen of the forest, received high praise from the sage, not only for her gifts to men but also for her charm. Forests should be green with trees and plants. Oshadhi Sukta of Rig-veda addresses to plants and vegetables as a mother, 'O Mother! Hundreds are your birth places and thousands are your shoots.' [69]. The plants came into existence on their earth before the creation of animals [70]. Chandogya Upanishad elaborates generated plants which in turn generated food [71]. The Atharvaveda mentions certain names of Oshadhis with their values. Later this information became an important source for the Ayurveda. The Rig-veda instructs that forests should not be destroyed [72]. The Atharvaveda talks about the relation of plants with earth, 'The earth is the keeper of creation, a container of forests, trees and herbs.' Plants are live. There is an important quotation in a Purana which says, 'One tree is equal to ten sons.' The Atharvaveda prays for continuous growth of herbs, 'O Earth! What on you, I dig out, let that quickly grow over.' And another prayer says, 'O Earth! Let me not hit your vitals. The 'Avi' element referred in the Atharvaveda, as the cause of greenness in trees, is considered generally by Vedic scholars as 'Chlorophyll.' The term 'Avi' is derived from the root 'Av' and thus gives the direct meaning of 'protector.' Hence, plants were studied as a part of the environment and their protection was prescribed by the Vedic seers.

3.11   The Concept of Sacrifice (Yajna)

The sacrifice 'Yajna' is regarded as an important concept of Vedic philosophy and religion but when we study it in its broader sense, it seems to be a part of Vedic environmental science. Yajurveda and Rigveda describe it as the 'navel (nucleus) of the whole world.' It hints that Yajna is regarded as a source of nourishment and life for the world, just as navel is for the child. Vedas speak highly of 'Yajna.' Through it, seers could understand the true meaning of the Mantras. All sorts of knowledge were created by Yajna. It is considered as the noblest action. In simple words, Yajna signifies the theory of giving and take. The sacrifice simply has three aspects: Dravya (material), Devata (deity) and Dana (giving). When some material is offered to a deity with adoration, then it becomes Yajna. Pleasing deity returns the desired material in some different forms to the devotee. This Yajna is going on in the universe since the beginning of the creation and almost everywhere for production and, also for keeping maintenance in the world. Even the creation of the universe is explained as Yajna in the Purusha Sukta. Thus, the concept of Yajna seems to be a major principle of ancient environmental science. In the environment, all elements are inter-related, and affect each other. Sun is drawing water from the ocean through rays. Earth gets rain from the sky and grows plants. Plants produce food for living beings. The whole process of nature is nothing but a sort of Yajna. This is essential for the maintenance of environmental constituents. The view that Yajna cleans atmosphere through its medicinal smoke, and provides longevity, breath, vision etc., is established in Yajurveda. Few scholars have attempted to study the scientific nature of the Vedic Yajnas. Undoubtedly, they have never been simple religious rituals, but have a very minute scientific foundation based on fundamental principles. According to Vedic thought, Yajna is beneficial to both individual and the community. Yajna helps in minimizing air pollution, in increasing crop yield, in protecting plants from diseases, as well as in providing a disease-free, pure and energized environment for all, offering peace and happiness of mind. Moreover, Yajna serves as a bridge between desire and fulfilment.

3.12   Coordination Between All-Natural Powers

Ancient seers knew about various aspects of the environment, about cosmic order, and about the importance of co-ordination between all-natural powers for universal peace and harmony. When they pray for peace at all levels in the 'Shanti Mantra' they side by side express them believe about the importance of coordination and interrelationship among all-natural powers and regions. The prayer says that not only regions, waters, plants trees, natural energies but all creatures should live in harmony and peace. Peace should remain everywhere. The mantra takes about the concord of the universe space of the sky, peace of mid-region, peace of earth peace of waters, peace of plants, peace of trees, peace of all-gods, peace of Brahman, peace of universe, peace of peace; May that peace comes to me.

From the above detailed discussion, some light is thrown on the awareness of ancient Vedas about the environment, and its constituents. The Vedic vision is to live in harmony with the environment was not merely physical but was far wider and much comprehensive. The Vedic people desired to live a life of a hundred years and this wish can be fulfilled only when the environment will be unpolluted, clean and peaceful. The knowledge of Vedic sciences is meant to save human beings from falling into an utter darkness of ignorance. The unity in diversity is the message of Vedic physical and metaphysical sciences. The essence of the environmental studies in the Vedas can be put here by quoting a partial Mantra of the Ishavasyopanishad 'One should enjoy with renouncing or giving up others part'. The Vedic message is clear that the environment belongs to all living beings, so it needs protection by all, for the welfare of all.

Imagine a smart community who has deeply imbibed the Vedic view of environment discussed above in many generations and is organised to create environment resource governance and management, then their goal that is to be realised is well postulated in various Vedic texts and their actions will be well guided.

Buddhism have their own view, non-conformal to many Vedic views on the environment. The lengthy preoccupation of Vedas on cosmology, and the Brahmanical cult of knowledge custodian, monopolist, protector and keepers were rejected in

Buddhism which wants to simplify, make it easy to understand by common people and directly take the entire humanity to the realm of people, their mental makeup and their actions that create a desirable environment. Buddhism was meant for all and not for any castes such as Brahmins. Buddhism today is world religion and its spread has been mentioned in the early part of the chapter.

4 Buddhist View of Environment

There have been many studies on Ecology and Hinduism [73], Ecology and World Religions [74] and Religions and Deep Ecology [75], Ecology and Christianity, Judaism and Buddhism [76–79]. Buddhists understand "Nature as life" that is inter-related and interdependent. Nature, they believe, is alive and at least partly conscious. It is neither sacred and perfect nor evil and to be conquered. The deep reality of Nature is not separate from our fully enlightened nature (the Buddha-nature). Buddhists understand 'Nature' as useful without any unique, intrinsic reality of its own. It can also be understood as the living web that interconnects individual beings, both sentient and non-sentient, in interdependence. What is ultimately real about that web is its Buddha-nature, and its Buddha-ness. When we purify our minds, we experience the true nature of Nature, and then we see that we are living in a Pure Land or Buddhaland. That Buddhaland is not somewhere else, but right here. The Sixth Chan Buddhist Patriarch the Venerable Master Huineng quoted the Buddha as saying: "As the mind is purified, the Buddhaland is purified". From the Buddhist viewpoint, humans are not in a category that is distinct and separate from other sentient beings, nor are they intrinsically superior. All sentient beings have, the potential to become fully enlightened. Buddhists do not believe in treating non-human sentient beings as objects for human consumption. Enlightened beings do not harm sentient life. If they did, they would not be enlightened beings. They have compassion for unenlightened beings, who are attached to our polluted world, filled with pain and suffering, and who do not experience themselves as living in a pure Buddhaland. By looking inward, within one's own body-mind, one gradually realizes that there is no ultimate division between inside and outside, that the patterns of the natural environment are not separate from the patterns of our own body-minds. The experience of those patterns is not considered an ultimate truth or the goal of Buddhist practice, but awareness of them is an important aspect of the Path that leads to enlightenment.

4.1 Nature and the Buddhist Path to Enlightenment

Nature as wilderness is important to Buddhists because it provides a place where rapid progress in Buddhist practice, or self-cultivation, can be made. Nature grounds us and can soothe us. Unspoiled natural locations, usually place in the wilderness where the natural energies are peaceful, are the ideal places for Buddhist practice.

Here are what some traditional Buddhist sources tell us about the benefits of the natural environment as a place for Buddhist practice. The Buddha said, 'I am pleased with that bhikkhu's [monk's] 'dwelling in the forest'. 'And when he lives in a remote abode his mind is not distracted by unsuitable visible objects. He is free from anxiety, he abandons attachments to life; he enjoys the taste of the bliss of seclusion. The more desolate and distant the place is from human habitation, with wild beasts roaming freely about, the more prepared is the mind to soar up from the abyss of defilements, being always like a bird about to fly. The defilements are still there in the depths of the mind, but in such an environment the power of the mind is greatly developed and appears to have gotten rid of hundreds of defilements, with only a few remaining. 'This is the influence of environment which gives encouragement to an aspirant at all times'.

4.2 Nature, Karma and Buddhist Ethics

The core of Buddhist karma-based ethics is respect for life, particularly sentient life. On the everyday level of understanding, Nature changes according to the karma (patterns of intentional causal activities and their consequences) of all sentient beings. Mental pollution causes environmental pollution, and environmental pollution fosters mental pollution. The starting place for understanding just about anything about Buddhism is karma. Karma is the causal network of intentional actions, both mental and physical, that is the foundation of Buddhist ethical understanding. The foremost principle of Buddhist karma-based ethics is ahimsa, the principles of non-harming and of respect for life. This does not only refer to respect for human beings, but also for every manifestation of life on the planet, especially sentient life. As one's mind is purified, one's actions are purified. As a result, not only do mental attitudes that are dissonant or harmful to Nature disappear, but one's new mental states lead directly to more enlightened actions in relation to Nature and more enlightened influence on others about Nature. There is also influence from action to mind. As we act more responsibly towards life in Nature or life as Nature, the more our actions will purify and clarify our minds. Consideration of our actions and their consequences will lead us to more environmentally responsible ethical behaviour. Buddhist monks and nuns vow to follow moral precepts that prohibit harming the environment. There are vows for protecting the purity of the water; for not killing sentient beings who live in the earth; for not killing insects, birds, and animals; for not starting forest fires; and for respecting the life of trees, particularly ancient ones. In the contemporary world, Buddhist monastic communities are developing new ways of applying ancient Buddhist principles to their own environments. For example, in the Dharma Realm Buddhist Association, monks, nuns and lay people are getting involved in recycling; in teaching temple residents and the supporters of temples not to pollute their air, earth, and water; and in reforesting temple properties. While performing the ancient Buddhist rite of rescuing birds and animals originally consigned to death and liberating them, they are developing a new ecologic concern for making sure that those sentient beings are released into environmentally suitable habitats. The principles of compassionate ecology are also being taught in the Association's Buddhist schools. Buddhism stresses the importance of the preservation of the unsullied natural world as a place for practice on the Path to enlightenment. The Buddha said this about what he himself had experienced: There I saw a delightful stretch of land and a lovely woodland grove, and a clear flowing river with a delightful forest so I sat down there thinking, Indeed, this is an appropriate place to strive for the ultimate realization of that unborn supreme security from bondage, Nirvana.

Professor Stephanie Kaza is now retired and moved to Oregon was a professor of Environmental Studies at the University of Vermont and a scholar-practitioner of Zen Buddhism. The following Buddhist view on the environment is taken out of writings and lectures. Her books on Buddhist environmental thought include: Dharma Rain: Sources of Buddhist Environmentalism (co-edited with Ken Kraft, 2000), Hooked! Buddhist Writings on Greed, Desire, and the Urge to Consume (2005) Mindfully Green: A Personal and Spiritual Guide to Whole Earth Thinking (2008). She believed, once you take up the green practice, we see that environmental caregiving become a lifelong task. We cannot do this work alone but with the community. We need the encouragement and inspiration of others to help us find our way [80–83]. What is presented below in the wisdom of environment in Professor Kazak's own words which is brief to the point and motivate environmental action the basis of the smart environment.

4.3   Buddhist Teaching on Environment According to Professor Stephanie Kaza [15]

According to Professor Kaza, there are 7 different views of nature that causes conflicts.

a. Nature as a resource for human use
b. Nature as spiritual home
c. Nature as a set of ecosystem functions
d. Nature as evolutionary legacy
e. Nature as an enemy to be conquered
f. Nature as our collective unconscious
g. Nature as property to be owned

These views of the Natural Environmental leaves people with uncertain challenges and struggle. The scope and scale of these challenges can be overwhelming.

a. economic insecurity, unstable global markets
b. the sense of threat from multiple disasters and pervasive toxins
c. excess stimulation, too much information, cyber overload
d. difficult to share our fears and concerns
e. debilitating emotional states in response to cumulative concern

f. helplessness, despair, fear, frustration, depression, discouragement

There can be much suffering due to

a. Hard to grasp the full scale of human impact
b. Global forces shaping future for all local systems
c. Fear, hope, anxiety add to physical hardships
d. Difficult to discuss and find appropriate action and a tangled web
e. Economic, social, environmental systems intertwined
f. Conflicting values, religions, governance structures
g. Consequences of past choices are still unfolding

How can we find a way forward amid so many challenges?

(1) Stable or smaller world population

- Education for women, access to family planning options
- Increase family security and literacy

(2) Elimination of mass poverty

- Cooperative support from developed nations
- Stabilization of government structures

(3) Environmentally benign technologies

- Cradle to cradle thinking
- Green building design
- Public transport (ex. BRT, bus rapid transit)

(4) Environmentally full-cost pricing

- Include externalities
- Reform GDP measures
- Green taxes

(5) Sustainable consumption

- Product certification, green labelling
- Corporate accountability for production
- Safe chemicals in packaging
- Eat local, eat less meat

(6) Green knowledge and learning

- Campus sustainability movement
- Environmental majors
- Green jobs training

(7) Global environmental governance and cooperation

- Work with China and India to reduce carbon emissions
- Meet new EU standards for toxics and e-waste

(8) Transformation of consciousness

- Valuing quality of life and well-being
- Seeking common ground for a paradigm shift

How can religion help meet this challenge? It Creates Five capacities

(1) Engage members of faith-based groups
(2) Moral authority offers ethical guidelines, religious leadership
(3) Provide meaning by shaping worldviews consider new paradigms of well-being
(4) Share physical resources i.e. retreat centres, temple grounds, schools
(5) Build a community to support sustainability practices

Basic Buddhist teachings and practices common to all traditions are;

(1) Moral guidelines based on non-harming
(2) Central law of interdependence and causation
(3) Belief in liberation from suffering through insight
(4) Practices that strengthen intention and compassion

What teachings can Buddhism offer on the environment?

(1) Develop skilful means:

- Cultivating mindfulness
- Non-ego-based action
- Practicing equanimity

(2) Follow ethical guidelines

- Non-harming
- Practicing restraint
- Caring for other as self
- Taking the deep view

(3) Take up new paradigms for the well-being

- Relational thinking
- Practice path approach
- Seeking green wisdom

(4) Build a community for the shared purpose

I. Skilful Means, *Upaya*, in Sanskrit = using a range of appropriate techniques to relieve suffering or share teachings with different audiences and situations. This is the heart of compassionate response, because it is based on what is needed and what can be received.

Setting one's intention to be skilful.
Responding to the call

- What is needed? What can I do?

- What is effective action?
- What is meaningful?

(1) Cultivating mindfulness, medicine for the suffering of the world

  - developing the capacity to be with each moment as it is
  - mitigating chaos with calmness
  - restoring the "natural inward measure"

Mindfulness in action
Examples:

- mindful eating
- mindful leadership training (Shambhala Institute)
- mindfulness-based stress reduction training
- mindfulness practice in the schools
- Practicing awareness builds resilience in a world of challenges.

(2) Non-ego-based action

  - understanding interdependence of self and other
  - becoming aware of triggers and personal needs
  - observing the influence of ego and power
  - refraining from polarizing around the difference
  - practicing self-reflection as part of the action
  - Selflessness in action
  - Centre for Whole Community diversity and conservation work
  - mindful Occupy (Buddhist Peace Fellowship)
  - Reducing self-interest results in stronger shared outcomes.

(3) Practicing equanimity

  - being with the suffering of the world while still taking effective action
  - developing the capacity to stay cantered in changing conditions
  - being prepared for environmental impacts on social support systems
  - Stability and calmness increase the possibility for insight and kindness.

II. Living by an ethical compass

Choosing restraint based on awareness of others and commitment to social stability.
Rethinking priorities:

- choosing where to invest time, energy, money, relationships
- Considering how to care for the environment as self, self as the environment
- examining abuse of all beings, human and other
- considering cultural conditioning that condones environmental harming
- Investing time in regular self-reflection to ask: What is important? What really matters?

(1) Non-harming
     Taking the Bodhisattva vow

- reduce harm by practicing kindness
- acting with restraint and awareness of others

Choosing contentment

- support infrastructures of well-being
- the state of being not dissatisfied, the absence of craving
- recognize personal and social indicators of contentment

Examples

- food choices based on the degree of harm
- mobility choices based on the degree of harm
- parenting choices for a non-toxic household

Creating ecological, economic, ethical sustainability for the long

(2) Practicing restraint

Desire = grasping or craving after something, identifying with the craving, being "hooked" by addictive needs, the focus is on short term gratification.
     Three types of desire:

- The desire for more of something (greed)
- The desire for less of something (aversion)
- The desire for illusory options, fantasies (delusion)

Shenpa = "that sticky feeling" (Tibetan) that makes us insecure, uncomfortable, wanting to escape the world that's always changing; the urge for relief, for comfort food, alcohol, drugs, sugar, shopping, entertainment.
     Shenlok = refraining from acting on that urge, turn shenpa upside down, break open self-limiting patterns.
     Self-awareness of desire and the practice of restraint can be personally and socially transformative, a foundation for ethical behaviour.

(3) Caring for other as self

Buddhist precepts provide these primary guidelines:

- Not harming life
- Not taking what is not given
- Not participating in abusive relations
- Not speaking falsely
- Not using intoxicating substances or behaviours

These have environmental implications for personal and social behaviour.
Ethical people create an ethical society; an ethical society creates ethical people.

(4) Cultivating a Deep View

Observing time to practice a shift in perception

- the vast eons of time that created the earth and ocean, plants, and animals in contrast with the very short period of human life
- understanding that the past is linked to the present,
- a correction for short-term thinking

Ethics based on the long view of time and the human place in that long unfolding develops these capacities:

- patience
- perspective
- humility
- endurance
- serenity

III. Paradigms for well-being

- Individual = good health, satisfying work, supportive relationships, and sense of internal control in your life
- Social = safe and civil society, appropriate governance and market structures to support; community well-being
- Global = free from war, poverty, resource depletion; engaged in collaborative support for; planetary well-being

Spiritual well-being

- right relationship with self and community
- right relationships with the natural world
- ethical clarity, capacity for restraint from harming
- tolerance, respect for other paths

(1) Practicing Rational Thinking

- Experiencing flux-balance, the flowing ch'i of life force
- All beings relating to time and space and to others around them
- Perceiving from a systems perspective
- Human life as relational, both socially and ecologically

(2) Understanding self-more clearly

- Self as a process rather than an object
- Seeing ourselves in others, self-realization
- Ecological self, "interbeing"
- Self as an agent in the web of life

(3) Taking practice path Approach

- Orienting toward well-being for all

- Starting where you are, going from there
- Reaffirming intention
- Practicing ethical guidelines with others
- Deepening from beginner to novice to lifeway, as life allows

(4) Seeking Green wisdom

- From green mentors, trees, places, animal and plant beings
- Finding guidance and support, strengthening our intention
- Receiving the teachings of all beings, people, and places
- Following paths laid down by others before us
- Practicing wild mind, the original face of life arising

IV. Building a mindful society is a community effort.

Three jewels in Buddhist Teaching

- Buddha = the teachers
- Dharma = the teachings
- Sangha = the community of friends practicing together

Investing in community strengthens shared purpose.

- Living in a place, being part of the local ecosystem, seasons, weather
- Supporting local governance and business, cultivating social networks
- Reclaiming time, caring for food, the healing work of gardening
- Reducing consumption, living simply with more direct experience
- Cultivating friendships, deepening awareness practice

Ethical guidelines stabilize community relations.

- Growing a new civility based on kindness and respect
- Influencing each other toward more sustainable social norms
- Choosing to be in the web of relations as an active agent
- Teaching children to develop social values based on non-harming

New paradigms for well-being generate contentment.

- Recalibrating what matters through mindful awareness
- Celebrating the richness of experience and relationship
- Bringing full presence of mind and body to our interactions
- Sharing green wisdom with others on the practice path
- Taking the deep view of the place, health, nature, society
- Aiming with intention toward a mindful society, based on a deep view of mind and nature.

4.4 The Synthesis

The Vedic and Buddhist view on the environment clearly postulates a way of life which teaches how to live with an environment in a harmonious way that results in a sustainable environment. Once it happens in a smart city then it can be called smart living for the smart environment of the integrated smart city system of six components discussed above. The smart living advocated in this chapter is based on the Vedic and Buddhist view as two examples of the environment which differs from current practice but leads to smart living. The main difference is the Vedic view of the environment is based on aesthetic and scientific perception and appreciation of the environment which has in turn generated the protocol of environmental interaction-culture as part of Smart Vedic way of life that creates a smart environment. The Buddhist Environmental View is based on mind cultivation or mind-culture which is how the mind can be changed and cultivated with respect to the view of the environment that leads to the protocol of environmental interaction which is the basis of smart living based on the eight-fold path and three jewels which creates a smart environment as discussed above. Three jewels in Buddhist teaching are Buddhas (all humanity are Buddhas in making) are the teachers of environmental wisdom, Dharma the process of the teachings and practicing of environmental wisdom and Sangha, the community of friends which in the smart city terminology smart community practicing together environmental wisdom.

The cultural system has imbibed the environment aspect of faith discussed above which has been passed over from one generation to another in many centuries. If one analyses the environmental legislations which are the backbone of environmental regulation of the country, undoubtedly the cultural value system is the fundamental basis of formulation of these environmental legislations. This fact can be easily identified and verified if one makes one aspect to one aspect study of practicing religion and its belief and the environment legislation in India even if it is a secular and democratic country. The smart community is the implementation arm for all. environmental action based on awareness of the environment discussed under Vedic and Buddhist teachings. It is more efficient, cost effective than the current administrative system that exists and regulates the environment in India today. The concept of smart community has been well shaped by the concept of Sangha in the Buddhism discussed above. Buddhism shaped the concept of Sangha based on earlier concepts of Republics followed by ancient republics in India before the advent of Buddhism some 2500 years ago. During the colonial era when economic development of the country decelerated and degenerated rapidly and famines took place, and environment cultural norms of ancient India was replaced by a colonial administrative system and codes that was the beginning of the cultural, environmental and economic destruction of India having a great adverse impact even today. The colonial power was governed by greed, favouritism that divide and rule and corruption to extract and transport resources to their country which degraded the Indian environment. The greed and corruption in India has not disappeared from India after independence and so also the British administrative system that helped colonial rule. The colonial rulers tried to destroy ancient cultural practices as related to the environment. There was a need to detoxify India out of colonial rule, but it never took place from 1947 until 1991. The failed socialism of USSR became a model of development for India with their draconian and alien institutions up till 1991 with its mixed economy concept where the local initiative was suffocated and nullified by administrators, which further strengthened the alien colonial Indian Administrative System a name sake change from Indian Civil Service of British coloniser. It failed to resurrect what we have practiced in India thousands of years ago in the republics and Buddhist Sangha.

The influence of the smart city was positive in smart environmental management and Governance. Environmental resources management and Governance became fully constantly updating database using IOT and ICT system, use of environment modelling in the computing cloud for E-Democracy and E-Governance related to the environment using apps based on Spatial Decision Support System, by the smart community through well-defined E-Democracy and E-Governance System discussed in this chapter. A smart community was required to continue their never-ending skilling and education to use effectively IOT and ICT system and implement E-Democracy and E-Governance for the environment more effectively. Even today the 100 smart city programs have not done anything worthwhile for nurturing smart communities for the smart environment.

5 Conclusion

1. Smart Environment has been defined in this book. "Smart Environment is a knowledge based environment that develops extra ordinary capabilities to be self-aware, how it functions 24 h and 7 days a week and communicate, selectively, in real time knowledge to citizen end users for a satisfactory way of life with easy public delivery of services, comfortable mobility, conserve energy, environment and other natural resources, and create energetic face to face communities and a vibrant urban economy even at a time there are National economic downturns". This is similar to the definition of a smart city used in this book.

2. Just like Smart City system, it is a six-component integrated system such as Smart People, Smart Mobility, Smart Economy, Smart Governance, Smart Environment, and Smart Living; Smart Environment can also be considered as a six-component system namely for example for water element of environment, Smart water community, Smart Water Mobility, Smart Water Economy, Smart Water Governance, Smart Water Environment and Smart Water Centred Living. The same type of smart environment system can be designed for say Smart Forest.

3. Three types of design need to be generated for the smart environment. They are ICT and IOT system specific to the local environment needs, E-Environment Democracy and Environment Governance. All these three designs shall have linkages with six Smart Environment System.

4. IoTs are a network of technologies which can monitor the status of physical objects, capture meaningful data, and communicate that data over a wireless network to a computer in the cloud for software to analyse in real time and help determine action steps. Typically, each data transmission from a device is small, but the number of transmissions can be frequent. Each sensor will monitor temperature, pressure, rainfall or water quality and a natural part of the environment such as an area of ground to be measured for moisture or chemical content (pollution).

5. As many IoT nodes are spread across the study area, the setup requires a robust communication technology that covers a wide geographical area and can handle a huge amount of data traffic. For the unconstrained IoT nodes we use the traditional LAN, MAN, and WAN communication technologies such as WIFI, optic fibre, Ethernet, broadband.

6. In a smart environment system, there are design components and protocol component. While design can be generic, the protocol can be specific to environment issues.

7. The environment is the eco system. The Scientific study of the eco system has progressed considerably around the world which may be called scientific ecology. We have now a basket of descriptive models of the eco system, diagnostic models and predictive models which can be calibrated using mathematical tools. We have analytic tools to determine the environmental impacts. In other words, the knowledge base of the environment is available but requires more development within the university system of research to answer emerging environment issues which is ever changing. These models have remained within the confines of universities but have not come down to the field for applications where it is required to sort out environmental issues on a day to day basis. All means available knowledge should be used to share these environment knowledge bases based on the most up-to-date data.

8. The environment or eco system become smart when environment knowledge base, or ecosystem knowledge base is available at the place and to people it is required for acting. This calls for developing analytics for a common man based on these models available in cloud computing system accessed by apps in the smart phone and can be used to sort out emerging issues with or without human intervention using automata. This can only be executed at household and community level faced with identical problems addressed in the apps. The job of creating apps for common men based on the latest environment information and analytical models is the unfinished or never attempted jobs of universities or business. The other unfinished or never attempted job of the university and Non-Governmental Agencies is to strengthen community and households to be smart with minimal but up to date knowledge base of the eco system and analytic apps that enable the household/community to sort out the issues using apps in smart phones.

9. The domain of knowledge base required to manage the eco system issue at one location cannot be found in one department of studies in these universities alone which make this operation of linking university, developing ICT and IOT

infrastructure and smart phone apps for the field more complex. The best remedy is community concentrate first on the most pressing environmental issues and then add the next one to the list subsequently. For example, if a community has water related and sanitation related issues with no interconnections with each other and water issue is the most pressing issue, then start with water issue and a water community who subsequently embrace other issues.

10. There is a continuous training requirement for the community to make them capable of intervening scientifically and based on most current information to face the environment issues which can be an important task of the local university or business. Here the university/business train people with differing education level to face the issues to tackle the problems using mobile computing.

11. Environmental issues and required timely intervention require an appropriate ICT and IOT infrastructure to be developed in the study area for effective interventions again by the university/business which they are capable of. The technical management of these systems to tackle environment issues can also be designed by the local university/business. Internet based technical protocol for managing the ICT and IOT system can also be developed by the local university. Alternatively, this can be a livelihood of young university graduates doing related business.

12. In addition to ICT and IOT design and deployment for issue-based environment management at community and household level, there is a need to develop E-Environment Democracy Applications and E-Environmental Governance Applications as per specific needs.

13. E-Environment Democracy is the rightful role of citizens to participate in the environment management of smart cities as per the constitutions of the country using electronic means such as high-speed fibre optic internet network, mobile computing as well as Internet of Things that can transform the community to smart community with dominant role of citizen, and technology in smart cities. Majority of environment resources such as the air we breathe, the water we use, the public domain urban space and property we use every day are all common property resources and E-Democracy is the only way it can be used efficiently, responsibly and rationally.

14. The issues E-Environment Democracy faces are spatial environment issues which can only be solved by the domain eco system knowledge base. It requires spatial analytics using Geographic Information System based well-developed suites of Spatial Decision Support Systems mobile applications to intervene effectively. There is not much development taking place in this area and smart cities can form smart cities association and jointly develop such applications with local universities and share freely.

15. E-Environment Democracy ideally should follow Deep Ecology principles stated in this chapter integrated with scientific ecology to create a smart environment. This calls for a change of view on environment emanated from the first industrial revolution that advocates that the human and environment resources need to be conquered to meet the never-ending greed resulted in colonialism and slave traffic and greed-based living. These approaches should perish and

replace with thoughts discussed in this chapter based on a Vedic view of the environment and Buddhist view of the environment or any other religion being practiced.

16. E-Democracy requires directive principles which are stated in this chapter.

17. Most of the E-Governance applications are not based on water or other natural resources management using existing legislations. They are mostly centred around office management of clerical operations such as issuing birth and death certificate or caste certificate or land parcel GIS. The absence of E-Environment mobile applications to monitor and stop point source pollution or non-point source pollution is not there and the existing implementation of environment governance is far from satisfactory.

18. E-Environment Governance and E-Environment Democracy can only be implemented by the smart community including households since they know the issues and they want a remedy. A registration under Indian Societies act or similar in other countries makes this community a legal entity in India which can access Commercial Bank loans and are accountable to Government. These smart community shall be armed with technology, domain knowledge base, fully operational E-Democracy, E-Governance and continuing education set up to be effective.

19. A smart city should not be considered as a technological artefact. It is an ecological and cultural system. The ecological system defines the environment issues and can be diverse at every location within the city and may differ from other cities. A cultural system that interacts with the environment arises out of the way of life which is influenced by religious practices.

20. The manifestation of the cultural system in a city is determined by people who live there with their differing religions and caste compositions in a country that follows secular constitutions like India. These religions influence their followers to adopt a distinct world view of the environment as per their scriptures which are discussed in this chapter for two sample religion the Hinduism and Buddhism. They also have environment ethics on how to intervene with the environment. The Spatial Decision Support System Mobile applications should consider these facts while formulating applications as a robust value system based on religious belief.

21. Smart Community is the creator and sustained of the smart environment. Their cultural system is the key to community action which cannot be dictated from outside. The environmental view of the religion they practice is that one which directs their action.

22. The Environment view of Vedas gives how Veda view the environment and establish a protocol for environmental intervention by people and same with Buddhism. This is discussed in this chapter.

23. The way many schools of Buddhism such as Mahayana, Theravada, Chan and Zen Buddhism and Buddhist Tantric practices view and intervene with nature in a different way reinforces the cultural system. This is discussed in this chapter. Buddhism consider all elements of the environment as equal and even non-living. Humans are not superior to these elements. The conditioning of mind

and practice of mindfulness is the basis of intervention with the environment as Buddhist practice.

24. Religious and caste leader has a role in shaping the intervention of local community with nature and creating a smart environment as much as the practitioners of scientific ecology.

25. Most commonly used e-tools such as e-discussion, e-initiative, e-petition, e-consultation, e-feedback, e-complaints, e-polls, e-voting, e-campaigning, e-budgeting, e-meetings, e-democracy games, e-award and web casting may be used for e-democracy.

References

1. Batchelor M, Brown K (eds) (1992) Buddhism and Ecology. Cassell Publishers, London. One of five introductory volumes on the five major world religions and the environment, this set of essays addresses Buddhist teachings and practice and how they apply to ecological issues
2. Boston Research Centre for the 21st Century (1997) Buddhist perspectives on the Earth Charter. Boston Research Centre, Cambridge. A range of reflections on the potential for the Earth Charter to support Buddhist values in contemporary challenges to sustainability and peace
3. Gross R (1998) Interdependence and attachment: toward a Buddhist Environmental Ethic (pp 75–93) and Buddhist values for overcoming Pro-Natalism and Consumerism" (pp 108–124). In: Soaring and settling. Continuum, New York. Thoughtful and challenging reflections on difficult environmental subjects, bringing fresh interpretations of fundamental Buddhist principles
4. Habito RLF (1993) Healing breath: Zen Spirituality for a wounded earth. Orbis Books, Maryknoll, N.Y. An introduction to Zen practice in the context of personal and planetary woundedness, approached here through the Christian frame of love as well as the Buddhist emphasis on awareness
5. Harris I (2002) Buddhism and Ecology. In: Keown D (ed) Contemporary Buddhist Ethics. Curzon Press, Richmond, Surrey, pp 113–136. The author provides historical and philosophical evidence for and against the existence of a Buddhist environmental ethic
6. Ryan PD (1998) Buddhism and the natural world: toward a meaningful myth. Windhorse Publications, Birmingham. An in-depth study of the Buddhist version of a creation story, interpreted from an environmental perspective
7. Macy J (1991) Mutual causality in Buddhism and general systems theory. State University of New York Press, Albany. Analysis of Buddhist views of reality showing convergences with modern systems theory, including self as a process, co-arising of knower and known, body and mind, doer and deed, self and society
8. Bahuguna S, Shiva V, Buch MN (1992) Environmental crisis and sustainable development. Natraj Publishers, Dehradun
9. Beck R, Kolankiewicz L (2001) The environmental movement's retreat from advocating U.S. population stabilization (1970–1998), the first draft of history, 'Causes 3 & 4', June. See www.mnforsustain.org/beck_environmental_movement_retreat_long3_causes2.htm
10. Bookchin M (1996) The philosophy of social ecology: essays on dialectical naturalism. Rawat Publications, New Delhi
11. Doyle T, McEachern D (1998) Environment and politics. Routledge, London
12. Fox W (1989) The deep ecology-ecofeminism debate and its parallels. Environ Ethics 11(1), Spring
13. Fox W (1984) Deep ecology: a new philosophy of our time? Ecol 14(5–6)
14. Macy J, Brown MY (1998) Coming back to life: practices to reconnect our lives, our world. New Society Publishers, Gabriola Island, B.C. Practical experiential exercises to introduce deep ecology and Buddhist principles into environmental perception and organizing

15. http://www.greenfaith.org/files/buddhist-environmental-slides-stephanie-kaza/at_download/file
16. Washburn D, Sindhu U, Balaouras S, Dines RA, Hayes NM, Nelson LE (2010) Helping CIOs understand "Smart City" initiatives: defining the smart city, its drivers, and the role of the CIO. Forrester Research, Inc., Cambridge, MA. Available at http://public.dhe.ibm.com/partnerworld/pub/smb/smarterplanet/forr_help_cios_und_smart_city_initiatives.pdf
17. Giffinger R, Fertner C, Kramar H, Kalasek R, Pichler-Milanovi N, Meijers E (2007) Smart cities: ranking of European medium-sized cities. Centre of Regional Science (SRF), Vienna University of Technology, Vienna, Austria. Available at http://www.smart-cities.eu/download/smart_cities_final_report.pdf
18. Glaeser EL, Berry CR (2006) Why are smart places getting smarter? Taubman Center Policy Briefs, PB-2006-2. Available at http://www.hks.harvard.edu/rappaport/downloads/policybriefs/brief_divergence.pdf
19. Hall RE (2000) The vision of a smart city. In: Proceedings of the 2nd international life extension technology workshop, Paris, France, Sept 28. Available at http://www.osti.gov/bridge/servlets/purl/773961-oyxp82/webviewable/773961.pdf
20. Harrison C, Eckman B, Hamilton R, Hartswick P, Kalagnanam J, Paraszczak J, Williams P (2010) Foundations for smarter cities. IBM J Res Dev 54(4). https://doi.org/10.1147/jrd.2010.2048257
21. Rios P (2008) Creating "the smart city". Available at http://dspace.udmercy.edu:8080/dspace/bitstream/10429/20/1/2008_rios_smart.pdf
22. Partridge H (2004) Developing a human perspective to the digital divide in the smart city. In: Proceedings of the biennial conference of Australian library and information association. Queensland, Australia, Sept 21–24. Available at http://eprints.qut.edu.au/1299/1/partridge.h.2.paper.pdf
23. Hartley J (2005) Innovation in governance and public services: past and present. Public Money Manage 25(1):27–34
24. Torres L, Pina V, Royo S (2005) E-government and the transformation of public administrations in EU countries: beyond NPM or just a second wave of reforms? Online Inf Rev 29(5):531–553
25. Vinod Kumar TM, Associates (ed) (2014) Geographic information system for smart cities. Copal Publishing Group, New Delhi
26. International Centre for Communications, San Diego State University with direction from the California State Department of Transportation (1997) Smart communities guidebook. How California's Communities can thrive in the Digital Age
27. Bryce M (2003) Innovation, broadband and the smart community technology park, Perth, 03 05 05 Keynote on the Innovation Festival
28. The Government of Canada, Ministry of Industry (1999) Connecting Canadians, Smart Communities
29. Smart Community International Network (SCIN) www.scin.org
30. Smart Communities Guidebook, Pete Wilson, Governor of California (1997)
31. European Union, Telecities, www.telecities.org (2003)
32. The European Union, Euro cities, www.eurocities.org (2003)
33. Smart Valley, Inc. (1998) Smart Valley
34. Arup (2010) Smart cities transforming 21st century city via creative us of technology, 1st edn published September 2010. London, Hongkong, San Francisco, Sydney. www.Arup.com
35. Shiva V (2002) Water wars; privatization, pollution, and profit. South End Press, Cambridge, MA
36. Vinod Kumar TM (ed) (2015) E-Governance for smart cities. Springer, Singapore
37. Vinod Kumar TM (ed) (2016) Smart economy in smart cities. Springer, Singapore
38. Vinod Kumar TM (ed) (2017) E-Democracy for smart cities. Springer, Singapore
39. Rifkin J (2013) The third industrial revolution: how lateral power is transforming energy, the economy, and the world, Kindle edn
40. Rifkin J (2015) The zero marginal cost society: the internet of things, the collaborative commons, and the eclipse of capitalism. Kindle edn

41. Batty M, Axhausen KW, Giannotti F, Pozdnoukhov A, Bazzani A, Wachowicz M, Ouzounis G, Portugali Y (2012) Smart cities of the future. Eur Phys J Spec Top 214:481–518
42. Vinod Kumar TM (2015) E-Governance for smart cities. In: Vinod Kumar TM (ed) E-Governance for smart cities. Springer, Singapore
43. Dahiya B (2015) Shifting focus to the mobility of people. China Daily – Asia Weekly, 6–12 February. Available at https://www.academia.edu/10556084/Shifting_focus_to_mobility_of_people_China_Daily_-_Asia_Weekly
44. UN-HABITAT (2013) Planning and design for sustainable urban mobility: global report on human settlements 2013. Routledge, Abingdon, Oxon
45. Griffith RTH, Ferqus JW, Keith AB (10 January 2017) The Samhitas of the Rig, the Yajur, the Sama and the Atharva Vedas, Single Volume Unabridged, Createspace publishing independent platform full Sanskrit text and English translation distributed by Amazon
46. Panchamukhi AR (1998) Socio-economic Ideas in Ancient Indian Literature. Rashtriya Sanskrit Sansthan, Delhi, p 467
47. Roy RRM (1999) Vedic physics, scientific origin of Hinduism. Golden Egg Publishing, Toronto, p 58
48. Yajurveda 7.5
49. Rigveda 1.164.33
50. Rigveda 1.159,160
51. Atharvaveda, 12.1.12
52. Rigveda 7.49.2
53. Taittiriya Aranyaka 1.24.1-2
54. Atharvaveda 1.6.4
55. Atharvaveda, 4.27.4
56. Chandogya Upanishad 7.10.1
57. Shatapatha Bra. 1.9.3.7
58. Manusmriti 4.56
59. Roy RRM (1999) Vedic physics, scientific origin of Hinduism. Golden Egg Publishing, Toronto, p 84; Jaiminiya Bra. 1.192; Yajurveda 1.24; Shatapatha Bra. 8.7.3.10
60. Taittiriya Upanishad, 2.4
61. Rigveda, 1.37.2
62. Ibid 10.186.1
63. Taittiriya Upanishad 1.6.1; 1.5.1
64. Yajurveda 5.43
65. Ibia, 34.1-6
66. Rigveda 10.90.8
67. Yajurveda 19.20, 3.37
68. Atharvaveda 11.2.24
69. Chandogya Upanishad. 6.2.4
70. Rigveda 10.146
71. Rigveda, 10.97.2
72. Ibid, 10.97.1
73. Chapple CK, Tucker ME (eds) (2000) Hinduism and ecology. Harvard Center for the Study of World Religions, Harvard University Press
74. Barnhill DL, Gottlieb RS (eds) (2001) Deep ecology and world religions. State University of New York, Albany, NY
75. Barnhill DL (2001) Relational Holism: Huayan Buddhism and deep ecology. In: Barnhill DL, Gottlieb RS (eds) Deep ecology and world religions. State University of New York Press, Albany, NY, pp 77–106
76. Tucker ME, Williams DR (eds) (1997) Buddhism and Ecology. Harvard Center for the Study of World Religions, Harvard University Press
77. Tucker ME, Williams D (eds) (1997) Buddhism and Ecology: the interconnection of Dharma and Deeds. Harvard University Press, Cambridge. First serious scholarly collection of papers in the emerging field of Buddhism and Ecology. Includes critiques, field studies, and reinterpretation of classic texts

78. World Wide Fund for Nature: Christianity and Ecology (1992) Judaism and Ecology, Buddhism and Ecology, Hinduism and Ecology, Islam and Ecology, a series of volumes. Cassell Publishers, London
79. Hunt-Badiner A (ed) (1990) Dharma Gaia: a harvest of essays in Buddhism and Ecology. Parallax Press, Berkeley. An early collection of provocative articles and poetry addressing the intersection of Buddhist and environmental thinking and perception
80. Kaza S (1993) The Attentive Heart: conversations with trees. Ballantine, New York. Meditative nature writing essays from a Zen deep ecological perspective, with a particular interest in meeting individual trees as sentient beings
81. Kaza S (2000) To save all beings: Buddhist environmental activism. In: Queen C (ed) Engaged Buddhism in the west. Wisdom Publications, Boston, pp 159–183. An overview of Buddhist environmental history and initiatives in the United States with a summary of key philosophical principles used by people in the field
82. Kaza S, Kraft K (eds) (2000) Dharma rain: sources for a Buddhist environmentalism. Shambhala, Boston. Foundational anthology of classic and modern texts, including advice and reflection from Buddhist environmental activists
83. Kaza S (ed) (2005) Hooked! Buddhist writings on greed, desire, and the urge to consume. Shambhala, Boston. Essays by leading Buddhist teachers, scholars, and practitioners on Buddhist ethics related to modern issues of consumerism

Part I
Hong Kong

Smart Environment for Smart and Sustainable Hong Kong

Sujata S. Govada, Timothy Rodgers, Leon Cheng and Hillary Chung

Abstract Smart City is about incorporating a 'Smart Thinking' in the city's urban design and planning with a focus on People Place and Planet. This chapter will explore how Smart and Sustainable Environment is achieved when applying 'Smart Thinking' to the city's development with Hong Kong as the case study of the principles. Hong Kong has more recently implemented environmental policies and initiatives in enhance the quality of the environment. By studying the various government policies at a strategic level that improve the city's environment and analysing the public, private, institutional, academic and community initiatives, it provides a better understanding of the recent efforts in this regard. By assessing and reviewing Hong Kong's effort in creating a smart and sustainable city, this chapter enables us to conclude with the overall quality and future positioning of the city's environment in achieving a smart and sustainable Hong Kong.

Keywords Smart and sustainable city · Smart environment · Urban design · Planning · Green building · Zero carbon building · Pedestrianisation · Resource management · Technology · Open space · Harbour conservation · Pollution · Water management · Energy efficiency

S. S. Govada (✉)
School of Architecture, Institute for Sustainable Urbanisation, Chinese University of Hong Kong, UDP International, Hong Kong, China
e-mail: sujata@udpcltd.com

17/F, Tesbury Center, 28 Queen's Road East, Wan Chai, Hong Kong, China

S. S. Govada · T. Rodgers · L. Cheng · H. Chung
Institute for Sustainable Urbanisation, UDP International, Hong Kong, China
e-mail: tim@udpcltd.com

L. Cheng
e-mail: leon@udpcltd.com

H. Chung
e-mail: hillary@udpcltd.com

© Springer Nature Singapore Pte Ltd. 2020
T. M. Vinod Kumar (ed.), *Smart Environment for Smart Cities*, Advances in 21st Century Human Settlements, https://doi.org/10.1007/978-981-13-6822-6_2

1 Smart and Sustainable Cities

Resilience in cities has been the trending idea in the 21st Century, and rightfully so. More and more people are realising that climate change and global warming is no longer a discussion of something that is decades away but that it is beginning to impact our lives now. The recent publication by the Intergovernmental Panel on Climate Change (IPCC) has suggested that since the pre-industrial era, there has been an increase in the emission of anthropogenic Green House Gases (GHG) due to economic and population growth, which is the dominant cause of climate change. In 12 years' time, the world is looking at an increase in 2 °C, which will impact the bio-diversity and ecosystem with risks of rising sea-level, forest fire, species extinction, melting of ice sheets, and more. Take Asia as an example, the rise in temperature will induce higher risk of flooding causing damage to infrastructure, livelihood and settlements; heat-related incidents will increase human mortality; as well as droughts leading to water and food shortage. It is clear that humans have influence on climate change, leading to widespread impact on human and natural system. The report also warned that if we continue the current emission of GHG, further warming and long-lasting changes in all components of climate system, thus increase the likelihood of severe and irreversible impacts for people and eco-systems. Therefore, mitigations of environmental impacts, complemented by adaptations to climate change, are necessary to reduce and manage the risk. Substantial emission reduction over the next few decades would promote a better prospect for effective adaptation and contribute to climate-resilient pathways for sustainable development. The effectiveness of the strategies depend on some factors including whether there is effective institutions and governance, innovations and investments in environmentally sound technology and lifestyle choices [1].

Cities around the world have taken a conscious effort in adopting smart city concepts and strategies to shape better cities for our environment. However smart and sustainable city concepts need to be viewed together by integrating good planning and design into their physical and social fabric with a focus on people, place and planet along with information and communication technologies (ICT) to improve the quality of life of the city and its people. Through technology, cities can function more efficiently and adapt to their new environments better by utilising real-time data monitoring, assessment and sharing as information is received and can optimise decision making to better understand and improve the status quo and quality of city life. The concept of "Big Data" is increasingly popular because it allows for more timely decision making with more informed and accurate data. However, it is important for "Big Data" to be open and transparent to the public because it can foster and influence other opportunities for improving cities, including better planning and design. Equally, the objectives of educating, empowering and engaging with community and integrating technology into cities should be well thought out as the guiding principles of smart and sustainable cities.

To begin formulating principles for smart cities, first the definition of what constitutes as a smart city must be established. There are many different definitions, frameworks and approaches when it comes to understanding smart cities. In many of the smart city initiatives, there is a heavy emphasis on technology/ICT as the main component of these frameworks and models. Although technology has without a doubt played a significant role in the functioning, development and growth of cities around the world, nevertheless in some cases it is relied too heavily upon such that the overarching principles of how to effectively create a smart and sustainable city is lost.

We believe a smart and sustainable city should be discussed together, and should focus on encompassing three core values—people, place, and planet, while highlighting planning and design with 'smart thinking' as the central themes of developing smart and sustainable cities. Our framework consists of six elements in city development including Smart Living, Smart Environment, Smart Mobility, and Smart Infrastructure, Smart Governance and Smart Economy. People, Place, and Planet are taken at a higher level to establish its significance and contribution to smart and sustainable development. This re-defined positioning on smart and sustainable cities no longer places technology at the centre, but rather as a tool to enable the underlying concepts to flourish so that the fundamentals of smart and sustainable city development remains. These three core values combined with the central themes forms a holistic view for existing and future smart city developments. Our framework defines smart thinking between people, place, and planet as encouraging people to adopt an environmentally aware, sustainable and healthy lifestyle. With this new mind-set, cities can be planned and designed at a human scale for its people, to create a place with regards to the needs and aspirations of the people. In addition it is also critical to form a climate change adaptive and resilient city for the planet by considering the natural resources and built environment together to create a smart environment that is sustainable as well.

2   Smart and Sustainable Environment

Using our Smart City framework, we further develop the concept of sustainable development to achieve Smart Environment. Our key strategies for sustainable development are concerned with the natural environment of the city. A smart environment implements smart resource management for public open spaces to create a place where the people as well as the natural ecology and biodiversity can coexist in the dense urban environment in balance to provide a stimulating milieu for people to live, work and spend leisure time. Open spaces should be plentiful and abundant in greenery, as these spaces provide a place for social interactions and leisure activities, facilitating an inclusive and cohesive society that will provide physical, psychological and social health benefits for individuals and communities. A smart environment should have natural environment near to the built environment, promote and encourage Green Building designs and sustainable neighbourhoods, implement energy saving techniques, utilize sustainable materials, and manage waste, water and electricity usage efficiently. The public sector should take on the responsibility to lead by examples with a clear vision and encourage private sector, NGOs, institutes and community to follow, which can in turn influence social behavioural changes and have significant impact on the environment. The Government of a smart environment should also substantially advocate for environmental protection and play a leading role in educating the public on the dangers of unsustainable living, and also in introducing policies and regulations to safeguard the natural environment from excessive development or detrimental emissions, such as pollution control and management. The general public and the community needs to be more educated, aware and be more proactive and smart about protecting and safeguarding the environment. It is time to realise that it is our duty to take care of the planet and inspire others by our responsible behaviour.

3 Overview of Hong Kong's Built Environment

Hong Kong is one of the densest and most compact cities in the world with a population of over 7 million people residing within the overall land area of just over 1,104 km^2. Hong Kong's development is compact with a built up area of only 24% and the remaining 76% of the land consists of countryside, natural greenery, barren land, and water bodies including 6% of land allocated to agriculture. Thanks to the Government's urban development policy, Hong Kong has 24 country parks and 11 natural reserves cover an area of 443 km^2, almost 40% of total land area. These areas are prohibited from development so as to protect the biodiversity and natural habitats of these areas. The country parks and reserves make Hong Kong unique from other places. It is rare for a city with such a dense population to be able to experience the nature at such close proximity. Many travellers visit Hong Kong for hiking or mountain biking to explore the natural wonders of Hong Kong's country parks and reserves. Given the rich mountainous topography and a vast coastline, it is hard to experience a city similar to Hong Kong anywhere else of the world.

Hong Kong has adopted a mixed-use, high density, Transit Oriented Development (TOD) approach with the railway as the backbone of the city's public transport system to utilise its limited developable land resources for smart and sustainable development. The transit nodes created from the rail network and smart growth planning, as a result, formed the basis to allow the city to expand upwards rather than outwards to create the high dense environment Hong Kong has today. Hong Kong with its compact vertical high density development on either side of Victoria Harbour with a natural mountain backdrop provides an attractive city skyline and a memorable view of the city from the Peak. With the combination of high urban densities and high quality public transit services, over 90% of passenger trips in Hong Kong are made through public transport, which is one of the highest levels of transit usage in the world [2]. This has also driven down the cost of motorized travel and increased mobility and ridership [3].

The TOD model used by Hong Kong is known as the Rail plus Property model. Each railway station forms a community as the surrounding area of railway stations, providing residential, commercial offices, shops, schools, open spaces and other public facilities within close walking distance. Housing is kept within 500 m of subway (Mass Transit Railway—"MTR") stations to create seamless connections to the public transport system. MTR, as the sole provider of Hong Kong's railway system, is also granted the land development rights above, below and in direct proximity of transit stations [4]. This allows the corporation to benefit as a transit provider and a property owner, which has led to high density mixed-use developments around the stations that typically include high-rise commercials and residential, large-scale housing estates, shopping malls, as well as civic and public amenities and facilities.

The idea of TOD can also be found in the development of new towns in the New Territories in Hong Kong. The development of new towns started in the 1970s to decentralize development away from the urban core due to immigration influx from China. Rural farmlands were developed into New Towns together with the MTR lines, using the TOD including high density mixed use development with a pedestrian oriented approach. There are in total 9 new towns in the New Territories with TOD features accommodating over half of Hong Kong's population. The good connectivity with transit nodes enables residents to enjoy high mobility. Along with the high cost of having private cars, Hong Kong has an extremely low car ownership in comparison to other developed countries. Apart from motorised transit linkages, non-motorised options are also provided in new town development to facilitate the use of more environmentally friendly transport. All new towns have cycle track networks fitted with cycle parking facilities in order to promote safe and sustainable infrastructure. However, the poorly designed cycling track and lack of parking space affects connectivity and leads to illegal parking. Most importantly, cycling is viewed as a leisure activity instead of a daily mode for commuting. Therefore, it is essential to modify the design guidelines for cycle track and facilities to provide a more suitable environment for sustainable transport as well as to change people's mind-set in treating cycling as a mode of transit and a way to commute daily.

Hong Kong's successful public transit system relies on MTR besides public buses, mini buses and tram but very limited water transport for a city with such a vast coastline. The integration of TODs in Hong Kong has provided MTR with increased catchment for enhanced ridership but led to gentrification of the neighbourhoods as property prices shoot up as a result of the MTR developments. This mode of development has allowed the city to reach a population density of 6,603 persons/km^2, with an urban population density of 26,317 persons/km^2 as majority of the people reside in the urban areas.

Over the years the trend to maximize station related development, these TODs have transformed into expansive developments becoming Development Oriented Transit (DOTs) with little integration with surrounding areas. However, recognizing issues with such massive podium development, there is an effort to have more sensitive station related development on the newer MTR lines in the older urban districts and more towards Transit and Pedestrian Oriented Developments (TPOD). While it is an impressive achievement for the city to house such a high population and at the same time preserve its nature and countryside, it has come at the expense of the living environment and quality of life within urban areas. This has resulted in very poor living space standards, high density developments the norm with shrinking apartment sizes with the average apartment size as low as 20 m^2 due to high costs and affordability is a huge issue. Making housing more affordable and increasing the space standards of apartments and an opportunity to ownership is very important for Hong Kong people. There is also a lack of quality open and public space, as well as poor pedestrian space impacted by poor roadside air and noise pollution. Although Hong Kong has many country parks the open space is especially low within the city given that it is an extremely high density environment. According to the Hong Kong Planning Standards and Guidelines, the standard for the provision of open space in urban areas is a minimum of 2 m^2 per person, apportioned as 1 m^2 per person for district open space and 1 m^2 per person for local open space. Also this is quite low compared to other international cities such as New York or London and regional cities like Singapore and Shanghai. Raising this standard significantly higher should be a consideration for future city planning decisions to create a smart and sustainable environment.

Another challenge arose to accommodate the high population is the use of reclamation of flat land for development. Over 6% [5] of land including new towns and the airport were created through reclamation. It is a common practice in Hong Kong to retrieve land as reclamation is a cost effective way to acquire flat piece of land for development. However, it has serious impact on the environment. During construction and after completion, it destroys the habitat and well-being of marine animals, disrupts the ecosystem of the ocean and leads to water pollution [6]. These negative and harmful effects were especially obvious on the endangered Chinese White Dolphins. Their numbers have been dropping drastically at a rate of 2.5% each year [7] and remain critically low.

In view of such negative impacts, Hong Kong has become increasingly aware of the environmental problems in the urban environment as government, private developers, institutes, academia and community groups are all looking for innovative ways to improve the sustainability and liveability of the city. The lack of land resources has often been criticised as the root of the problem, as the government established the Land Supply Task Force to review and evaluate land supply for Hong Kong in the long run. Aside from studying additional land resources, others have considered how more thoughtful and innovative planning and design can improve the urban environment to cater to local population needs. There are several bottom-up initiatives which advocate for more community collaboration to provide better quality, sustainable public spaces by transforming existing urban environment through engaging the local community. Central and Western Concern Group, Designing Hong Kong, Hong Kong Public Space Initiative and Very Hong Kong are some examples of community groups that formed to influence more positive urban design and planning decisions. More recently Hong Kong's efforts at Sustainable Development through the Sustainable Development Council (SDC), Harbourfront Planning with the Harbourfront Commission (HC), and Green Building with Hong Kong Green Building Council (HKGBC) and the Energizing Kowloon East Office (EKEO) ensuring more local level planning in Kowloon East and recent initiatives from the Ministry of Environment Bureau and the Environment Protection Department (EPD) have made some contribution in raising awareness towards making Hong Kong a more environment friendly city. Moreover, many community groups have put efforts in raising the city and community's awareness towards green environmental strategies in recent years, from recycling, energy saving to waste and water management. The joint efforts of public and private sectors, institutes, academia and community groups have created great opportunities for Hong Kong to move forward to improve the environment for a more sustainable and liveable future. However, Hong Kong still has a long way to go in becoming environmentally sustainable and making people more environmentally aware.

4   Hong Kong's Strategic Position in Enhancing the Environment

As part of the Greater Bay Area (GBA), Hong Kong plays a significant role as the major international trading hub between China and the rest of the world when it returns to China in 1997. Since then, Hong Kong has transitioned itself from an industrial manufacturing industry to a more financial and service oriented industry due to cheaper land and labour cost in the Mainland. While industrial factories moving out of the city should mean improvements to the environment, pollution continues to be a concern for the city because of marine vessels emissions as well as pollution that is being carried from the northern breezes. Studies have shown that as much as 77% of pollutant particulates came from China. The lack of a regional cooperation between Hong Kong and China on environment protection has come at the expense of Hong Kong's people suffering from pollution further impacting their quality of life. As a GBA regional goal, Hong Kong should be communicating and working together with other cities within the region to address pollution for a better future.

To make matters worse, local participation on tackling environmental issues such as energy saving and waste management is also lacking as the public does not care enough about these issues. This is especially true for recycling. Despite the government's effort in promoting recycling through incentives and even a development of an Eco-Park to facilitate and educate, only few choose to recycle. The private sector is not helping the cause. Recycling is rarely offered as the option for building residents or tenants; many commercial buildings are producing a strong carbon footprint from overused air conditioning and lighting; buildings are designed with poor insulation that creates more burden on energy usage.

Compared to other global cities, Hong Kong is behind in addressing the city's environmental problems. Hong Kong is now focusing and making efforts in improving the environment at a strategic level. These strategies and initiatives are typically published in the form of consultations and city plans which are made open to the public for comments. The Smart City Blueprint, Hong Kong 2030+: Towards a Planning Vision and Strategy Transcending 2030 ("Hong Kong 2030+") and Hong Kong Climate Action Plan 2030+ are three of the city level strategies in creating a smart, innovative, liveable and sustainable city for the future. Victoria Harbour is recognized as one of the important natural assets within the city's urban environment, which has led to the establishment of the Harbourfront Commission as well as environmental community groups such as the Society for Protection of the Harbour which safeguard against any harbour reclamation in the future.

4.1 Smart City Blueprint

The Smart City Blueprint is a consultation study which aims to integrate technology to address the challenges in the urban environment. By embracing innovation and technology, the Government of the Hong Kong Special Administrative Region, China (the Government) has targeted to build a world-famed Smart Hong Kong, characterised by a strong economy and high quality of living[i]. Development plans include Smart Mobility, Smart Living, Smart Environment, Smart People, Smart Government and Smart Economy. The focus on Smart Environment aims to change how the Government manages Hong Kong's built and natural environment in the interest of improving Hong Kong people's everyday life and quality of living. Potential short-term, medium-term and long-term projects have been proposed, referring Table 1.

These projects and corresponding initiatives, supported by technological solutions, will help drive systemic and cultural changes in city-wide pollution and waste management, as well as regulatory changes that will encourage the incorporation of more sustainable design, standards, and management of buildings. These in turn will

Table 1 Potential projects for Smart Environment in HK

Aspect	Potential project
Green and intelligent buildings	• Promoting green and intelligent buildings in construction and maintenance
Smart grid	• Advancing the electrical grid with smart technology for better energy management by interconnecting various sensors, meters and appliances to allow remote monitoring of energy usage, and for users to manage their own energy demand and shift peak hours of energy usage
Intelligent waste management	• Exploiting the value of waste through improving Hong Kong's waste management practices to maximise landfill diversion and increase recycling and increasing efficiency in the overall waste management processes
Pollution management	• Enhancing pollution management by using remote sensing technology to monitor and reduce pollution
LED lighting	• Improving energy efficiency in commercial settings, e.g. in neon signs and illuminated signboards

Source PWC (2017) Report of Consultancy Study on Smart City Blueprint for Hong Kong. https://www.smartcity.gov.hk/report/

foster greater integration and optimisation across urban planning and development to improve efficiency and minimise impacts on the environment. The strategies and initiatives on Smart Environment revolve around decreasing the output of energy and pollution including alternative energy sources for coal-fired electricity, green building designs and carbon emission targets. Additionally, it recommended a potential waste management strategy by implementing a charging scheme on household, commercial and industrial waste. While it is critically important to address energy saving and pollution reduction in the natural environment, like many Smart City initiatives, the overarching principles relies too heavily on technology and does address the needs of the people enough. A major component that the Smart City Blueprint should focus on is how to improve the built environment and quality of life in Hong Kong, where there should be recommendations on how to improve and increase the public open space for people as well.

4.2   Hong Kong 2030+

As a strategic plan for Hong Kong, Hong Kong 2030+ aims to guide the planning, development, and the built and natural environment of the city. One of the three key building blocks of the strategic plan is planning Hong Kong as a liveable high density city. Recognizing Hong Kong's built environment as a compact city, the plan intends to capitalize on the city's blue and green assets to form an open space network. Victoria Harbour is expected to become the major open space within the urban environment of the city through exploring revitalization works for recreational eco-use and climate resilient use. At the same time, the plan looks to bring our natural environment closer by improving the connectivity to rural areas and country parks with better public transport such that the city's urban and natural environments are linked together for social interactions and leisure activities. At a smaller scale, the plan proposed public spaces should be reviewed and reinvented with better urban design concepts to create better quality of life that is more fitting to the urban environment. Future urban developments are also to adopt urban design concepts utilizing development opportunities with mixed-use as well as forming a connection with the open space and blue-green network.

Aside from improving the urban environment of Hong Kong through bringing the country parks and reserves closer to people for enjoyment, there are other assets in Hong Kong where the plan has neglected. One example would be the outlying islands, such as Lamma Island and Cheung Chau. The outlying islands are some of the few suburban areas of Hong Kong where people can escape from the busy city life and enjoy as a getaway paradise. They can be treated as the city's public space. Instead of capitalizing the existing natural assets, Hong Kong 2030+ also proposed the development of an East Lantau Metropolis (ELM) as a new central business district through reclamation. While there is a general consensus that Hong Kong lacks space to cater the city's future growth, the ELM has more potential to showcase best practices for Hong Kong's dense urban environment rather than a new

CBD. Hong Kong's harbour is uniquely positioned to be the CBD, with Central, Island East and Island West on the Hong Kong island side and Kowloon, Kowloon East and Kowloon West on the Kowloon Peninsula side, together creating a Hong Kong CBD which will be unmatched anywhere else in the world.

The Hong Kong 2030+ also fails to mention how Hong Kong can involve at a regional level to strategically position the city within the context of GBA. As it is apparent that the liveability and sustainability of Hong Kong is also influenced by our counterparts in Mainland to put in their effort to push for environmental protection, collaboration is necessary. There needs to be a forward thinking for Hong Kong to take initiatives to work together with cities in the GBA leading to a more smarter environment and a sustainable region.

4.3   Hong Kong Climate Action Plan 2030+

This action plan published in 2017 set a target for carbon emissions reduction in 2030. In order to meet the target, the action plan has suggested a number of strategies.

From the mitigation perspective, sources of electricity supply will be changed from coal to natural gas, non-fossil fuels and renewable energy. Low carbon transportation will be promoted through the use of rail as the backbone for low carbon transport, control private car growth as well as promote walking and cycling. With other measures such as Green Building, Building Energy Code and Green procurement, the plan hopes to reduce energy intensity by 40% by 2025.

In terms of adaptation, there are four key areas including infrastructure, planning, water security as well as conservation and biodiversity. For infrastructure, there will be coastal protection such as flood walls for sea-level rise. Also, there will be drainage and flood management including the construction of porous pavement, rain garden, green roof, water harvesting, retention lake, bio-swale and retention tank to improve the flood resilience ability of the city. For planning, smart city development and urban climate adaptive planning are promoted to strengthen the urban fabric. It is essential as Hong Kong has many ageing buildings which pose difficulty to cope with the impact of climate change. Moreover, urban micro-climate and air ventilation considerations are being incorporated to alleviate heat island effect to considerably improve air ventilation within the urban environment. The third key area is water security, which aims to explore new source of water supply such as reclaimed water, desalinated water as well as recycled grey water and harvested rainwater, while at the same time conserve water. The final key area is conservation and biodiversity, aspiring to enhance country parks and special areas, promote sustainable farming and fisheries as well as enhance biodiversity in urban environment. Also a Biodiversity Strategy and Action Plan has been implemented.

To transform Hong Kong into a more resilient city to climate change, the action plan has pointed out a few key points including the need to prepare for emergencies and deal with extreme heat as well as the importance to raise community awareness and implement regulations to control consumption-based emission such as waste charging, in hope to change citizens' lifestyle choices and adjust consumption demand [8].

4.4 Harbourfront Protection

Victoria Harbour is one of Hong Kong's rich blue assets. Situated between Hong Kong Island and the Kowloon Peninsula, the 73 km long harbourfront can enjoy views of the urban core of the city from either side of the harbour.

However, Victoria Harbour has changed drastically over the years, not only the change in views of the skyline, but also due to land reclamation. Since the 1980s, 2,800 ha of land reclamation has occurred in Victoria Harbour. Such destructive activities led to choppy waters in the harbour and also water pollution, which is detrimental to human health. Therefore, water activities such as water polo and swimming are forbidden in the Victoria Harbour. Pollution not only affects human activities, but also the ecosystem and marine life.

Realizing the severity of land reclamation could potentially destroy Hong Kong's important blue assets, the Society for the Protection of the Harbour advocated against further reclamations against the Harbour. Their involvement has led to the formulation of the Protection of the Harbour Ordinance. Other community and civil society also voiced out against reclamation of the Harbour. In 2004, the Chief Secretary formed the Harbourfront Enhancement Committee after listening to Citizen Envisioning @ Harbour, a group formed by sixteen organizations, in support of making sure harbourfront planning is undertaken in a collaborative manner. The Committee provided input for planning, designs and development issues of the harbourfront as well as explored a framework for the management of the Harbour. In 2010, the Harbourfront Enhancement Committee was transformed into today's Harbourfront Commission (HC), which acts as an advisory and advocacy role for all planning, design, development, management and operations related to the harbourfront area, in hope that the committee creates a long term sustainable strategic plan for the Harbour [9].

HC has set up task forces on Harbourfront Developments focusing on three specific parts of Hong Kong which are (1) Hong Kong Island; (2) Kowloon, Tsuen Wan and Kwai Tsing; and (3) Kai Tak. The Harbour Planning Principles and Guidelines prepared by the HC are taken into consideration by the task forces when assisting the HC in advising on harbourfront development proposals and strategies. The Commission also has set up a working group on the Protection of the Harbour Ordinance to review the implementation of the ordinance, including diagnosing existing issues, proposing new recommendations and investigating on the feasibility of any possible solutions. However, the effectiveness of the Commission remains unclear since the Commission does not have any statutory or executive rights and the government can reject their proposals [10]. There were plans to create a Harbourfront Authority but this seems to be on the back burner now as currently affordable housing and land supply are a priority.

Victoria Harbour has finally regained its attention. In the past, the development of the harbour was more commercial-oriented. It was not used for recreational purposes. Now, the government wants to strike a balance between recreational and commercial use. Though more effort is required, the harbour is now in the process of transforming into a more pleasant place for everyone to enjoy.

5 Government Policies and Environmental Initiatives

Hong Kong has been pursuing a holistic vision of building a low-carbon and more sustainable city in recent years. The Government has embarked on numerous green initiatives and projects which energize and encourage the community and businesses to work side by side towards achieving more sustainable living and working practices. The following section outlines some examples of initiatives which strive to achieve a sustainable future in five different environmental aspects.

5.1 Air Quality Management

Ensuring good air quality safeguards public health and is therefore a vital component of a liveable environment. In Hong Kong, the major challenges faced by air quality management are poor roadside air quality, emissions from local power plants, and regional pollution. Over the past two decades, multiple solutions have been implemented to cope with the challenges. To reduce roadside air pollution, stricter vehicle emission standards have been adopted to tighten the controls on vehicle emission such that new vehicles will produce less pollution on the road. Existing vehicles are also encouraged to be retrofitted with particulate reduction devices through grants and other incentive programs. Aside from roadside air pollution, the Air Pollution Control Ordinance was formed in 2014 to ensure local power plants and industrial activities follow the control and regulations by mandating the approval and license acquired from the EPD. The efforts in reducing roadside emissions has been progressing as major air pollutants such as respirable suspended particles (RSP), sulphur dioxide (SO_2), and nitrogen oxides (NO_x) have successfully decreased by 50% or more since 1999. More work still needs to be done as nitrogen dioxide (NO_2) levels continue to remain steady over the years.

In 2013, the Government has taken another major step forward by releasing "A Clean Air Plan for Hong Kong". The plan not only looks to reduce roadside emission output by replacing with cleaner energy, but also reducing the source of pollution—the number of vehicles on the road. Initiatives such as bus route rationalization and forming a cycling network are some of the strategies to improve public transport and connectivity in the city. The plan also recognizes that urban planning and design has a significant role in improving the city's air quality, with increased vegetation and planting to create better quality of streets, as well as improving connectivity for seamless a pedestrian network to improve walkability and reduce car dependency. In addition, the plan looks at pollution at a regional level by assessing Hong Kong with the rest of the Pearl River Delta (PRD) Region. It extends air quality management as a collaborative effort with Hong Kong and Guangdong in the long run as our Guangdong counterparts also needs to tighten their air quality standards for the benefit of the region.

Since the release of "A Clean Air Plan for Hong Kong" in 2013, the government also published a progress report in 2017 comparing roadside air quality, marine emission, and power plant emission from 2012. Improvement on roadside air quality was seen through retiring old diesel commercial vehicles (DCV) and subsidizing liquefied petroleum gas vehicle owners to encourage replacement of their catalytic converters. Franchised buses are also replacing with new and low emission technology to further reduce roadside pollution. At the same time, the government has been actively promoting walking and cycling in urban areas to reduce car dependency and number of vehicles on the road. Further reduction is projected in the future. Aside from improving roadside air quality, marine emission also has been reduced since 2012 despite Hong Kong still remains as one of the busiest ports in the region. Stricter regulation on fuel switch for ships at birth has been effective, and will be further promoted through cross-boundary, regional collaboration with our counterparts in PRD to establish the Domestic Emission Control Areas (DECA) in the future. Meanwhile, electricity sector has played its part in reducing pollution from power plants by continually replacing coal with natural gas and non-fossil fuel to produce cleaner energy. New and higher standards are also set up under Air Pollution Control regulations to control emission from machinery.

As part of the Paris Agreement, Hong Kong targets to reduce its carbon emission by 65–70% by 2030 using 2005 as a base. Phasing down coal in the fuel consumption of Hong Kong's electricity generation is one of the city's main focuses in reaching the goal as electricity is responsible for 55% of Hong Kong's energy-end-use (Fig. 1).

Overall, Hong Kong recognizes the urgent need to work hard to improve air quality. The plan demonstrates the government's determination in creating good ambient air quality. It introduces new alternatives, proposes scaling up the existing measures,

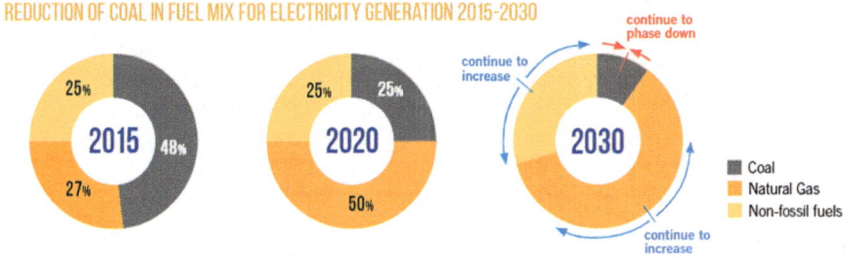

Fig. 1 Hong Kong's plan to reduce coal in fuel mix by 2030. *Source* Hong Kong Climate Action Plan 2030+

and adopts a cross-boundary approach to proactively cooperate with regional stake-holders.

5.2 Noise Pollution Control

Being a compact and densely populated city, noise pollution is an issue in Hong Kong. There are different types of noise pollution within the urban areas of the city, including traffic, construction and aircraft. Traffic noise is one of the main contributors as around 960,000 people, mostly in the older urban areas, are affected by excessive traffic noise alone. As a compacted city, Hong Kong is constantly working on new developments, which resulted in many on-going constructions that build on the noise level on the street. Aircraft noises used to have a significant impact in the urban areas as the Kai-Tak airport was located at the Victoria Harbour surrounded by residential buildings. It was not until the airport was moved to Chek Lap Kok in 1998 when it becomes less impactful. Apart from noises produced from vehicles and constructions, buildings also generate their respective noise pollution from ventilation systems as air conditioning is almost a necessity due to Hong Kong's hot and humid climate and environment. The noise pollution can be ascribed to poor planning and dense development in the past.

To alleviate noise pollution, the EPD has been working closely with developers and the private sector. They looked into various innovative building materials and designs. Window design for a student hostel of the Hong Kong Polytechnic University is a good example of noise pollution control. Designed with sound absorption materials in the frame, noise can be reduced even with the window opened. Besides exploring new designs and imposing stringent controls over construction, the EPD also encourages sharing of good practice by setting up the Quiet Construction Working Group and the launch of internet platforms. From a planning perspective, EPD is also giving its input in planning for new towns and development. A more conscious effort is placed to separate residential buildings from major roads. A buffer zone is in place between roads and developments to reduce the effect of traffic noises. For areas where buffers are not preferable, noise barriers are installed on the sides of major roads to minimize the exposure of the noise, however they are not visually pleasing. While improvements have been made, the roadside condition remains poor for local roads because vehicles and pedestrians are compacted into the limited space available. There needs to be strategies and initiatives adopted to improve the street environment, such as landscaping and widening of pedestrian space as buffers so that street noises do not travel directly to people. This may be challenging as the road network ROW within the city is rather limited for several streets which are too narrow to accommodate streetscape improvements. The provision and design of open space is also a strategy to be considered so it can become an active space for communities and be used as a buffer from noise.

5.3 Water Quality Improvement and Management

Water quality is vital to our health and well-being. Government has been implementing a number of water pollution control programs over the past decades to ensure clean and healthy water, which has led to cleaner beaches and less polluted rivers. Among which, the Harbour Area Treatment Scheme (HATS) has widespread success in tackling water quality problems in Hong Kong. Being the largest environmental infrastructure project in Hong Kong, it reduces key pollutants and improves water quality in the Victoria Harbour. Although it is rather late, the first stage of HATS commenced in 2001 and treats 75% of sewage generated from the Victoria Harbour [11]. Second stage was fully commissioned in 2015 providing treatment to the remaining 25% of sewage from the northern and south western parts of Hong Kong Island [12]. All sewage is conveyed to Stonecutters Island Sewage Treatment Works for treatment and disinfection, which is then discharged into the harbour. During the review stage, it was proven that the scheme has adequate capacity to handle projected sewage flow. It is also in compliance of the Water Quality Objectives (WQOs) which is an objective and scientific benchmark to protect and ensure marine waters are suitable for marine life growth and different human uses in a sustainable manner. After the implementation of the scheme, overall the water quality has enhanced [13] (Fig. 2).

Besides maintaining high quality marine water, much effort has been put into cleaning the shorelines and achieving excellent beach water quality. One of the ini-

Fig. 2 Coastal cleanup at Silverstrand Beach after Typhoon Mangkhut. *Source* https://www.worldcleanupday.org/news/2018/09/27/hong-kong-community-joins-hands-to-clean-up-post-typhoon-mangkhut

tiatives is "Clean Shorelines", which is an inter-departmental working group set up in 2012 to review and formulate measures to improve the cleanliness of the shore- lines. EDP, as the secretariat of the group, is responsible for assessing and rating the cleanliness condition and the effectiveness of the enhanced cleaning work through the use of a five-level shoreline Cleanliness Grading System. Immediate follow- up by relevant departments is essential if any site is graded as fair, unsatisfactory or poor. Other departments including the Food and Environmental Hygiene Department, Leisure and Cultural Services Department, Marine Department and Drainage Ser- vices Department are responsible for the cleanliness of ungazetted beaches, gazetted beaches, sea surface and the provision of wastewater and storm water drainage services. Apart from government departments, the community and the NGOs are actively involved in the initiative through organising various campaigns. For exam- ple, over 150 volunteers and representatives from different departments carried out a clean-up event after the Typhoon Mangkhut at Silverstrand Beach in 2018 [14]. Such collaborations are valuable in raising public awareness and promoting the importance of shorelines protection. Aside from marine and beach water, river water is another key water source that requires special attention. Sampling stations and water Quality Index are used to evaluate and monitor the pollution status and the changes in water quality [15].

Apart from improving the water quality, the government has started to manage water resources in a more sustainable way through the use of reclaimed water. Reclaimed water is highly treated to make it clear in appearance and odourless, and safe to reuse. The use of reclaimed water is an important long term in ensuring sufficient water resources for sustainable development, water conservation and pro- tection as it conserves drinking water, reduces the overall demand for fresh water and treated effluent discharged into the aquatic environment and conserve drinking water [16]. There are multiple ways to use reclaimed water such as cleaning roads and vehicles, irrigate parks and sport fields, flushing toilet and fire fighting. For example, the scheme launched in Ngong Ping and Shatin Sewage Treatment Work in 2006 and 2011 respectively are actively promoting the use of reclaimed water.

Another strategy in water management is the Sewage Services Charging Scheme. It was implemented in 1995 to cover the cost of sewage services and promote long term sustainability. Types of charges include Sewage Charge and Trade Effluent Surcharge. It applies to all users whose premises are linked to public sewers [17]. The use of polluter-pays principle links the charges to the amount of water used, hopes to raise people's awareness of the price of wastewater production [18].

5.4 Resource Management

Municipal solid waste (MSW) is another hotly discussed environmental prob- lem in Hong Kong given its densely populated character. The Government has announced plans to further reduce MSW disposal in the "Blueprint for Sustain- able Use of Resources 2013–2022". Three significant actions are "The Environmental Levy Scheme on Plastic Shopping Bags (PSB Levy Scheme)" "Mandatory Producer Responsibility Scheme (PRS) on Waste Electrical and Electronic Equipment (WEEE)", and "PRS on Glass Beverage Containers" initiated by the EPD [19]. These programs are designed to reduce generation of waste and encourage materials recycling and reuse.

The first PRS is the PSB Levy Scheme. It is a direct economic initiative imposed to minimise the excessive use of plastic shopping bags (PSBs). The first phase took place from 2009 to 2015, which has significantly reduced the amount of PSBs disposed into the landfills by 80%. However, it only covered 3,000 retail outlets and the number of PSBs in the retail sector has increased. Due to the confined coverage, the effectiveness of the scheme was restricted. Therefore, the PSB Levy scheme was amended to cover all retail outlets from 2015 [20]. Along with the levy scheme, a campaign "Bring your own Bag (BYOB)" was set up to promote the use of reusable bags to replace plastic bags. The scheme hopes to change people's habit and reduce the usage of PSBs to a sustainable level in the long run [21].

In terms of WEEE, there was a new proposal for a new PRS on WEEE. The reasons are twofold. First, internationally there was a trend in tightening the control over WEEE and a proper mechanism to treat and recycle WEEE was needed in Hong Kong. Second, the practice of exporting most WEEE to other countries was no longer sustainable due to the decrease in demand for second-hand items in other places. As a result, the new PRS on WEEE was implemented in 2018. All suppliers of regulated electrical equipment (REE) are required to register with the Environmental Protection Department before distribution of REE and arrange free removal service for their customers who want to dispose REE. By the end of 2018, all REE will not be allowed to be abandoned at the landfills or other disposal facilities. Sequentially, WEEE will be treated at The Waste Electrical and Electronic Equipment Treatment and Recycling Facility, which has advanced equipment and technologies on transforming waste into reusable items [22].

The third initiative of PRS is on Glass Beverage Containers. It creates circular economy through the provision of practicable solutions. In the past, due to the low commercial value in recycling glass containers, most of them ended up in the landfill even though they can be transformed into building materials, concrete and paving applications. EDP appointed contractors for the collection and treatment services in 2017–2018, and is preparing for the implementation of the PRS to be submitted to the Legislative Council for subsidy, hoping to implement a more practicable scheme.

The government also adopts a bottom-up and community approach to advocate the idea of "Use Less, Waste Less" and "Clean Recycling" [23]. The idea of "Use Less, Waste Less" is to cherish the limited resources on the planet and minimise landfill disposal through reusing and recycling waste, or transforming them into energy. On the other hand, the idea of "Clean recycling" is proposed due to improper recycling. The recycling bins are stuffed with other refuse including food waste and unfinished drinks, which contaminates other recycled items in the bins, which increases the operational cost and makes other items non-recyclable, which might cause those "recycled items" into the landfill. By educating the public the importance of clean recycling and waste separation at source, the effectiveness and value of recycling can be enhanced. Combining the two ideas, the government plans to set up Community Green Stations with both operational and educational purposes in 18 districts in Hong Kong. The green stations vehicles are designated to collect materials from the community and ensure all materials are properly treated and recycled.

5.5   Energy Saving

Energy saving is critically important as one can save precious natural resources and also help cut down on pollution by using energy more efficiently. The government therefore has legislation and regulations to monitor the efficient use of energy, including the setting up of the Electrical and Mechanical Services Department in 1994, the imposition of the Building Energy Efficient Ordinance and the Building Energy Codes [24]. As the commercial sector accounted for the largest share of energy consumption, the government continues to mobilize the industries to act through implementing various schemes. The leading scheme is the Building Environmental Assessment Method (BEAM) Plus for New and Existing Buildings. The Beam Plus is an independent assessment scheme, which started in 1996 and revised in 2012, to evaluate building sustainability [25]. There are 6 aspects included in the scheme, namely site aspects, material aspects, energy use, water use, indoor environmental quality, as well as innovations and additions. The 6 aspects cover multiple stages of a building project from demolition, planning, to construction and commission [26]. Through affordable means, the scheme hopes to minimise the environmental impact of buildings and alleviate the internal and external environmental quality, hence benefiting the health and wellbeing of residents.

As the pioneer, the government took the lead to demonstrate possible ways to save energy including the construction of the Zero Carbon Building and the T-Park. The Zero Carbon Building is one of the successful low-carbon sustainable buildings as it is 45% more energy-efficient compared to other buildings. Located in Kowloon Bay, it is the first Zero Carbon emission building in Hong Kong. The use of renewable energy such as biodiesel and solar power and the implementation of passive design and green active system help save energy by 20–25% [27]. This building aspires to promote low-carbon living and exchange information and knowledge in low-carbon design and technologies.

Another project is the T-Park, which is a place where waste is converted into energy. Sewage sludge is used as fuel to produce heat energy during incineration, which is converted to electricity. Fluidised bed incinerator, which is a high-tech thermal technology to process the sludge with high efficient combustion, reduces 90% of waste to be disposed of in the landfills and the emission of greenhouse gases. Also, the particulates and pollutants in the flue gas are removed through a gas cleaning system. A Continuous Emission Monitoring System is in operation to assure the cleaned flue gas complies with the international emission standards. Moreover, T-Park is a place with educational and recreational purpose. Its vision is to change people's behaviours and attitudes towards waste and resource management (Fig. 3).

Fig. 3 T-Park. *Source* Hong Kong Climate Action Plan 2030+

The architectural design of T-Park also promotes energy saving. The streamlined and wave-form design with green roofs help to get the best use of sunlight and ventilation, which reduces energy consumption [28].

To better promote energy saving in Hong Kong, the Government has therefore published the first strategic plan named the "Energy Saving Plan for Hong Kong's Built Environment 2015–2025+" in 2015 [29]. It provided a summary of the current energy use in Hong Kong and pledged a new target of reducing energy intensity by 40% by 2025, mainly through the use of three strategies [30]. The achievement to be made requires widespread support from the cohesive community and the industries though the government will take the lead. The first strategy is to improve the energy efficiency of all buildings including new and existing ones since buildings account for over 90% of electricity usage in Hong Kong [25]. The second strategy is to change the current lifestyle of Hong Kong people and promote energy saving practices. The final strategy is to encourage the use of more energy efficient products.

Since collaboration of professionals and the government with public participation and education is essential for energy saving, the government has partnered with different organisations such as HKGBC to promote better built environment in Hong Kong [31]. Besides the efforts from private sector, the Government also emphasizes the vitality for the whole community. The community are encouraged to take a more proactive role in achieving the new target in energy saving.

6   Smart Environment Projects in Hong Kong

Improving the environment should not depend only on the government. As we go over the different case studies, there is a place for everyone to involve, from academia to institutes and social advocacies, Smart Environment can be brought forward even from grassroots initiatives. Through "Smart Thinking" with considerations of People, Place, and Planet, sensitive ideas can be formed to achieve smart and sustainable developments for the city.

6.1   Case Study: Energizing Kowloon East

Energizing Kowloon East is an initiative aiming to promote Kowloon East to be the next Commercial Business District of Hong Kong, boost economic growth and competitiveness through improving connectivity, design and branding [32]. The conceptual masterplan for the area has been revised from time to time to achieve the best outcome. The most updated master plan focuses on 5 aspects including walkability, Green CBD, Smart City, vibrancy and creation [33], in which the strategies for Green CBD actively contribute to the idea of smart environment. Though the project is still underway, the following completed actions and strategies illustrate the successfulness of the initiative in enhancing the environment.

Kwun Tong Promenade, located in Kowloon East, is converted from the former inaccessible Kwun Tong Public Cargo Working Area that handled recycled paper. It is one of the strategies and actions taken under the Kai Tak Development and Energizing Kowloon East. The project is a response to the community as they urge for more waterfront recreational space and sub-regional development in Kwun Tong. The promenade was completed in 2 stages. The first stage including 200 m waterfront open space was opened in 2010. During the second stage in 2015, the promenade was extended by 750 m. Around 1 km in length, the promenade with waterfront boardwalk allows visitors to enjoy close-up views of different landmarks of East Kowloon including the Kai Tak Cruise Terminal and the Runway. It also provides a spectacular night view of Hong Kong Island East, Victoria Harbour and Lei Yue Mun [34]. The aesthetic quality of the pedestrian environment is also enhanced with landscaping. The promenade is now one of the most popular landmark in Hong Kong and becomes a place with diversifies activities for everyone, through the use of architectural intervention. In the later stages of the initiative, the promenade will also be linked with other parks and gardens in the masterplan, which would further enhance the vibrancy and diversity of the waterfront.

Energizing Hoi Bun Road is an initiative to transform the area near Hoi Bun Road into places for leisure and cultural activities for locals and visitors. Hoi Bun Road was originally a road next to the public cargo working area. Through revamping Hoi Bun Road Park and implementing the "Fly the Flyover Operation", Hoi Bun Road is now a vibrant and aesthetically appealing (Fig. 4).

Fig. 4 Kwun Tong Promenade. *Source* http://www.hongkongextras.com/whatsnew.html

Hoi Bun Road Park will be renovated and equipped with a variety of amenities and landscaping elements. Originally, it had limited green coverage with inadequate facilities, a place-making approach will be incorporated to energise the park. Grey infrastructure including pumping station, refuse collection point and dry weather flow inceptors will be transformed into more aesthetically appealing infrastructure through the use of attractive urban design and landscape design. Also, there will be additional and renovated facilities provided in the park such as soccer pitch, elderly fitness equipment, sitting-out areas, environmental features and thematic features. The park now has larger green coverage with increased bio-diversity in planting species [35].

The "Fly the Flyover Operation" aims to make use of the space beneath Kwun Tong Bypass for art and cultural activities to complement the attractive Kwun Tong Promenade. Various other facilities including gallery, outdoor open spaces, multi-purpose rooms, an open stage, urban farms, a restaurant, food kiosks and a pop-up store are provided to bring vibrancy and create new synergy to the area [36]. It was once used as informal open space for different events and activities. The scheme is now in partnership with a NGO, HKALPS Limited. It adds vibrancy into the area by bringing a wide range of activities and events into the area. Other improvements include the provision of roadside parking space, pedestrian linkages between Hoi

Bun Park and the promenade, landscaping elements, lighting features and streetscape enhancement [35] (Figs. 5 and 6).

Revitalising green space is crucial in creating a smart environment. In EKEO, one of the main themes is to improve the physical appearance of the surroundings including key streets and pedestrian pathways with greenery. The main strategies

Fig. 5 Place-making beneath Kwun Tong Bypass in Phase I. *Source* http://www.hongkongextras. com/whatsnew.html

Fig. 6 Place-making beneath Kwun Tong Bypass in Phase II. *Source* http://www.hongkongextras. com/whatsnew.html

include the revitalisation of Tsun Yip Street Playground and the creation of Blue-green infrastructure Tsui Ping River.

Tsun Yip Street playground has turned into an aesthetically appealing park with an industrial culture theme showcasing the industries in Kwun Tong in the past such as textile and garments. It is now an inviting place with multi-purpose lawn and stage for performance or exhibition. It provides flexible green leisure space for different activities. There are art installation and water features, enhancing the overall environment of the park. Located at the centre of the former Kwun Tong industrial area, the park has become an important node in the pedestrian network and a catalyst of the area that brings the community together and illustrates the history of the Kowloon East [37]. Phase 1 was completed in 2014 and has transformed the area from an ordinary sitting place to a well-designed green space with art gallery for exhibition, leisure activities, community engagement and public art. It is the first project illustrating how creativity can be incorporated into public space. Through the conversion of industrial containers into shelters and exhibition pavilions, it shows small public area can be redesigned and adapt to new uses. The interactive design of the display panels opens new ways of interaction between the audience and the exhibition. The playground is now an important node and catalyst in the area bringing the community together and illustrating the history of the Kowloon East.

Tsui Ping River, currently known as the King Yip Nullah, is situated at the centre of Kwun Tong. Making use of its central location, the project aspires to revitalise the river through the provision of water elements, landscaping, ecological and water quality enhancement as well as green riverine corridor. The riverine corridor will incorporate a number of public spaces including the re-provision of Shing Yip Street Rest Garden, which will be renamed as Tsui Ping River Garden, at a temporary public carpark nearby [38]. Tsui Ping River will become a landmark where everyone can enjoy the riverine scenery. Moreover, the walkability and connectivity along the river will be enhanced with walkways along the river, cross-river walkways and landscaped decks.

All buildings are encouraged to reduce carbon emission through the construction of Green Buildings and the adaptation of low carbon design to achieve green community in EKEO. Green Buildings refer to buildings that have achieved BEAM Plus New Buildings with a rating of Gold or above (Fig. 7). Apart from green buildings, there are Green Check Points set up to provide information regarding environmentally friendly facilities to the public. These check points can be Green Buildings or environmental facilities such as Energizing Kowloon East Office and Kowloon Bay Waste Recycling Centre.

The most influential example of Green Building and Green Check Point is the Zero Carbon Building. The concept of Zero Carbon Building, mentioned in the above chapter, is the first project completed in the Energizing Kowloon Initiative [39]. It is located at the centre of Kowloon Bay, like an oasis in the compacted urban environment. It is 3-storey in height including facilities such as eco-home, eco-office and eco-cafes. The landscape area is designed to mitigate heat island effect by providing high greenery ratio. It is also home to Hong Kong's first urban native woodland that promotes biodiversity. Apart from architectural design, ZCB

Fig. 7 EMSD Headquarter, an example of BEAM Plus Building in EKEO. *Source* Hong Kong Climate Action Plan 2030+

uses different eco-building technologies, some of which are used for the first time in Hong Kong, ZCB's integrated design is based on the principle of energy saving and eco-efficiency, which reflects consideration of the linkage and relationship between the built environment and the nature. Its success in innovation and environmental performance was shown from the awards and recognition the project received from around the world.

6.2 Case Study: Green Deck (GD) Project

A smart environment implements smart resource management for public open space, as a balance to the urban city and to provide a stimulating milieu for people to live, work and spend leisure time. The Green Deck (GD) Project proposed to construct an upper deck over the Cross Harbor Tunnel plaza, a busy and highly congested road segment in Hong Kong, to tackle the problems of poor air quality, overloaded pedestrian footbridge, poor connectivity within the district, and lack of open space in that area. Proposed by The Hong Kong Polytechnic University (PolyU), the GD project aims not only to renew the area by providing more green space, but also to foster sustainable development and smart environment in Hong Kong.

Hung Hom is one of the oldest districts in Hong Kong, as well as the interconnection point between the Peninsula Island and the Hong Kong Island. The Cross Harbour Tunnel Plaza, where the GD is proposed, is one of the most traffic-heavy spots in Hong Kong as the cross harbor tunnel is the major connector serving on average 115,190 vehicles per day [40] and many of them are popular bus routes. Additionally, the Hung Hom station is a major railway terminal for the East Rail Line serving New Territories and Kowloon, as well as for the connector trains to mainland China. As a result, the Hung Hom station and Cross Harbour Tunnel Plaza becomes a key transportation node in Hong Kong that suffers heavy air and noise pollution, especially during the traffic jam.

The existence of the cross-harbour tunnel not only leads to environmental impact to the surrounding area, but also creates mobility issue that bars that walkability. The plaza consists of two levels, which the ground level area is used while waiting for the buses, and the upper level is a pedestrian footbridge connecting Hum Hong station to the Hong Kong Polytechnic University. The pedestrian flow is therefore limited to the urban form. The undesirable nature of the surrounding context makes the Cross Harbour Tunnel Plaza a functional space that forms part of the barrier. Due to the close proximity to the Hong Kong Polytechnic University and major transportation node, the pedestrian footbridge sometime was transformed into an informal place where people perform and conduct informal activities. The convenient positioning of the plaza also determines its potential of transforming into a community place and gathering spot in the district.

Considering the unrevealed opportunities and constraints of the site, PolyU proposed to transform the plaza into a GD that provides high quality space with assistant of environmental technologies.

A number of research projects have been undertaken by PolyU experts with a view to achieving low carbon consumption and emissions. An air filtration and purification system including electrostatic precipitators for use to filter and purify the polluted air is being included in the scheme. Reduction of Heat Island Effect will be achieved through the increase in biomass and vegetation. The deployment of environment-friendly items including photovoltaic cells, wind catchers, solar lighting, recycled grey-water, recycled glass and DeNOx agent are being explored. Table 2 show examples of some researchers associated with smart environment in case of GD.

The Hung Hom Area, according to a study by Civic Exchange, an independent Hong Kong public-policy think tank, in 2017, is undersupplied regarding the open space per person [48]. The shortage of open space could be resolved with the provision of the 43,000 m^2 from GD. With the aid of the abovementioned technologies, the GD will provide a green open space in the city centre for students, residents and visitors. Utilising the vertical spaces, the GD will place-make this placeless site into an active gathering zone for the area and offer spaces for happenings. It will create a unique natural vista and green lung in the urban jungle that supports a sustainable lifestyle.

More than a community open space, the GD will also become an attraction for tourists given its strategic location. It will ripple with other attraction points, such

Table 2 Potential initiatives of GD

Aspect	Potential initiative
Air quality	In GD situation, PM10 concentration and carcinogenic compounds, has reduced or improved than the value measured with non-green deck situation [40]
Noise	It is proposed to use louvres together with micro-perforated sound absorbers and powerful sound absorbers as the noise mitigation measure [41]
Temperature and thermal comfort	It shows that the temperatures increase from the center to the boundary of the parks by 1 °C. Thermal comfort and acceptability of the thermal environment were observed, so were the benefits of the reduced morbidity and morbidity, as well as increased productivity and recreational value [42, 43]
Renewable energy	It will be a "net-zero-energy consumption" deck if the roofs or facades of other buildings developed together with the deck are allowed for PV installation; however, the initiative is limited due to green areas and other facilities on the Deck [44]
Solar PV pavement	To explore more possible areas for solar power generation, the walkable solar PV floor tile is proposed for installation on pavements and cycling tracks, which receive a lot of sunshine every day [45]
Recycled materials	It is estimated that approximately 143 tonnes of waste glass can be recycled by the construction of the GD, paved with an architectural glass screed (mortar) and coated with a thin layer of photo-catalyst [46]
3D Spatial analysis	It is noted that there is a potential to have minor Plot Ratio/Building Height relaxation on buildings in nearby area to increase the living and working spaces to meet the increased population flow whilst minimizing its environmental impact [47]

as the Hong Kong Coliseum, the East Tsim Sha Tsui area, the Hong Kong Science Museum, and the Hung Hom Waterfront, to establish a tourism network adding value to the surrounding lands. As the quality of space will be enhanced by the green technologies, the GD will support a comfortable journey for walking and biking between these key spots. The walkability of the district is then improved with a well-designed network. Mobility with a pedestrian oriented design is therefore fostered.

The GD would be a city park and also serve many functions of satisfying the public demand and creating a better city, as an integrated infrastructure in one of the busiest areas of Hong Kong. In contrast to view Smart Environment as a single space, GD aims to achieve Smart Environments with multiple technologies, to improve the performance based on the experience on many environment elements, from diversified groups of inhabitants, companies and researches in order to continuously expand

and improve the expected benefits of Smart Environment over a more comprehensive scope.

6.3   Case Study: Walk DVRC Initiatives

Des Voeux Road Central (DVRC) is one of the main arterial roads in Hong Kong's Central Business District (CBD). DVRC is also the home of Hong Kong's world famous tramway, also referred to as 'Ding Ding' by locals. The road has a long history and witnessed the transformation of Hong Kong's Central district for over a century. It is currently very congested by public transport and private vehicles, which contributing to poor air quality, as well as noise pollution. Pedestrians are also limited in space as individuals often find themselves fighting for space on the streets. Recognising the poor roadside conditions and lack of space, professional institutes, namely Hong Kong Institute of Planners (HKIP), began study the possibility to bring change to the historic road for a better street environment at a human scale in 2000. They proposed to the Hong Kong Government to transform DVRC into a car-free, pedestrian and tram shared public space. The concept was not well received and was put at halt since then. In 2014, HKIP, in conjunction with several other institutes, university and transport consultancy, revisited the idea of pedestrianizing DVRC. The study brought other civic groups together and began to promote the concept as Walk DVRC Initiative.

By 2016, through continuous efforts and negotiation with various departments, the Government agreed to trial for a one-day road closure in 2016. The road closure resulted in a turnout of 14,000 visitors with participants joined for different activities on the street. The tram continued in operation during the road closure to simulate how the street environment becomes when turned into a shared space. The success for the event has led to the establishment of Walk DVRC Limited to steer the Walk DVRC Initiative into realisation. The one-day trail also has derived the current plan for DVRC, which is to leave one lane for vehicular flow and continue the operation of tram while the other lanes will be used for pedestrians and other activities. As a compensation or attraction for car users, the pedestrian environment will be enhanced with landscaping such as seating and greenery.

The Walk DVRC Initiative advocates that the city should prioritize the need of pedestrians over vehicles, hoping that this idea would become a core principle when planning the CBD. Moreover, street design prioritising people creates a more attractive and safe city as it encourages walking and interaction. The long term goal of this initiative is to make Hong Kong a more connect, healthy and smart city.

6.4   Case Study: Smart Cultural Triangle Precinct

Central District is one of the first communities, and urban districts, built in the early colonial days of Hong Kong. Government institutions, European merchants, Indian businesses and local Chinese settlements were concentrated around the streets of Central. Over time, the area continued to evolve as Hong Kong's main business district; shaping what we see today. However, rapid economic expansion from 1970s onwards redefined Hong Kong's cityscape. During this time of expansion, building modern skyscrapers, and fast-moving urbanisation, took precedence over preserving buildings and/or landmarks of historical and cultural significance. As a repercussion to urban renewal and redevelopment, local community networks were also disrupted and/or displaced due to over the years. In recent years, a public outcry emerged to preserve the local culture and heritage sites and ensure Hong Kong's heritage and cultural artifacts are preserved. Through years of discussions, the government, along with its appointed public bodies and committees, identified potential heritage sites for further conservation of cultural heritage in Hong Kong. In 2009, the Chief Executive on Hong Kong announced the "Conserving Central" programme, which intends to adaptively reuse existing vacated buildings within Central District. The Police Married Quarters (PMQ) on Hollywood Road, the Central Market (CM) and the Central Police Station (CPS) are three of eight buildings indicated for adaptive reuse within the programme. These three buildings, after much public debate, are currently in various stages of implementation. The Hollywood Road PMQ has been adaptively reused for the creative industries since 2014; the CPS, being redeveloped by the Jockey Club, has turned into a centre for heritage and arts and is only recently open to public in early 2018; the CM is being developed by the Urban Renewal Authority (URA) to form the "Central Oasis" with constructions currently underway. Other heritage revitalisation projects include Pak Tsz Lane, now with Pak Tsz Lane park completed, and Yu Lok Lane historic building revitalisation, which is underway as part of the controversial and much debated long-term redevelopment plan for Graham and Peel Street market area.

The district is a place of contrast between new and old; oftentimes the old is hidden underneath the veil of new development around. The many new hip restaurants and bars that make up much of the SOHO area, which runs along the Mid-Levels Escalator, compete daily with the longstanding 'Dai Pai Dong' style restaurants and famous food stalls, such as the popular Lan Fong Yuen on Gage Street; these areas will be nicely complemented by the CM redevelopment. All across the district, there are still traditional businesses operating next to the new, including an old-style Chinese barbershop hidden away next to the PMQ, traditional paper printing shops, and key and watch makers, etc.

Furthermore, the district is host to many art galleries, drawing both local and international interest, an important aspect positioning Hong Kong as a creative centre in Asia. Creativity is apparent as you walk along Hollywood Rd, towards the PMQ and Sheung Wan area. There are also hints of a growing street art culture; interesting and quirky paintings, graffiti, art works can be discovered, including a local graffiti

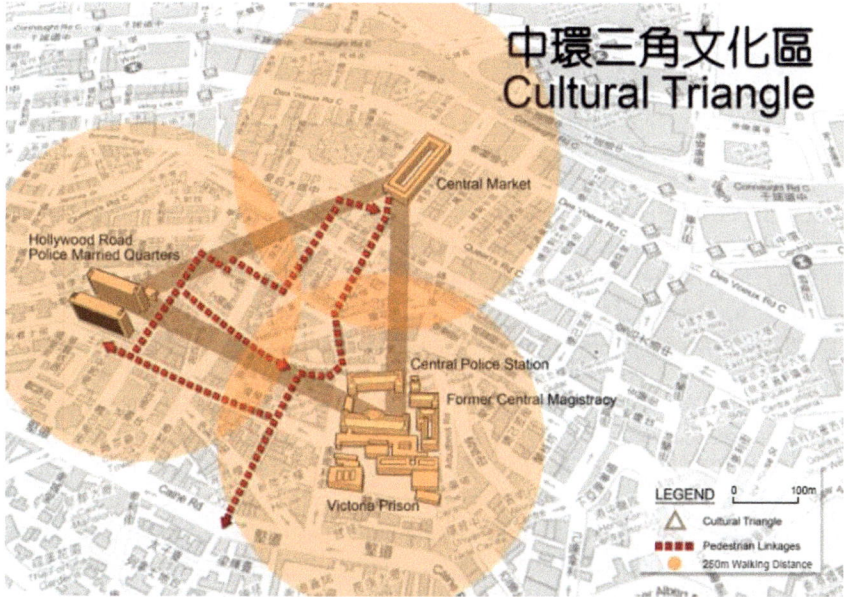

Fig. 8 The location of the three anchor buildings in Cultural Triangle. *Source* Institute for Sustainable Urbanisation

artist's interpretation of the Kowloon Walled City commissioned by G.O.D's central shop and geometric shapes painted on the entire side of a building. Moreover, there is unique urban greenery, for example the remarkable growth of trees on the wall of the PMQ. These tree-walls are unique, serving as a metaphor for the constant play between man and nature as the roots slowly take over the walls, resulting in a beautiful and contemplative sight.

Given their location and close proximity within Central, the Institute for Sustainable Urbanisation (ISU) proposed that the three key developments form the three cultural nodes or the key anchors at the vertices of the Smart Cultural Triangle Precinct (SCTP) in Central (Fig. 8). Within this context, the Mid Levels Escalator, Staunton Road, Hollywood Road, Pak Tsz Lane, Peel and Graham Street markets could potentially become key linkages and arteries between these three important cultural nodes in Central. By bringing together these three important sites and working collaboratively, the cultural linkages between the sites could be further strengthened (The SCTP's "trails") and be the catalysts to re-energise the area within, giving the potential for the SCTP to be one of Hong Kong's most vibrant areas filled with character, antique shop houses, street markets, studios, workshops, 'Tong Lau' buildings and other buildings/structures of heritage value; a place to be for locals and visitors.

In addition to promoting the historic Central District through three key Government-based conservation projects, ISU proposed the precinct can be further enhanced through urban design interventions, place-making and traffic calming initiatives to improve the pedestrian environment with high quality public realm, which can be a game-changer to the community. In Hong Kong's compact and highly dense CBD, public open space is scarce and hard to come by. The conservation of CPS, PMQ and CM are able to preserve what are the few remaining space from turning into high-rise developments. The Smart Cultural Triangle Precinct can utilize the buildings as open spaces with better connectivity and accessibility. This in return can act as a catalyst to re-energize the area as an integrated, sustainable, cultural and heritage destination for the people of Hong Kong and tourists.

The importance of community engagement and public's support is well recognized. ISU has continued to promote the Smart Cultural Triangle Precinct concept to various stakeholders, including URA, District Council, and community groups. Earlier in 2018, ISU had organized CityWalk, a walk tour to raise awareness and educate the public on the rich cultural and heritage assets within the district. Collaborating with local community groups, ISU is looking to continue with various community engagement activities to involve the local community to participate in bringing the culture and history of Central District to locals, visitors, and beyond.

7 Conclusion—Hong Kong's Future Positioning

Studying the government policies, it is apparent that Hong Kong has a long way from achieving Smart and Sustainable Environment. Re-visiting the core values of the Smart City Framework discussed in the beginning of the chapter, the people of Hong Kong needs to elevate to develop their mindset with "Smart Thinking" in order to realize the healthy lifestyle achieved through Smart and Sustainable Environment. There is a lack of public awareness in environmental protection, varying from recycling, water and energy saving, waste management and more. However, it is not to say Hong Kong completely lacks "Smart Thinking", as we see many initiatives from the grassroots and NGOs such as the Walk DVRC are putting in efforts to improve the urban environment. The government should invest hard in educating more of the public so that they become more conscious about the environment and taking more interests in the bottom-up initiatives that are happening in the city.

Apart from the people, Hong Kong as a place has infinite potential developing Smart and Sustainable Environment. The extreme dense environment of the urban areas made it possible for public transport to run efficiently with closer destinations. All the while Hong Kong's urban environment is blessed with the Victoria Harbour that has the opportunity to turn into a world-class waterfront for the public's enjoyment. As we have seen in the case study on Kowloon East, the city has been aware of the potential the Harbour can bring with installation of the Kwun Tong promenade with more similar public space along the harbourfront to follow. However, relying only on the harbourfront should not be the only objective of the city's environmental plan. More open space within the city through initiatives such as the Green Deck in Polytechnic University should be made available within the urban environment and developing an open space network that connects to the Harbour would be a more holistic approach the government should consider. Also, unused or vacant space in the urban area can be transformed into flexible open space for different activities like the flyover project in Energizing Kowloon East to bring vibrancy and creativity into the area. In terms of Hong Kong's initiatives for the planet, the city's planning has allowed an abundance of green space is preserved as country parks and reserves occupied 40% of the city's land area. However, due to lack of affordable housing there is considerable pressure to develop these country park and green belt areas, which will not be a good thing.

Furthermore, the city needs to focus on dealing with other environmental issues which are harming the natural environment, such as air quality, water quality and waste management. Hong Kong should avoid coming up with non-statutory bodies and piecemeal strategies which lacks an overarching plan with coordination. Improving the environment quality cannot solely fall into the hands of Hong Kong but rather joint collaborative efforts between cities of the Greater Bay Area. Similarly in "A Clean Air Plan for Hong Kong", the city government should proactively engage with Mainland cities to derive a regional plan to collectively improve and preserve the natural environment in the region.

Hong Kong, as an international city, requires huge effort in raising public awareness and implementing its policies to improve its environment further. First, public education is essential to raise awareness and consciousness about the importance of environment and resiliency and being more proactive. The community should also make an effort in realising the existence of different initiatives and the importance of environmental protection and become equal partners in ensuring a smart and sustainable environment. Realising that Hong Kong has the potential to create smart environment, more work is needed from all parties. Also, apart from conserving the current green space such as the country parks and natural reserves, it is crucial to create more open space within the urban area with its compact high density development to develop a network of open spaces from the hinterland to the harbour a more liveable environment and city for everyone to enjoy. Although it is good to contain and concentrate development in the urban areas, more open space is needed in the inner city area. Since the standard for the provision of open space per person in Hong Kong is lower compared to other developed cities, the standard can be refined to create a smart and sustainable environment. Finally, long term environmental policies are fundamental in creating a better environment for Hong Kong. Policies shall be well-organised and coordinated with a forward-looking and sustainable perspective in the long run.

References

1. IPCC (2014) Climate change 2014: synthesis report. Contribution of working groups I, II and III to the fifth assessment report of the intergovernmental panel on climate change [Core Writing Team, Pachauri RK, Meyer LA (eds)]. IPCC, Geneva, Switzerland, p 151
2. Transport and Housing Bureau (2017) Public transport strategy study. [Online]. Available: https://www.td.gov.hk/filemanager/en/publication/ptss_final_report_eng.pdf
3. Suzuki H, Cervero R, Iuchi K (2013) Transforming cities with transit: transit and land-use integration for sustainable urban development. The World Bank
4. Parker D, Wood A (eds) (2013) The tall buildings reference book. Routledge
5. CEDD (2018) CEDD—1 introduction. [Online] Cedd.gov.hk. Available at: https://www.cedd.gov.hk/eng/about/organisation/int.html. Accessed 20 Dec 2018
6. Nissim R (2011) Land administration and practice in Hong Kong, vol 3. Hong Kong University Press
7. Duong I (2017) Are the benefits of land reclamation worth the environmental impact they cause?. [Online] Yp.scmp.com. Available at: https://yp.scmp.com/over-to-you/op-ed/article/108181/are-benefits-land-reclamation-worth-environmental-impact-they-cause. Accessed 20 Dec 2018
8. Environment Bureau (2017) Hong Kong's Climate Action Plan 2030+. [Online] Enb.gov.hk. Available at: https://www.enb.gov.hk/sites/default/files/pdf/ClimateActionPlanEng.pdf. Accessed 20 Dec 2018
9. Lewis R (2010) Hong Kong real estate. Responsible Research
10. Lee EW, Chan EY, Chan JC, Cheung PT, Lam WF, Lam WM (2013) Public policymaking in Hong Kong: civic engagement and state-society relations in a semi-democracy. Routledge
11. Drainage Services Department (2009) Caring for our harbour. Available via https://www.dsd.gov.hk/EN/Files/publications_publicity/publicity_materials/leaflets_booklets_factsheets/HATS.pdf. Accessed 24 Jan 2018
12. Environmental Protection Department (2015) Water quality objectives. Available via https://www.epd.gov.hk/epd/wqo_review/en/wqo.htm. Accessed 29 Oct 2018
13. Hong Kong Government (2018) Harbour Area Treatment Scheme. Available via https://www.gov.hk/en/residents/environment/water/harbourarea.htm. Accessed 29 Oct 2018
14. Environmental Protection Department (2018) Working group on clean shorelines. Available via http://www.epd.gov.hk/epd/clean_shorelines/node/12.html. Accessed 24 Jan 2018
15. Hong Kong Government (2017) River water quality. Available via https://www.gov.hk/en/residents/environment/water/riverwater.htm. Accessed 29 Oct 2018
16. Drainage Services Department (2018) Available via https://www.dsd.gov.hk/EN/Sewerage/Environmental_Consideration/Reclaimed_Water/index.html. Accessed 29 Oct 2018
17. Environmental Protection Department (2018) Available via http://wqrc.epd.gov.hk/en/sewerage/sewage-charges.aspx. Accessed 29 Oct 2018
18. Drainage Services Department (2017) Available via https://www.dsd.gov.hk/EN/FAQ/Sewage_Charges/index.html. Accessed 29 Oct 2018
19. Environmental Protection Department (2018) Producer responsibility schemes. Available via http://www.epd.gov.hk/epd/english/environmentinhk/waste/pro_responsibility/index.html. Accessed 24 Jan 2018
20. Ibid
21. Environmental Protection Department (2015) Available via https://www.epd.gov.hk/epd/psb_charging/en/psb_charging/charge.html. Accessed on 29 Oct 2018
22. Environmental Protection Department (2018) Producer responsibility schemes. Available via https://www.epd.gov.hk/epd/english/environmentinhk/waste/pro_responsibility/index.html. Accessed on 29 Oct 2018
23. Environmental Protection Department (2014) Community Green Station. Available via http://www.epd.gov.hk/epd/english/resources_pub/videos/green_station_eng1.html. Accessed 24 Jan 2018

24. Tong KW (ed) (2014) Community care in Hong Kong: current practices, practice-research studies and future directions. City University of HK Press
25. Brebbia C, Sendra J (2018) Sustainability and the city. WIT Press, Southampton
26. Hong Kong Green Building Council (2018) BEAM Plus New Buildings. https://www.hkgbc. org.hk/eng/NB_Intro.aspx. Accessed 30 Oct 2018
27. Vinod Kumar TM (2017) Smart economy in smart cities" international collaborative research: Ottawa. St Louis, Stuttgart, Bologna, Cape Town, Nairobi, Dakar, Lagos, New Delhi, Varanasi, Vijayawada, Kozhikode, Hong Kong. Springer, Singapore Google Scholar
28. Environmental Protection Department (2018) T - PARK. [Online] Tpark.hk. Available at: https://www.tpark.hk/en/. Accessed 20 Dec 2018
29. Environment Bureau (2015) Energy Saving Plan for Hong Kong's Built Environment 2015–2025+. Available via http://www.enb.gov.hk/sites/default/files/pdf/ EnergySavingPlanEn.pdf. Accessed 25 Jan 2018
30. Hong Kong Government (2015) Energy Saving Plan for the Built Environment outlines roadmap to reduce energy intensity by 40 percent by 2025 (with photo/video). [Online] Info.gov.hk. Available at: https://www.info.gov.hk/gia/general/201505/14/P201505140408. htm. Accessed 20 Dec 2018
31. Hong Kong Green Building Council (2018) About HKGBC. Available via https://www.hkgbc. org.hk/eng/Abouthkgbc.aspx. Accessed 24 Jan 2018
32. Development Bureau (2012) Vision and mission. Accessible via https://www.ekeo.gov.hk/en/ vision/index.html. Accessed on 30 Oct 2018
33. Development Bureau (2012) Conceptual Master Plan 5.0. Accessible via https://www.ekeo. gov.hk/en/conceptual_master_plan/master_plan_5/index.html. Accessed on 30 Oct 2018
34. Leisure and Cultural Services Department (2014) Kwun Tong Promenade—introduction. [Online] Lcsd.gov.hk. Available at: https://www.lcsd.gov.hk/en/parks/ktp/index.html. Accessed 20 Dec 2018
35. Legislative Council Panel on Development (2017) Improvement of Hoi Bun Road Park and adjacent area. [Online] Available at: https://www.ekeo.gov.hk/filemanager/content/public/en/ dev20170425cb1_817_6_e.pdf. Accessed 20 Dec 2018
36. Development Bureau (2012) Energizing Kowloon East—Energizing Hoi Bun Road. [Online] Ekeo.gov.hk. Available at: https://www.ekeo.gov.hk/en/award/HKIP_2017_Publication_ Content/index.html. Accessed 20 Dec 2018
37. Development Bureau (2015) Converting Tsun Yip Street Playground as Kwun Tong Industrial Culture Park. [Online] Available at: https://www.ekeo.gov.hk/filemanager/content/public/en/ 20151028_Converting_Tsun_Yip_Street_Playground_as_Kwun_Tong_Industrial_Culture_ Park_eng.pdf. Accessed 20 Dec 2018
38. Development Bureau (2016) Reprovisioning of Shing Yip Street Rest Garden as Tsui Ping River Garden. [Online] Available at: https://www.legco.gov.hk/yr15-16/english/fc/pwsc/papers/p16- 11e.pdf. Accessed 20 Dec 2018
39. Construction Industry Council (2018) Overview of ZCB. Accessible via http://www.cic.hk/ eng/main/zcb/ZCB_experience/Overview_of_ZCB/. Accessed on 30 Oct 2018
40. Transport Department (2018) Road Tunnels and Control Areas—Traffic Statistics—Cross-Harbour Tunnel I Annual Transport Digest 2017. [Online] Available at: https://www.td.gov.hk/ mini_site/atd/2018/en/section4_10.html. Accessed 20 Dec 2018
41. Lee SC (2014) Effect of the Green Deck on local air quality. [Online] Available at: https://www.polyu.edu.hk/cpa/greendeck/pdf/3_Prof_SC_Lee_Effect_of_the_Green_ Deck_on_local_air_quality.pdf. Accessed 20 June 2017
42. Tang SK (2014) Effect of the Green Deck on local noise environment. [Online] Available at: https://www.polyu.edu.hk/cpa/greendeck/pdf/2_Prof_SK_Tang_Effect_of_the_ Green_Deck_on_the_local_noise_environment.pdf. Accessed 20 June 2017
43. Chau CK (2016) Investigating the effects of greenery on temperature and thermal comfort in Urban Parks. [Online] Available at: https://www.polyu.edu.hk/cpa/greendeck/pdf/11_Prof_ CK_Chau_Investigating_the_effects_of_greenery_on_temperature_and_thermal_comfort_ in_urban_parks.pdf. Accessed 20 June 2017

44. Chan EHW (2016) Costs and benefits analysis on the thermal effect on the Green Deck to the surrounding outdoor environment. [Online] Available at: https://www.polyu.edu.hk/cpa/greendeck/pdf/10_Prof_Edwin_Chan_Costs_and_benefits_analysis_on_the_thermal_effect_of_the_Green_Deck_to_the_surrounding_outdoor_environment.pdf. Accessed 20 June 2017
45. Yang HX (2014) Renewable energy applications on the Green Deck. [Online] Available at: https://www.polyu.edu.hk/cpa/greendeck/pdf/4_Prof_HX_Yang_Renewable_energy_applications_on_the_Green_Deck.pdf. Accessed 20 June 2017
46. Yang HX (2016) Research and development of solar PV pavement for application on Green Deck. [Online] Available at: https://www.polyu.edu.hk/cpa/greendeck/pdf/14_Prof_HX_Yang_Research_and_development_of_solar_PV_pavement_for_application_on_Green_Deck.pdf. Accessed 20 June 2017
47. Poon CS (2014) Maximise the use of recycled glass in cement-based construction materials for the Green Deck. [Online] Available at: https://www.polyu.edu.hk/cpa/greendeck/pdf/6_Prof_CS_Poon_Maximise_the_use_of_recycled_glass_in_cement-based_construction_materials_for_the_Green_Deck.pdf. Accessed 20 June 2017
48. Shen G (2017) A study of the development potential in Tsim Sha Tsui East using 3D spatial analysis technologies. [Online] Available at: https://www.polyu.edu.hk/cpa/greendeck/pdf/16_Prof_Geoffrey_Shen_A_Study_of_the_Development_Potential_in_Tsim_Sha_Tsui_East_Using_3D_Spatial_Analysis_Technologies.pdf. Accessed 20 June 2017

Part II
Ahmedabad–Gandhinagar

Automation Based Smart Environment Resource Management in Smart Building of Smart City

Jignesh G. Bhatt, Omkar K. Jani and Chetan B. Bhatt

Abstract Globally, many smart cities have been observed developing in different countries to offer high quality life and excellent working-living environment to their citizens. Being powerhouses of potential and skilled workers, smart cities contribute immensely to the overall development of the society and nation. As Smart home or smart building contributes at the core of effective smart city realization as an important and basic building block, for long term sustainable growth, it becomes quite imperative to monitor critical environmental parameters of building to make the life quite liveable. Such smart buildings monitored by Building Automation Systems (BAS), which have started demonstrating rapid growth potential on account of rising energy costs, stringent scarcity of fossil fuels for power stations and continuous abnormal-unpredictable climate changes, etc. Smart buildings with energy efficiency are need of the day and frequent terrorist attacks and rising security concerns worldwide, security and surveillance have been major focused areas today, where BAS are providing solutions. This chapter keeps its major emphasis not only on automatic monitoring of critical parameters, but also suggest technological approaches. Optimized utilization of energy usage, integration of renewables as well as with smart grid (energy backbone of smart city) shall also be covered in broad perspectives. The chapter shall suggest useful guidelines and recommendations for smart building

J. G. Bhatt (✉)
Department of Instrumentation and Control Engineering, Faculty of Technology,
Dharmsinh Desai University (DDU), Nadiad 387001, Gujarat, India
e-mail: jigneshgbhatt@gmail.com

O. K. Jani
Kanoda Energy Systems Pvt. Ltd, 501, Sheraton House, Opp. Ketav Petrol Pump,
Polytechnic Road, Ambawadi, Ahmedabad 380015, Gujarat, India
e-mail: omkar.jani@kanoda.com

C. B. Bhatt
Government MCA College, Maninagar (E), Ahmedabad 380008, Gujarat, India
e-mail: chetan_bhatt@yahoo.com

J. G. Bhatt
B-303, Siddhashila Apartment, Opp. Prernatirth Derasar, Jodhpur, Satellite,
Ahmedabad 380015, Gujarat, India

© Springer Nature Singapore Pte Ltd. 2020
T. M. Vinod Kumar (ed.), *Smart Environment for Smart Cities*, Advances in 21st Century
Human Settlements, https://doi.org/10.1007/978-981-13-6822-6_3

automation for smart cities with interesting discussions of example case studies and implemented proof of concepts.

Keywords Building automation · Smart building · Smart city · Smart environment resource management · Smart grid

List of Abbreviations

BAS	Building Automation System
CO_2	Carbon Dioxide
CPP	Critical Peak Pricing
CT	Communication Technology
DDU	Dharmsinh Desai University
DR	Demand Response
DSM	Demand Side Management
GERMI	Gujarat Energy Research and Management Institute
GoG	Government of Gujarat
GoI	Government of India
GUI	Graphical User Interface
HAN	Home Area Network/Home Automation Network
ICT	Information and Communication Technology
IITR	Indian Institute of Technology Roorkee
IOT/IoT	Internet of Things/Internet of Things
IT	Information Technology
LED	Light Emitting Diode
LF	Load Forecasting
PDPU	Pandit Deendayal Petroleum University
PM	Particulate Matters
Prosumer	Producer + Consumer
PV	Photo Voltaic
RTP	Real Time Pricing
SCADA	Supervisory Control And Data Acquisition
SG	Smart Grid
TDS	Totally Dissolved Solids
TSS	Totally Suspended Solids
ToU	Time of Use
VOCs	Volatile Organic Compounds
WSN	Wireless Sensor Network

1 Introduction

With objectives to cater high quality lifestyle and excellent environment to citizens globally, smart cities have been developed in different countries. Highly skilled and learned citizens living-working in such smart cities are expected to contribute immensely on account of their experience and domain expertise to the development of society and nation at large. As mentioned in [1], globally going on urbanization is estimated to urbanize 500 million people by 2030 resulting in approximately 60% of the world's population living in cities. Urbanization is also contributing significantly to climate change as 20 largest cities consume 80% of the world's energy and urban areas generate 80% of greenhouse gas emissions worldwide. Climate change, energy scarcity, environmental pollution, etc. have been some of the major challenges of rapid urbanization that need to be resolved by efficient urban planning for effective and sustainable development without putting pressure on resources.

Smart environment resource management with low carbon electricity ecosystem should therefore necessarily be the essential segment of urban planning to avoid future sources of greenhouse emissions, while developing more livable and efficient urban spaces. It could also alleviate population pressure on natural habitats and biodiversity thus reducing the risks to natural disasters. Smart buildings with low carbon footprint powered by smart grid would help immensely to smart prosumers in smart environment resource management and much positive impact on overall optimized energy consumption.

A basic building block of such a smart city is a 'Smart Building' which contributes at the core to realize the transformation of the smart city. As the smart citizen of smart city shall live and work in smart building environment, it is the need of an hour to closely monitor various critical parameters associated.

Referring to our earlier contributions, in [2] detailed technical review of smart grid along with identification of critical applications and parameters, while in [3] e-governance of rooftop-based solar photo voltaic rooftop system has been covered with special focus Gandhinagar solar city project. Next in [4], smart grid pilots along with interesting applications have been discussed including various initiatives of UGVCL and GERMI. Last in [5], interesting details of smart metropolitan region development of Ahmedabad-Gandhinagar twin city metropolitan region with smart grid installation with Naroda area at the focus was presented.

In line with our above mentioned earlier contributions, this chapter begins with conceptual explanations of associated terminologies and later identifies key critical parameters for smart environment monitoring in smart building. Next, the chapter presents relevant technological approaches involved and put forth case study examples for the same purpose. This chapter indicates relevant issues and challenges and suggests necessary recommendations for citizens' participation at the end.

2   Smart City Evolution: Smart Grid, Smart Building and Smart Environment

Transformation of legacy cities into smart cities necessitates subsequent transformation of conventional unidirectional non-smart electrical grids into bi-directional smart grids with plethora of e-applications for different domains.

Describing a smart city as a sustainable and efficient urban center that provides a high quality life to its inhabitants through optimal management of its resources with energy in particular, Calvillo et al. [6] identified energy management is one of the most demanding issues owing to the complexity of the energy grid systems and their vital role. To achieve resolution of such an issue, Khansari et al. [7] mentioned that in a smart city by having the right information at the right time, citizens, service providers and city government, etc. would be able to make better energy switching-scheduling-consumption related decisions to provide quality life to urban residents with overall sustainability of the city. Further zeroing in [8], seek to analyze recent shifts in goals concerning domestic energy uses by investigating various domestic practices such as lighting, heating and cooling spaces, cooking/eating and leisure activities, cleaning, etc. with relevant appliances as well as their timing relevant characteristics for flexible timing-of-use.

Moreno-Munoz et al. [9] outlined evolution of 'Smart Energy Communities' that would allow the active participation of the 'prosumers' in a genuinely open market. Outlining continuously rising energy demands from rapidly increasing IoTs in smart city applications, Ejaz et al. [10] presented unifying framework for energy efficient optimization and scheduling of IoT based smart cities along with an interesting case study of energy-efficient scheduling in smart homes in smart cities.

Homes and buildings are constituent and basic elements of any city, therefore, in such a situation of revamping, without converting homes and buildings into smart ones, the overall transformation of cities and citizens would remains incomplete. Therefore, to fulfil the overall objectives of 'smart city', both the citizen and the home or building in which the citizen enjoys his/her life and works are necessarily be 'smart'. Original concept of 'Smart Home' has been further expanded into 'Smart Building' to make it applicable to large residential/commercial/institutional complexes such as apartments, societies, offices, hotels, hospitals, community centres, etc. Generally, buildings equipped with duly automated subsystems for routine operation, safety, security, communication, entertainment, etc. are known as 'Smart Buildings'. Smartness is evolved by deployment of different sensors and actuators for such automation along with smart configuration of control systems.

Smart environment is the real world physical environment for living and/or working inside such smart buildings duly equipped with special sensors, displays, actuators and controlling devices for continuous monitoring and control. The environment resources include ambience, air quality, water quality, etc.

3 Critical Parameters for Smart Environment Monitoring [11–14]

In recent years there has been an increase in public awareness about the effects of the indoor environment on people's health, comfort and work efficiency. Indoor work environment has also been considered as a crucial factor in the context of productivity. Parameters such as ambient temperature and humidity, illumination, barometric pressure, vibration and acoustic noise, air quality, water quality, safety and security have been considered important critical parameters for smart environment and monitored closely.

3.1 Ambient Temperature and Humidity [11–14]

Room temperature is a key ambient parameter. Very high or very low values of temperature have direct impacts on work efficiency and expectancy of lifespan. Humidity affects digestion and temperament of a person. Especially, cities on sea shores and beaches, it becomes important. Humidity is also dependent upon rainfall and it has impacts upon mood swings as well as work efficiency. Environment with controlled ambient temperature and humidity can provide thermal comfort to the occupants.

The comfort level of a human being results into his/her health and work efficiencies. Ambient temperature and humidity are closely interrelated parameters that directly affect comfort and therefore, health, mood and temperament of persons staying in the building. Generally, it is recommended to maintain temperature within range of ~19.5–20.5 °C and humidity within range of ~25–80% RH.

Environments with very low temperature and very low relative humidity causes dry respiratory, while those with very high temperatures and very high relative humidity result into heat stroke. Similarly, environments with very high temperature with very low humidity causes dehydration, while those with very low temperature with very high humidity rheumatism.

3.2 Illumination

Natural illumination is preferred, however, in absence of natural illumination, artificial illumination is provided by lighting. Illumination has direct effects on fatigue and boredom and hence, on work efficiency. Environments with controlled illumination can provide visual comfort to the occupants.

As mentioned by Falchi et al. [15], excessive artificial lighting also results into environmental pollution, which is observed one of the most rapidly rising recently with its levels growing exponentially over the natural nocturnal lighting levels provided by starlight and moonlight.

3.3   Barometric Pressure

Barometric pressure affects blood pressure and health of the person. It becomes very important parameter to monitor for cities in hilly terrains in particular. Abrupt variations in barometric pressure cause uneasiness and discomfort to the citizens as it not only affects their digestion system, it also swings mood and temperament.

3.4   Vibration and Acoustic Noise

Vibration is very important parameter from safety point of view, especially for the cities established in seismic zones and/or in hilly areas such as Himalayan terrains including Haridwar, Nainital, etc. in India. Larger amplitude earth vibration measurements could result into earthquakes and spread threat of life in citizens. Places with vibration and acoustic noise within limits can provide acoustic comfort to the occupants.

Huang et al. [16] mentioned that neutral sound pressure level for aural comfort in typical air-conditioned offices have been found in the range of 45–70 dB, with a mean of 57.5 dB. Undesirable sounds outside this range have been considered as Noise which could cause concentration failures, if the workspaces have poor acoustic isolation properties (Fig. 1).

3.5   Air Quality [11–14]

Clean air for breathing is fundamental necessity of everyone's life. Without stringent monitoring of air quality, survival of citizens could be in danger. Breathing in poor quality air shall quickly affect health of citizens and the entire city can get into health hazards or serious diseases. Particulate Matters (PM), Volatile Organic Compounds (VOCs) and Carbon dioxide (CO_2) contents have monitored for determination of air quality. Stringent biological, chemical and physical monitoring is required for overall air quality maintenance. Natural and artificial both types of ventilation should be employed for consistently maintaining air quality.

3.5.1   Particulate Matters (PM)

Considered as most hazardous air pollutants, particulate matters (PM 2.5 and 10) are two major air quality (purity) parameters which represent presence of particles with diameters 2.5 and 10 μm. Such particles can cause serious problems of respiratory and breathing such as cardiovascular problems, asthma attacks, etc.

(a)
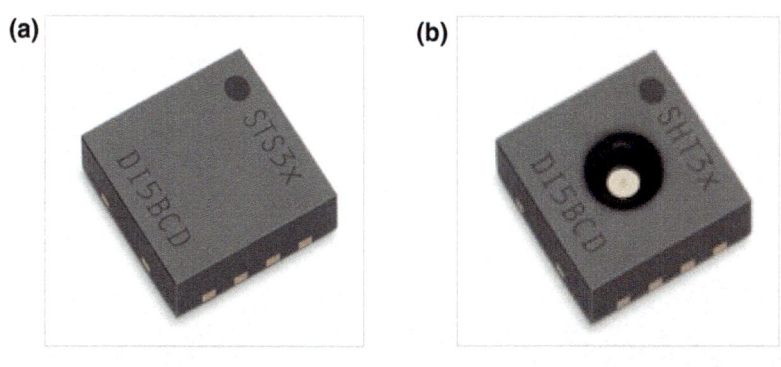

(b)

(c)

Fig. 1 **a** Temperature sensor, **b** Humidity sensor [12] and **c** Vibration and acoustic noise sensor [13]

3.5.2 Volatile Organic Compounds (VOCs)

These are indoor pollutants due to increased usage of certain products and building materials in construction. Such pollutant materials severely affect health and productivity, causing problems like headaches, dizziness, eye irritation, etc.

(a) **(b)** **(c)**

Fig. 2 **a** Particulate matters sensor, **b** CO_2 sensor and **c** Gas sensor [12]

3.5.3 Carbon Dioxide (CO_2)

Being an important indicator of indoor air quality, CO_2 content is keenly monitored as it directly affects the productivity and well-being of the building residents. To maintain CO_2 under specific limits, active ventilation is required (Fig. 2).

3.6 Water Quality [17]

Clean and pure drinking water is basic need for everyone. Without close monitoring of water quality, survival of citizens could lead to danger. Consumption of contaminated water can make serious and adverse effects on citizens' health and spread waterborne contagious diseases. Chemical parameters like pH, Conductivity, Total Dissolved Solids (TDS), Physical parameters such as solid concentrations such as Totally Suspended Solids (TSS) and turbidity are mainly monitored for ensuring water quality (Fig. 3).

Fig. 3 Water quality sensor [17]

3.7 Safety and Security

Safety and security against thefts, fire and gas leaks, unauthorised access to premises, etc. are necessary not only for young and physically fit citizens, but also for children, differently enabled persons, ladies and elderly citizens.

4 Technological Advancements: HAN-BAS, WSN and IoT

4.1 HAN-BAS

Home Area Network (HAN) is the network within the premises of a house or building, enabling devices, and electrical loads to communicate with each other and dynamically respond to externally sent signals (i.e. price, capacity utilization, scheduling information, etc.). This type of network could be characterized by low data rate requirements and provides necessary communication infrastructure for the energy meter. As reported via Reuters at [18], from $5.77 billion in 2013, the global market of home automation and security control is likely to reach $12.81 billion by 2020 with projected CAGR of 11.36% between 2014 and 2020. This clearly exhibits the potential and demand at international levels.

Building Automation System (BAS) is a data acquisition and control system that incorporates various functionalities provided by central control system of a building. Modern BAS is a computerized, intelligent network of electronic devices, designed to monitor and control the lighting, internal climate, and other systems in a building for creating optimized energy usage, safety and security, information and communication, and entertainment facilities. BAS reduces building energy consumption and, thereby, reduces operational and maintenance costs as compared to an uncontrolled building. BAS core functionality keeps building climate within a specified range, provides light to rooms based on an occupancy schedule, monitors performance and device failures in all systems, and provides malfunction alarms to building maintenance staff. A building equipped with BAS is often referred to as a Smart Building or Intelligent Building.

4.2 WSN

Wireless Sensor Network (WSN) refers to a group of spatially dispersed and dedicated sensors for monitoring and recording the physical conditions of the environment and organizing the collected data at a central location. WSNs measure environmental conditions like temperature, humidity, vibration, sound, pollution levels, wind, etc. (Fig 4).

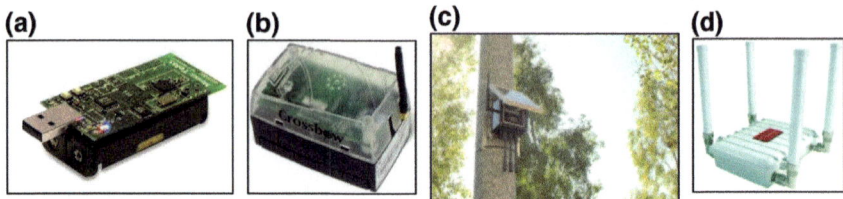

Fig. 4 **a** WSN mote, **b** WSN mote in enclosure [28], **c** Solar powered WSN mote [17] and **d** Wireless data aggregator unit

4.3 IoT

Wikipedia defines Internet of Things (IoT) as the network of physical devices, vehicles, home appliances, and other items with embedded electronics, software, sensors, actuators, and connectivity which enables these things to connect and exchange data, creating opportunities for more direct integration of the physical world into computer-based systems, resulting in efficiency improvements, economic benefits and reduced human intervention. Referring to market researcher Gartner German news agency FAZ vide its website article at [19], last year 8.4 billion networked devices were in use, which were 31% higher in numbers than previous year; with an estimated quantity to reach 20.4 billion devices by 2020.

In [20], after providing vision and role of IoT in smart city applications, with smart home applications in particular, along with communication technology related details.

The main attractive feature of IoT devices is they are accessible-modifiable over internet and consume less power due to sleep-sniff abilities as well as their high security data communication with good latency performance.

5 Participation of Smart Citizens and Initiatives

Smart citizen living-working in smart building powered by smart grid of smart city, shall not merely remain a consumer like legacy citizen, but also be a producer of electricity. Thus, the smart citizen shall be a 'Prosumer' (Producer + Consumer) of electrical energy. Smart buildings shall be equipped with small size wind turbines and/or rooftop solar Photo Voltaic (PV) cells to produce electricity which shall be adjusted-credited against the consumption bills.

Smart citizen without compromising convenience, comfort or liking, shall configure/schedule various appliances to run at their best efficiency during off-peak hours to serve the need of service as well as to consume least. Smart citizen shall adapt schemes of dynamic pricing and participate actively in Demand Side Management (DSM) and Demand Response (DR) based on intelligent Load Forecasting (LF) suggestions sent from utility company. Novel concepts such as Real Time Pricing (RTP), Time of Use (ToU), and Critical Peak Pricing (CPP) shall serve as major guidelines for effective, efficient, and timely decision-making for optimized use of electricity.

Describing various Government of India initiatives on smart cities developments, Upadhyaya [21] suggested that the focus should be sustainable and inclusive development. The authors explained basic idea of smart cities with its components and applicability in Indian cities.

Presenting smart grid evolution as energy independence and environmentally sustainable economic growth, Vijayapriya and Kothari [22] presented basic components and model setup of smart grid network along with a model for smart home.

Focusing upon role of IoT in renewable resources integration to electric grid, [23] presented conceptual implementation of architecture in smart cities and proposed consumer communication area network framework.

6 Issues and Challenges

(i) Smart energy systems could attain a sustainable future by tackling challenges and issues related to production, processing, and end use of energy.

(ii) Electricity theft has been identified as one of the major challenge vide [24] along with communication availability and reliability as some of the major issues.

(iii) Significant amount of novel approaches are required to fulfil the objectives such as decarbonisation as well as reduction in impacts of climate change.

(iv) Mir and Ravindran [20] identified challenges for IoTs such as pending technological standardization, managing and fostering rapid innovation, privacy and security, absence of governance, vulnerability to internet attack, etc.

(v) Return over investment, break-even period, capital and maintenance expenditures, requirements of highly skilled manpower, etc.

7 Case Study Examples

Focusing on urban IoTs, Zanella et al. [25] presented an interesting case of practical proof-of-concept implementation in the city of Padova, Italy, various services (smart city applications) have been characterized along with network types, traffic rates, tolerable delays, energy sources and feasibility.

Reviewing concepts, motivations and applications of smart cities, Arasteh et al. [26] described IoT technologies for smart cities including practical experiences of various installations across the world and challenges faced.

Design and implementation of working models of BAS based on wired technology has been presented in [27]. With necessary customization, these models could be suitably modified to serve need based application requirements. Using similar approach, existing buildings can also be converted into smart buildings with suitable modifications.

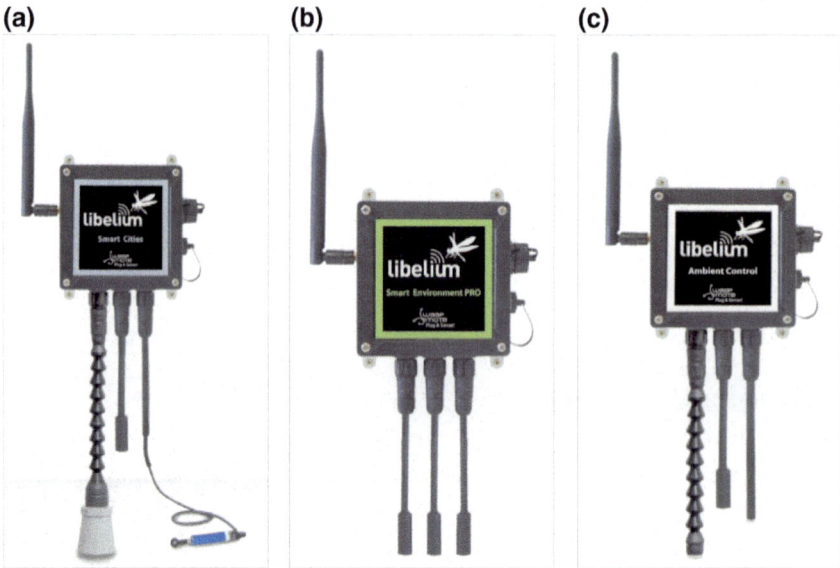

Fig. 5 **a** Wireless smart cities sensor, **b** Wireless smart environment sensor and **c** Wireless smart ambient control device [17]

Fig. 6 View of central monitoring system [27]

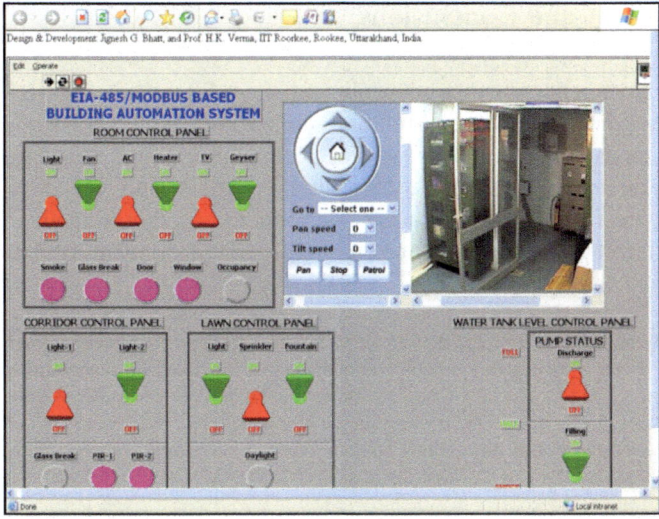

Fig. 7 Snapshot of remote access of GUI via internet [27]

Standard modules packaged with various relevant sensors are available as shown in Fig. 5, which could be employed as per customized needs. Furthermore, a central monitoring system should be evolved as shown in Fig. 6 from where entire geographic spatial region should be continuously monitored, citizens could be alerted against abnormal possible occurrences and necessary records could be maintained for future use (Fig. 7).

8 Guidelines and Recommendations

(i) Each smart building in a smart city can differ in terms of requirements of applications and therefore that of communication. Therefore, there cannot be a single model which could be made universally applicable. However, with necessary customization, suitable model could be developed and implemented to serve need based requirements. New buildings should be constructed duly equipped with smart technologies to be smart buildings since inception, while existing buildings should be converted into smart buildings with suitable modifications.

(ii) Usage of smart building technologies along with smart plugs, LEDs and climate sensors have been advised to utilize for reduction in energy consumption, CO_2 reduction and spreading awareness.

(iii) Designers, maintenance staff as well as citizens should be imparted necessary training and awareness for easy adaptability.

9 Summary and Conclusions

This chapter started with explanation of importance of smart home/smart building as core and mandatory element of smart grid serving as energy backbone of smart city. Next, critical parameters for smart environment monitoring and citizen participation have been discussed. Relevant technologies along with useful case study examples have been presented along with earlier contributions of authors. Finally after identifying applicable issues and challenges, the article ends with making useful guidelines and recommendations.

Acknowledgments The authors take this opportunity to express their sincere thanks to Editor Prof. T. M. Vinod Kumar and his entire team, bulletin editors and Springer staff members for their valuable guidance, excellent co-operation and timely help extended. Co-operation received in development of this book chapter from the faculty members, management and office bearers of affiliating organization of the authors is acknowledged with thanks. Useful contributions and cooperation received from all the cited sources of references are also gratefully acknowledged.

References

1. Bansal N, Shrivastava V, Singh J (2015) Smart urbanization—key to sustainable cities. In: Real corp 2015, vol 2. pp 551–560
2. Bhatt J, Shah V, Jani O (2014) An instrumentation engineer's review on smart grid: critical applications and parameters. Renew Sustain Energy Rev 40:1217–1239. https://doi.org/10.1016/j.rser.2014.07.187
3. Bhatt J, Jani O (2015) E-governance for photo voltaic powergrid: solar city Gandhinagar, Gujarat, India. In: Kumar TMV (ed) E-governance of smart cities. Springer, pp 177–230
4. Bhatt J, Jani O (2016) Smart grid: energy backbone of smart city and e-democracy. In: E-democracy for smart cities. Springer Singapore, pp 319–366
5. Bhatt J, Jani O (2018) Smart development of Ahmedabad–Gandhinagar twin city metropolitan region, Gujarat, India. In: Kumar TMV (ed) Smart metropolitan regional development. Singapore, Springer Nature, pp 313–356
6. Calvillo CF, Sanchez-Miralles A, Villar J (2016) Energy management and planning in smart cities. Renew Sustain Energy Rev 55:273–287. https://doi.org/10.1016/j.rser.2015.10.133
7. Khansari N, Mostashari A, Mansouri M (2014) Impacting sustainable behavior and planning in smart city. Int J Sustain L Use Urban Plan 1(2):46–61
8. Smale R, Vliet B van, Spaargaren G (2017) When social practices meet smart grids: flexibility, grid management, and domestic consumption in The Netherlands. Energy Res Soc Sci 34:132–140. https://doi.org/10.1016/j.erss.2017.06.037
9. Moreno-Munoz A, Bellido-Outeirino FJ, Siano P, Gomez-Nieto MA (2016) Mobile social media for smart grids customer engagement: emerging trends and challenges. Renew Sustain Energy Rev 53:1611–1616. https://doi.org/10.1016/j.rser.2015.09.077
10. Ejaz W, Naeem M, Shahid A et al (2016) Efficient energy management for internet of things in smart cities
11. Sensirion Inc., Four key indoor environmental parameters for indoor air quality. https://www.azosensors.com/article.aspx?ArticleID=995. Accessed on 26 Dec 2018
12. Sensirion Inc., Environmental sensors | Sensirion. https://www.sensirion.com/en/environmental-sensors/. Accessed on 26 Dec 2018
13. NoiseMOTE – Envirowatch Ltd. http://www.envirowatch.ltd.uk/noisemote/. Accessed on 29 Dec 2018

14. Treacy M (2013) 10 environmental sensors that go along with you | TreeHugger. https://www. treehugger.com/clean-technology/environmental-sensors.html. Accessed on 29 Dec 2018
15. Falchi F, Cinzano P, Elvidge CD et al (2011) Limiting the impact of light pollution on human health, environment and stellar visibility. J Environ Manage 1–9. https://doi.org/10.1016/j. jenvman.2011.06.029
16. Huang L, Zhu Y, Ouyang Q, Cao B (2012) A study on the effects of thermal, luminous, and acoustic environments on indoor environmental comfort in offices. Build Environ 49:304–309. https://doi.org/10.1016/j.buildenv.2011.07.022
17. Plug & Sense! Sensor Networks made easy—models | Libelium. http://www.libelium.com/ products/plug-sense/models/. Accessed on 29 Dec 2018
18. Reuters Research and Markets, "Research and Markets: Global Home Automation and Control Market 2014–2020—Lighting Control, Security & Access Control, HVAC Control Analysis of the $5.77 Billion Industry." https://web.archive.org/web/20160505124414/; https://www. reuters.com/article/research-and-markets-idUSnBw195490a%2B100%2BBSW20150119. Accessed on 26 Dec 2018
19. Rudiger Kohn F ne, Große internationale Allianz gegen Cyber-Attacken. https://www.faz.net/ aktuell/wirtschaft/diginomics/grosse-internationale-allianz-gegen-cyber-attacken-15451953-p2.html?printPagedArticle=true#pageIndex_1. Accessed on 26 Dec 2018
20. Mir MH, Ravindran D (2017) Role of IoT in smart city applications : a review. Int J Adv Res Comput Eng Technol 6(7):1099–1104
21. Upadhyaya AV (2016) Smart cities: a vision for development of Indian cities. Imp J Interdiscip Res 2(10):700–710
22. Vijayapriya T, Kothari DP (2011) Smart grid: an overview. Smart Grid Renew Energy 02(04):305–311. https://doi.org/10.4236/sgre.2011.24035
23. Al-Ali AR (2016) Internet of things role in the renewable energy resources. Energy Procedia 100:34–38. https://doi.org/10.1016/j.egypro.2016.10.144
24. Shukla P (2015) Smart cities in India
25. Zanella A, Bui N, Castellani A, Vangelista L, Zorzi M (2014) Internet of things for smart cities. IEEE Internet Things J. 1(1):22–32. https://doi.org/10.1109/JIOT.2014.2306328
26. Arasteh H, Hosseinnezhad V, Loia V, Tommasetti A, Troisi O, Shafie-Khah M, Siano P (2016) Iot-based smart cities: a survey. In: EEEIC 2016—International Conference on Environment and Electrical Engineering. https://doi.org/10.1109/EEEIC.2016.7555867
27. Bhatt J, Verma H (2015) Design and development of wired building automation systems. Energy Build. https://doi.org/10.1016/j.enbuild.2015.02.054
28. Connect with new age wireless device—telosb sensors«telosb sensors|telosb motes|telosb. https://telosbsensors.wordpress.com/2013/12/03/connect-with-new-age-wireless-device-telosb-sensors/. Accessed on 29 Dec 2018

Part III
Chandigarh

Smart Urban Green Spaces for Smart Chandigarh

Prabh Bedi, Mahavir and Neha Goel Tripathi

Abstract The principles of sustainable development necessitate that a balance be maintained between environment and development to ensure a sustainable future. In India, growing population and rapid urbanisation is resulting in significant land being used for settlements resulting in decrease in open spaces across many cities. It has long been established that the presence of natural areas in and around urban settlements contributes to a quality of life by providing important ecological, social and psychological benefits to humans. Amongst the rare exceptions are cities established after India's independence, such as Gandhinagar and Chandigarh, where the urban greenery was pre-integrated in the City Master Plans at the initial design phase. However, the recent trends and analysis indicates that Chandigarh region has been seen struggling to maintain a balance between economic, environmental and social sustainability. Today, the periphery of Chandigarh is characterised by unregulated construction and rapid urbanisation. The Urban Green Spaces framework is not yet integrated in regional planning in a systematic way. In this research, the authors have assessed the organization and implementation of the existing framework and structure of the urban green spaces and its planning in Chandigarh Region Area. Though the green spaces in the city were integrated with the master plan, the periphery spaces have not been successfully addressed. This research aims to develop a comprehensive conceptual framework for urban green space for Chandigarh Region based on spatial planning and ecological principles. Attempt has been made to propose a smart urban green strategy for the Chandigarh region. Built upon the geospatial technological tools the framework will help in identifying and protecting the green spaces in the region.

P. Bedi (✉)
Sushant School of Planning and Development, Ansal University, Gurgaon, India
e-mail: prabhb@gmail.com

Mahavir · N. G. Tripathi
School of Architecture and Planning, 4, Block-B, I. P. Estate, New Delhi 110002, India
e-mail: mahavir57@yahoo.com

N. G. Tripathi
e-mail: nehagtripathi@gmail.com

© Springer Nature Singapore Pte Ltd. 2020
T. M. Vinod Kumar (ed.), *Smart Environment for Smart Cities*, Advances in 21st Century Human Settlements, https://doi.org/10.1007/978-981-13-6822-6_4

Keywords Chandigarh · Urban green spaces · Urban periphery · Green infrastructure

1 Introduction

Chandigarh, 'The City Beautiful' was planned as capital city of undivided Punjab. It is located near the foothills of the Shivalik range of the Himalayas in Northwest India and is independent India's first planned. Planning along the lines of Garden City concept, Le Corbusier envisioned the periphery to be a large green space whose agriculture would support the city and the city in turn would generate employment for the local people. In the last four decades, the towns of Panchkula and Mohali have grown around Chandigarh not only in terms of population but have spread in area too turning it into a vast and economically vibrant region. Even though the green cover within the city limits of Chandigarh have been maintained, the rapid urbanisation of the periphery especially since after 2008 has been impacting the periphery area.

Climate change and increasing urbanization will exacerbate the temperatures in the urban areas creating heat islands. Green spaces can play a significant role in the complex urban ecosystems. Green spaces in urban areas in form of parks, green roofs and walls and road-side trees can reduce the temperature.

Smartness of a space lies in its aesthetics and attractiveness, healthy environment, ability to create social cohesion and economic viability. These spaces benefit the communities directly and indirectly from aesthetics, environment, recreation and economic standpoints. At the same time urban areas can accrue the benefits of mitigating air pollution and revive the fast depleting biodiversity habitats.

2 Evolution of Chandigarh and Its Region

Chandigarh derives its name from the temple of Goddess 'Chandi'[1] and 'garh'[2] that lay beyond the temple. The gently sloping plains where present day city of Chandigarh exists, was in the ancient past, a wide lake ringed by a marsh [3]. The fossil remains found at the site indicate a large variety of aquatic and amphibian life. The area was also known to be a home to the Harappans during the Indus Valley Civilisation. Since the medieval through modern era, the area was part of the large and prosperous Punjab Province, which was divided into East and West Punjab during partition of the country in 1947. Established immediately after Independence, the city was conceived not only to serve as the capital of Punjab, but also to resettle thousands of refugees who had been uprooted from West Punjab (in present day Pakistan) [3].

---

[1]Goddess of power as per the Hindu mythology [1].
[2]Fort [2].

In March 1948, the Government of the then erstwhile Punjab, in consultation with the Government of India, approved a 114.59 km^2 (11,459 ha) area at the foothills of Shivaliks as the site for the new capital. The site selected by Dr. M. S. Randhawa, the then Deputy Commissioner of Ambala was a part of the erstwhile Ambala district as per the 1892–93 Gazetteer of District Ambala [4]. Chandigarh was designed for a population of five lakhs and density of approximately 4400 persons per Sq. km (44 persons per hectare) [5]. Subsequently, with the reorganization of the erstwhile State of Punjab in 1966 into present day Punjab, Haryana and Himachal Pradesh, the city of Chandigarh assumed the unique distinction of being the capital of two states, Punjab and Haryana. The city was declared as a Union Territory and put under the direct administrative control of the Central Government.

Chandigarh is symbolic as the first city that was planned in independent India. Planned on modernist thinking,[3] the then Prime Minister, Pt. Jawaharlal Nehru envisioned the new city to represent free India. Albert Meyer, an American planner and architect, was hired in 1949 to formulate the plan of Chandigarh city. Mayer developed a superblock[4] based-city threaded with green spaces which emphasized the cellular neighborhood and traffic segregation. The site had certain inherent landscape feature, which were fully retained as well as integrated into the plan like the gentle grade was used to promote the drainage and rivers to orient the plan [8] (Refer Fig. 1).

The natural features of the site consisted of the panoramic range of the Shivalik Hills, the gently sloping land form from the North-East to the South-West, seasonal rivulets on its north-western and south-eastern flanks; and an eroded valley with a small nallah [5] running through the centre [8].

The conceptual design was further detailed by young architect, Mathew Nowicki in Mayer's team, who gave it a touch of artistry and monumentality as he synthesised it into a simple pattern 'The leaf', bringing in the element of sensitization towards the environment and ecology from the inception into the plan (Refer Fig. 2).

Due to untimely death of the Nowicki, Le Corbusier, a swiss-french architect and urban planner, was appointed to complete the design for the capital city. He had a repute of being the forefather of the Modernist architectural. Corbusier's approach to planning was to create a scientifically rational and comprehensive solution to urban problems in a way that would both promote democracy and quality of life [11]. Based on the principles of scientific rationalism, efficiency, and social improvement through design, new ideas of mass production and democracy inspired his vision of the built environment. These manifested in planning and architecture through the use of minimal ornamentation, repetitive units, high rise structures, and separation of land use and zoning.

---

[3]Modernism in urban planning was based on the principles of clearly distinguishing everything including the four main principals of human activities: living, working, recreation so as to bring order in the city and conceive it as a machine that works according to specific rules [6].

[4]Superblock is an area of urban land bounded by arterial roads that is the size of multiple typically-sized city blocks [7].

[5]Nallah is a runnel or a brook [10].

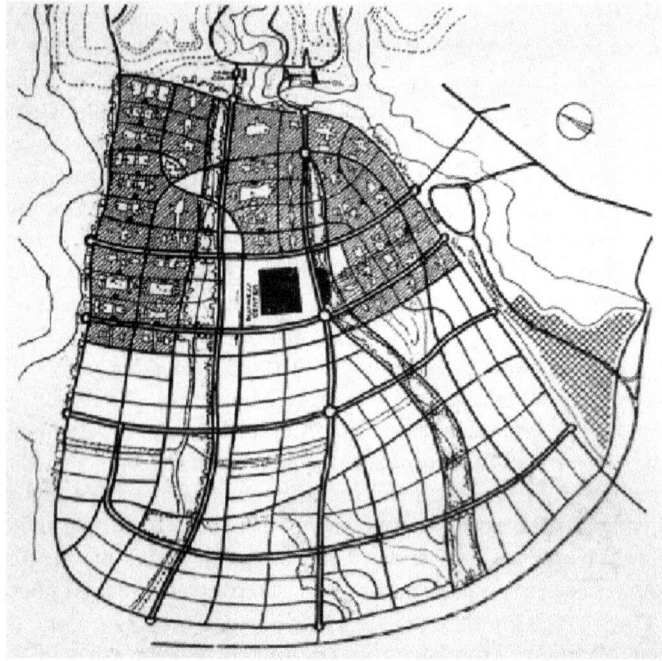

Fig. 1 Chandigarh city plan by Albert Mayer. *Source* Mayer [9]

City plan was modified from one with a curving road network to rectangular shape with a grid iron pattern for the fast traffic roads, besides reducing its area for reason of economy. The city plan was conceived as post war Garden City[6] wherein vertical and high rise buildings were ruled out, keeping in view the socio economic-conditions and living habits of the people.

The city was conceptualized as an urban organism with its major functions: living, working, care of the body and spirit and circulation, being equated with different parts of human body. Neighborhood called a sector was rectangular in shape of about one square kilometer each which defined its basic unit. Le Corbusier planned the Capitol Complex at the top of Chandigarh city resembling the head; the intellectual base, reflecting his conviction that governance should begin here as the head rules the body. Residential sector constituted the living part whereas the Capitol Complex, city centre, Educational Zone (Post Graduate Institute, Punjab Engineering College and Punjab University) and the Industrial Area constitute the working part. The Leisure Valley, Gardens, Sector Greens and Open Courtyards were designed to represent the care of body and spirit. The circulation system comprised of 7 different types of roads known as 7 Vs. Subsequently, pathways for cyclists called V8 were added to

---

[6]Method of urban planning proposed in 1898 by Sir Ebenezer Howard in the United Kingdom were in the urban areas were planned to be self-contained communities surrounded by greenbelts, containing proportionate areas of residences, industry and agriculture [12].

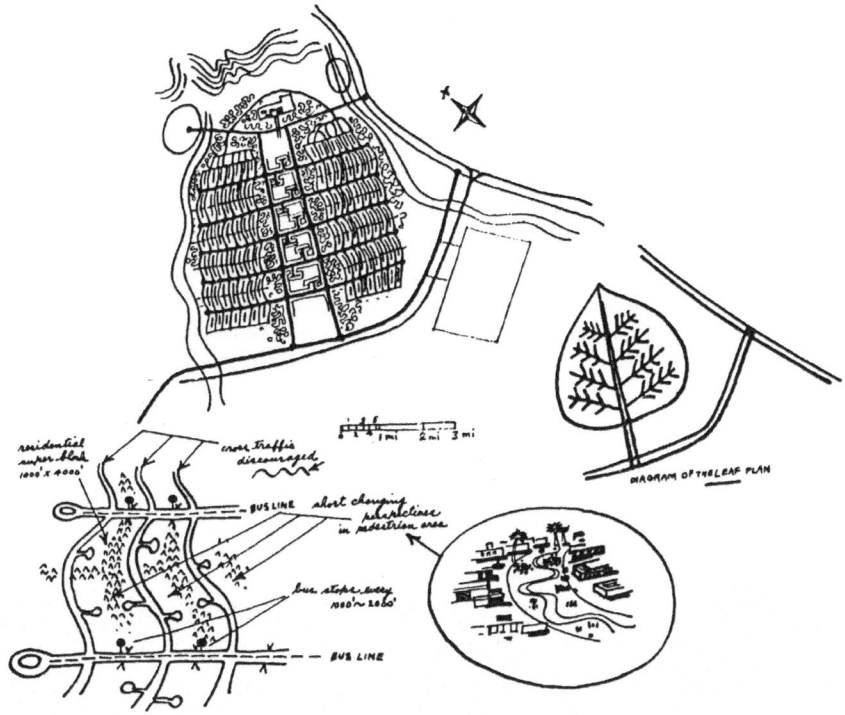

Fig. 2 The 'leaf plan' of Chandigarh conceptualised by Mathew Novicki. *Source* Mayer [9]

the circulation system [8]. Combining modernity with ecology, 8 km long linear-park was planned running through the city from its north eastern tip to its south-western end forming the city level urban green.

Corbusier's design of an integrated system of seven roads was to ensure efficient traffic circulation. The city's vertical roads run northeast/southwest known in local language as 'path' and the horizontal roads run northwest/southeast known as 'marg'. These intersect at right angles, forming a grid of network for movement. This arrangement of road-use has led to a hierarchy of movement, which also ensures that the residential areas are segregated from the noise and pollution of traffic [8].

The work place of the city, the industrial area comprised of 2.35 km^2, is located in the extreme southeastern side near the railway line. The location of the industrial areas was in conjunction with the prevalent wind direction of the area, which is north, north west. The area has been allocated and planned for non-polluting and light industry and is directly connected to the civic centre by a V-3 road. A wide buffer of fruit trees has been planted to screen off the industrial area from the rest of the city. Tree plantation and landscaping has been an integral part of the city's Plan. Twenty-six different types of flowering and twenty-two species of evergreen trees

[3] all over the city help to ameliorate the harsh climate of the region, especially the hot and scorching summers.

3 Natural Conditions of Chandigarh

Chandigarh experiences cold dry winder, hot summer and sub-tropical monsoon. Evaporation usually exceeds precipitation and the weather is generally dry. The city experiences four seasons: summer or hot season from mid-March to mid-June, rainy season from late June to mid-September, post monsoon Autumn from mid-September to mid-November and winter from mid-November to mid-March. The dry spell of summer is long but with occasional drizzles and thunderstorms. May and June are the hottest months of the year with mean daily minimum and maximum temperatures being about 40 and 25 °C, respectively. Southwest monsoons bring high intensity showers in late June and July. The weather at that time is hot and humid. January is the coldest month with mean maximum and minimum temperatures being around 24 and 1.8 °C respectively [13].

3.1 Temperature

The hot season lasts for two and a half months, from April end to beginning of July, with an average daily high temperature above 36 °C. The hottest days of the year are in May, with an average high of 41 °C and low of 27 °C.

The cool season too lasts for two and a half months, from December to middle of February, with an average daily high temperature below 23 °C. The coldest days of the year are in January, with an average low of 8 °C and high of 20 °C [13].

3.2 Clouds

The average percentage of cloud cover, that is the sky covered by clouds, in Chandigarh varies significantly by season. The clearer part of the year in begins towards the end of August and lasts for about three and a half months until middle of December. The month of September experiences the clearest skies. The cloudier part of the year begins around middle of December and lasts for about eight and a half months. The month of July is the cloudiest day of the year.

3.3 Precipitation

A wet day is one with at least one millimeter of liquid or liquid-equivalent precipitation. The chance of wet days in Chandigarh varies very significantly throughout the year. The wetter season lasts three months, from middle of June to middle September. The wettest month being July. The drier season lasts nine months, from middle of September to middle of June, November being the driest month.

3.4 Rainfall

Chandigarh experiences extreme seasonal variation in monthly rainfall. The rainy period of the year lasts for 10 months, from middle of December to middle of October. Most rainfall is during July and August, averaging at 220 mm. The rainless period of the year lasts for two months, from middle of October to middle of December. The least rain falls in November, with an average total accumulation of 5 mm [8].

3.5 Sun

The length of the day varies significantly over the course of the year in Chandigarh. The shortest day and the longest day coincides with the winter and summer solstice. Summer solstice giving the region the long daylight time of 14 h and winter solstice giving the shortest daylight time of only 10 h.

3.6 Humidity

Chandigarh experiences extreme seasonal variation in the perceived humidity. The comfort level of humidity is dependent on the dew point, determined by the evaporation of the perspiration from the skin. Lower dew point feels drier and higher dew point feels humid. The humid period of the year lasts about three and a half months from middle of June to beginning of October, August being the most humid.

3.7 Wind

The average hourly wind speed in the city experiences significant seasonal variation. The windier period is from middle of January to end of June with average wind speed of 10 km per hour. The calmer month experience about 6 km per hour of wind speed [8].

The predominant wind direction is north and northwest. Seven months of the year experience the north wind and three months experience the west wind.

3.8 Topography

Chandigarh lies at an elevation of 1053 feet (321 m) above mean sea level. The topography within 3 km has modest variations in elevation with maximum elevation change of 50 m. This change within 15 km is about 100 m. This variation is significant within 100 km at about 3000 m.

90% of the area within 3 km of Chandigarh is covered in artificial surface, within 10 km is 65% of the area is under agricultural land and 20% under artificial surface. This increases to 76% under agricultural land and 12% under trees within 100 km [8, 13].

While planning the city of Chandigarh, the natural and climatic conditions of the region have been accounted for. The placement of the principal functions has been based on the climatic conditions and the gradient of the land, like the industries have been located considering the prevalent wind direction. The alignment of the roads, which is in the grid form is in conjunction with the prevalent wind direction as well.

4　Land Use Through the Development Plans

The economic constraints led the master plan to be implemented in two phases, catering to a total population of half a million (five lakhs). Phase-I consisting of 30 low density sector spread over an area of 36.42 km^2 (Sector 1–30) for 150,000 people whereas Phase-II consisting of 17 considerably high density sectors (Sectors 31–47) spread over an area of 24.28 km^2 for a population of 350,000 [8].

Chandigarh city is under the administration of Union Territory of Chandigarh. The planning functions of the city are undertaken by the Urban Planning Department and the Municipal Corporation of Chandigarh manages the provision and maintenance of services and amenities within the municipal area.

The area of the Municipal Corporation is 79.74 km^2 out of 114 km^2 [14]. With regards the existing land use (including the extension areas of the city) about 38% of area is categorized as residential and 5% as commercial and a significant 8% each is assigned towards forests. On the other hand, the proportion of public and semi-public use land constitutes 11% and transportation land constitutes only 7% of the total area (Table 1).

While planning the city of Chandigarh, every emphasis has been laid for providing green spaces, throughout the city. The open space per capita in the city was 82 m^2 in the early stages of development [8] With the subsequent increase in population, the

Table 1 Existing and proposed land use classification, 2014 and 2031

Land use type	2014		2031	
	Area (km²)	Percent	Area (km²)	Percent
Residential	43.19	37.88	43.98	38.58
Commercial	5.42	4.76	5.99	5.26
Transport	8.28	7.26	8.49	7.45
Industrial	5.37	4.71	6.52	5.72
Public/semi-public	12.01	10.54	13.86	12.15
Recreational	9.83	8.62	11.41	10.00
Agriculture	0.00	0.00	2.72	2.39
Public utility	1.22	1.07	1.38	1.21
Railway	1.28	1.12	1.28	1.12
Defence	6.37	5.58	6.37	5.58
Forest	8.55	7.50	10.38	9.10
Vacant land	12.48	10.94	0.39	0.34
Reserved	0.00	0.00	1.25	1.10
Total area	114.00	100.00	114.00	100.00

Source Chandigarh Master Plan 2031 [8]

per capita green spaces in the city came down to 56 m² in 2010, which is still quite high as compared to 21.43 m² per capita in the Delhi. Whereas, in Gandhinagar, another post-independence planned city of India the green cover is 164 m² per capita [15]. In the new cities that have been recently planned, Naya Raipur and Amravati, the green cover has been designated at 27 and 53% of the total area [15].

The city of Chandigarh was planned with the city at the core and agricultural area in the periphery. In order to enforce the vision of the Plan, Punjab New Capital (Periphery) Control Act was enacted in 1952 to ensure that no haphazard development in the periphery of the planned city.

5 Chandigarh Periphery Control Act

Howard Ebenezer, in his model of Garden City, attempted to plan for both the town and country. The agricultural green belt was an essential element of the Garden City concept. Realizing that development is continuous, Ebenezer also envisioned satellite cities developing around a central city. The satellite cities were contained cities with their own green belts. He advocated the idea of ensuring the continuity of the rural green character of the countryside around the urban, grey landscape of a city despite

Fig. 3 The Garden City
concept. *Source* Howard [16]

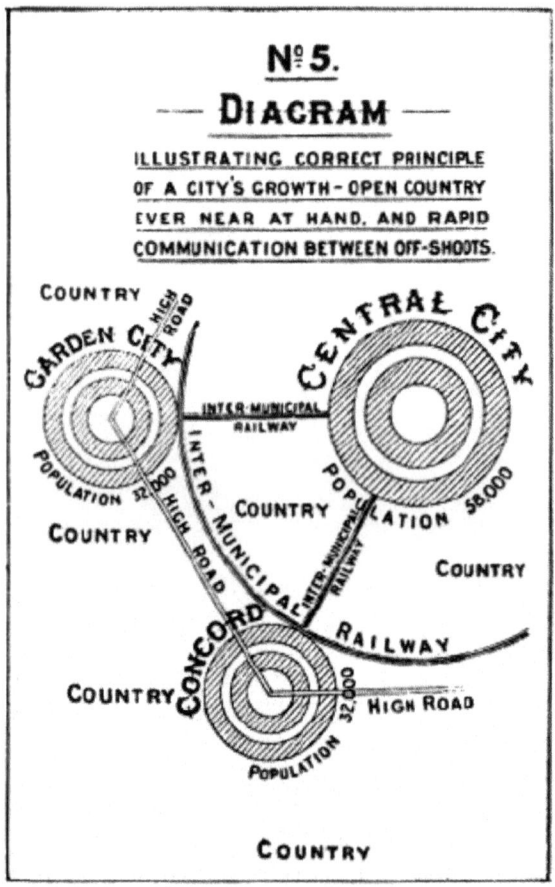

controlled developments. His ideas found form in the shape of the Greater London
plan of 1944 which provided for a green belt of 8 km all around the city to contain
the sprawl of London [12].

Planning along the same lines so as to maintain the aesthetic of the new City,
Le Corbusier envisioned the periphery to be a large green space whose agriculture
would support the city. He emphasized that the agriculture on the periphery would
support the city and the city in turn would generate employment for the local people
(Fig. 3).

The city of Chandigarh as planned was spread over 114 km^2. The city that formed
the core was to be surrounded by agriculturally dominant area spread over 1500 km^2.
This was a 16 km buffer extending from the outer boundary of the area acquired for
Chandigarh (Refer Fig. 4).

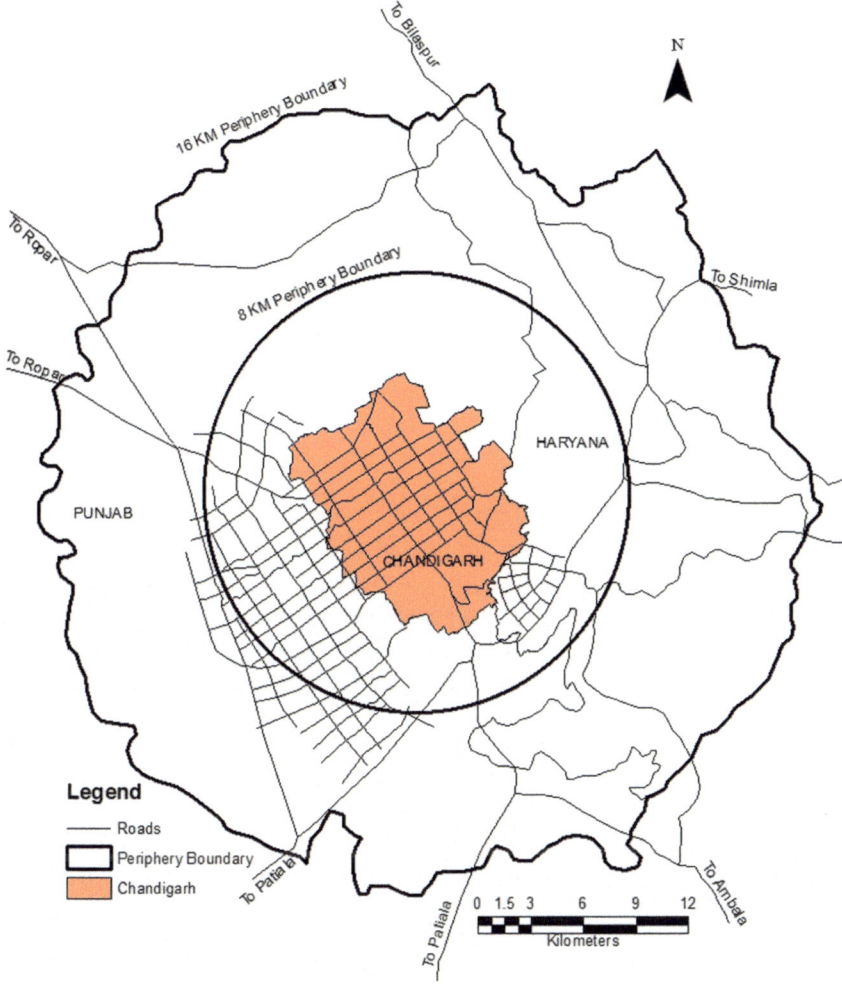

Fig. 4 Chandigarh capital periphery boundary. *Source* Developed by authors from Chandigarh Master plan 2031

5.1 The Act

In order to enforce the implementation of the vision of an agriculturally dominant periphery, the Punjab New Capital (Periphery) Control Act, 1952 popularly known as the Chandigarh Periphery Controlled Act was established in 1952. Its objective was to prevent mushrooming of unplanned construction around the new city. The Periphery Control Act was created to regulate the use of land and prevent unauthorized urbanization within 8 km. A controlled periphery around the new capital was proposed to serve as much as a green envelope, as a buffer separating the new capital from the indigenous urban expression all around [17].

5.2 The Purpose

At the time of conceptualization, the rationale of the said Act was to protect the surrounding rural community from getting urbanised, to prevent the growth of slum like inhabitations, to freeze the land use in the demarcated boundary, and to stop conversion of agricultural[7] land into uses other than agriculture or subservient to agriculture [17]. Another purpose of the Periphery was to provide locally available building material (from brick kilns) for the construction of the city [17].

5.3 Violations to the Act

However, there emerged gaps between what was visualized and what came up as the Chandigarh grew. Development is largely guided by the political and economic factors that are highly unpredictable and dynamic in nature. The time gap between planning and development leads to unforeseen realities which has been the case of Chandigarh and its Periphery.

Violation to The Punjab New Capital (Periphery) Control Act, 1952 started in 1962 with the establishment of Army Cantonment at Chandimandir, Air Force Station and Hindustan Machine Tools factory within the periphery. In the same year, the periphery control area was extended from 8 to 16 km through an amendment to the Act. These establishments were considered essential and were not deemed to be violations of the Act as such.

In 1966 the state of Punjab was divided into Punjab, Haryana and Himachal Pradesh. Chandigarh continued to be the capital of the states of Haryana and Punjab and was made the Union Territory under the direct administrative control of the central government. The Union Territory of Chandigarh comprised of 114 km^2, 70 km^2 under the city limits and 44 km^2 in the Periphery area. 1021 km^2 of the periphery area came under Punjab and 295 km^2 under Haryana. At the time of state reorganisation, in 1966, only 3% of the controlled area came under Chandigarh, 23% went with Haryana and 74% fell under Punjab control. It had a population of 1.04 million in 2001 spread across 458 villages and 12 urban centres [5].

Soon after the reorganization of the states, townships were established by respective state governments in areas adjoining Chandigarh. The state of Punjab set up a 22 km^2 town of Mohali towards south west while Government of Haryana set up a 20 km^2 urban area of Panchkula towards south east. The establishment of these

---

[7]Agriculture here includes horticulture, dairy farming, poultry farming and the planting and upkeep of orchards [17].

towns was in violation to the Punjab New Capital (Periphery) Control Act, 1952. The towns of Mohali and Panchkula have over the years developed a unique relation with Chandigarh city at the core. Panchkula bordering Chandigarh on the southeast functioned as a big service center. In Mohali, industrial impetus was given which has led to the growth of present day town spread across 22 km^2.

In 1975, a high powered co-ordination committee was constituted under the then Ministry of Urban Development which in 1977 led to the formulation of Chandigarh Urban Complex Plan consisting of Chandigarh, Mohali and Panchkula comprising of 330 km^2. The three cities are popularly known as Tri-City.

In 1984, Chandigarh Interstate Regional Plan was prepared by Town and Country Planning Organisation, New Delhi for an area of 2431 km^2 for a population of 25 lakhs [8] (Refer Fig. 5).

In 1990, Punjab Government declared an area of 40 km^2 to be a Free Enterprise Zone (FEZ) in Dera Bassi Tehsil of Patiala District [18] (now part of SAS Nagar District), where the setting up of industries was to be permitted. This further undermined the power of the Periphery Act.

In 1998, Punjab Government decided to pass a regulation for an over-arching regularization of all existing unauthorized constructions, as well as permitting the future construction of educational and health institutions [19]. This move by the government resulted in mushrooming of educational institutes on the outskirts of the urban area along the main roads and highways. The states of Punjab and Haryana had themselves violated the green buffer the by developing the townships of Panchkula, Mohali and Zirakpur. In 1999, Chandigarh Interstate Metropolitan Regional Plan was to be prepared for a 50 km radii [8].

In order to capitalise on the new wave of urbanisation, especially in the tier II and III towns, the Government of Punjab notified Greater Mohali Area Development Authority Regional Plan 2056 covering an area of 1021 km^2 which is the periphery area of Chandigarh falling within the jurisdiction of Punjab. Subsequently, Haryana government added approximately 6 km^2 of area to Panchkula Development Plan [8, 20].

There have been numerous violations to the Act. In the last four decades, the towns of Panchkula and Mohali have grown around Chandigarh not only in terms of population but have spread in area too turning it into a vast and economically vibrant region. These developments by the state government took place in the city's greenbelt, an area under green activities like agriculture and trees. The controlled area gradually became uncontrolled. Punjab and Haryana governments instead of protecting the green belt established satellite towns adjacent to Chandigarh to benefit from the economic opportunities provided by it. The Periphery Control Act which was notified in order to ensure the sustainability of rural landscape around Chandigarh, due to economic and developmental pressures failed to do so. Increasing urbanization and sprawl impacted the environment of the region.

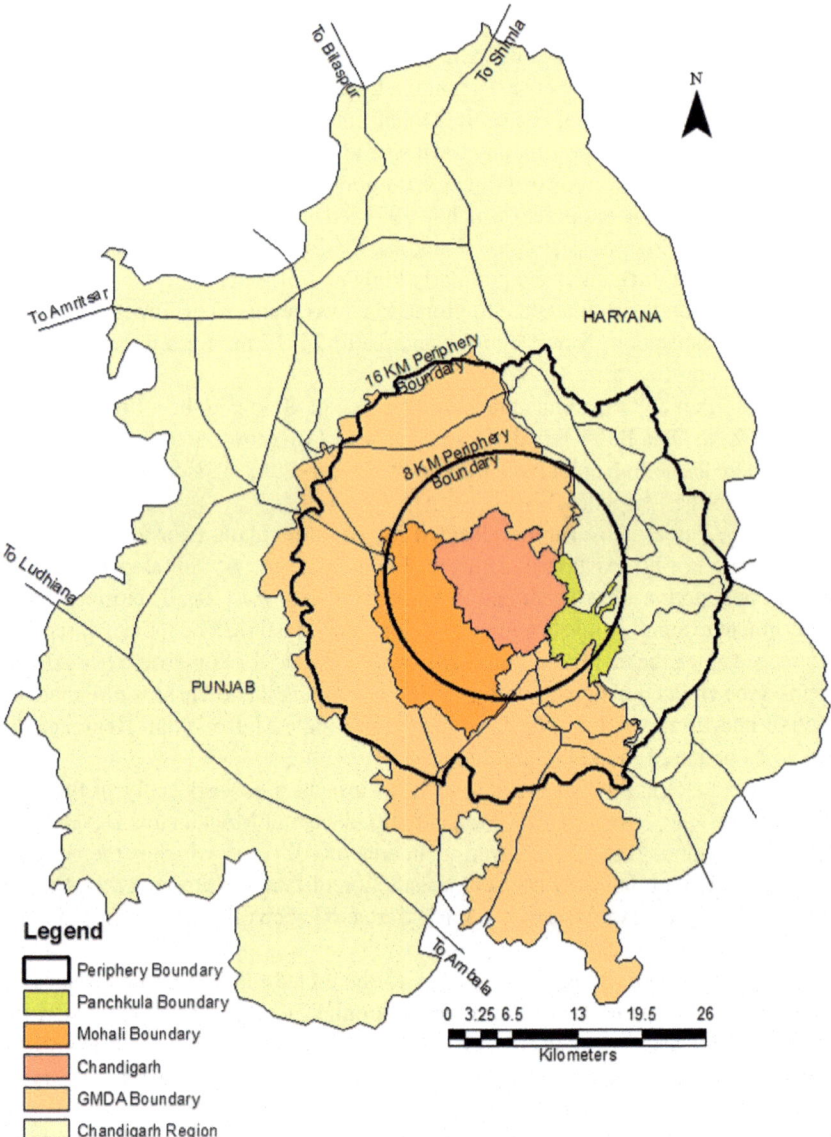

Fig. 5 Spatial spread of Chandigarh metropolitan region. *Source* Developed by authors with inputs from Chandigarh Interstate Regional Plan

6   Green Spaces in Urban Area

Urban green space is a component of 'green infrastructure', that comprises of all urban land covered by vegetation of any kind. This covers vegetation on private and public grounds, irrespective of size and function, and can also include small water bodies such as ponds, lakes or streams which are known as the 'blue spaces' [21]. In the context of Chandigarh, all of the area within the Periphery Boundary as per the Punjab Capital Periphery Control Act. 1952 is being considered as the study area, hence includes the area under agricultural activities as Green Space.

Green spaces in cities are considered a crucial factor for realizing environmental quality goals, which are within the SDG Goal 11 of making cities and human settlements inclusive, safe, resilient and sustainable. A target of the Goal is to provide universal access to safe, inclusive and accessible, green and public spaces, in particular for women and children, older persons and persons with disabilities [22]. The SDGs have been formulated very recently, in 2016. The importance of Green Spaces in and around the city which emanated from the need of improving the health and providing better housing conditions during the industrial era, has been realized a s a prerequisite by the Planners from the most initial stages.

6.1   Benefits of Green Spaces

Benefits of Green Spaces are numerous. Direct benefits of green spaces are reduction in energy consumption, ground water recharge, protection of lakes and streams from polluted runoff, reduced soil erosion, rainfall retention, water quality protection, reduced heat island effect and lesser overall temperatures.

An indirect benefit of urban green areas is a decrease in the crime rates of the area which has been substantiated by a study conducted in Philadelphia where it was found that areas where grass, trees and shrubs were found, crime rates, especially robberies and assaults were lesser [23]. This has been corroborated by another study that the presence of large street trees was most important in reducing rates of crime [24]. Green cover in an area is an indicator of the neighborhood being cared for and hence creating a perception in the mind of the criminal that he may be watched.

It has been observed by the UW-Madison Department of Urban and Regional Planning [25] that the urban green spaces are helping in regulating the climate and air quality as well as reducing the energy consumption and in turn are recharging the groundwater. Wolf observed in 2004 that the trees in a parking lot can reduce on-site heat buildup, decrease runoff and enhance night time cool downs. Tests in a mall parking lot in Huntsville, Alabama showed a 31° difference between shaded and unshaded areas. There are evidences that green cover modifies temperatures and in turn reduces the amount of fossil fuel used for cooling and heating. Properly placed deciduous trees reduce house temperatures in the summer, allowing air conditioning units to run 2–4% more efficiently [26].

6.1.1   Rooftop Green Spaces Reduce Urban Heat Island Effect

The urban heat island effect is a term used to describe the hotter temperatures found in cities compared to rural areas within the same region. Human activities increase the temperature within cities, and with a reduction in greenery there is little natural cooling. An excellent way to combat this is to turn rooftops into green spaces or green/living roofs. Planted roofs are exceptionally effective at reducing environmental heat as well as provide greater energy-efficiency within the building itself.

6.1.2   Greater Energy Conservation

Not only does an increase in green space directly affect the building it surrounds or is on (in the case of green roofs), but they also contribute to energy conservation to the city as whole. By reducing temperatures naturally and helping to bring down air temperature, buildings' HVAC systems can run at a slightly lower power. This is especially true for residential neighborhoods which may see up to a 4% increase in AC efficiency when trees are properly planted to block sunlight in the hottest parts of the day [27].

6.1.3   Natural Reduction in Noise Pollution

Noise pollution in the form of nearby industrial buildings or busy roads and highways can naturally be buffered by green spaces. Strategic planting of shrubs and trees are a way to not only enhance the surrounding areas but also help protect people from unpleasant, disruptive noise.

Enclosed green spaces may not actually be such a bad idea, especially for green spaces around office buildings where the area functions as a place for employees to relax. Noise pollution in densely populated areas within a city can be quieted by the use of insulated concrete walls surrounding green spaces, even if only as a single wall barrier alongside a busy highway adjacent to the space [27].

6.1.4   Improved Air Quality

Air pollution is a serious issue in many big cities that are densely populated. It is fairly common knowledge that plants clean the air, a primary reason why so many homeowners enjoy their houseplants. On a larger scale, green spaces with trees and shrubs are an ideal way of naturally cleansing the air.

6.1.5   Erosion and Runoff Control

Water and wind erosion cause sediment to seep into bodies of water, storm drains, and roads. When erosion occurs in an urban setting it can easily lead to flooding, mudslides, and dust storms. A densely planted green space is the natural way of reducing erosion. A well-planted green space will also help prevent nutrient loss in the soil, which can still occur in poorly planted spaces.

6.1.6   Promotes a More Active Lifestyle

Numerous studies have concluded that access to more green space also increases the population's activity level. A major factor in the struggle of obesity is regular exercise, a problem only made worse within cities with little access to parks and similar open spaces. Simple observational research shows people living in greener areas are more active. This is especially true for children, as majority of the physical activity in childhood outside of the school playground is in parks [27].

6.1.7   Effectively Helps Prevent Disease and Ease Mental Illness Symptoms

Not only does being in a green space help with the healing process, both physically and mentally, but it also works towards preventing diseases like Alzheimer's and Dementia. It comes as no surprise that more and more hospitals and similar healthcare buildings are creating "healing gardens", green spaces where patients can visit and relax. Green spaces also help with mental conditions like ADHD (particularly in children) as well as insomnia, depression, anxiety, and chronic stress [27].

There is a plethora of different benefits associated with adding more greenery to cities. As the concern over climate change and human adaption to these changes continues, the practice of bringing more of nature back into our urban areas is need of the hour. Empirical studies indicate multiple approach towards green Infrastructure (Table 2).

6.2   Urban Green Spaces Around the World

There are a number of cities around the world which have shown that planning of urban areas can be environment friendly. The most livable cities are as known for their open space, have a large core area as open green public space. Hyde Park in London, Central Park in New York, the Bukit Timah Nature Preserve in Singapore, Phoenix Park in Dublin are some such examples functioning as the lungs of the cities. Curitiba, a Brazilian city of a population of 1.8 million people consumes 23% less

Table 2 Infrastructure approaches to environmental sustainability

Approaches	Meaning as defined	Examples
Green infrastructure (strategic conservation)	Strategically planned and managed networks of natural lands, working landscapes and other open spaces that conserve ecosystem values and functions and provide associated benefits to human populations	Hubs and corridors of natural areas
Green infrastructure (water specific)	Green infrastructure is an approach to wet weather management that is cost-effective, sustainable, and environmentally friendly. Green infrastructure management approaches and technologies infiltrate, evapo-transpire, capture and reuse storm water to maintain or restore natural hydrology	Green roofs; bioswales, rain gardens, disconnected storm water systems
Ecological infrastructure	Naturally occurring systems that provide ecosystem services	Wetlands, forests, aquifers, floodplains, watersheds, etc.
Natural infrastructure	Term used to describe natural capital or assets with regards to the services they provide, with particular emphasis on habitat creation and watershed management	Watershed planning; landscape-scale conservation development planning
Greening infrastructure	Infrastructure development with minimized environmental impact	Vegetated swales, pervious paving
Green infrastructure	Infrastructure that supports more sustainable or 'green' communities	Renewable energy sources, smart grids, infrastructure to promote conservation & efficiency
Sustainable infrastructure	Design and construction of all infrastructure systems that maximize triple bottom line benefits while still serving their intended purpose	Recycled content; energy and water efficiency; low-waste processes; material sourcing; ecologically-cognizant design
Smart infrastructure	Use of Geospatial technologies encompassing remote sensing and GIS for the purpose of monitoring of the green spaces should be incorporated in planning	Monitored through sensors and geospatial technology models, remote sensing

Source adapted from Greening the Grey: *An Institutional Analysis of Green Infrastructure for Sustainable Development* [28, 29]

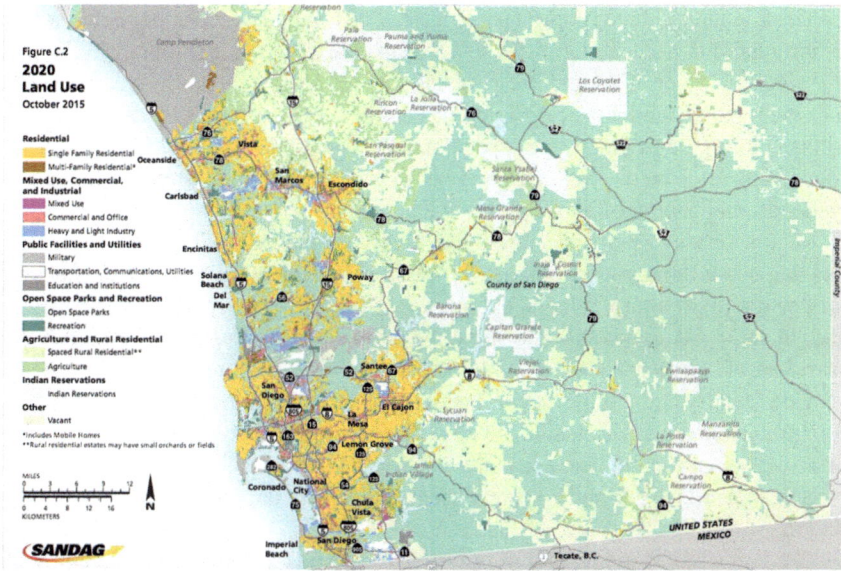

Fig. 6 Proposed land use 2020 San Diego region. *Source* Regional plan SANDAG [30]

fuel per capita than the Brazilian national average. The city has 16 parks, 14 forests, and over 1000 green public areas shared by its residents [28].

Indian examples are Sanjay Gandhi Van in Mumbai, The Ridge in Delhi, and Nandan Van in Naya Raipur apart from cities of Bangalore, Gandhinagar and Jaipur.

6.2.1 San Diego Forward: The Regional Plan (2019–2050)

The goals of the Regional Plan of San Diego are to provide innovative mobility choices and planning to support a sustainable and healthy region, a vibrant economy, and an outstanding quality of life for all. The Regional Plan identified smart growth and sustainable development as important strategies to direct the region's future growth toward compact, mixed-use development in urbanized communities that already have existing and planned infrastructure, and then connecting those communities with a variety of transportation choices. In 2008, the San Diego region included about 3.1 million people forecasted to increase by 1.25 million people by 2050 [30].

The San Diego region has made significant changes in the approach to planning in planning for more compact, higher density, and walkable development in order to protect the open spaces in the region. The natural environment in the San Diego region includes three general geographic areas: the coast, the mountains, and the desert (Fig. 6).

The San Diego Region open space strategy has been developed on the principles of sustainability and smart growth. The Sustainable Community Strategy (SCS) land use pattern also protects and preserves about 1.3 million acres of land, more than half the region's land area. SANDAG has defined regionally significant open space to include: region-defining open space, natural resource areas, region-serving open space, and rural lands [30]. These areas include parks, steep slopes, floodplains, wetlands, and habitat of native plants and animals.

Region-serving open space are areas lightly developed with activities or facilities that serve the region as unique or outstanding recreational, safety or managed production (agriculture, mineral extraction). These areas are retained as open space and, in some cases, increased to serve the region's expanding needs. Additionally, corridors of open space within and between communities are retained in order to provide identity and a sense of community, and to link significant open space areas.

Rural lands are areas outside the identified urban area that are planned to remain in a low intensity, rural land use pattern. These areas provide a contrast to complete urbanization and result in the visual appearance and feeling of more openness in the region.

The Farmland Mapping and Monitoring Program (FMMP), administered by the Division of Land Resource Protection at the California Department of Conservation, produces maps and statistical data to analyse impacts to California's agricultural resources [30]. To characterize existing and potential farmland, agricultural lands are rated according to soil quality and irrigation status FMMP maps are updated every two years using aerial photographs, a geographic information system, public review, and field reconnaissance.

The goal is to assure that adequate quantities of diverse habitat types are maintained, and that the plants and animals found in these habitats are less likely to become endangered. The regional habitat conservation plans in the San Diego region are designed to provide an umbrella of protection for multiple species by conserving their habitats and the linkages that allow them to travel between habitats [30].

The State of California has initiated a habitat conservation planning process that concentrates on the conservation of large parcels of land and emphasizes planning for environmental systems. The Natural Community Conservation Planning (NCCP) program pilot effort focuses on the coastal sage scrub habitat of the coastal California gnatcatcher in Southern California. The program is designed to address a habitat type of several species.

As part of the SANDAG participation in planning for the conservation of regional habitats, SANDAG developed a database of conserved lands in 2010 (SANDAG Conserved Lands database, 2010). This database, which is regularly updated and available to the public, serves as the basis for monitoring habitat conservation [30].

The SCS land use pattern incorporates finalized habitat plans as well as the conservation of other sensitive resource lands as reflected in plans by local jurisdictions. These local and regional plans ensure the conservation of plant and animal species, and natural habitats through low density zoning, conservation easements, and land purchases.

Table 3 Hierarchy of open spaces

Service level	Hierarchy of services	Population
Housing cluster	3–4 local parks and playgrounds	5000
Neighbourhood	3–4 local parks and playgrounds	15,000
Community	2–3 community level parks and open spaces	100,000
District/zone	1 district level park, sports complex, maidan	500,000
Sub-city centre	1 city level park, sports complex, botanical garden, maidan	1,000,000

Source URDPFI Guidelines, 2014
an open space in or near a town, used as a parade ground or for events such as public meetings [32]

6.3   Green Spaces as Per URDPFI Guidelines

URDPFI Guidelines [31] define the open spaces as recreational spaces, organised greens and other common spaces such as vacant lands/open spaces including flood plains, forest cover in plan areas. The Guidelines prescribe 10–12 m^2 per person of open space. These norms are over and above the protected zones, ecological conservative areas in the hilly regions. A hierarchy of green spaces is prescribed consisting of parks, play fields, other open spaces (Refer Table 3).

In case of Chandigarh, The Green Space have been planned at each sector level in the form of neighbourhood parks and playgrounds. There are a series of community spaces and city level parks live Rose Garden and Rock Garden. The hierarchy of services as stated in the Guidelines has been followed in case of Chandigarh.

6.4   Green Spaces in Chandigarh City

The concept of urban green parks can be traced back to 19th century for promoting well-being in new industrialised cities. Importance of green spaces for the well-being of humans has been well established through many researches, especially after the World War II. In concurrence with thought, Corbusier kept the provision of large number of open spaces, green belts, city parks and neighborhood parks in Chandigarh right from the beginning. There exists well-structured and order in the hierarchy of open spaces in the city ranging from the neighborhood level to the city level. The major principle which guided the formulation of Chandigarh's landscape design was that its original topography has to be retained and integrated [8]. Provision of green spaces in each sector were basically meant to provide lungs to the area and

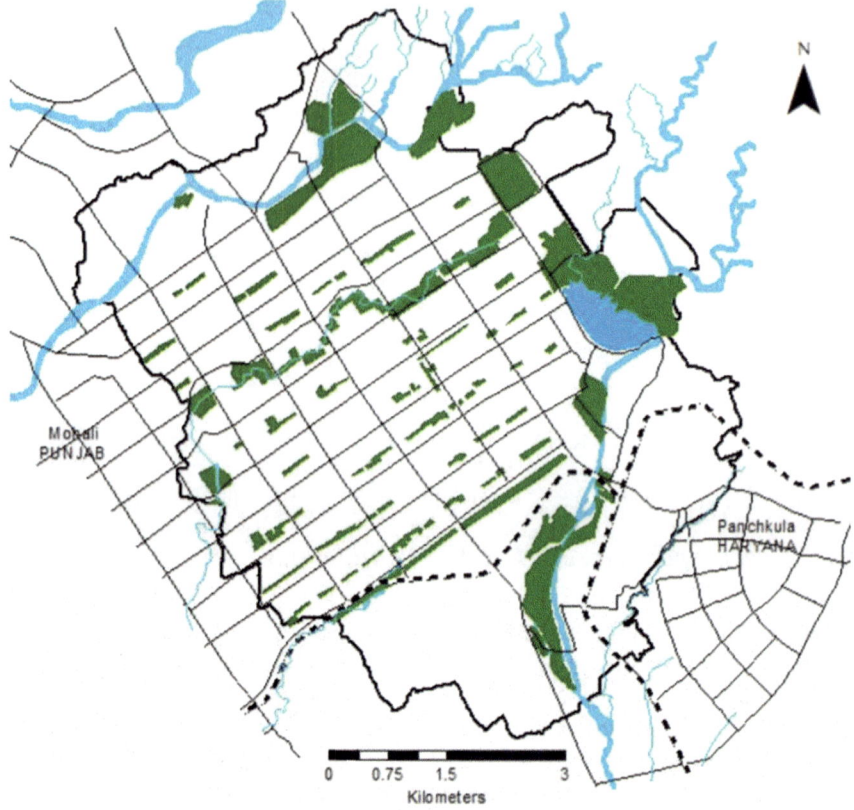

Fig. 7 Green Spaces in Chandigarh city. *Source* Chandigarh Master Plan 2031 [8]

for developing areas of leisure. Unlike other Indian cities the green areas in the Chandigarh city are planned within each sector, which is the neighbourhood level (Fig. 7).

Additionally, green open spaces are available either in the form of private housing, schools, colleges or other institutional large campuses such as Punjab Engineering College, Punjab University; over and above the planned green areas in the city layout, contribute as much as the overall city greenery and open spaces.

Another legacy of the city's green identity is the establishment of long clear cut vistas created in the first phase of the city connecting the residents to the skyline of the hills through continuous green spaces such as sector greens, Leisure Valley and Sukhna Lake promenade. As tree plantation and landscaping has been an integral part of the city's Master plan, the city's green cover rose from 21% in 1991, 29.3% in 2003 to 35.7% in 2007 as per Forest Survey of India's Report 2007 [33]. The per capita green space availability in the city presently is around 54.45 m^2 which is better than most of the European cities [34].

The most peculiar feature of the city's landscaping is the tree planting along roads, open spaces, green belts and around building complexes. The road plantations were based on study of the movement of sun in relation to direction, scale, size and architecture along avenues. To minimize glare along the avenues running north-east to south-west, dense foliage evergreen trees like Ficus infectoria[8] and Schleichera oleosa[9] were planted to form green tunnels. The shopping streets were planted with flowering species like Cassia fistula,[10] Cassia javanica[11] and Jacaranda[12] species [35].

6.5 Green Spaces in Chandigarh's Periphery

Prior to the siting of Chandigarh, the Periphery was characterized by an agrarian landscape of rural villages, farmland and forested areas [20]. Le Corbusier sought to maintain this functioning rural landscape by creating an agricultural greenbelt circumscribing the urban core of Chandigarh. As stated in Sect. 6 the Punjab New Capital (Periphery) Control Act, 1952 was enacted soon after the preparation of the Master Plan, to help implement this vision by restricting urban development, but despite its existence, the Periphery has seen rapid land use change.

The landscape of the Periphery is made of four different geographical features; the Himalaya, the gravel upland (Dun), Shivalik Hills, the rolling plain at the base of the foothills (Ghar), and the Chandigarh alluvial plain [35] (Refer Fig. 8).

Each of these features of the natural environment contributes to the rural way of life. This interconnected ecosystem has been significantly impacted by the development on the Periphery. The natural areas are at risk of depletion under the threat of urbanization, therefore impacting the rural lifestyle.

The majority of the Periphery is made up of Chandigarh Alluvial Plain, located at the foot of the Shivalik Hills on a gentle southwestern slope. The plains are an important feature of the Periphery as they provide fertile soil for agriculture. The Shivalik Hills are located to the northeast of Chandigarh, running parallel to the Himalayan Mountains. Patiala-Ki-Rao and Suhkna Choe are the main seasonal streams flowing through the Periphery, previously referenced in Sect. 3 were defining element of the Mayer Plan.

There are two reserve forests functioning to protect the Sukhna Lake catchment area, which fall within portions of Punjab, Haryana and the UT. Sukhna Lake is an artificial lake located to the southwest of the Shivalik Hills. The lake was created in 1958 by damming the Sukhna and Kansal Choes, seasonal streams flowing from the Hills [3]. This catchment area is located within portions of Punjab, Haryana and

---

[8]Common name: white fig, locally known as *pilkhan.*

[9]Common name: cylone oak, locally known as *kusum.*

[10]Common name: golden shower tree, locally known as *amaltas.*

[11]Common name: apple blossom tree, locally known as jangli dalchini.

[12]Common name: blue jacaranda, locally known as neeli gulmohar.

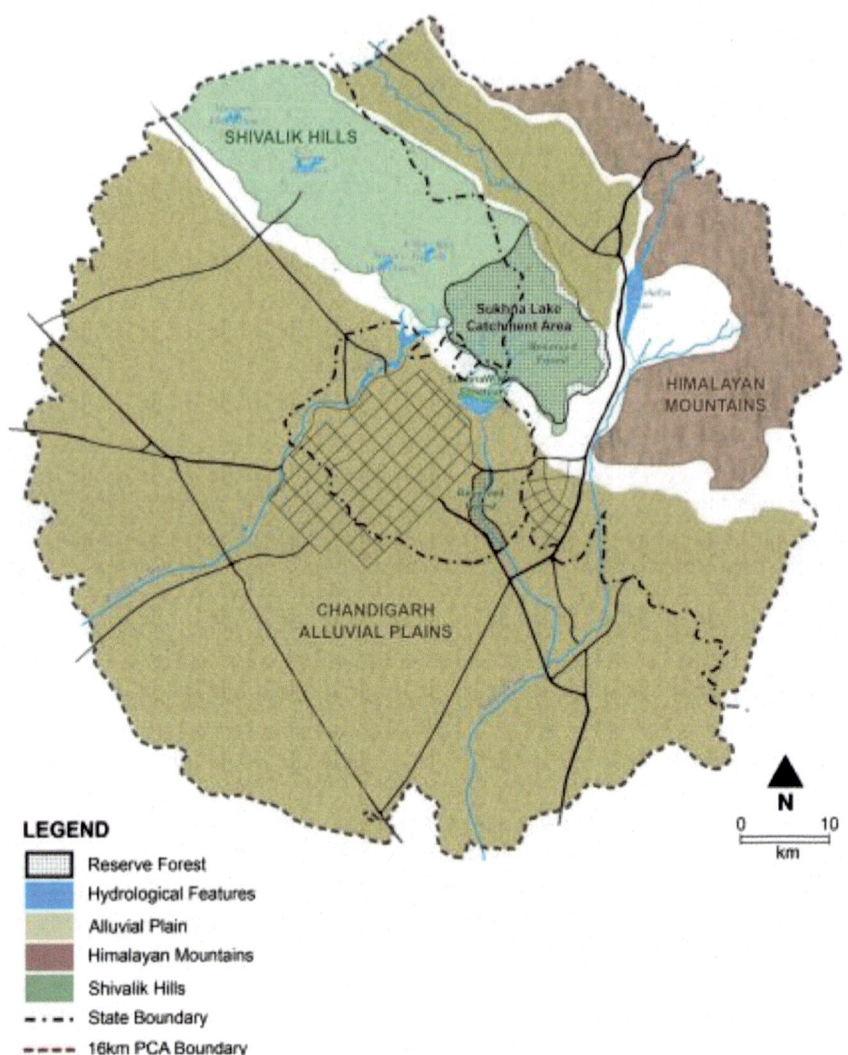

Fig. 8 Natural features in Chandigarh periphery area. *Source* Weber Abigail, within the edge: a revised approach to urban containment within Chandigarh periphery [35]

the UT. The lake was a design feature in Le Corbusier's Master Plan of Chandigarh (Refer Fig. 9).

Rural land within the Periphery Control Area was permitted to serve one of two uses intended to meet the day to-day needs of the residents of the urban core; agriculture or the production of building materials [20]. Land was designated as 'agricultural and afforestation zone', or a 'brick field zone' which included brick and lime kilns.

Fig. 9 Drainage network of Chandigarh. *Source* Developed by authors from Survey of India Toposheet sheet and Google Map

Active agriculture and the rural village remain, but as urban development encroachments have been pushed further out into the Periphery. Villages in close proximity to Chandigarh have changed significantly, becoming visibly more urban in density and commercial activities.

The development of the satellite towns Panchkula in Haryana and Mohali in Punjab paved the way for construction to proceed within the Periphery even before Chandigarh had reached its full capacity [20]. Increasing land values in Chandigarh drove people to these neighboring townships, or further out into the Periphery where urban services were not located. Unauthorized development began emerging in response to the rapid growth and demand for services, In Punjab, the Greater

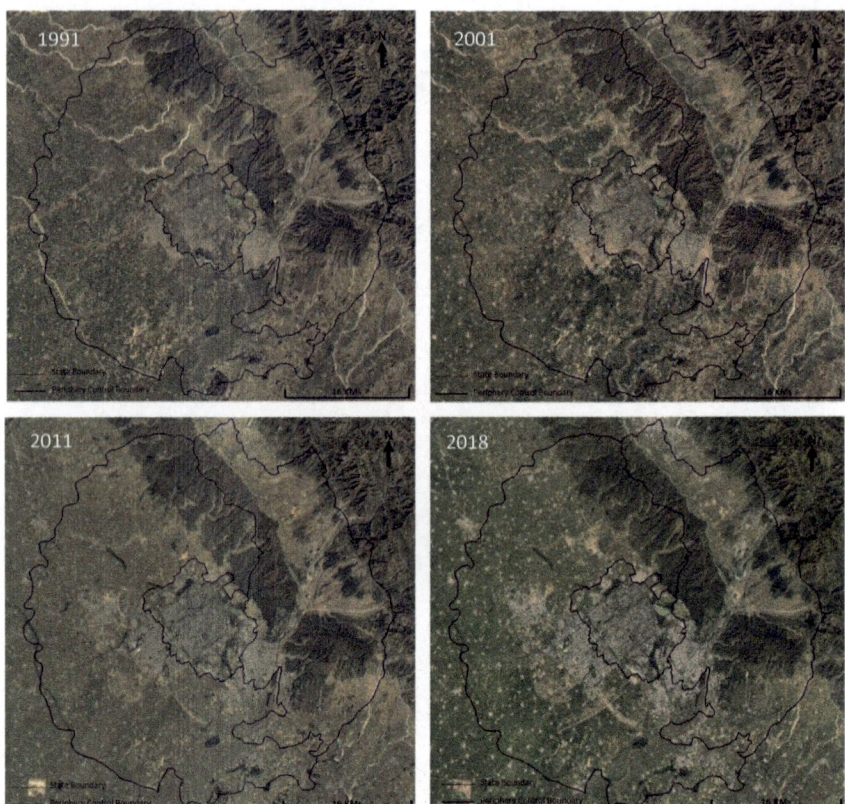

Fig. 10 Temporal comparison of built up in chandigarh periphery area. *Source* Developed by authors using Google Earth

Mohali Area regional plan creates seven integrated Economic Hubs, while Haryana's regional growth strategy incorporates five planned settlements [3].

This trend of master planned developments has emerged over the last decade. Throughout the Periphery township development, commercial establishments, mixed-use and academic campus came up. The magnitude of land use change is evident from the study of historical and contemporary Google Earth maps (Refer Fig. 10). New developments have been permitted due to relaxed interpretations of the Punjab New Capital (Periphery) Control Act, 1952 and jurisdictions continuing to justify new projects in the public interest. This rapid development reflects the large-scale shift from a primarily rural Periphery of agricultural activities to a Periphery of a mixed urban land uses and activities.

The built up area in the Periphery Control Area increased by more than 40% over the decade between 2000 and 2012 (Refer Table 4).

In Punjab sub region of Chandigarh Periphery Area, nearly all of the area has been notified under the Greater Mohali Metropolitan Area Regional Plan.

Table 4 Increase in the built up in periphery control area of Chandigarh

Year	Built up area (km^2)	Rural component
2000	151.52	1348.48
2006	164.47	1335.53
2012	215.67	1284.33

Source: Saini 2011 [36]

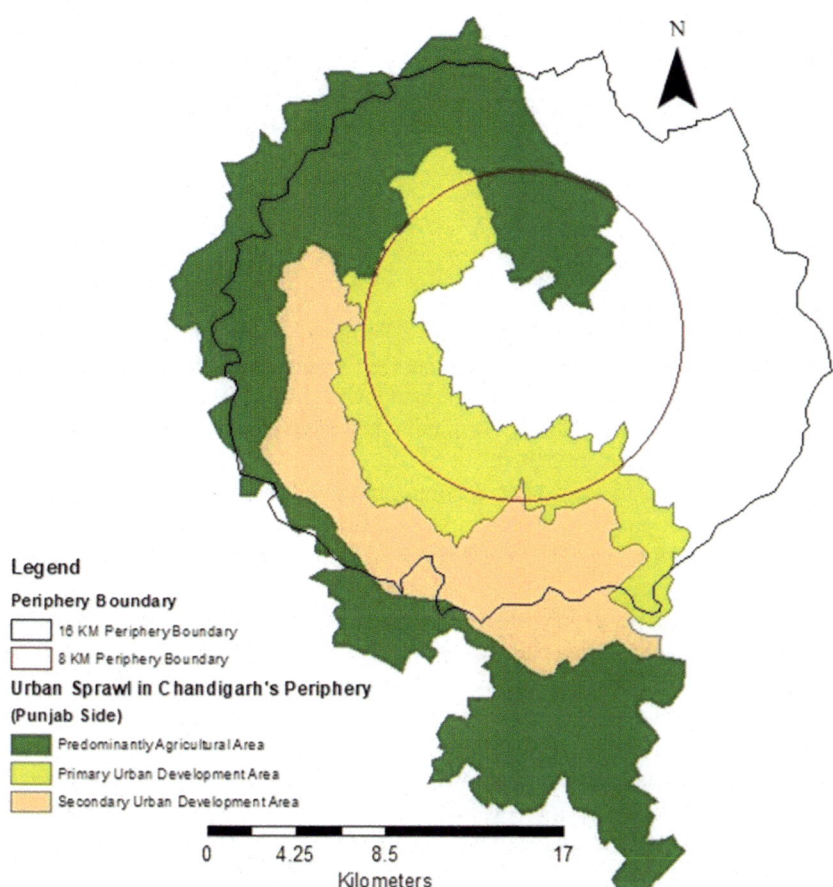

Fig. 11 Proposed urbanisation by 2058 in Punjab sub region of periphery control area. *Source* Adapted from Greater Mohali Regional Plan 2058

Even though three areas have been demarcated within the sub-region, namely Primary Urban Development Area, Secondary Urban Development Area and Agriculture/Tourism/Rural Area, the whole sub-region is likely to grow into a huge urbanized area by 2058 (Refer Fig. 11).

Table 5 Land cover in the Haryana sub-region of Chandigarh periphery control area

Land cover type	Area in km^2				
	1972	1990	Percent change	2008	Percent change
Agriculture	128.00	113.30	−11.48	90.64	−20.00
Built-up	2.58	27.07	949.39	52.80	95.04
Reserve forest	103.17	103.17	0.00	103.17	0.00
Open vegetation	19.30	11.20	−41.98	8.57	−23.45
Special area (cantonment)	18.57	18.57	0.00	18.57	0.00
Water bodies	22.29	18.25	−18.11	15.28	−16.26
Mining	1.09	3.09	183.43	5.10	64.95
Slums	–	0.35		0.87	148.46
Total	295	295		295	

Source Saini, 2011 [36]

In Haryana sub region of Chandigarh Periphery Area, the built up area has increased at nearly 50 percent decadal rate (Refer Table 5 and Fig. 12).

The on-the-ground reality is that the periphery of Chandigarh is developing in a manner directly counter to the act originally intended to preserve it as a permanent, functioning agricultural greenbelt.

The green cover that has been declining in the Chandigarh Periphery Area due to rapid urbanisation is leaving an impact on Chandigarh. There has been a notable increase in water logging within the city as well as increase in the surface temperature of the region [37].

Chandigarh has per capita urban green space of 54.45 m^2 [34], having more than 35% of its geographical area under forest and tree cover, making it one of the greenest cities of India [3]. Being a planned city, the green spaces are well integrated in the master plans. Similar strategies have been adopted in the satellite towns of Panchkula and Mohali.

Chandigarh is one of the fastest growing cities in the country, with a decadal growth rate of 40.30%, has the potential to become the I.T. city of Punjab and Haryana [13]. The city is witnessing a rapid expansion, industrial boom, increased trade opportunities coupled with high population growth rate accelerated due to migration from neighboring states as well as from U.P. and Bihar. Chandigarh is beset with increasing urban environmental problems due to the growth of unsustainable economic activities, increase in vehicular transport, in spite of a public transport system. The most pertinent concerns include air and noise pollution, depleting ground water and contamination of water bodies and lakes, destruction of natural and manmade resources of cultural value and improper management of domestic and industrial effluents. This situation is leading to depletion of green cover, groundwater table and rising levels of air and noise pollution both in the core and periphery area.

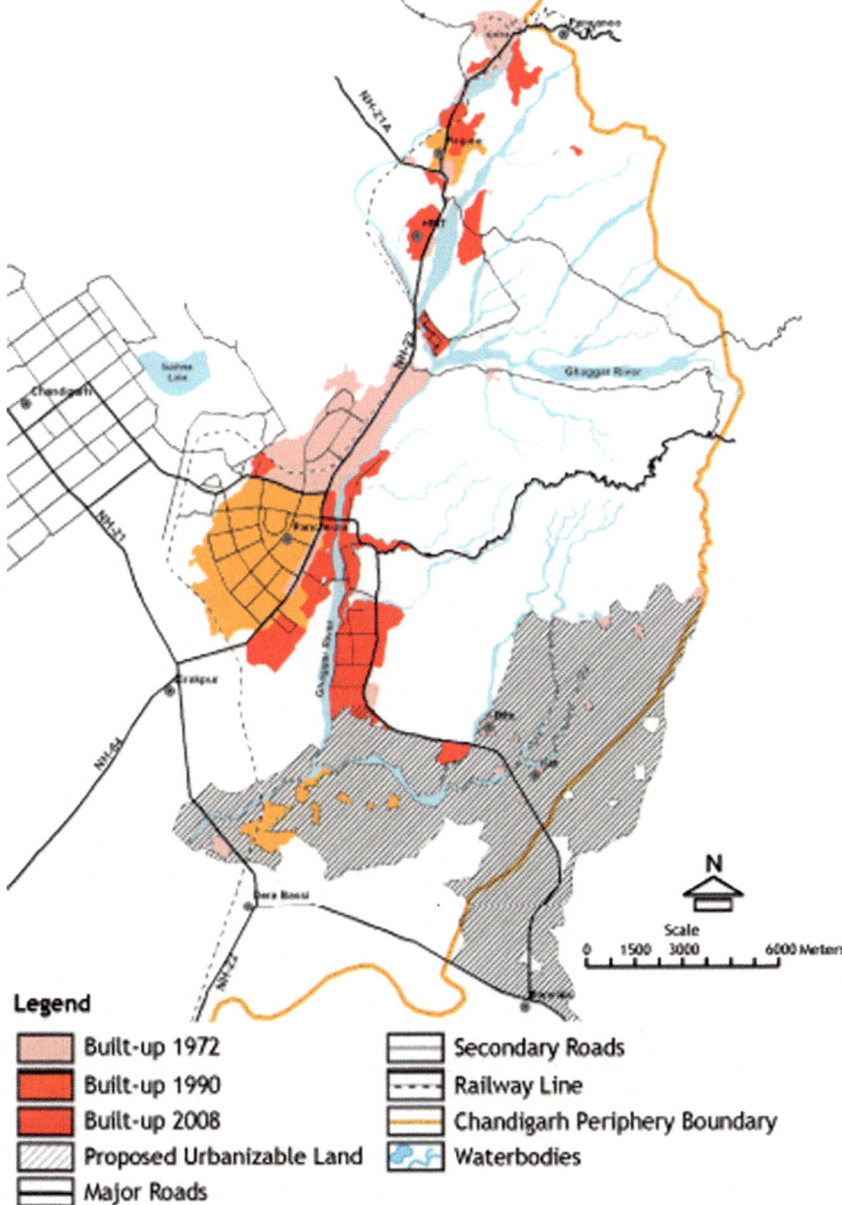

Fig. 12 Proposed urbanisation in haryana sub region of periphery control area. *Source* Land use changes in Haryana sub-region of Chandigarh controlled area: a spatio-temporal study [36]

In the periphery area the forest cover is confined to hardly 14 villages out of a total of 458, four each in Punjab and Chandigarh Sub-zones of the Periphery Zone and six in Haryana Sub-zone. Most of these villages are located along different choes[13] flowing through the Periphery Zone Forest land in Haryana Sub-zone is more than two times that of Punjab and Chandigarh Sub-zones put together. Bir Ghaggar village in Haryana recorded 98% of its total area as forest. The number of villages with more than half of their land under forest cover declined from 11 in 1971 to 4 in 2001. The actual decrease here was from 8729 to 2105 ha [39].

Mohali and Panchkula are not satellite towns in its true essence as there needs to be an essential green belt between the core town and its satellites. However, in both the cases, there is a continuum of built up with the city of Chandigarh at its core (Fig. 13).

7 Challenges for Chandigarh Periphery Region

The region lacks a coordinated vision for Development as well as Protecting Open Spaces. In the case of Chandigarh, while Corbusier visualised the green spaces within the city in detail, however the same consideration was not extended to the periphery region by the states of Punjab and Haryana within whose aegis the area lies.

Watershed-scale planning is critical as environmental systems are best analysed within a watershed framework, rather than within municipal boundaries. Water logging in Chandigarh city, but also in Mohali is a result of encroachment of development on seasonal 'choes' in the periphery region.

Multiplicity of governance as the state government's efforts to go for coordinated planned development were lacking. To resolve the multiplicity of governance the Government of India constituted a Coordination Committee, in 1966 to ensure continuity in policy toward the Periphery Zone. The Coordination Committee appointed two working groups one for preparation of the Regional Plan for Chandigarh Inter-State Region and the other for the framing of requisite legislation for its effective enforcement and implementation. However, none of these could be effective in preventing development in the Periphery Area.

Stress on natural resources in the region that is the two most noticeable features in the land use change in the Periphery Zone were the decrease in the forest area and increase in the share of land not available for cultivation. The forest cover got denuded by six per cent points and land not available for cultivation' increased by 7% points during 1971–2001. The forest cover was completely wiped away from 42 out of 50 villages where it existed in 1971. This phenomenon was more typical of the villages falling in the Punjab and Chandigarh Sub-zones. The decline in the net sown area and cultivable wasteland was marginal during 1971–2001.

---

[13] Seasonal streams [38].

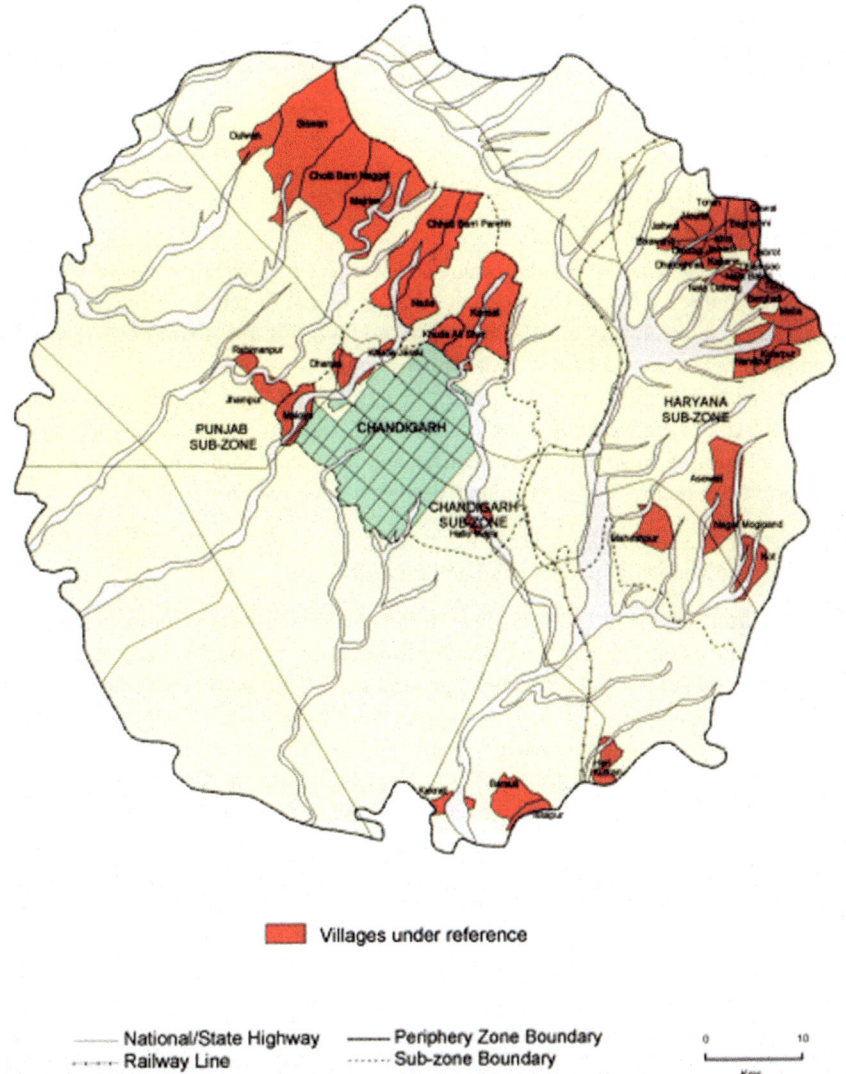

Fig. 13 Forest depleted villages. *Source* Land use changes in Haryana sub-region of Chandigarh controlled area: a spatio-temporal study [36]

Mobility challenges in form of regional transport system accentuate the issues in the region. There exist links to provide an easy access to the people living in the periphery Area to commute to the city Core.

8 Smart City Chandigarh

Chandigarh was selected for the Smart City project in May 2016 out of 98 cities which are under the Smart City Project of Government of India. The central idea is to make use of technology and the advanced modes of lifestyle to optimize the living conditions in the selected cities under this program.

The vision statement for Chandigarh Smart City has been defined as to create a Vibrant and Unique Regional Centre. The City beautiful is envisioned to become leader in livability, sustainability, equality, and innovation [40]. The city proposal is to improve public life and safety, and liveability using two initiatives:

- Smart Integrated e-Governance
- Intelligent Multi-modal command and control center

The administration has decided on carrying out area-based development in sectors 17, 22, 35 and 43 on a pilot basis. Under the Smart projects-Smart water meters, smart grid, wi-fi spots, intelligent traffic management system, pedestrian pavements is what all the sectors would be provided with. Apart from these, there would be smart pipes that would be connected with a wireless processor to help detect leakage in real time, with a system of online monitoring of the entire distribution in place everywhere [40].

Footpaths and cycling tracks will also be laid on roadsides all along the inter-sector roads as well as those leading to markets to encourage people to switch over to non-motorised transport. Traffic-sensing lights or smoother flow of traffic and lesser pollution, e-surveillance with CCTV cameras and sensors at every point to help in better traffic management is being planned for the city.

Chandigarh is among four cities which the French government has identified to invest in as part of the Indian government's smart city project and has allocated as much as two billion Euros to be spent over the next few years.

9 Challenges for Integrating Smart Regional Environment

One of the major problems facing the city is the enormous population growth both of the city and that of the adjoining areas. There is also rapid increase in the floating population, which come to work, and avail of the services in the city.

Root cause of Chandigarh's problems has been that Chandigarh planning and development has been seen in isolation compared to the development of its periphery and region. While Chandigarh is a planned city its peri-urban areas are neither rural nor urban. Le Corbusier's tool of greenbelt for Urban containment was not effective.

Le Corbusier viewed the Punjab New Capital (Periphery) Control Act, 1952 as a critical feature protecting his plan. In year 1966, after re-organization of state of Punjab, when Chandigarh became Union Territory (a centrally administered unit) with capital of two states of Punjab and Haryana and its periphery area got divided into three parts with majority of share going to state of Punjab and Haryana.

10 Smart Green Open Spaces Strategy

In the past few decades Chandigarh region have been seen struggling to maintain a balance between economic, environmental and social sustainability. It needs to be acknowledged that new challenges like affordable housing, economic generation have existed and dictated the current development pattern. The economic and developmental pressure is a reality, which the Punjab and Haryana Government has tried to cater to by developing new satellite towns. However, it cannot be denied that the overall vision for the region has been lacking and the development has taken place in phases due to which the green spaces in the region have been affected. The approach now, that needs to be advocated is preserving the existing green spaces in the region. This involves not developing new areas in the region that doesn't have existing infrastructure service and preserve the open spaces in the region for ecological functions.

The smart green space strategy for the region should be based on the principles of growth management and green infrastructure (Refer Table 2). Additionally, keeping the regional challenges in mind, the open spaces in the region should be planned and sensitive lands identified should be protected. There is a need to achieve a balance between the protection of agricultural lands and open spaces and the urban development to promotes smart and responsible growth.

Overall strategy for open spaces needs to cover the following:

- Prepare the smart open green strategy keeping in mind the present situation and constraints
- Implement the strategy through a central authority with representatives of all involved governments and stakeholders
- Allocate responsibilities to respective state governments and local departments
- Monitor the progress against the set targets using smart tools
- Review action plan periodically
- Review and update the strategy if required.

10.1 Chandigarh Region

Chandigarh city has been designed as greenfield city in India. The periphery of the City was visualized as buffer as a functional landscape that contains the growth of the city. However, the development of the satellite towns of Panchkula in Haryana and

Mohali in Punjab in Periphery area started even before Chandigarh reached its full potential of development. Increasing lands values in Chandigarh city and affordable options in the periphery drove this development. Resultantly, the rural villages in close proximity to Chandigarh have undergone significant changes. The rapid growth in the rural population can be attributed to in-migration between 1971–2001 due to economic opportunities being provided by Chandigarh city. Subsequently, Mohali and Panchkula, the satellite towns became the magnet.

The path of urbanisation in the region may not be reversible, however, measures can be taken by the authorities to ensure that the region is planned as a single entity maintaining the basic character and ethos of the Plan initially laid out by Le Corbusier.

To achieve the Green Space vision a four–point strategy has been proposed that is adopting a coordinated approach towards the planning of the Periphery and the Core; conserving and retaining the water and tree cover; adopting modern techniques of greening the region and basing the planning for the region on the principles of watershed.

The smart growth principles need to acknowledge the Periphery area as an integral feature of the historic master plan. Yet, at the same time the developmental pressure and sound economic opportunities for the region cannot be denied. Hence a new periphery area should be delineated keeping in mind to preserve and maintain the visualized heritage of Chandigarh. Strategies should discourage sprawl and further conversion of natural and agricultural land for urban development.

For meeting the development challenge two-fold approach should be adopted—one to use the existing developed area more efficiently and second to minimize the new urban centres in Chandigarh's green belt. The periphery area needs to be visualized as an entity comprising of open spaces, agricultural lands, village settlements, urban development and planned accordingly.

10.2 Protection of Sensitive Lands in the Region

A lot of effort went in selecting the site for Chandigarh city. The natural features of the selected site were incorporated in the design right from beginning. The ecological setting of the city was duly selected keeping in mind the vision for the city and the region. The landscape of the periphery comprises of distinct geographical features:

- The Himalaya,
- Shivalik Hills to the northwest that provided a backdrop, the gentle southern slope provided efficient drainage
- The Chandigarh alluvial plain [35] at the base of the foothills
- Hydrological features

Each of these natural features plays an important ecosystem services. In the past, the infrastructural demands like road and rail transport networks, water treatment, sewage disposal, educational institutions, health centres and recreation sites have pressure on the land in the Periphery Zone thereby converting the area under cultivation, cultivable wasteland and forest, and bringing these under the urban development.

The seasonal water 'choes' in the region help in reducing runoff volumes and reduces peak flows by utilizing the natural retention and absorption capabilities of vegetation and soils. Due to development, some of them have encroached upon hampering the ecological function played by them. Conserving and retaining the water and tree cover; adopting modern techniques of greening the region and basing the planning for the region on the principles of watershed is recommended.

The open spaces in the region should be planned and sensitive lands identified should be protected as an integrated and connected system with different open spaces supporting one another to achieve multiple benefits. As stated by Kimmel [29], planning and decision making for vibrant and environmentally sustainable communities requires a systems perspective that integrates green and grey infrastructure. Additionally, there has to be a financing mechanism in place that funds the conservation plans and sensitive land protection in the periphery zone. Without funds and collective ownership, the area to be preserved will always face the threat of development.

As stated the Urban Green Areas do not have to be limited to parks on the ground but can extend to agricultural land as well as green roofs and walls. 'Greens on Building' that is green roofs and walls can increase the green cover specially in the core and dense urban areas with negligible land area for vegetation. These Greens in a Building not only play an important role in thermal insulation of the building itself, but also increase the water recharge and reduce runoff of precipitation into urban sewers.

Some strategies for maintaining the green character of the region could be through adoption of smart laws that enforce 'Greens on Buildings' by creating green walls and roofs. The impact of this can be monitored through sensors and geospatial technology models can be generated to monitor the heat island effect. This will help the authorities in taking corrective measures to reduce the heat island effect.

10.3 Promote Collective Governance and Responsibility

The governance structure in the Chandigarh region is complex. The reorganization had negative implication for the periphery in terms of multiplicity of governance. Due to lack of interstate coordination, competing interest and differing interpretations of the language of the law, the reorganization made it particularly difficult to enforce a unified vision of the rural landscape within the Periphery [20].

There are multiple government in charge of different spatial segments of the region. In 1975, in response to the uncontained urban growth in the periphery the Central government formed a coordination committee for balanced development of the Chandigarh Region. Based on the recommendation of this committee, in 1984 'Inter State Chandigarh Regional Plan' for 2001 was formulated. The Committee held numerous meetings and also formally obtained the comments of relevant Departments of the Government such as Revenue, Industries, Housing and Urban Development and Local Government [19]. The Committee lacked Statutory power and has remain ineffective as a result, the development taking place in periphery is lacking coordination and vision.

Green open space strategy cannot be separated from governance, as the process of planning involves a wide range of stakeholders and individual citizens. There is a need for emergence of a central regional authority with a vision as well as clear legal status, technical and professional capacity to plan and a robust administration to implement and enforce [41]. Additionally, not only State level coordination but there is a need to identify and support existing inter-agency and inter-municipal collaborations.

10.4 Use of Smart Tools Which Essentially Include Geospatial Technologies

Smart tools which essentially include geospatial technologies encompassing remote sensing and GIS for the purpose of online monitoring of the green spaces should be incorporated in planning. It is pertinent that not only the extent of the green spaces, both in the city limits of Tri City but also in the Periphery area but also the quality be monitored and upgraded from time to time. Close monitoring. Some strategies for maintaining the green character of the region could be through adoption of smart laws that enforce 'Green on Buildings' by creating green walls and roofs. The impact of this can be monitored through sensors and geospatial technology models can be generated to monitor the heat island effect. This will help the authorities in taking corrective measures to reduce the heat island effect.

It must be noted that the city and its environs are to be considered as ecological space. The authors emphasize that Smart Green Spaces for Chandigarh and its Periphery has to be based on technology within the framework of sustainability. It is pertinent that through smart green spaces the city builds on its social capital. Planning and implementing for the smart green spaces will reduce the adverse impact of urbanization and at the same time maximize the intangible benefits and build on the aspects of culture in turn boosting the economy of the area.

11 Conclusions

Chandigarh has been one of the most discussed city in India. It was designed to represent the modern India post-independence. However, it is important that modern urban planning should not only focus on the improvement of how a city works, but also how the open spaces or green spaces are areas within a city and the region. Realizing the number of different benefits associated with increased green spaces in urban settings will help us build healthier, vibrant and sustainable settlements.

The challenges that are arising in the planned city of India can be addressed through smart tools which essentially include geospatial technologies encompassing remote sensing and GIS for the purpose of online monitoring of the green spaces. It is pertinent that not only the extent of the green spaces, both in the city limits of Tri City but also in the Periphery area but also the quality be monitored and upgraded from time to time. The authors emphasize that Smart Green Spaces for Chandigarh and its Periphery has to be based on technology within the framework of sustainability. Close monitoring of these aspects will help the authorities in improving the green index and per capita ratio.

It must be noted that the city and its regions are to be considered as ecological space. It is pertinent that through smart green spaces the city builds on its social capital. Planning and implementing for the smart green spaces will reduce the adverse impact of urbanization and at the same time maximize the intangible benefits and build of the aspects of culture in turn boosting the economy of the area.

References

1. Shabdkosh (2018) Meaning of Chandi. https://shabdkosh.raftaar.in. Accessed on 12 June 2018
2. Shabdkosh (2018) Meaning of Garh. https://shabdkosh.raftaar.in. Accessed on 12 June 2018
3. Chandigarh the city beautiful: the official website of the Chandigarh administration (n.d.) history http://chandigarh.gov.in. Accessed on 10 June 2018
4. Government of Haryana (1993) Haryana district gazetteers: Ambala District, Haryana gazetteer organisation, revenue department, Chandigarh
5. Census of India (2011) District census handbook: village and town directory, Government of India
6. Modernism in Urban Planning (2015) https://archiobjects.org. Accessed on 24 June 2018
7. Super Blocks (2018) https://www.thedesignresponse.com. Accessed on 23 August 2018
8. Chandigarh Master Plan 2031 (2016) Chandigarh administration, Chandigarh
9. Mayer AA, Papers, [Box 18, Folder 28–29], Special Collections Research Center, University of Chicago Library
10. Shabdkosh (2018) Meaning of Nallah. https://shabdkosh.raftaar.in. Accessed on 12 June 2018
11. Almeida T (2013) Le Corbusier: how utopian vision became pathological in practice. https://orangeticker.wordpress.com. Accessed on 24 Aug 2018
12. Howard E (1902) Garden cities: a solution of the housing problem. J Sanitary Inst 23(4):670–674
13. City Development Plan (n.d.) Chandigarh Administration, Chandigarh
14. Municipal Corporation of Chandigarh (2018) Basic information. http://mcchandigarh.gov.in/. Accessed on 26 June 2018
15. Chaudhary P, Bagra K, Singh B (2011) Urban greenery status of some Indian cities: a short communication. Int J Environ Sci Dev 2(2)
16. Howard E (1898) Garden cities of tomorrow. http://urbanplanning.library.cornell.edu/DOCS/howard.htm. Accessed on 30 June 2018
17. Law Archives (2018) Punjab new capital periphery control act 1952. https://archive.india.gov.in/allimpfrms/allacts/1848.pdf. Accessed on 20 Aug
18. Punjab Urban Development Authority (2018) Notification on free enterprise zone. http://www.puda.nic.in/. Accessed on 25 Aug 2018
19. Greater Mohali Area Development Authority (2018) Report of the state-level committee to a policy framework for the Chandigarh periphery controlled area and regulating constructions therein. http://gmada.gov.in. Accessed on 01 Nov 2018

20. Chalana, M (2014) Chandigarh: city and periphery. J Plan Hist. 4(1):62–84
21. WHO (2017) Urban green spaces: a brief for action, World Health Organisation, Denmark
22. United Nations (2018) United nations-disability. https://www.un.org. Accessed on 19 Nov 2018
23. Wolf KL (2004) University of Washington; http://www.cfr.washington.edu. Accessed on 02 Dec 2018
24. Donovan GH, Prestemon JP (2010) The effect of trees on crime in Portland, Oregon. Environment and Behavior
25. Department of Planning and Landscape Architecture (2018) Urban green spaces in Madison. http://urpl.wisc.edu. Accessed on 02 Dec 2018
26. Turf Grass (2017) Environment. http://turfgrassod.org. Accessed on 02 Dec 2018
27. The Environmental Blog (2018) Benefits of green spaces in urban areas. https://www.theenvironmentalblog.org. Accessed on 09 Dec 2018
28. Healthy Parks Healthy People Central (2018) Urban planning and the Importance of green spaces in cities to human and environmental health. http://www.hphpcentral.com. Accessed on 19 Nov 2018
29. Kimmel C (2013) Greening the grey: green infrastructure for sustainable development. http://narc.org. Accessed on 16 Nov 2018
30. SANDAG (2015). San Diego forward: the regional plan. San Diego. http://www.sdforward.com. Accessed 05 December 2018
31. Government of India (2014) Urban and regional development plan formulation and implementation guidelines, Ministry of Urban Development, Government of India, New Delhi
32. Shabdkosh (2018) Meaning of Maidan. https://shabdkosh.raftaar.in. Accessed on 12 July 2018
33. Narula S (2009) Chandigarh Gardens and Greens
34. Chaudhry P, Tewari VP (2011) Urban forestry in India: development and research scenario. Interdiscip Environ Rev 12(1)
35. Weber A (2014) Within the Edge: A revised approach to urban containment within Chandigarh Periphery (unpublished thesis), University of Washington
36. Saini S (2011) Land use changes in Haryana sub-region of Chandigarh controlled area: a spatio temporal study. Inst Town Planners India J 8(4)
37. Nimish G, Chandan MC, Bharath HA (2018) Understanding current and future landuse dynamics with land surface temperture alterations: a case study of Chandigarh. ISPRS Annals of the Photogrammetry, Remote Sensing and Spatial Information Science, Volume IV-2, India
38. Shabdkosh (2018) Meaning of Choes. https://shabdkosh.raftaar.in. Accessed on 12 Sep 2018
39. Krishan G (2006) Chandigarh peripheral zone 2020 a study in futuristic geography (unpublished thesis). Punjab University, Chandigarh
40. Government of India (2017) The smart city challenge 2—Chandigarh, Ministry of Urban Development, New Delhi
41. Indian Institute of Remote Sensing (2013) Evolving a regional perspective on 'Greater Chandigarh Region' (GCR) Using RS and GIS (Graduate)

Smart Environment Through Smart Tools and Technologies for Urban Green Spaces

Case Study: Chandigarh, India

Kshama Gupta, Kshama Puntambekar, Arijit Roy, Kamal Pandey, Mahavir and Pramod Kumar

Abstract Urban Green Spaces (UGS) are an integral part of urban environment and act as lungs for rejuvenating the urban environment and improving the quality of life and health of residents. UGS assist in regulating urban microclimate, biodiversity conservation, alleviating floods, enhancing air quality, and also promotes physical and mental wellbeing of urban populace. They also provide spaces for improved social environment and are considered highly beneficial for physical, social and cognitive development of urban children. UGS exists in diverse shape, size, vegetation cover and types and includes parks, gardens, railway corridors, road side green, derelict monument sites, etc. [1] defined UGS as urban land that consists of unsealed, permeable, soft surfaces such as soil, grass, shrubs and trees. Chandigarh is the first planned city of modern India and is known for its uniformly distributed and ample UGS within its boundaries. However, only quantification of amount of UGS is not

K. Gupta (✉) · P. Kumar
Urban and Regional Studies Department, Indian Institute of Remote Sensing, 4, Kalidas Road, Dehradun 248001, India
e-mail: kshama@iirs.gov.in

P. Kumar
e-mail: pramod@iirs.gov.in

K. Puntambekar
Department of Planning, School of Planning and Architecture, Neelbad Road, Bhauri, Bhopal 462030, India
e-mail: kshama@spabhopal.ac.in

A. Roy
Disaster Management Studies, Indian Institute of Remote Sensing, 4, Kalidas Road, Dehradun 248001, India
e-mail: arijitroy@iirs.gov.in

K. Pandey
GIT&DL, Indian Institute of Remote Sensing, 4, Kalidas Road, Dehradun 248001, India
e-mail: kamal@iirs.gov.in

Mahavir
Department of Physical Planning, School of Planning and Architecture, 4-Block-B, Indraprastha Estate, New Delhi 110002, India
e-mail: mahavir57@yahoo.com

© Springer Nature Singapore Pte Ltd. 2020
T. M. Vinod Kumar (ed.), *Smart Environment for Smart Cities*, Advances in 21st Century Human Settlements, https://doi.org/10.1007/978-981-13-6822-6_5

sufficient to harness the full range of benefits from UGS for smart urban environment. The UGS should be accessible, uniformly distributed and maintained for its daily use by urban population. Although, there are restrictions on development within the Chandigarh, the high pace of urbanization, population increase and economic development in surrounding area is creating a pressure on infrastructure of Chandigarh as well. Chandigarh is also facing the issues of traffic congestion, air pollution and environmental degradation which were unheard before. The multi-faceted issues and benefits of UGS call for smart approaches for their effective utilization and management. Geospatial technologies integrated with Information and Communication Technology (ICT) tools provide a useful and smart tool in the hand of planners for quantification, assessment and evaluation of UGS for smart management. They can help in identifying the vulnerable areas as well as to assess the accessibility and distribution of UGS. They can be used for quantitative as well as qualitative analysis of UGS by applying a range of remote sensing data sets and geo-analysis. Many indices have been developed using these datasets for evaluation and monitoring of UGS. With the growing technological advancements in smart web-based tools, these technologies can also be used effectively for monitoring and management of UGS through integration of ICT tools and citizen centric services. This study demonstrates various innovative geospatial and ICT tools for evaluation and monitoring of UGS.

Keywords Urban green spaces · Geospatial technologies · ICT · Chandigarh · Mobile app · Carbon sequestration

1 Introduction

Urban Green Spaces (UGS) can be a significant part of sustainable development and an important contributor for improved quality of life in urban areas. Development of UGS need to consider interdisciplinary and integrative approaches such as economic, political, social, cultural, management and planning aspects to improve existing urban green space's facilities and services, and to optimize urban green space policies.

UGS does not include only those spaces, which are planned and managed for use by the urban population, it also includes recreational areas, natural overgrown forests and gardens, historical sites, road side avenue vegetation, railway corridors, community greens and like. UGS are defined as "land that consists predominantly of unsealed, permeable, 'soft' surfaces such as soil, grass, shrubs and trees" [1] (the emphasis is on 'predominant' character because of course green spaces may include buildings and hard surfaced areas). It is the umbrella term for all such areas whether or not they are publicly accessible or publicly managed and includes all areas of parks, play areas and other green spaces specifically intended for recreational use, as well as other green spaces with other origins. UGS includes all public and private open spaces in urban areas, primarily covered by vegetation, which are directly (e.g. active or passive recreation) or indirectly (e.g. positive influence on the urban environment) available for the users. In consists of all semi-natural areas, managed parks and gardens, supplemented by scattered vegetation pockets associated with roads and incidental locations as well.

UGS create opportunities for recreational activities, which contributes towards improving people's health, well-being and quality of life, essentially providing environment that help to alleviate stress [2]. UGS provide numerous benefits to urban residents by acting as urban lungs-absorbing pollutants and generating much needed oxygen to balances city's natural urban environment [3]. These areas also function as a visual screen by avoiding too much spatial uniformity and act as noise barriers [4]. Social and health benefits of UGS include spaces for physical activity for physical, mental and cognitive health, improved social cohesion and educational opportunity, which results in improved mental health, decreased risk of cardiovascular ailments, obesity and diabetes [5]. UGS are also extremely effective in restoring psychological health and stress alleviation, increased physical activity and reduced exposure to air pollutants [6]. Studies reveal that even availability of few trees and grass outside a high-rise building leads to improved ability to concentrate and enhanced capacity to cope with surmounting daily stress of life [6, 7]. The physical, social and cognitive development of children is positively impacted by the access to UGS [8]. Apart from providing recreational areas for residents, UGS also helps to enhance the beauty and environmental quality of neighborhoods. Hence, UGS are a necessity for healthy urban population [9] rather than a luxury.

The environmental benefits of UGS are innumerable, such as combating pollution, decreasing air and noise pollution, biodiversity conservation, flood alleviation and management and sink areas for carbon di-oxide [10, 11]. UGS also helps in regulating urban micro climate by reducing thermal stress and energy consumption through countering the warming effects of paved surfaces [12]. It plays a pivotal role for recharging ground water supplies and by reducing run-off. The economic benefits of UGS range from on-site benefits such as direct employment and revenue generation to less tangible effects on property values, retention of businesses and attracting tourists. If the total benefits of UGS could be transformed in terms of economic cost, its benefits will outweigh the cost incurred for its development, management and monitoring [13]. UGS plays an important role in improving the environmental value and quality of life in cities and hence, their optimum provision in urban areas is must for maintaining the smart and livable urban environment. Smart cities movement is centered on the use of Sensors and Internet of Things (IOT) technology. However, environmental sustainability is also one of its core principles. It is believed that new digital technologies can enhance the wellbeing of urban residents and will be able to achieve environmental sustainability. UGS are critical for smart environment and making the cities habitable, healthy and energy efficient [3]. A well planned and managed urban green can enhance the ecological, social and economic value of a city, with a consequently positive effect on its inhabitants [14, 15].

The world is becoming urbanized and it is projected that by 2050, 70% of the world population will live in urban areas. Recent trend shows that now more urbanization will take place in developing world especially in Asia, Africa and Latin America. India is also experiencing exponential growth in urban population due to opportunities offered by urban areas. The urban areas are indispensable as they are hubs of economic growth and innovation as well a symbol of better opportunities and life. While urban areas are important and will continue to grow, due to current patterns of economic growth, they influence urban environment significantly by consumption of resources and energy. The increasing demand due to population growth and changing consumption patterns lead towards poor air quality, waste management and resource management. The situation is exacerbated with the challenges posed by changing climate scenario, increasing CO_2 emissions and air pollution, inefficient supply and disposal systems, security and crime, etc. Polluted urban environments have significant undesirable impacts on health and wellbeing of urban populace. The developing economies are hard pressed for resources and it is necessary for the resource planners to think ahead and plan properly the existing and future areas to enhance quality of living in urban areas. With a promise to achieve this objective, the smart city term has come into use by the end of 20th century and its main emphasis was laid on use of web based technology including IoT and sensor network for improved management of urban spaces like solid waste, urban utilities, crimes, traffic, UGS and so on. The concept is progressive and resource-efficient while providing at the same time a high quality of life. This encompasses many terms such as information city, digital city and sustainable city [16]. In last one decade, this term has gained importance and many government initiatives have been launched worldwide for development of smart cities [17, 18]. Despite the discussion about its concept, there is a lack of consensus on what a smart city actually is [19, 20]. However, it is generally understood that smart cities make use of information and communication technology (ICT) extensively with an emphasis on sensor driven models [21, 22] for sustainable and energy efficient future. It is also defined as "a conceptual model where urban development is achieved through the use of human, collective and technological capital" [23]. The term smart city is, therefore, an umbrella concept that contains a number of sub-themes such as smart urbanism, smart economy, sustainable and smart environment, smart technology, smart energy, smart mobility, smart health, and so on [24, 25]. The planners' community consider it as a comprehensive model addressing the sustainable development. For smart cities to be truly smart, the issues needs to be addressed in an integrated manner and aspires to achieve the sustainable development. It has to be a comprehensive strategy for maintaining smart environment protection, transportation, infrastructure and management, etc. Management and monitoring of environmental issues by applying smart tools and technologies helps in achieving the smart environment.

The Chandigarh is the first planned city of modern India and has a total area of 114 km^2. Chandigarh has distinct status, because it is the capital of two provinces, namely Punjab and Haryana, hence it is also known as an administrative city. Chandigarh is a planned city, with a high standard of civic amenities. The city's decennial population growth for decade 1991–2001 was 44.33%, which has declined in the previous decade of 2001–2011 to 17.19%. It holds a population of 9,00,914 in 2001 census and 10,54,686 population as per 2011 census (http://censusindia.gov.in/2011-prov-results/data_files/chandigarh/Provisional%20Pop.%20Paper-I-Chandigarh%20U.T.pdf). Initially, the city was planned to house only 5 lakh people, however, it now houses more than double the originally planned population. With the growing population in last three decades, Chandigarh also faces the issues of slums and squatter settlements, traffic congestion, air pollution and environmental degradation. The city was planned on green concept and still boasts of well planned, integrated and uniformly distributed UGS but the high pace of development in Chandigarh and surrounding regions is creating mounting pressure on existing infrastructure. It calls for not only quantification of amount of UGS but also an assessment of its quality to harness the full range of benefits from UGS for smart urban environment. The situation and the multi-faceted issues faced by city calls for smart and integrated approaches for effective utilization of resources and management of smart environment in the city and to retain its image of city beautiful.

Geospatial technologies integrated with ICT tools provide a useful tool in the hands of planners for quantification, assessment and evaluation of UGS for smart management. They can help in identifying the vulnerable areas as well as to assess the accessibility and distribution of UGS [8, 26]. They can be used for quantitative as well as qualitative analysis of UGS by applying a range of remote sensing data sets and geo-analysis. Many indices have been developed using these datasets for evaluation and monitoring of UGS. Remote Sensing (RS) technology provides essential data and useful monitoring tool by visualizing the various factors of city fabric and their changes in relation to policy scenario. Remote Sensing data helps in an accurate estimation of UGS as well as reliable relationship among various land covers at micro level. The red-edge band in RS data can be an important data for assessment of vegetation health and subsequent quality of green in urban areas. The integration of this information with ICT tools improves the perception ability of environment information, level of environment protection, provides the necessary information base for informed decision making as well as can provide participatory mechanisms to involve citizens for improved resource efficiency. With the growing technological advancements in smart web-based tools, geospatial technologies along with ICT tools can provide an effective tools and mechanisms for smart monitoring and management of UGS.

This chapter tries to discuss the use of geospatial technologies and ICT tools for evaluation and monitoring of urban green spaces and also dwell over the issues faced for the management of UGS with a case study of Chandigarh.

2 Tools and Technologies for Smart Urban Green Spaces

The smart city agenda has a strong emphasis towards use of innovative technologies for achieving the sustainable development. As per Gary Grant, it is important that smart cities be as much about nature, health, and wellbeing as traffic flows, crime detection, and evermore efficient provision of utilities. There are themes and initiatives which could bring nature into the planning and operation of smart cities. Well planned and accessible UGS are critical for smart environment by reducing the production of carbon dioxide and by providing numerous benefits to urban populace.

Tools for measuring the green and quantifying the benefits of UGS will be very much helpful in assessing the benefits provided by the UGS and its ability to enhance the smartness of the cities.

2.1 Geospatial Technologies

Geospatial technologies relates to the collection and processing of data that is associated with geographic location. This is a broad term and encompasses an amalgamation of technologies namely remote Sensing, GIS and Global Navigation Satellite System (GNSS). The "Geo" prefix (in Geospatial) alludes to a Greek word which implies Earth and "Spatial" identifies with space. Remote Sensing is concerned with acquisition of images, information and data about an object without coming in direct contact. GNSS provides the satellite based geo-location and is being extensively used in public domain for navigation and geo-location in various walks of human life. The GIS is integrated with remote sensing data and GNSS to produce meaningful information and analysis of data for host of applications for natural resource management. The geospatial technology likewise contain georeferenced data procured from surveying and mapping and photogrammetry technique, and convergence of data with upcoming techniques like Internet of Things (IoT) and Location Based Services (LBS) (Fig. 1).

Today, geospatial technologies is being used in various walks of life for the betterment of human kind, be it governance, environment or societal development. It has also been used considerably in various aspects of urban planning, research and management studies, such as urban monitoring [27, 28], urban suitability analysis

Fig. 1 Components of geospatial technology

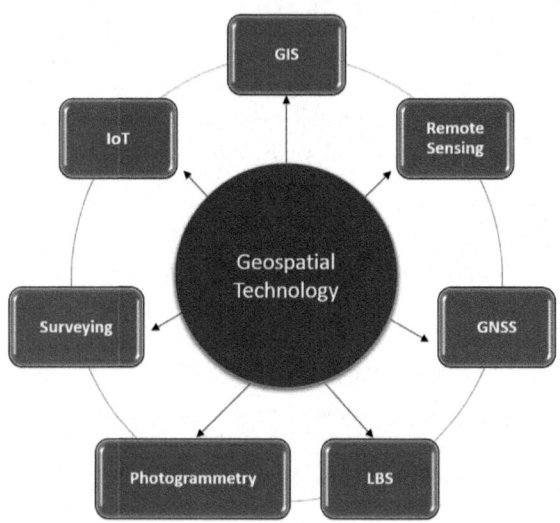

[29], site selection studies, multi-criteria analysis, study of urban growth [30] and urban sprawl, urban hazard, urban climate and assessment of UGS [31, 32]. These technologies together have huge potential for development of tools and methods for assessment and evaluation of UGS for effective management and smart environment. The growing availability of very high resolution satellite data sets as well as open source data sets for example Landsat, Sentinel-1, 2 and 3 provides ample opportunities for studying Earth surface features, vegetation and land cover. Coarse to medium resolution remote sensing datasets have been utilized for broad assessment of green cover at city level whereas the high resolution few meter to sub-meter datasets is being utilized for evaluation of UGS at neighborhood and sub-city scales [26]. Remote sensing datasets have been utilized by researchers for deriving many indices which can provide useful information on UGS quantity, density, spatial distribution, vegetation stress, carbon sequestration, impact of UGS on micro climate and many more [8, 32]. When combined with ICT or web based tools, it can also be utilized for management and monitoring of UGS for achieving smart environment.

2.1.1   Role of Remote Sensing in Urban Environment Monitoring

Remote Sensing techniques offer unique perspectives to study the urban as it provides numerous benefits over conventional methods in terms of synoptic coverage, geographic location, repetitive view and multiple scales. Remote Sensing forms the primary data source in Geospatial information studies. It is the process of gaining information about an object, place or thing without being in physical contact with it. Remote Sensing data is accessible from a range of sources and information collection strategies. Such information are procured with gadgets (e.g. cameras) situated on ground or the sensors or cameras set on aircraft/spacecraft. The objects on the Earth's surface reflect, emit or backscatter varying amount of electromagnetic energy (EMR) in different wavelengths depending on their physical and chemical properties. The estimation of these energies through terrestrial/airborne/space-borne sensors forms the basis for understanding the characteristics of Earth's surface features through remote sensing.

Remotely sensed satellite data is acquired in two modes, namely passive and active. Passive remote sensing is based on electromagnetic radiation produced by the sun and reflected off the Earth's surface features. Active remote sensing depends on sensors that send and record the reflected or backscattered pulse of energy. Most of the readily available remote sensed data is passively collected and is limited to energy not absorbed by the Earth's atmosphere (Fig. 2).

Remote Sensing data obtained in optical, thermal or microwave spectral region have utilities in studying UGS. The optical remote sensing systems are grouped into four categories, depending on the number of spectral bands and bandwidth used in imaging process: (i) Panchromatic (1 band), (ii) Multispectral (3–10 bands), (iii) Super-spectral (10–50 bands), and (iv) Hyper-spectral (50–300 bands). Satellite images acquired in optical region of EMR have high utility in understanding the extent, type, vigor and density of UGS [33]. Especially, the mapping in near infrared

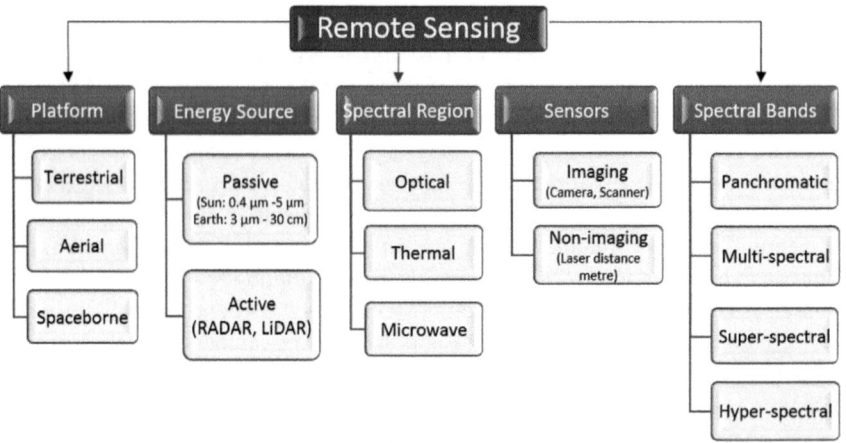

Fig. 2 Remote sensing data types for studying urban green spaces

and red-edge region is widely used for understanding and mapping the vegetation characteristics worldwide.

Thermal remote sensing is based on premise that there is a strong relationship between the amount of radiant flux radiated and the temperature of an object. These systems are capable of sensing the changes in land surface temperature (LST) associated with changes in vegetation proportion in an urban set-up [2, 34].

Microwave remote sensing has the advantage of all-weather capability, unlike the satellite images acquired in optical region of EMR. Microwave remote sensing data acquired with multi-temporal and multi-frequency Synthetic Aperture Radar (SAR) data is useful in estimation of stand density, area delineation, species separation, tree height estimation, detection of plant stress, etc. [35]. The biomass dependence of radar backscatter changes as a function of radar wavelength, polarization and incidence angle. The synthetic aperture radar (SAR) data in microwave remote sensing can be utilized to evaluate above-ground standing biomass. The interferometric SAR (InSAR) based on microwave signals also relate the canopy height with backscatter which depends on density, wavelength and polarization of incident energy.

Light Detection and Ranging (LiDAR) technique is a remote sensing method that uses light in the form of a pulsed laser to quantify distances. LiDAR data is quite useful in measurements of canopy height which in turn can also provide information about biomass, age etc. and used extensively for biomass studies [36]. The photogrammetry methods using optical stereo satellite data are also employed to measure tree height in sparse canopies. It can be used for three-dimensional visualization of UGS or a Digital Elevation Model (DEM) can be created to visualise or analyse the urban built-form and UGS. Unmanned aerial vehicle (UAV) remote sensing has great potential for UGS mapping in a complex cityscape due to their capability to capture images at low altitudes with super spatial resolution up to 2–4 cm. This data is capable of providing highly detailed information of UGS with each and every individual tree.

Remotely sensed data has high utility in studying the vegetation type, distribution, growth pattern, or the change in physiology and morphology of UGS. The reflectance, albedo, leaf water content, fraction of photo-synthetically active radiation (fapar), etc. are the few measurements that can be made using remotely sensed data and useful to analyse the UGS extent, vigour, density and biomass contents. Such data measured over time is helpful for the monitoring the UGS systems and GIS technique helps in multi-layer driven modeling and decision making.

2.1.2 Geographic Information System (GIS)—A Processing Platform for Urban Planning

Urban or any other thematic analysis requires that remote sensing imagery to be converted into perceptible information for use in conjunction with other data sets. Geographic Information System (GIS) integrated with remote sensing have been widely applied as an effective tool in urban analysis and modelling [37]. Remotely sensed data is converted into thematic layers in GIS and further analyzed for extracting the relevant information. GIS lets us visualize, analyze, map, query and interpret the data to understand underlying processes and trends. GIS is integral to remote sensing. The large volumes of data generated by remote sensing cannot be analyzed without GIS.

GIS is used to capture, store, manipulate, analyse, manage, and present the geographically-referenced data. It helps to inter-relate urban form and the UGS for better understanding of their spatial patterns and relationships. It is useful in analysing the composition and function of UGS as well as the understanding of inter-related processes between UGS and surrounding environment. GIS can be used as tool in both problem solving and decision making, as well as for visualisation of UGS data in a spatial environment. The data in a GIS framework is generally stored as raster images or vector layer. While the discrete data is well represented as vector layer, it is advantageous to represent continuous data as raster images. Typically, the satellite images are available as raster images while the analysed maps are available both as raster images or vector layers. The vector data is stored as point, line or polygon layers. Typically, an isolated tree can be mapped as point and the UGS as polygons. GIS technology empowers urban planners with enhanced visibility and helps them to study UGS extent, composition and vigour fluctuation over time when integrated with RS data. GIS can be effectively employed to predict effects of UGS on the urban environment and vice versa.

2.1.3 Global Navigation Satellite System (GNSS)

Global Navigation Satellite System (GNSS) refers to a constellation of satellites providing signals from space that transmit position and time data to GNSS receivers available at users' end. Presently, the GNSS include Europe's Galileo, the USA's NAVSTAR Global Positioning System (GPS), Russia's Global'naya Navigatsionnaya Sputnikovaya Sistema (GLONASS) and China's BeiDou Navigation Satellite System with GPS as the oldest GNSS system. GPS was invented with the need to have an independent military navigation system and with time, it was opened up for public use. GNSS technique is useful to geo-reference the satellite data and to enable them as GIS-ready or to capture varied information needed for multi-criteria modeling of UGS and surrounding environment in a GIS domain.

India has launched its own Indian Regional Navigation Satellite System (IRNSS), also known as NavIC, which is designed to provide position information to users in India as well as the region extending up to 1500 km from its boundary. The system is designed with a constellation of 7 satellites and a vast network of ground systems and expected to provide a position accuracy of better than 20 m in the primary service area. India has also launched the GPS Aided Geo Augmented Navigation-GAGAN project as a Satellite Based Augmentation System (SBAS) for the Indian Airspace. The objective of GAGAN is to establish, deploy and certify satellite based augmentation system for safety-of-life civil aviation applications in India but it also has position location-specific applications in urban planning (https://www.isro.gov.in/applications/step-towards-initial-satellite-based-navigation-services-india-gagan-irnss).

2.2 Information and Communication Technologies (ICT)

ICT has played a major role in the development process of humankind. Modern era is considered as a technology era, there has been tremendous growth in this field and it has changed the way people pursue quality of life. Computing and communicating devices have witnessed miniaturizations in every aspect. From desktop to palmtop and now embedded chips have revolutionized computing paradigm, there has been a landslide shift from wired to wireless communication with the advent of mobile phones. Information has now transformed to Geo-enabled information. Today, GIS, GNSS, Digital Media and mobile phones allow smart data collection and quickly transforming it to smart information and knowledge.

Internet of Things (IOT) and Cloud Computing are the latest additions to the modern ICT architecture, which allow the offering in term of services or more commonly known as Service Oriented Architecture (SOA). This kind of architecture is capable of providing on-demand services for various applications as an optimized solution. Geo-enabled IoT and Web 2.0 ensures participatory and contributory information sharing. IPv6, XML and Application Programming Interface (API) are the major implementation paradigm of the service-oriented architecture. SoA has made the availability of big satellite data even in mobile phones without compromising the quality of data and internet bandwidth. These amalgamations of hardware and software tools have enabled citizens to provide their point of view and feedback on government schemes designed and developed for the masses. Today, a mobile phone consists multiple inbuilt sensors like proximity, GNSS and light sensor; camera, barometer, magnetometer, accelerometers, etc. With all these inbuilt sensors, the scope of location-based services has widened and leading towards improved citizen engagement, information dissemination, knowledge sharing, increased efficiency of service delivery and enhanced productivity.

The integration of these technologies such as IoT, cloud computing, mobile GIS with responsive websites provide huge potential for effective management and governance of urban areas. These tools have immense capacity to provide citizen-centric tools and can be deployed for developing effective participatory mechanism in order to achieve smart management and monitoring of UGS. It can be applied effectively for monitoring of the services and functions provided by urban local government by obtaining the feedback from users for functional assessment of UGS as described in later sections.

3 Study Area Profile

The Union Territory of Chandigarh is the first planned city of modern India and has a total area of 114 km^2. Chandigarh has distinct status, because it is the capital of two state governments, Punjab and Haryana, hence it is also known as an administrative city. Chandigarh is a planned city, with a high standard of civic amenities. Its name has been derived from Chandi Mandir, an ancient temple devoted to the Hindu goddess Chandi, near the city in Panchkula District. The city is located near the foothills of Himalayas in north-west India (Fig. 3) and is drained by two seasonal rivulets, Sukhna Choe in the east and Patiala-Ki-Rao Choe in the west. The city enjoys a sub-tropical humid climate with very hot summers, mild winters, unreliable rainfall and great variation in annual temperature (-1 to $46\ °C$). The average annual rainfall is 1110.7 mm. As per Forest Survey of India report, Chandigarh has 38.8% of total green cover mostly deciduous trees. The city's decennial population growth for decade 1991–2001 was 44.33%, which has declined in the previous decade of 2001–2011 to 17.19%. It holds a population of 9,00,914 in 2001 census and 10,54,686 population as per 2011 census (http://censusindia.gov.in/2011-prov-results/data_files/chandigarh/Provisional%20Pop.%20Paper-I-Chandigarh%20U.T.pdf). The union territory is reported to be one of the cleanest and also heads the list of Human Development Index in 2015.

3.1 Location

Chandigarh is located near the foothills of the Sivalik range of the Himalayas in north-west India. The geographic co-ordinates of Chandigarh are 30.74°N and 76.79°E. It has an average elevation of 321 m (1053 ft.) [38]. Chandigarh is bordered by the state of Punjab to the north, the west and the south, and to the state of Haryana to the east. It is considered to be a part of the Chandigarh capital region or Greater Chandigarh, which includes Chandigarh, and the city of Panchkula (in Haryana) and

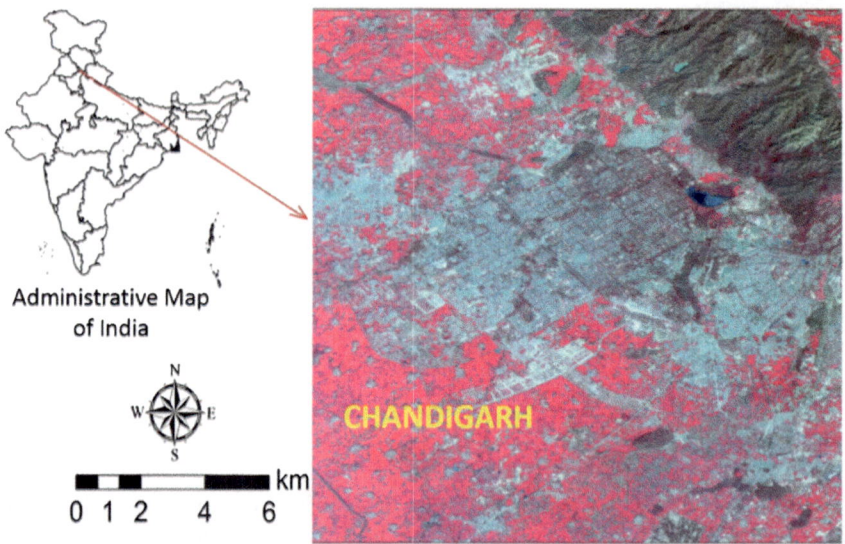

Administrative Map
of India

Fig. 3 Study area Chandigarh

cities of Kharar, Kurali, Mohali, and Zirakpur (in Punjab). It is located north of New Delhi, southeast of Amritsar and just southwest of Shimla.

3.2 Planning of Chandigarh

The city was conceived as a living organism by Le Corbusier with close semblance of functions. The administrative area situated in the north houses major state function comprising the Secretariat, High Court and Assembly Chamber, is the head, the Educational Centers in the north-east are limbs, the Chief Commercial and Civic Centre is the heart whereas the residential area are analogous to the trunk of the organism. The network of roads and footpaths through which the circulation of traffic and population takes place is the circulatory system. The spacious parks, green belts and other open spaces act as lungs of the city (http://chandigarh.gov.in/cmp2031/physical-setting.pdf). The basic unit of planning of the city is sector of the size of 800 m * 1200 m with an area of 246 acres. Each sector is planned as an independent unit consisting of facilities such as shopping centres, schools, hospitals, places of worship, recreational centers, play grounds, plenty of open spaces, etc. within a 10-min walking distance of the residents. Most of the sectors are following a structure style with green space along the middle of sector while the built-up (commercial, institute, super market, etc.) are arranged around the green centre. Along the road is green trees system (Fig. 4). Initially, city was planned for 47 sectors which has now increased to 63 sectors. The city has been built in three phases: sectors 1–30 in

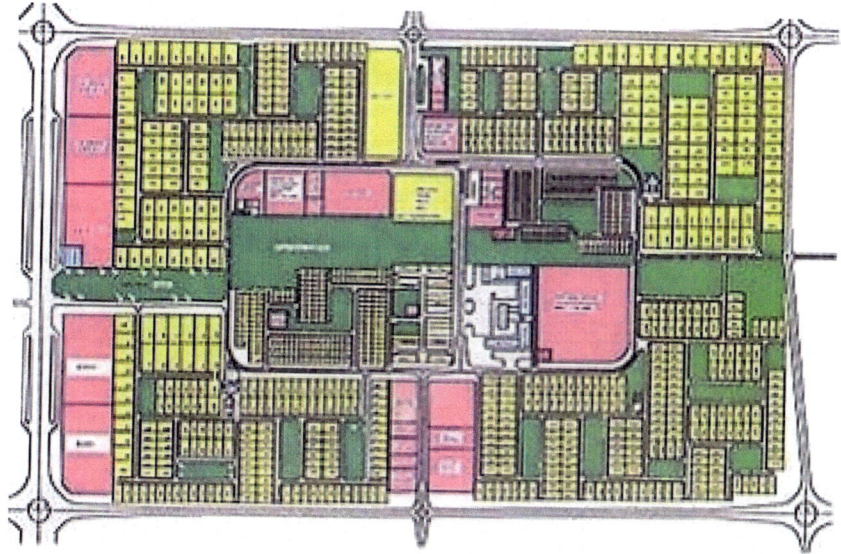

Fig. 4 Layout of one sector. *Source* Chandigarh Master Plan 2031

first phase, sectors 31–47 in the second phase and sectors 48–63 in third phase of development to accommodate the growing population.

3.3 Roads

The road and vegetation cover system in Chandigarh is very specific and planned. There are many sectors and they all have fix road system such that roads are divided in seven categories from V1 to V7; V-1 are the fast roads which connect Chandigarh to other towns, V-2 are Arterial roads, V-3 are fast vehicular sector dividing roads, V-4 denotes the meandering shopping streets, V-5 are sector circulation roads, V-6 are access roads to houses, and V-7 are foot paths and cycle tracks as shown in Fig. 5.

3.4 Urban Green Spaces in Chandigarh

Chandigarh is one of the planned cities in post-independence India and known for its immaculately planned UGS. The master plan of the city was prepared by French Architect Le Corbusier who has given strong emphasis on provision of well-connected, planned and networked UGS [40]. The tree plantation in Chandigarh was planned as an integral part of master plan. Le Corbusier not only provided a con-

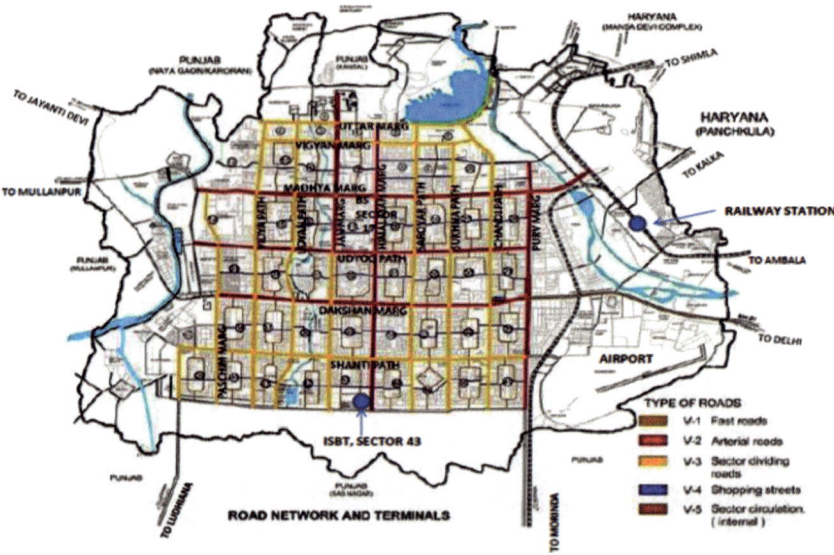

Fig. 5 Chandigarh road network. *Source* Chandigarh Master Plan 2031

ceptual framework for tree plantation but also indicated shapes of trees and there planting arrangement in relation to the scale of road and the position of sun. Based on such elaborate conceptual guidelines, plantation schemes were suitably evolved for each area. The selection of species has to be perfectly planned in Chandigarh with respect to location and site conditions like soil, ground water table, climatic conditions like rainfall, temperature, ornamental and aesthetic requirement, pollution abatement, distance of the plantation site from the residential building, public utility services, etc. Plantation of fruit bearing species such as mango, black plums, guava, tamarind, Indian gooseberry, cluster fig in dedicated plots was carried out within the city.

Most of Chandigarh is covered by dense banyan and eucalyptus plantation. It also hosts many popular gardens, e.g. Zakir Hussain Rose neighborhood greens. The city is surrounded by urban forests with total area of 3245.30 ha that include Rock Garden, Terraced Garden, Bougainvillea Garden, Shanti Kunj and many others public green spaces. According to Forest Survey of India, total green cover (forest cover and tree cover) of Chandigarh is 54 km^2 which form 38.8% of total geographical area [38].

3.5 Present Situation

The tri-city of Chandigarh-Mohali-Panchkula along with Zirakpur forms a large urban agglomeration with high spurt of urban growth in its surroundings and converting the total area in a large metropolitan region. The rapid urbanization, population increase and economic development in surrounding region of Chandigarh is creating a pressure on infrastructure of Chandigarh as well. The city beautiful is now facing multiple issues on infrastructure front mainly water and electricity shortage, traffic congestion, lack of efficient public transport leading to growth in number of vehicles and cleanliness. Chandigarh has been ranked worst in India after Delhi in terms of waste management in India (https://www.quora.com/What-are-the-problems-faced-by-residents-of-Chandigarh).

The city of Chandigarh was built on the concept of green city with abundant open spaces, access to sun, space and verdure by every dwelling unit. The city is known for its uniformly distributed and ample UGS within its boundaries (http://chandigarh.gov.in/knowchd_redfinechd.htm). But only quantification of amount of UGS is not sufficient to harness the full range of benefits from UGS for smart urban environment. The UGS should be accessible, uniformly distributed and maintained for its daily use by urban population. Chandigarh is also facing the issues of traffic congestion, air pollution and environmental degradation which were unheard before. The urban forest on the outskirts of city boundary is continuously under threat of reduction of foliage density and unable to manage original quality. Besides, the dead and diseased trees planted in first phase of development needs to be replaced with newer plantation. Unfortunately, emphasis is more on the plantation of ornamental trees rather than local varieties which requires more effort to nurture and maintain (http://chandigarh.gov.in/greencap/gcap-2009/gap-protect-imp2009.pdf). Local authorities are hard pressed for resources and planning to hand over the management of neighborhood park to resident welfare associations which needs to be monitored on a regular basis for efficient use of resources provided to them. The situation and the multi-faceted issues faced by city calls for smart and integrated approaches for effective utilization of resources and management of smart environment in the city and to retain its image of city beautiful.

4 Quantitative Evaluation and Spatial Distribution of Urban Green Spaces in Chandigarh

Geospatial Technologies can play a major role for quantitative evaluation and assessment of spatial distribution of UGS. To determine the vegetation conditions, various remote sensing-based vegetation and water indices are used. A few of them, which are used regularly are presented in Table 1. The indices are calculated by using surface reflectance of the spectral bands of data by satellite sensors. For example, the spectral signature curves as shown in Fig. 6, detects the various phenomena such as green and dry vegetation, and soil by the effective reflectance. The reflectance of vegetation are of two types i.e. green/wet (i.e., photosynthetic) and dry (i.e., non-photosynthetic). Figure 6 shows reflectance of green vegetation expressing unique spectral signature which helps it to distinguish it from other types of reflectance. It is found to be low in both the blue and red regions of the electromagnetic spectrum

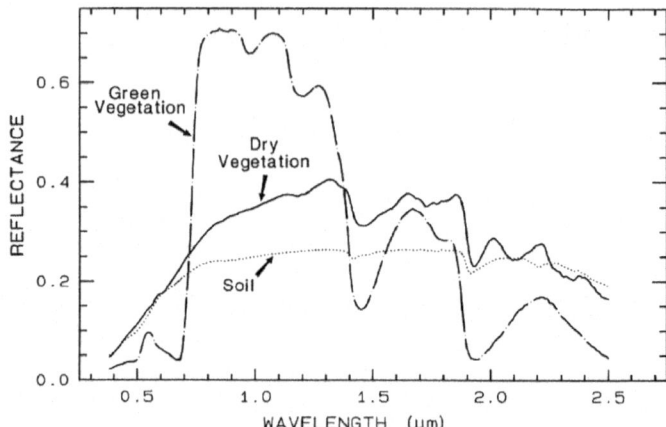

Fig. 6 The spectral reflectance curves of green and dry vegetation and soil along with the spectral wavelengths (after [39])

(i.e., 0.4–0.6 µm). This is because of absorption by chlorophyll for photosynthesis and has a small peak in green region (i.e., 0.5 µm). The reflectance of green vegetation is higher in the near infrared (NIR) region as compared to visible region because of cellular structure of the leaves. The absorption in the spectrum of dry non-photosynthetic vegetation occurs due to the cellulose, nitrogen and lignin. For the green vegetation, these absorptions are weak due to the presence of water bands.

Out of these indices, NDVI has found a wide application in vegetation studies as it has been used to estimate crop yields, pasture performance, and rangeland carrying capacities among others [41]. It is often directly related to other ground parameters such as percent of ground cover, photosynthetic activity of the plant, surface water, leaf area index and the amount of biomass. Generally, healthy vegetation will absorb most of the visible light that falls on it, and reflects a large portion of the near-infrared light. Unhealthy or sparse vegetation reflects more visible light and less near-infrared light. Bare soils on the other hand reflect moderately in both the red and infrared portion of the electromagnetic spectrum [42]. NDVI has been used extensively in urban areas for quantification of UGS.

4.1 Extent of Urban Green Spaces

The Green Index (GI) i.e., percentage of green for Chandigarh city was computed by utilizing Indian Remote Sensing Satellite data of Resourcesat-1 LISS-IV which is acquired on October 23, 2014. Resourcesat-1 LISS-IV is a high resolution multispectral camera operating in three spectral bands (Green, Red and Near Infrared). LISS-IV provides a ground resolution of 5.8 m at Nadir. A binary classification image of green and non-green classes were classified based on NDVI measurements. The

Table 1 Geospatial indices for vegetation studies

S. No.	Vegetation index	Equation	Reference
1	NDVI (normalized difference vegetation index)	$(\rho_{NIR} - \rho_R)/(\rho_{NIR} + \rho_R)$	[41]
2	SAVI (soil adjusted vegetation index)	$\frac{1.5(\rho_{NIR} - \rho_R)}{\rho_{NIR} + \rho_R + 0.5}$	[63]
3	NDWI (normalized difference water index)	$(\rho_{NIR} - \rho_{SWIR})/(\rho_{NIR} + \rho_{SWIR})$	[64]
4	MSI (moisture stress index)	ρ_{SWIR}/ρ_{NIR}	[65]
5	EVI (enhanced vegetation index)	$2.5(\rho_{NIR} - \rho_R)/(\rho_{NIR} + 6*\rho_R - 7.5*\rho_B + 1)$	[33]
6	SIPI (structure insensitive pigment index)	$(\rho_{800} - \rho_{445})/(\rho_{800} + \rho_{650})$	[66]
7	PRI (photochemical reflectance index)	$(\rho_{531} - \rho_{570})/(\rho_{531} + \rho_{570})$	[67]
8	REIP (red edge inflection point)	$700 + 40\left(\frac{\left(\frac{\rho_{670}+\rho_{780}}{2}\right) - \rho_{700}}{\rho_{740} - \rho_{700}}\right)$	[68]
9	LAI (leaf area index)		[69]
10	MCARI (modified chlorophyll absorption ratio index)	$((\rho_{700} - \rho_{670}) - 0.2*(\rho_{700} - \rho_{550}))*(\rho_{700}/\rho_{670})$	[52]
11	MCARI 1 (modified chlorophyll absorption ratio index 1)	$[1.2*[2.5*(\rho_{800} - \rho_{670}) - 1.3*(\rho_{800} - \rho_{550})]]$	[70]
12	MCARI 2 (modified chlorophyll absorption ratio index 2)	$\frac{1.5*(2.5*(\rho_{800}-\rho_{670})-1.3(\rho_{800}-\rho_{550})}{\sqrt{(2\rho_{800}+1)^2-(6\rho_{800}-5\sqrt{\rho_{670}}}-0.5}$	[70]
13	Red edge position index (REPI)	$(705 + 35*(0.5*(\rho_{780} + \rho_{660}) - \rho_{700})/(\rho_{710} - \rho_{700})$	[71]

ρ_x = reflectance at specific wavelength

NIR = Near Infrared band

R = Red band

G = Green band

B = Blue band

NDVI image was generated using Resourcesat-1 LISS IV data of study area. The negative values of NDVI measurement were classified as built-up area and positive values were classified as green class. 100 m × 100 m grid was overlaid over the binary image and percentage of green in each cell was calculated. Based on the percentage, each cell had been classified in four green qualify classes. The result showed that vegetated area occupies nearly 40% of total area i.e., 5034.8 ha. Vegetated area is distributed systematically in whole area (Figs. 7 and 8).

The analysis indicated ample UGS in Chandigarh with a good emphasis on its distribution. The highest amount of green was found in the sectors from 1 to 6 and in the sectors along the central green area. However, when sector map of Chandigarh was overlaid on the binary image, it was found that not all sectors has good amount of UGS (Fig. 9). The UGS area was variable among the sectors, where sector 14 has largest area with 113.0 ha, the second is sector 12 with 104.8 ha. Sector 64 with 0.3 ha

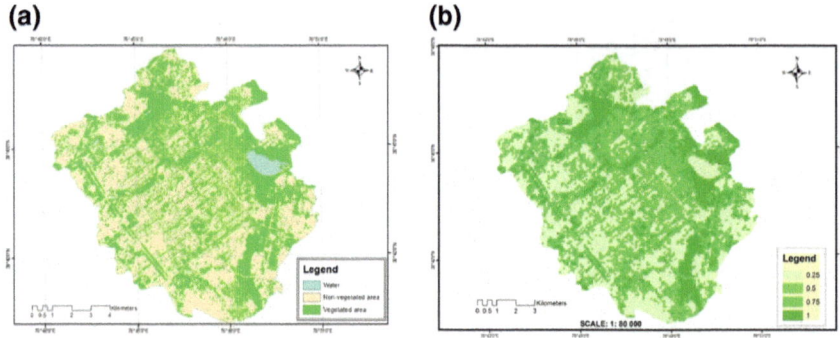

Fig. 7 Percentage of green distribution based on NDVI measurements **a** binary classification **b** percentage of green map

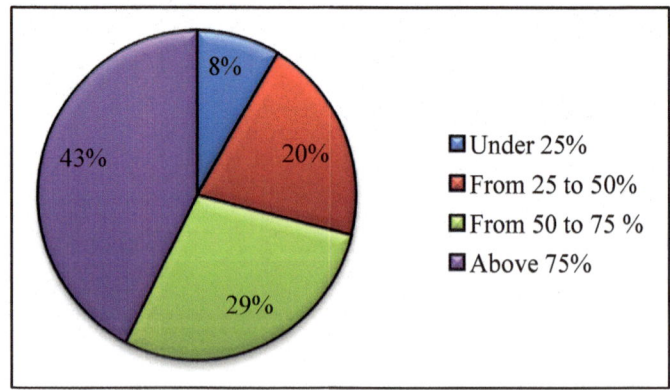

Fig. 8 Percentage of green in Chandigarh, 2014

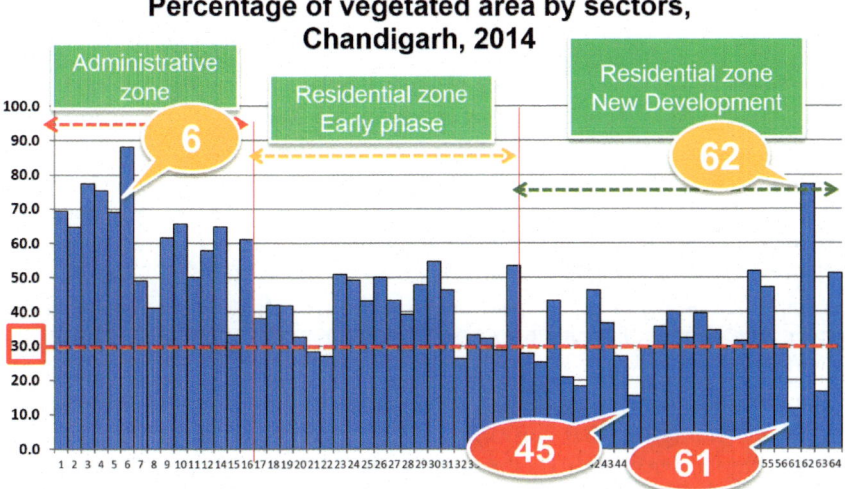

Fig. 9 Percentage of urban green spaces by sectors

and sector 62 with 0.9 ha has the lowest vegetation cover. However, when compared with total area of sector, there is significant change. The percentage of green space per unit area has larger significance for socio-environmental reasons rather than the absolute area of UGS. Most of the sectors in Chandigarh have percentage of green space greater than 30%. Sector 6 and sector 62 have the maximum percentage of green space (88.1 and 77.3%, respectively). There were some sector having percentage of green less than 20% such as sector 63, sector 61, sector 45 and sector 41. Specially, the percentage of green in sector 61 was lowest with 11.7% and most of area was occupied by residential. This analysis conveys an interesting development pattern of Chandigarh. The administrative zone has highest amount of UGS owing to institutional area while newly developed sectors in later phase of development has comparatively less amount of vegetation than the sectors developed in first phase of Chandigarh. This is due to the existence of village abadi area in this sector. It raises some pertinent questions to planning community such as why the planning norms are not applied to urban villages when they become part of urban extension. Why they have been left as totally unmanaged, unplanned spaces in the midst of planned urban areas although they are also effected by pressure of urbanization? The use of geospatial technologies provided a unique insight into the distribution of UGS and also assisted in understanding the development patterns.

4.2 Type of Green and Its Evaluation

Assessment of percentage or amount of Green based on NDVI measurement method do not provide information on the type of green space. The type of green is considered important by Jansson and Persson [43] as they stated that areas with dense vegetation offers more ecological benefits than grass lawns and they are more preferred by children to play as it provides the variety of play methods. Grahn and Stigsdotter [44] discuss that dense vegetation area provides more opportunities for stress restoration. When percentage or amount of green measure is applied for assessment of UGS, an important aspect of green space analysis is mixed. Is it right to equate a fully grown tree with a small shrub or a patch of grass? The qualities and ecological functions provided by different type of vegetation differs considerably. It is found by researchers that urban forest and urban trees provide more opportunities for heat and air pollution mitigation than grass lawns [45]. Besides, highly maintained lawns and trees sequester much less CO_2 than natural surfaces with little maintenance. With more lawn cover than tree canopy cover, the balance can actually shift to emitting CO_2. Hence, quantification of the type of green based on density and quality of vegetation is important than amount of green alone. Remote sensing data can be utilized effectively for classification of vegetation in various density classes. NDVI thresholding, pixel based classification and object based classification are few of the techniques which could be employed for assessment of type of green based on density classification. Since, object based image classification takes into account the shape, texture and other properties of various objects in the image along with spectral properties, it was employed on merged product of Resourcesat-1 LISS IV and Cartosat-1 (2.5 m spatial resolution) image for type of green classification. The basic type of urban green classes (forest dense vegetation, park and garden, grass or low vegetation, open spaces without vegetation, avenue trees, water and built-up areas) were identified as target classes. Hence, the final type of green map shows eight classes classified: forest, dense vegetation, park and garden, road side green, open space and sport complex, vacant, water and built-up. Further, the map was overlaid with 100 m × 100 m fishnet for deriving the weighted type of green map (Fig. 10). The result shows that sector 1, 2, 3, 4 or sectors along the central green (sector 10, 16, 23) has the highest values and outer sectors has lesser values.

4.3 Proximity to Urban Green Spaces

Using geographic area and percentage of green measures, a big patch of green available in the vicinity or within the residential area can be termed as area with high amount of green space. But the question arises, is it sufficient to have a big green patch in the vicinity? Or the green space should be properly distributed? Is the percentage of green is sufficient to measure the distribution of UGS? Optimal spatial distribution of UGS has significance in order to gain ecological and environmental

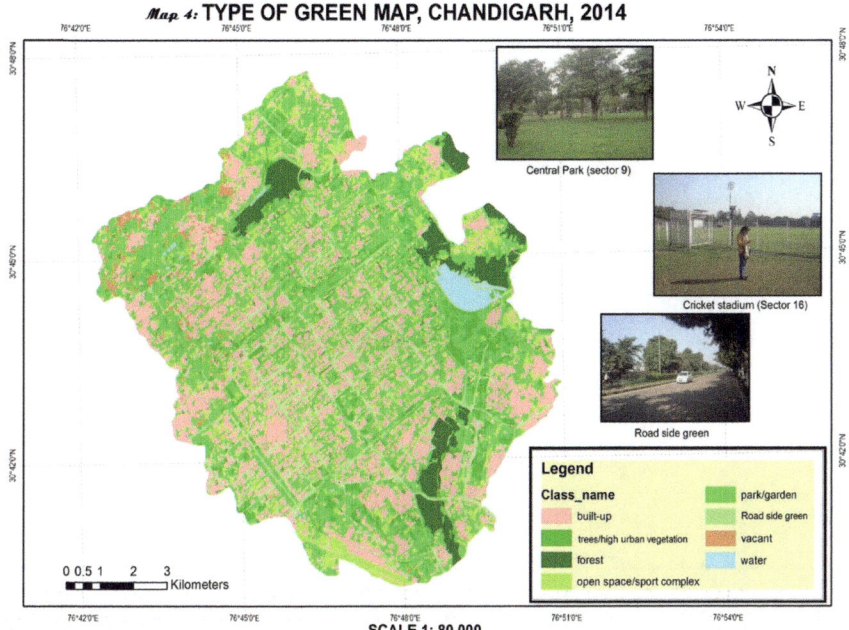

Fig. 10 Type of green map, Chandigarh, 2014

benefits from the UGS. Researchers have found that proximity to green space leads to an increase in the value of the property as it increases the environmental quality [46, 47]. Distributed and accessible UGS are required to enhance its use for active and passive recreation for urban populace as well as for regulating urban micro climate [48]. Toftager et al. [49] found that people living more than 1 km from green space have lower odds of using green space to exercise and keep in shape compared with persons living closer than 300 m to green space. Hence, proximity to green is an important parameter for assessing the spatial distribution of UGS.

Geospatial technologies play an important role for assessment of UGS by providing many tools and techniques for assessment of proximity to green. Most employed method is the buffer analysis. Meng et al. [50] had developed Urban Green Space Index (UGSI) to assess the proximal UGS in the vicinity of an individual building using buffer analysis. Gupta et al. [8, 26] also used buffer analysis in geospatial domain to assess the proximity to green in parts of Delhi. For the assessment of proximity to green in Chandigarh study area also, buffer analysis was performed. However, in this study, different weightages were given to different type of green with decreasing distance. Based on type of green map, "ring buffers" are created with distance in turn as 100, 200, 300, 400, 500 m. Every buffer ring was given varying weightage with respect to type of green class and distance from the UGS. For example, first 100 m buffer was given 1 weightage in the vicinity of forest area and dense vegetation but the same buffer distance was given lesser weightage of 0.8 in the

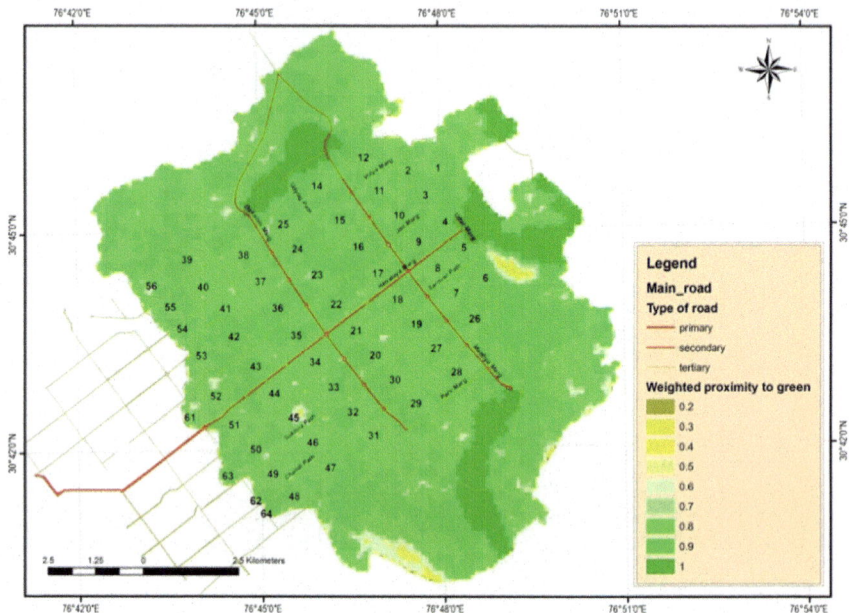

Fig. 11 Proximity to green, Chandigarh, 2014. *Source* Author

vicinity of grass lawns. Again, the grid of 100 m × 100 m was overlaid on generated buffers for each type of green and proximity to green map was generated (Fig. 11). The result showed that all the sectors in Chandigarh have high level of proximity to green (>0.7 on the scale of 0–1). The result indicates the integrated planning of Chandigarh with ample emphasis on networked green areas by Corbusier and that earns the Chandigarh the title of city beautiful.

4.4 Multi-criteria Evaluation of Urban Green Spaces

Remote Sensing integrated with GIS provides unique opportunities for understanding UGS by integrating characteristics of urban green and characteristics of built-up area. Urban Neighborhood Green Index [26] integrates the characteristics of urban vegetation, i.e., amount of green and type of green (basic type of urban vegetation, i.e., dense, low/grass vegetation, etc.) and characteristics of urban built-up at neighborhood level i.e., proximity to green, built-up density and height of structures using multi criteria evaluation in GIS. The similar approach was applied to study the UGS of Chandigarh and a Weighted Urban Green Space Index (WUGSI) was derived for multi-criteria evaluation of UGS [38] (Fig. 12). The result from WUGSI map illustrated that nearly 72% area of Chandigarh is covered by high and very high quality green. The area in low quality green only occupied small proportion with 66 ha

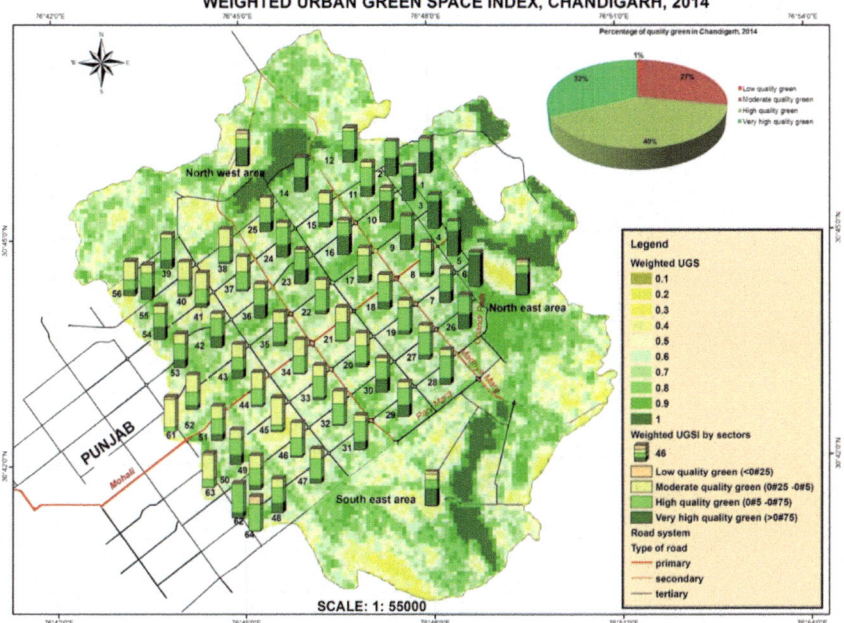

Fig. 12 Weighted urban green space index. *Source* Author

(0.57%). The analysis also reveals an interesting pattern for UGS in Chandigarh that sector 1 to sector 6 have very good quality of green as it comprised of mainly institutional area (Fig. 12). The sectors in first phase of development of Chandigarh also consists ample UGS which is more than 30% of sector area. However, few sectors in second phase of development has very low green percentage as low as 11%.

The mean and standard deviation of weighted percentage of green, weighted type of green, proximity to green and WUGSI were compared to bring out role and relationship among the parameters. The standard deviation (SD) values reflect the distribution of the UGS. Majority of sectors have good distribution of SD such as sector 45 and sector 41 have SD as 0.12; sectors 21, 24 and 33 have SD as 0.08. Some sectors have unequal distribution, such as sector 64 (SD = 0.28), sector 54 (SD = 0.2) and sector 62 (SD = 0.27) (Fig. 13). Sectors 64, 62, 53 have comparable mean percentage of green, however, has low WUGSI mean values that shows poor distribution of green. It also shows that Chandigarh is a planned city of modern India where urban green system is well arranged. Around the world, sustainable development has become a top policy discussion as countries struggle to maintain or enhance economic growth without compromising the future. One of the very important factors to resolve the problem is by improving the quality of urban environment, for which UGS plays a major role. Green space systems require improvement of the spatial pattern of urban green space. Green spaces need to be uniformly distributed throughout the city area, and the total area occupied by green spaces in the city should be large

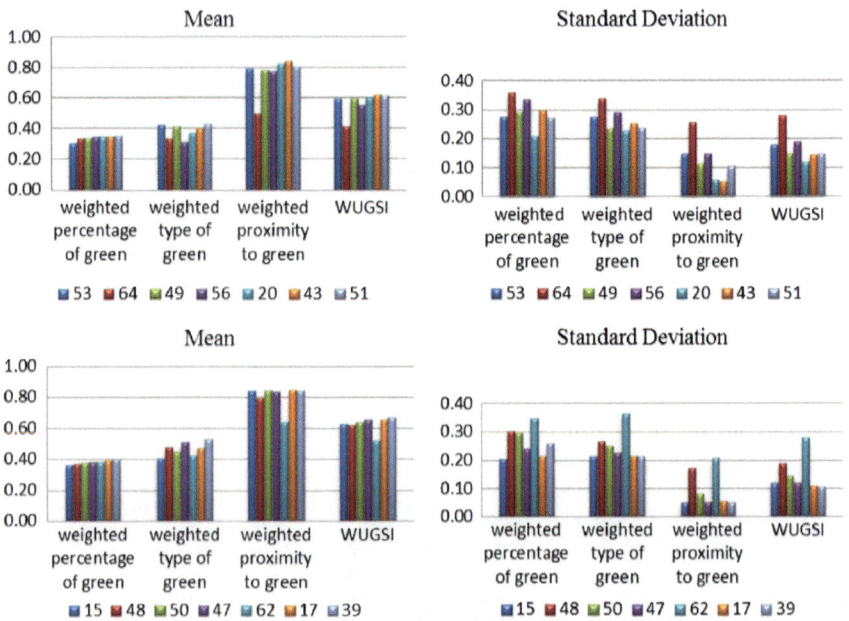

Fig. 13 Mean and standard deviation of some special sectors

enough to accommodate the city population needs [48]. The developed index can be a decision support tool to evaluate, quantify and compare distribution of green structure. Rather than merely measuring the overall percentage of green, i.e., GI, the WUGSI reflects the importance of distribution of green areas in specific area and environments.

4.5 Vegetation Health and Stress Monitoring

One of the important aspects of vegetation status other than the amount of green in an urban environment is the quality of green cover. Compared to the natural vegetation, urban vegetation is constantly under stress due to the impact of urban activities in its vicinity. Urban regions are source of a number of pollution apart from the other stress like lack of moisture and restricted growth areas [51]. Some of the important atmospheric pollutants are a result of the transportation sector and the industrial effluents. They include oxides of sulphur (SO_2), nitrogen (NO_x), ozone (O_3), etc. All these pollutants cause negative effect on the vegetation health due to their reactive nature. This causes the de-naturation of the protein, and leaching of the chlorophyll in the foliage resulting in overall decline in the vegetation health. UGS are meant to act as lungs for the city and the good health of the vegetation in the urban environment can help in mitigating the negative effects of air pollution if maintained and planted properly with pollution mitigating and local varieties. For ensuring quality services from UGS, it is important to monitor the urban green space for the spatial and temporal changes in the health of the green cover. Remote sensing with its capability of synoptic coverage and temporal repetivity provides an important tool for monitoring the status of the UGS using super-spectral and hyperspectral satellite data.

The spectral analysis of the vegetation provides an important source of assessing the urban vegetation health and the level of stress on the vegetation. As stresses occur, there is a change in the foliar chemistry and membrane structure, which is reflected in the changes in the vegetation spectral signatures. This aspect provides us with the physiological based spectral indices of vegetation stress and health. Some of the important spectral indices used for assessing vegetation health are NDVI, SAVI. NDWI, MSI, EVI, SIPI, PRI, REIP (Table 1). Apart from the different vegetation indices, some of the plant foliar biochemical indices like foliar nitrogen content are also used for assessing the vegetation health and stress in the vegetation (Table 1).

The increasing availability of super spectral and hyperspectral remote sensing datasets in commercial as well as open domain has huge potential for assessment and monitoring of vegetation health and stress in urban areas. The red edge band in remote sensing satellites such as Sentinel 2A, World view 2 and 3, RapidEye etc. provides unique opportunities for assessing the vegetation stress in urban vegetation as it is sensitive to stress induced changes in chlorophyll [51, 52, 72]. In this study, Sentinel 2A level 1C product, which is freely available, is utilized for computing the chlorophyll and Leaf Area Index (LAI) to assess the vegetation stress in Chandigarh.

Field samples were collected from the study area and chemical analysis was carried out to estimate the chlorophyll content in smaples. Sentinel-2A data of Chandigarh area was used to calculate MCARI, MCARI1 and MCARI2 indices (Table 1). These indices were compared with the chlorophyll content calculated by chemical analysis. The results of the analysis shows that MCARI2 index has better correlation with the field samples (Fig. 14). Therefore, MCARI2 index was applied to all the monthly dataset for the year 2017–18. The data obtained from OSM road buffer was used to further classify data in family wise, species wise and location (Sector) wise.

Figure 15 represents the variation of MCARI by plot type. It can be observed that neighborhood, central parks (except two plots) and plantations have relatively high values of MCARI, which indicates good vegetation health of those areas. The lowest values of chlorophyll content in avenue trees and urban forest signifies the stressed condition of these areas. The stressed condition of avenue tree may be due to traffic induced pollution while urban forest found to be stressed due to human interference and activities and lack of moisture in these areas. The study needs detailed investigation by taking into account the phenological cycle for conclusive results. Figure 16 represents family wise MCARI2 index for all the months. This index shows vegetation health of species for corresponding months. It can be observed that vegetation is healthiest in months of March and October. This improvement in vegetation health is due to pre monsoon season in October, and due to spring season in March. In the month of May, vegetation was highly stressed due to high insolation, low water availability. It also shows that Annonaceae and Myraceae have

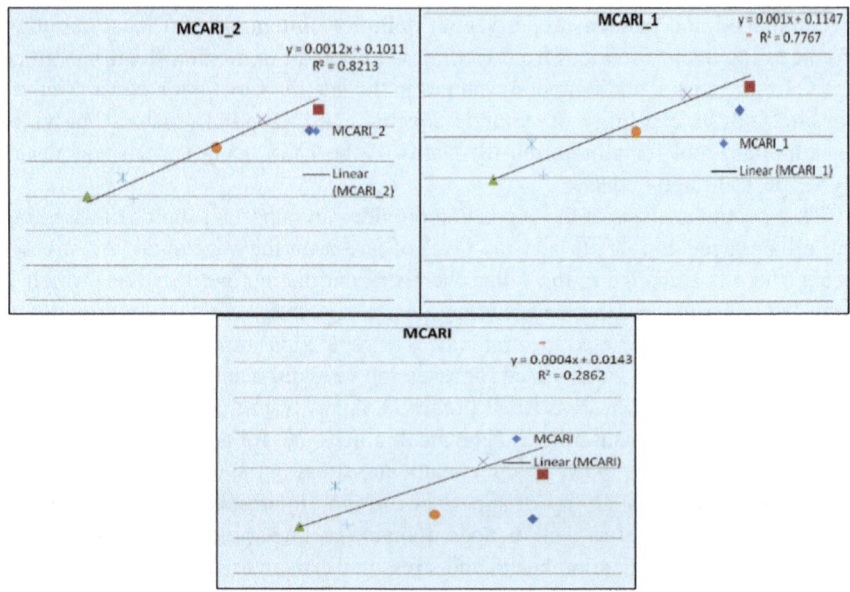

Fig. 14 Correlation between lab analysis and S2A data using MCARI equations

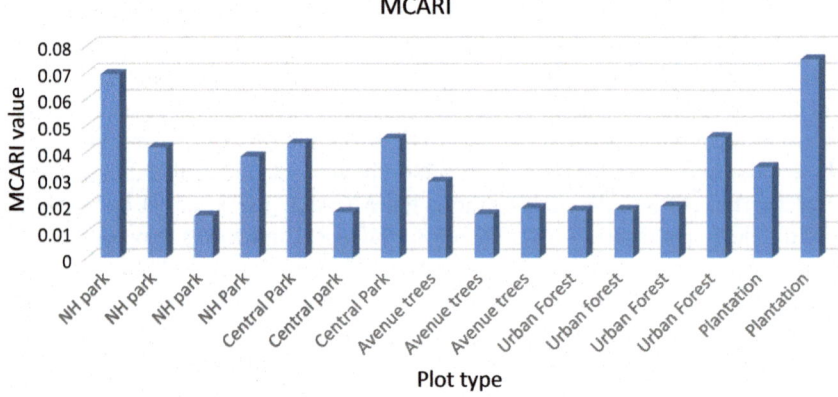

Fig. 15 Plot wise variation in MCARI values

highest index value where as lythraceae and barringtoniaceae have lowest index value. Remaining families have chlorophyll content between these two extremes. The study in under progress and further work is going on for developing a correlation between air pollutants and plant stress for improved assessment of UGS.

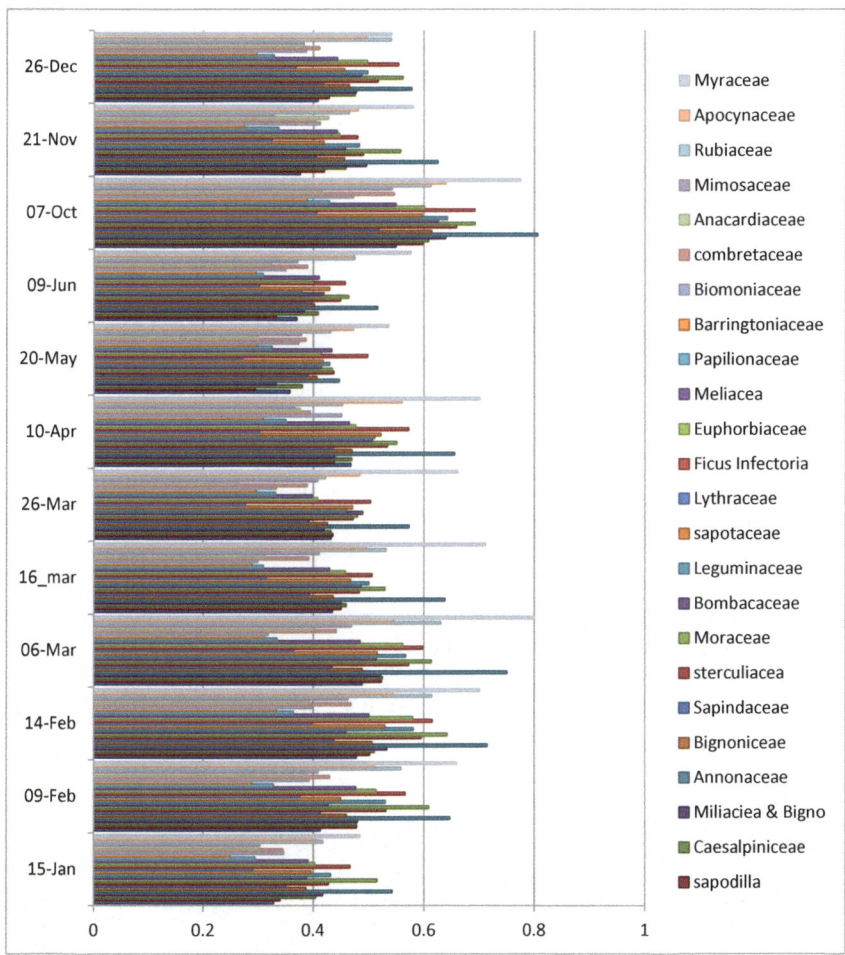

Fig. 16 Family wise chlorophyll variation

5 Web Based Tools and Mobile App for Functional Assessment of Urban Green Spaces

Satellite remote sensing provides an effective means for quantification of amount of green, spatial distribution of green and vegetation health and stress monitoring. However, regular and effective utilization of public UGS is also a management issue and it is dependent on various factors such as accessibility, maintenance, number of activities, feeling of safety and naturalness. The public UGS (PGS) are those UGS which are accessible to the urban population for social use and provide various functions to the urban populace for physical, social and psychological well-being [8]. The functionality of green can be defined as fulfillment of the function or purpose for which it has been defined. The UGS face several issue on the management and maintenance front as represented in Fig. 17.

The functional green spaces are required to be accessible, maintained and should be able to provide a variety of experiences and amenities to the urban populace for its daily and regular use and for maximization of social and health benefits (Fig. 18).

- *Accessibility*—This is a crucial aspect in determining everyday usage of a park. The location of the UGS should be within walking distance of user to enable this relationship to develop, since most of the users prefer walking to the green areas on foot. The walking distance varies according to the area/category of the park. For example tot lots (smallest unit of PGS as per Master Plan of Delhi 2021) are meant to be the play spaces for small age group of children. Hence, they should be within 50–150 m distance for its unhindered and regular use by children [8]. Similarly, neighborhood park meant to provide active and passive recreation to adult population also. A study done by [8] in parts of East Delhi found that 89% of park users reside within 5–10 min of walking distance. Hence, this hierarchy of UGS should be within 5–10 min of walking distance. On the other hand, city level greens are meant to provide weekend recreation for family and a large user group and they can be located within 1 hour of car distance [53].
- *Maintenance*—This factor greatly influence the social and economic value of a park. People feel more relaxed and comfortable in well maintained surroundings. Maintenance relates to cleanliness, presence and long term availability of facilities (like benches, footpaths, swings etc.), regular cutting of large trees, adequacy of water supply etc. A well maintained park always attracts larger user groups.

Fig. 17 Current situation of urban green spaces

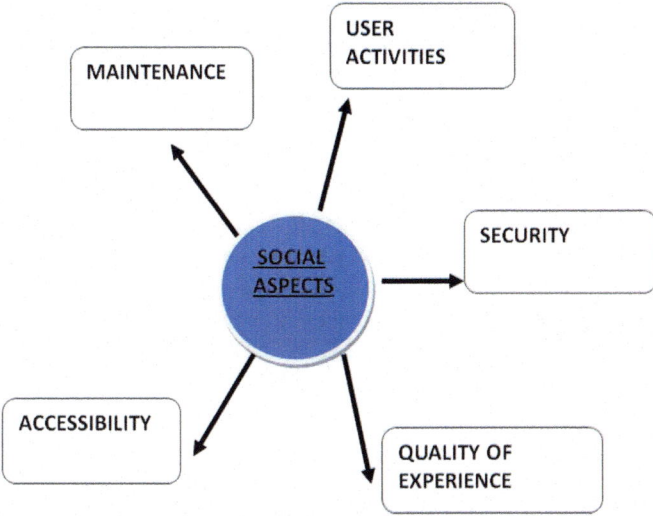

Fig. 18 Development of criteria for functional assessment

The study done in parts of Delhi [8] had also found that UGS with properly maintained footpaths, vegetation, facilities and amenities are preferred to be used on regular basis. In many cities, Resident Welfare Associations (RWAs) were given the responsibility to maintain the UGS within their jurisdiction with the financial support provided by the local government. The RWAs with active and co-operative members perform well and are able to maintain UGS efficiently. However, few of the RWAs are not that active and continuous monitoring of the fund utilization for maintenance of the parks is a big challenge for the local government.

- *Security*—It is an important issue that can influence the value of experience of individuals. Negative factors like fear of crime can affect the usage, since many user groups like females and children might avoid such spaces. The presence of notorious and anti-social elements, absence of guards within the UGS boundaries creates a negative impact and is detrimental to the use of UGS.
- *Quality of experience*—It refers to the ways in which UGS are interpreted by the user. Individual user interpret sites according to their needs and how the sites support their lifestyles [54, 55]. It mainly refers to the perceptions and emotional responses of the users within the park. Mostly, natural surroundings induce the feelings of well-being and relaxation.
- *User activities*—These refer to activities or the functions fulfilled by a green area. These differ at different levels [53]. Every level in the hierarchy of UGS system has different function and every level is complementary to each other. For example, large areas of forest in urban periphery may have utility for weekend recreation for whole urban population while small parks in inner city may have strong connection with everyday life. These are descriptive, what people actually do, walk, walk with a dog, observe nature, cycle etc. Hence, urban spaces are capable of multi-

functional outputs. While evaluating urban spaces (especially parks) based on social aspects, all the above key factors should be taken into consideration.

Chandigarh boasts of safe and accessible UGS, however, the local government finds it difficult to keep up with the increasing pressure for management of UGS. The major issues faced for the management of UGS in Chandigarh are replacement of dead, dying and diseased trees, planting for future replacement, transplantation of trees, biodiversity conservation and maintenance of neighborhood parks. The administration is of the view that maintenance and management of neighborhood parks may be transferred to Resident Welfare Associations (RWA). This strategy have been adopted in Delhi and many other cities. While this strategy proved to be quite advantageous whenever there is an active RWA, but ineffective if there is no monitoring of the UGS by the concerned local authorities.

The relation between parameters for assessing the functionality of UGS can be obtained only through user's perception survey. Various participation methods have been developed to facilitate involvement of people's personal experiences in urban woodland planning of Europe [53]. A range of methods including collaborative planning groups, field visits, public meetings and household surveys intended to involve local residents and other stakeholders in strategic planning of green areas has been used in planning and managing municipally owned woodlands and other green space in Helsinki since 1995 [56]. Although residents were satisfied with the opportunity to participate, the green area planning authorities felt that the process had been too costly and time consuming. The questions of how to involve 'silent' groups more in planning, and how to involve social information with other existing planning information was also raised.

Since, field based user opinion surveys are very time and resource consuming, a web based application was developed to collect user perceptions and to compute the functionality index of the park based on user responses. The web application (Fig. 19) for urban green space monitoring allows the user to give their valuable feedback about the green spaces by selecting the UGS from the geospatial map layer and then by answering a questionnaire related to the green space area on various aspects related to accessibility or distance from the residence, frequency of use, maintenance of UGS, feeling of safety and number of activities in the UGS through a predefined set of questions. Based on the received feedback and rating given to each questions by the authors, a functionality index mark (Fig. 20) is also calculated. The tool also provides an average index value for each UGS based on the feedbacks submitted till now by previous users. The developed tool can be an effective tool for grading the functional level of the UGS. It will enable the authorities to obtain the information about the functional status of a particular UGS. The collected feedback would be useful to take appropriate measure to maintain the green space area. This web application is being developed for Chandigarh to provide a citizen centric ICT based tool for monitoring and management of UGS in Chandigarh. It is fully developed by utilizing open source tools such as PHP, JavaScript, CSS, HTML, Leaflet for Front End Design and PostgreSQL, PostGIS in backend. ArcMap, Geoserver, OpenGeoSuit are also used for integration of green space shape file with this web application. This application

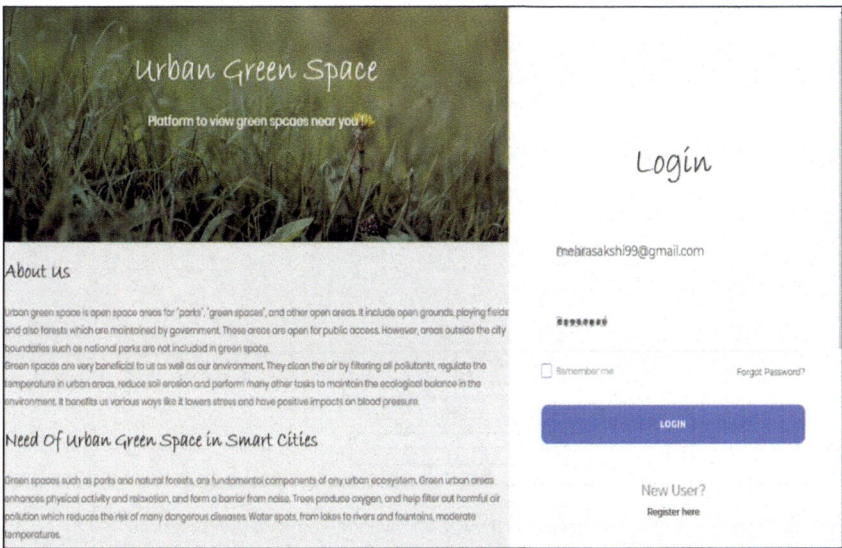

Fig. 19 Web based application for functional assessment of urban green space

is being converted into mobile app for wider application. This tool has the potential to be extended to include other Indian cities in future for providing an ICT based participatory tool for improving the management and maintenance of UGS. The mobile app can also be used for identification of tree database with their geo-location, their health status, threats and maintenance issues. The wider application of this tool can provide an effective mechanism in the hand of citizens for monitoring and participation for governance and management of UGS.

6 Urban Green Spaces and Urban Micro Climate

UGS reduces the temperature in the immediate surroundings due to evaporative cooling and helps in regulating the micro climate. It is well documented by many researchers that as one move away from UGS, the park cool island effect reduces and temperature increases. Urban areas are highly heterogeneous and it is resource consuming and nearly impossible to install ground based sensor network at multiple locations in urban areas to capture urban heterogeneity. Only remote sensing data has the advantage of synoptic, continuous and simultaneous view of the whole urban area which is very important to understand Land Surface Temperature (LST) patterns. A series of satellite and airborne sensors have been developed to collect data in thermal region of electromagnetic spectrum such as HCMM, Landsat TM/ETM+, AVHRR, MODIS and ASTER. These datasets have also been used for extracting emissivity data of different surfaces at varied resolutions and accuracies. Landsat

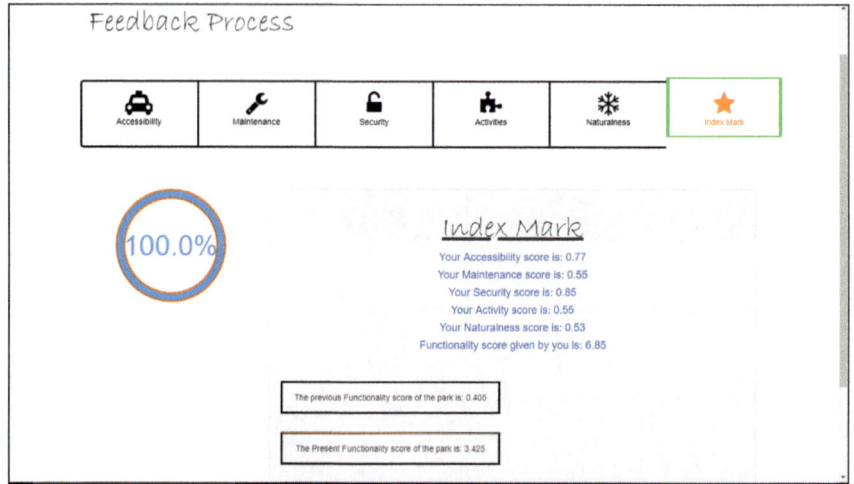

Fig. 20 Functionality index computed by web application based on user feedback

TIR and ASTER are two sensors, which have been widely used for urban studies due to their comparatively finer resolution (~30 m). In urban climate studies, LST and emissivity is mainly used to understand the LST patterns, its relationship with surface characteristics, assessing urban heat island and for relating LST with surface energy fluxes.

The Landsat 8 satellite image (30 m spatial resolution) obtained in thermal region of electromagnetic spectrum have been utilized to estimate Land Surface Temperature (LST) by using Planck's equation (Fig. 21). Landsat is the longest series Earth Observation Satellite (EOS) since 1972 and available freely from www.Earthexplorer.usgs. gov. The overall process for calculation of LST was performed in QGIS, which is an open source software. This software can also be downloaded and installed on any computer system and also provides many plugins for GIS operations. It can be observed from Fig. 21 that the city area exhibit moderate temperature due to uniform distribution of vegetation in Chandigarh city. The Land surface Temperature has negative correlation with Pervious Surface Fraction (PSF) which mainly consists of vegetation areas. Figure 22 shows the strong negative correlation of PSF with LST which is found to be −0.90. It shows that with the increase in amount of pervious area fraction, the Land surface temperature will decrease as the amount of built-up and impervious surface decreases in relation to pervious surface (Fig. 22).

Further, the line profile diagrams were used to analyse the change in temperature pattern when moving away from the vegetation area. In this analysis, the various lines of 200 m were drawn from the residential green park area and green belts towards the residential area and industrial area to see how the surface temperature pattern changes from the centre of park to built-up. Around 20 samples have been collected in this way and later on their average is considered according to the distance.

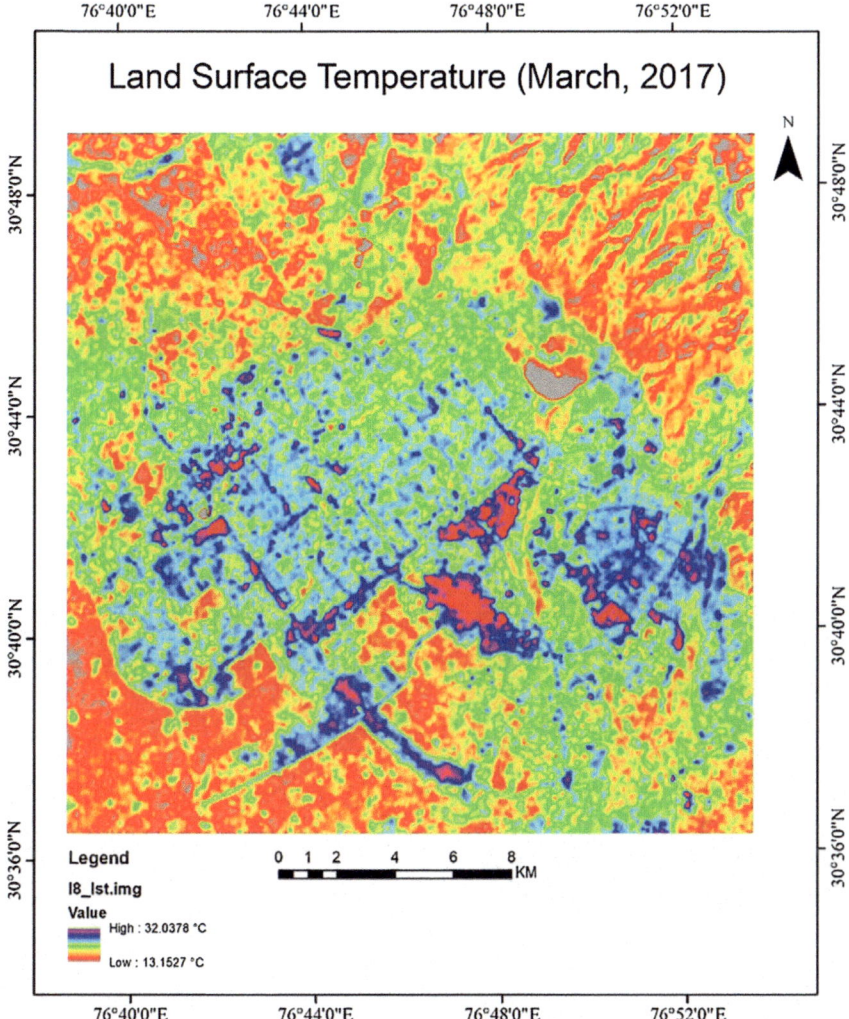

Fig. 21 Land surface temperature, March 2017

There is a significant regular increase of surface temperature with distance from the centre of green area of the order of 0.9 °C temperature change for every 100 m distance. However, it stabilizes approx. after 160–170 m distance (Figs. 23, 24, 25 and 26). It indicates that a regular and uniform distribution of UGS at every 120–150 m distance may help in regulating the urban micro climate.

Figure 27 shows the profile of LST in early phase of development and new development. It is observed that early phase of development has uniformly distributed UGS at every 150–160 m of distance which helps in regulating the microclimate and maintaining the LST through evaporative cooling in urban area. However, new devel-

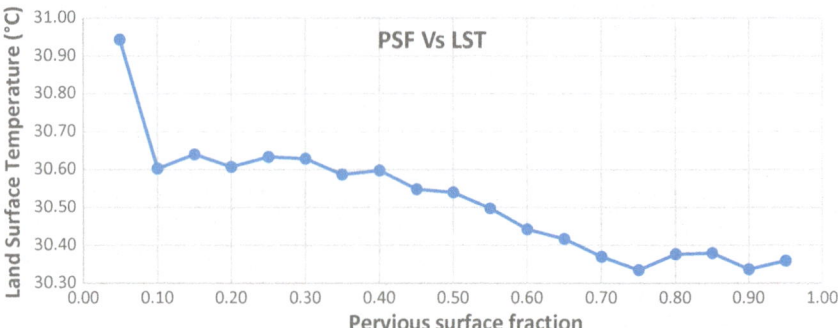

Fig. 22 Land surface temperature versus pervious surface fraction

opment do not have uniform distribution of UGS which results in higher temperatures as one moves away from the UGS.

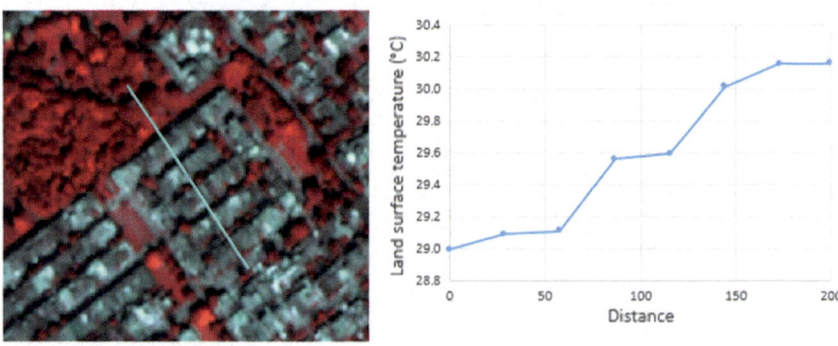

Fig. 23 Line profile (park to residential)

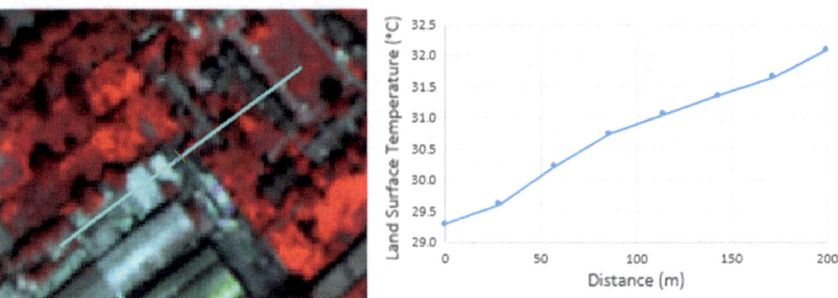

Fig. 24 Line profile (park to industrial)

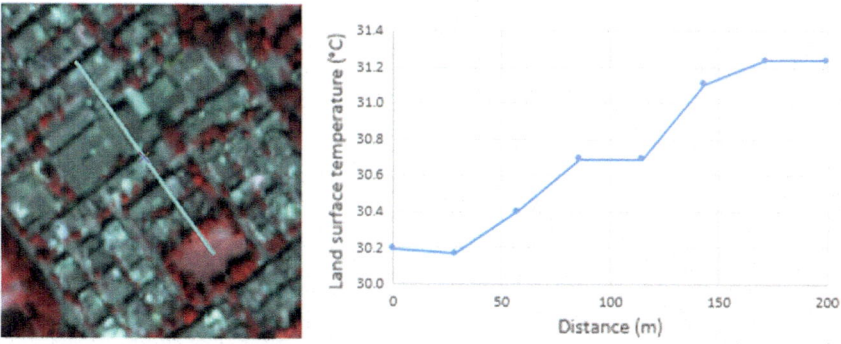

Fig. 25 Line profile (park to residential)

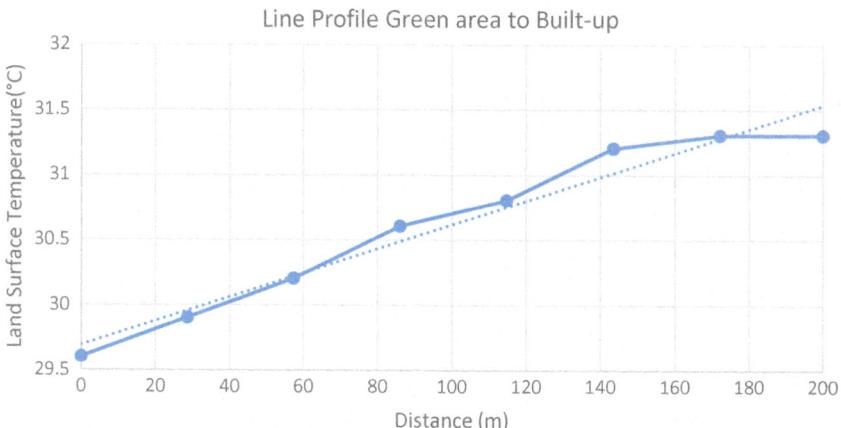

Fig. 26 Average line profile green area to built-up

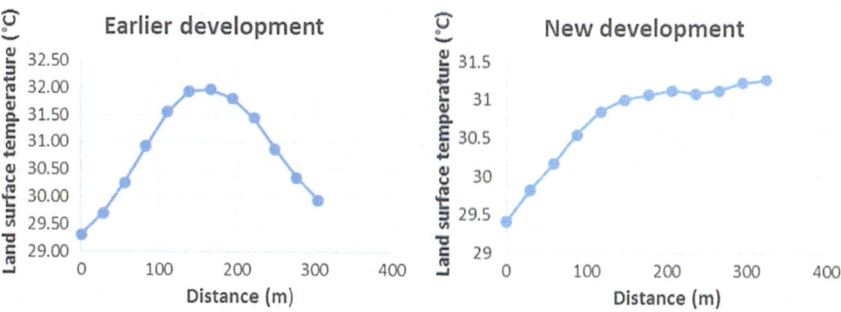

Fig. 27 Land surface temperature in early phase of development and new development

Table 2 Key carbon pools for forests and their definition

Carbon pool	Definition
Live trees	Live trees: above ground
	Live trees: below ground
Understory vegetation	Tree seedlings
	Shrubs, herbs, orbs, grasses
Standing dead trees	Standing dead trees: above ground
	Standing dead trees: below ground
Down dead wood	Down dead wood
	Stamps and dead roots
Forest floor	Fine woody debris
	Litter
	Humus
Soil carbon	Soil carbon
Wood products	Harvested wood mass

7 Estimation of Carbon Sequestration in Urban Green Spaces

Cities have strong impact on carbon cycles as the transportation and industries are considered as one of the major sources of CO_2 emission. UGS can play an important role in removing the atmospheric CO_2 by its sequestration. In this study, tree footprints are digitized from Google Earth image and information on tree height were derived from DSM. The sequestration of carbon was computed using empirical equations obtained from the literature. The computation was carried out for one sector of Chandigarh for demonstrating the methodology. The amount of carbon di oxide any tree can replace depends on its foliage, age, girth, height etc.

Carbon sequestration is storing atmospheric CO_2 that is removed from the atmosphere or before it enters the atmosphere. Carbon sequestering is one of the most effective and promising ways of geoengineering and mitigating the effects of global warming. The term "carbon sequestration" is used to describe both natural and deliberate processes by which CO_2 is either removed from the atmosphere or diverted from emission sources and stored in the ocean, terrestrial environments (vegetation, soils, and sediments), and geologic formations.

Terrestrial sequestration (sometimes termed "biological sequestration") is typically accomplished through forest and soil conservation practices that enhance the storage of carbon (such as restoring and establishing new forests, wetlands, and grasslands) or reduce CO_2 emissions (such as reducing agricultural tillage and suppressing wildfires) (Table 2).

On an average, a tree planted in tropical climate can sequester around 50 lb of carbon dioxide per year. The ability to sequester is maximum during the age of 20–50 years, which reduces with increase in age. Though not much research is done on the sequestration capacity of the trees in tropical climate, this study was an attempt to calculate the sequestration capacity of any tree in such regions. The procedure adopted was conceived as a learning activity for the students in National Computation Science Institute (NCSI). NCSI operates in partnership with the Education, Outreach and Training Partnership for Advanced Computational Infrastructure (EOT-PACI), The National Center for Supercomputing Applications, the University of Illinois at Urbana-Champaign, and many other institutions. Also, this method was used by trees for future to calculate the carbon sequestration in the afforested areas.

The procedure involved number of steps such as mapping the vegetation, determination of green weight and dry weight of the tree and carbon content, which will be further utilized to determine the carbon sequestrated by the tree. The step by step process for determining the carbon sequestration quantity in any area is discussed in the following section.

7.1 Determining the Green Weight

The green weight of the tree is calculated below and above the ground based on the studies by Georgia forestry commission. For this purpose, information on diameter of trunk and height of tree is needed. The coefficient of multiplication will vary as per the species, however, the equations given below could be considered as average of all the species.

$$\text{For trees with D} < 11$$
$$W = 0.25D^2HS \tag{1}$$

$$\text{For trees with D} \geq 11$$
$$W = 0.15D^2HS \tag{2}$$

where,

D   Diameter of the trunk in inches
H   Height of the tree in feet
W   Above-ground weight of the tree in pounds
S   Wood density (gm/cm^3)

Various studies consider the below ground weight of wood ranging between 15 and 20%. For this study, below ground root system weight of 20% of the total tree weight is considered. However, 15% could also have been considered for calculating the total weight of the tree.

Thus,

$$T = 120\% \text{ of } W$$

where,

T = total tree weight

To estimate the biomass of tree at species specific level, in tropics, mathematical equations has been developed and used by many researchers [57, 58]. In Maharashtra, Chavan and Rasal [59] has developed an equation where wood density is considered from the Poona district gazetteer of the Bombay State. The specific wood densities can also be taken from [60]. However, where ever the exact values of wood density are not available the standard average value 0.6 gm/cm^3 was taken. For this study, average value of wood density 0.6 gm/cm^3 is considered.

7.2 Determining the Dry Weight

Dry weight of tree is calculated on the basis of an extension publication from the University of Nebraska [59]. The dry weight of the tree is considered as 72.5% and moisture content is 27.5%. To get the value of dry weight, the total weight of the tree is multiplied by 72.5%.

7.3 Determine the Weight of Carbon

Yin et al. [61] suggested the use of carbon content which ranges from 45 to 60% of the total dry weight by using Carbon Storage Dynamic Analysis Method. In this investigation, an average of 50% of total dry weight is considered. The atomic weight of carbon is 12.00 and atomic weight of oxygen is 15.99. Thus, the weight of Carbon dioxide is $C + 2 * O = 43.99$. The ratio of CO_2 to C is $43.99/12.00 = 3.67$.

7.4 Calculating CO_2 Sequestrated in Chandigarh

For calculating carbon sequestrated by trees in Chandigarh, the study area selected is sector 16 as shown in Fig. 28. The crown of vegetation foliage was digitized using Google Earth. Few samples were taken for each tree typology to find out the crown trunk ratio. This generalisation was adopted on observing the trees full grown height and the trunk diameter in surroundings. It is not possible to know the exact age of the trees required to calculate the CO_2 sequestrated every year. The age range of 60 years was classified in three categories 1–20, 20–40 and 40–60 years. Figure 29 shows the distribution of trees according to age and Table 3 presents the species details and steps to calculate the CO_2 sequestration.

The estimation of the amount of CO_2 sequestrated every year for a small area of one square kilometer is 1653.523 metric tons. As per the World Bank in 2014,

Table 3 Calculating CO_2 sequestration

Common name	Botanical name	Total green weight (in pounds)	Dry weight (72.5% of total green weight) (in pounds)	Weight of carbon sequestrated (50% of dry weight) (in pounds)	Weight of CO_2 sequestrated (weight of carbon sequestrated 3.6663 (in pounds)	CO_2 sequestrated every year (weight of CO_2 sequestrated/age) (in pounds/year)
Cotton tree	*Bombax ceiba*	123,878,432.16	44,908,371	22,454,185.25	82,323,779.4	4,008,379.71
Golden shower tree	*Cassia fistula*	668,645,111.80	484,767,706.09	242,383,853.05	888,651,920.42	23,177,527.22
Putranjiva	*Putranjiva roxburghii*	40,042,182.15	29,030,582.06	14,515,291.03	53,217,411.50	1,297,536.66
African tulip tree	*Spathodea campanulata*	94,984,266.06	68,863,592.89	34,431,796.45	126,237,295.31	3,541,060.69
American mahogany	*Swietenia mahagoni*	115,322,397.54	83,608,738.22	41,804,369.11	153,267,358.46	4,429,440.13
Total		1,042,872,389.72	756,082,482.54	378,041,241.27	1,386,012,602.88	36,453,944.41
						16,535.23 tonnes/year

India was emitting 1.7 metric tons of carbon per person per year. Accordingly, it is sufficient for 972 persons.

This is a small demonstration of how to use the satellite data to measure the carbon sinks around us. This will facilitate to map and quantify the carbon sinks in smart cities and equating the sinks and sources.

8 Summary and Concluding Notes

Chandigarh is the first planned city in post-independence India and is known for its emphasis on green and holistic approach to planning. The city is also known as city beautiful. Initially, the city was planned for a population of 5 lakhs but it houses more than 10 lakh people as per 2011 census. Besides, the surrounding region of Chandigarh including the city of Mohali, Panchkula and Zeerakpur are going through a phase of rapid transformation and urbanization. The city beautiful is also facing the similar issues as faced by other rapidly growing towns such as proliferation of slums, environmental degradation, traffic congestion, increasing pollution levels etc. The multi-faceted issues needs to be addressed through integrated approach by employing innovative tools and technologies for smart management of urban environment. Geospatial technologies integrated with ICT tools provide many tools for efficient management of infrastructure resources and UGS.

Geospatial technologies especially satellite remote sensing provides a range of remotely sensed data at varying spatial and spectral resolutions in different wavelength regions for vegetation studies. The RS data in optical domain is demonstrated

Fig. 28 Part of Chandigarh city for quantifying carbon sequestration

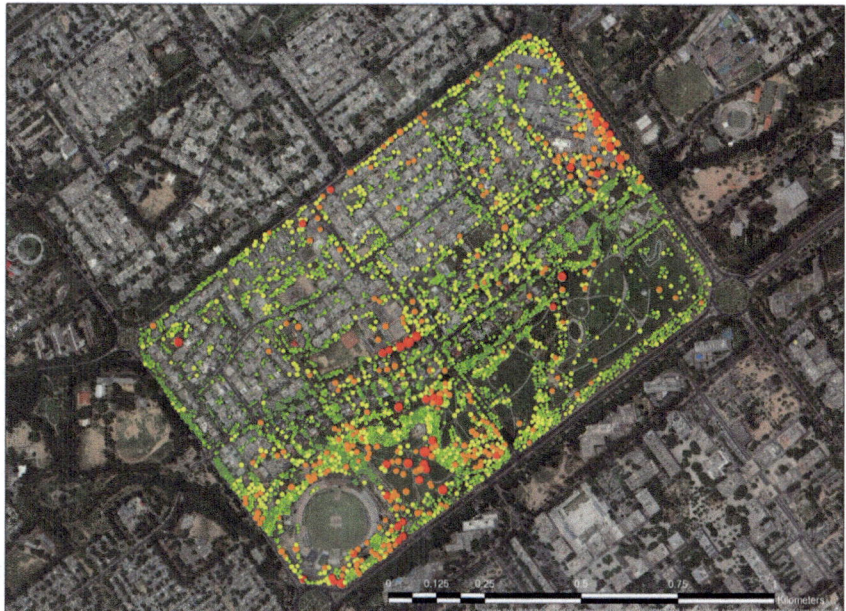

Fig. 29 Classification of vegetation according to age

in this study to assess the extent and type of UGS in Chandigarh. It is found that the city has good quality green as 43 out of 59 sectors have quality of green from high to very high (more than 50%). Only 3 sectors have low quality of green under 25%. The percentage of green analysis across sectors of Chandigarh presents an interesting trend which shows that early phase of Chandigarh's development has well managed and ample UGS, however, later phase of development has lesser amount of green in few sector. While studying the proximity to green to assess the distribution of UGS in Chandigarh through GIS tools, it is observed that all the sectors in Chandigarh has uniformly distributed UGS with more than 0.7 value at the scale of 0–1. The analysis of UGS through developed multi-criteria index i.e. WUGSI reveals that although Chandigarh has very good distribution of UGS but few of the sectors in second and third phase of development exhibits low quality of green.

The availability of red edge band in remote sensing satellites such as Sentinel 2A, World view 2 and 3, Rapid Eye, etc. and hyperspectral sensors provides unique opportunities for assessing the vegetation stress in urban vegetation as it is sensitive to stress induced changes in chlorophyll. The estimation of chlorophyll content in various parts in Chandigarh through Sentinel-2A data shows lower values of chlorophyll content in avenue tress and urban forest area as compared to neighborhood, central parks and plantations which indicates stressed condition of vegetation in these areas. The avenue trees are stressed due to increasing traffic and resultant air pollution while urban forest faces a threat due to human interference and cutting of trees.

Geospatial Technologies integrated with ICT technology has been demonstrated here for development of web based citizen centric tools and mobile app for monitoring and management of functional UGS. The tool is designed to compute a functionality index for UGS which can be utilized for grading of UGS, to identify management issues and to encourage authorities or Resident Welfare associations responsible for management to provide improved environment in UGS. The developed mobile app can be enhanced for collection of geo-tagged tree inventory data, to identify dead and diseased trees for felling and to obtain geo-tagged information on new plantation. These tools can help citizens to participate in the monitoring and management of UGS.

Remotely sensed data in thermal region has been extensively employed to establish relationship with surface characteristics, assessing urban heat island and for relating LST with surface energy fluxes. It has also been applied for assessing the park cool island effect and impact of UGS on regulating the urban micro climate. Since, Chandigarh has uniformly distributed UGS it is found that availability of UGS at every 150–160 m distance can help in regulating the urban micro climate.

The study also demonstrated the potential of high resolution remote sensing data for creation of tree inventory and further utilization for computation of carbon sequestration potential. It has been found that one sector in Chandigarh sequestered sufficient amount of CO_2 to cater to nearly 972 persons.

The study demonstrated the potential of geospatial technologies and ICT tools for improved assessment and management of UGS in Chandigarh through a number of studies. It has also highlighted the benefits and issue faced by UGS in Chandigarh. The study concludes that with the application of Geospatial technologies, the UGS analysis can be done effectively. It can be an effective and smart tool in preserving and monitoring green and open spaces in an urban area [62]. These tools are widely accepted in urban landscape planning as it can provide better understanding on the spatial pattern and changes of land use in an area. In future, spatial arrangement of green spaces can be captured in more refined manner with the help of 3D model using stereo pair/LiDAR data, which will provide additional height information to more realistically model the influence of neighboring buildings.

Acknowledgements The authors acknowledge the encouragement provided by Director, IIRS and Dean (Academics), IIRS for carrying out these studies. The authors are also thankful to all the students of IIRS and CSSTEAP, Dehradun and SPA, Bhopal who have worked under the guidance of authors for their support in data generation and processing.

References

1. Dunnett N, Swanwick C, Woolley H (2002) Improving urban parks, play areas and green spaces, urban research report, Department of Landscape, University of Sheffield, Department for Transport, Local Government and the Regions: London, http://publiekeruimte.info/Data/Documents/e842aqrm/53/Improving-Urban-Parks.pdf. Last accessed on 5 Nov 2018

2. Panagopoulos T, Gonzalez-Duque JA, Dan MB (2016) Urban planning with respect to environmental quality and human well-being. Environ Pollut 208:137–144
3. Nijkamp P, Leventa TB (2004) Urban green space policies: a comparative study on performance and success conditions in European cities. In: Proceedings of the 44th European congress of the European regional science association, 25–29 Aug 2004, Porto, Portugal. http://dare.ubvu.vu.nl/bitstream/1871/8932/1/20040022.pdf
4. Dole J (1989) Greenscape 5: green cities. Architects' J 61–69
5. Lee ACK, Maheswaran R (2010) The health benefits of urban green spaces: a review of the evidence. J Public Health 33(2):212–222
6. Ulrich R, Simons R, Losito B, Fiorito E, Miles M, Zelson M (1991) Stress recovery during exposure to natural and urban environments. J Environ Psychol 11:201–230
7. Grahn P, Stigsdotter U (2003) Landscape planning and stress. Urban For Urban Green 2:1–18
8. Gupta K, Roy A, Luthra K, Maithani S, Mahavir (2016) GIS based analysis for assessing the accessibility at hierarchical levels of urban green spaces. Urban For Urban Green 18:198–211
9. McMahon ET (1996) Green enhances growth. http://plannersweb.com/1996/04/green-enhances-growth/. Last accessed on 5 Nov 2018
10. European Commission. Directive, 2008, 2008/50/EC of the European Parliament and of the Council of 21 May 2008 on ambient air quality and cleaner air for Europe
11. World Health Organisation (2014) Burden of disease from ambient air pollution for 2012
12. De Ridder K (2003) Benefits of urban green space, EVK4-CT-2000-00041, Belgium
13. Mcpherson EG (1992) Accounting for benefits and costs of urban green space. Landsc Urban Plan 22:41–51
14. Coles R, Caserio M (2001) Social criteria for evaluation and development of urban green spaces. URGE-improving the quality of life in urban regions through urban green initiative, Project Deliverable 7. Pdf. Retrieved from http://www.urgeproject.ufz.de/PDF/D7SocialReport.pdf
15. Buff City Status Report (2003) Urban green structure. Baltic University Urban Forum City Status Report V. Retrieved from http://www.balticuniv.uu.se/index.php/component/docman/docdownload/229-urban-green-structure
16. Yigitcanlar T (2006) Australian local governments' practice and prospects with online planning. URISA J 18(2):7–17
17. Jong M, Joss S, Schraven D, Zhan C, Weijnen M (2015) Sustainable–smart–resilient–low carbon–eco–knowledge cities; making sense of a multitude of concepts promoting sustainable urbanization. J Clean Prod 109:25–38
18. Yigitcanlar T (2016) Technology and the city: systems, applications and implications. Routledge, New York
19. Angelidou M (2015) Smart cities: a conjuncture of four forces. Cities 47:95–106
20. Hortz T (2016) The smart state test: a critical review of the smart state strategy 2005–2015's knowledge-based urban development. Int J Knowl Based Dev 7(1):75–101
21. Yigitcanlar T, Baum S (2008) Benchmarking local e-government. In: Electronic government: concepts, methodologies, tools, and applications. IGI Global, pp 371–378
22. Caragliu A, Del Bo C, Nijkamp P (2011) Smart cities in Europe. J Urban Technol 18(2):65–82
23. Angelidou M (2014) Smart city policies: a spatial approach. Cities 41:S3–S11
24. Gudes O, Kendall E, Yigitcanlar T, Pathak V, Baum S (2010) Rethinking health planning: a framework for organising information to underpin collaborative health planning. Health Inf Manag J 39(2):18–29
25. Lara A, Costa E, Furlani T, Yigitcanlar T (2016) Smartness that matters: comprehensive and human-centred characterisation of smart cities. J Open Innov 2(8):1–13
26. Gupta K, Kumar P, Pathan SK, Sharma KP (2012) Urban neighborhood green index—a measure of green spaces in urban areas. Landsc Urban Plan 105(3):325–335
27. Jat MK, Garg PK, Khareb D (2008) Monitoring and modelling of urban sprawl using remote sensing and GIS techniques. Int J Appl Earth Obs Geoinformation 10(1):26–43
28. Gupta K (2013) Unprecedented growth of Dehradun urban area: a spatio-temporal analysis. Int J Adv Remote Sens GIS Geogr 1(2):6–15

29. Sarika B, Gupta K, Kumar P (2015) Smart planning through geo-enabling of digital database for planning authorities: a case study of regional park zone (RPZ) pockets in Navi Mumbai notified area. In: ISG 2015 conference at Jaipur, 16–18 Dec 2015

30. Chowdhury PKR, Maithani S (2014) Modelling, urban growth in the Indo-Gangetic plain using nighttime OLS data and cellular automata. Int J Appl Earth Obs Geoinformation 33(1):155–165

31. Voogt JA, Oke TR (2003) Thermal remote sensing of urban climates. Remote Sens Environ 86(2003):370–384

32. Vaidya G, Pawar AS, Gupta K (2017) Mapping urban green spaces for neighbourhood sustainability by using urban neighbourhood green index, Spandrel. Int J SPA Bhopal 1(12):27–39

33. Huete A, Didan K, Miura T, Rodriguez EP, Gao X, Ferreira LG (2002) Overview of the radiometric and biophysical performance of the MODIS vegetation indices. Remote Sens Environ 83(1–2):195–213

34. Wu Z, Zhang Y (2018) Spatial variation of urban thermal environment and its relation to green space patterns: implication to sustainable landscape planning. Sustainability 10:2249. https://doi.org/10.3390/su10072249

35. Lee W-J, Lee C-W (2018) Forest canopy height estimation using multiplatform remote sensing dataset. J Sens. https://doi.org/10.1155/2018/1593129

36. Dhanda P, Nandy S, Kushwaha SPS, Ghosh S, Krishna Murthy YVN, Dadhwal VK (2017) Optimising spaceborne LiDAR and very high resolution optical sensor parameters for biomass estimation at ICESat/GLAS footprint level using regression algorithms. Prog Phys Geogr 41(3):247–267

37. Lang S, Blaschke T (2006) Bridging remote sensing and GIS—what are the main supporting pillars? In: International archives of photogrammetry, remote sensing and spatial information sciences; ISPRS: Vienna, Austria, vol XXXVI-4/C42

38. ThiLoi D, Tuan PA, Gupta K (2015) Development of an index for assessment of urban green Spacesat city level. Int J Remote Sens Appl 5(1):78

39. Clark RN (1999) Spectroscopy of rocks and minerals and principles of spectroscopy. In: Rencz AN (ed) Manual of remote sensing, vol 1. Wiley, New York, pp 3–58

40. Singh D, Singh A, Sharma AK, Sodhi L (1998) Burn mortality in Chandigarh zone: 25 years autopsy experience from a tertiary care hospital of India. Burns 24(2):150–156

41. Rouse JW, Haas RH, Schell JA, Deering DW (1973) Monitoring vegetation systems in the Great Plains with ERTS. In: Third ERTS symposium, NASA SP-351 I, pp 309–317

42. Holme AM, Burnside DG, Mitchell AA (1987) The development of a system for monitoring trend in range condition in the arid shrublands of Western Australia. Aust Rangel J 9:14–20

43. Jansson M, Persson B (2010) Playground planning and management: an evaluation of standard-influenced provision through user needs. Urban For Urban Green 9:33–42

44. Grahn P, Stigsdotter UK (2010) The relation between perceived sensory dimensions of urban green space and stress restoration. Landsc Urban Plan 94:264–275

45. Hostetler M, Escobedo F (2016) What types of urban greenspace are better for carbon dioxide sequestration? WEC279, Wildlife Ecology and Conservation Department, UF/IFAS Extension. Retrieved from https://edis.ifas.ufl.edu/pdffiles/UW/UW32400.pdf

46. Jim CY, Chen WY (2009) Value of scenic views: hedonic assessment of private housing in Hong Kong. Landsc Urban Plan 91:226–234

47. Troy A, Grove JM (2008) Property values, parks, and crime: a hedonic analysis in Baltimore, MD. Landsc Urban Plan 87:233–245

48. Bao T, Li X, Zhang J, Zhang Y, Tian S (2016). Assessing the distribution of urban green spaces and its anisotropic cooling distance on urban heat island pattern in Baotou, China. Int J Geo Inf 5(12). https://doi.org/10.3390/ijgi5020012

49. Toftager M, Ekholm O, Schipperijn J, Stigsdotter U, Bentsen P, Gronbaek M, Randrup TB, Kamper-Jorgensen F (2011) Distance to green space and physical activity: a Danish national representative survey. J Phys Act Health 8(6):741–749

50. Meng Q-Y, Liu Y-Q, Li X-J (2014) Urban green space remote sensing retrieval with LIDAR and multi-spectral satellite data. In: Open GIS conference, Szekesfehervar, Hungary

51. Pitt R, Bannerman R, Clark S, Williamson D (2004) Sources of pollutants in urban areas (part 2)—recent sheetflow monitoring. J Water Manag Model, Jan 2005. https://doi.org/10.14796/jwmm.r223-24
52. Haboudane D, Miller JR, Tremblay N, Zarco-tejada P, Dextraze L (2002) Integrated narrow-band vegetation indices for prediction of crop chlorophyll content for application to precision agriculture. Remote Sens Environ 81:416–426
53. Van Herzele A, Wiedemann T (2003) A monitoring tool for the provision of accessible and attractive urban green spaces. Landsc Urban Plan 63(2):109–126
54. Norberg-Schulz C (1979) Genius loci: towards a phenomenology of architecture. Rizzoli, New York
55. Caserio M (2001) User perceptions of landscape in the Val Fontanabuona, research proposal. UCE, Birmingham, UK
56. Sipilä M, Tyrväinen L (2005) Evaluation of collaborative urban forest planning in Helsinki, Finland. Urban For Urban Green 4(1):1–12
57. Brown S, Gillespie AJR, Lugo AE (1989) Biomass estimation methods for tropical forests with applications to forest inventory data. For Sci 35:881–902
58. Negi JDS, Manhas RK, Chauhan PS (2003) Carbon allocation in different components of some tree species of India: a new approach for carbon estimation. Curr Sci 85(11):1528–1531
59. Chavan BL, Rasal GB (2010) Sequestered standing carbon stock in selective tree species grown in University Campus at Aurangabad, Maharashtra, India. Int J Eng Sci Technol 2:3003–3007
60. Brown S (1997) Estimating biomass and biomass change of tropical forests: a primer. FAO Forestry Paper-134. Retrieved from http://www.fao.org/docrep/w4095e/w4095e00.htm
61. Yin W, Yin M, Zhao L, Yang L (2012) Research on the measurement of carbon storage in plantation tree trunks based on the carbon storage dynamic analysis method. Int J For Res 2012:10 p (Article ID 626149). https://doi.org/10.1155/2012/626149
62. Ruangrit V, Sokhi BS (2004) Remote sensing and GIS for urban green space analysis—a case study of Jaipur city, Rajasthan, India. J Inst Town Plan India 1(2):55–67
63. Huete AR (1988) A soil-adjusted vegetation index (SAVI). Remote Sens Environ 25(3):295–309
64. Gao BC (1996) NDWI—a normalized difference water index for remote sensing of vegetation liquid water from space. Remote Sens Environ 58(3):257–266
65. Hunt ERJR, Rock BN (1989) Detection of changes in leaf water content using near-and middle-infrared reflectances. Remote Sens Environ 30:43–54
66. Penuelas J, Baret F, Filella I (1995) Semi-empirical indices to assess carotenoids/chlorophyll—a ratio from leaf spectral reflectance. Photosynthetica 31(2):221–230
67. Gamon J, Serrano L, Surfus J (1997) The photochemical reflectance index: an optical indicator of photosynthetic radiation use efficiency across species, functional types, and nutrient levels. Oecologia 112(4):492–501. Retrieved from http://www.jstor.org/stable/4221805
68. Guyot G, Baret F (1988) Utilisation de la Haute Resolution Spectrale pour Suivre L'etat des Couverts Vegetaux. In: Proceedings of the 4th international colloquium on spectral signatures of objects in remote sensing, Aussois, France, Jan 18–22 1988. Retrieved from https://www.researchgate.net/publication/234432105_Utilisation_de_la_Haute_Resolution_Spectrale_pour_Suivre_L'etat_des_Couverts_Vegetaux
69. Eschenbach C, Kappen L (1996) Leaf area index determination in an alder forest: a comparison of three methods. J Exp Bot 47(9):1457–1462. https://doi.org/10.1093/jxb/47.9.1457
70. Haboudane D, Miller JR, Pattey E, Zarco-tejada PJ, Strachan IB (2004) Hyperspectral vege-tation indices and novel algorithms for predicting green LAI of crop canopies: modeling and validation in the context of precision agriculture. Remote Sens Environ 90:337–352. https://doi.org/10.1016/j.rse.2003.12.013
71. Vogelmann JE, Rock BN, Moss DM (1993) Red edge spectral measurements from sugar maple leaves. Int J Remote Sens 14(8):1563–1575. https://doi.org/10.1080/01431169308953986
72. Daughtry C (2000) Estimating corn leaf chlorophyll concentration from leaf and canopy reflectance. Remote Sens Environ 74(2):229–239. https://doi.org/10.1016/S0034-4257(00)00113-9

73. http://censusindia.gov.in/2011-prov-results/data_files/chandigarh/Provisional%20Pop.%
    20Paper-I-Chandigarh%20U.T.pdf
74. http://chandigarh.gov.in/knowchd_redfinechd.htm
75. https://www.isro.gov.in/applications/step-towards-initial-satellite-based-navigation-services-
    india-gagan-irnss
76. http://chandigarh.gov.in/cmp2031/physical-setting.pdf
77. http://chandigarh.gov.in/greencap/gcap-2009/gap-protect-imp2009.pdf

Part IV
Gandhi Nagar

A Solar Intensive Approach for Smart Environment Planning in Gandhinagar, Gujarat

Asfa Siddiqui, Dixit K. Joshi, Sami Rehman, Pramod Kumar and V. Devadas

Abstract With time and technological development, human race is gradually elevating its dependence on machinery. The incessant usage of energy is therefore increasing manifolds. However, the realisation of reducing reliance on non-renewable energy sources was felt long back, no concrete steps are taken in day to day life. Solar Energy is considered to be one of the most abundant and sustainable forms of energy in a tropical country like India. It can thus be used to fulfil the demands risen due to dearth of urbanization. Gandhinagar was chosen to be one of the first Model Solar City after the Solar City Master Plan was prepared for the city. Considering the growing demand and need, the city was assessed using Remote Sensing (RS) and Geographic Information System (GIS) technique for its building footprints and solar rooftop potential was evaluated using standard insolation models. It was felt that geospatial techniques play a pivotal role in mapping the potential zones useful for resource tapping. It is expected that only 30% of rooftop area can reduce the dependence on non-renewable energy by 200% in residential sector. Planned cities and all planned developments within the urbanized regions can adhere to use technology

A. Siddiqui (✉) · P. Kumar
Urban and Regional Studies Department, Indian Institute of Remote Sensing, Indian Space
Research Organization, Department of Space, Government of India, 4, Kalidas Road,
Dehradun 248001, India
e-mail: asfa.aas@gmail.com; asfa@iirs.gov.in

P. Kumar
e-mail: pramod@iirs.gov.in

D. K. Joshi
20, Gayatri Nagar Society, Hanuman Tekri, Palanpur 385001, India
e-mail: dixitjoshi007@gmail.com

S. Rehman
Department of Electrical and Electronics Engineering, Energy Acres: Bidholi via Premnagar,
Dehradun 248007, India
e-mail: rehmansami08@gmail.com; srehman@ddn.upes.ac.in

V. Devadas
Department of Architecture and Planning, Indian Institute of Technology
Roorkee, Roorkee 247667, India
e-mail: devadasv59@gmail.com; valanfap@iitr.ac.in

© Springer Nature Singapore Pte Ltd. 2020 197
T. M. Vinod Kumar (ed.), *Smart Environment for Smart Cities*, Advances in 21st Century
Human Settlements, https://doi.org/10.1007/978-981-13-6822-6_6

to calculate the effect of usable rooftop area, height of the structure or building and insolation falling on the surface so as to reduce their energy demand. Such technologically sound and low cost solutions can help in driving developing nations' need of reliance on non-renewables. The study is an attempt to understand how efficiently geospatial technology can play a role in determining the solar potential and help in converting the city into a sustainable and green city by using the tool of "Smart Energy".

Keywords Technology · Green energy · Renewable · Non-renewable · Smart energy

1 Introduction

Urbanization today is soaring at a very rapid rate throughout the world. Not only in the urban sector, today, energy required in every front of livelihood. In this era of threat towards the non-replenishable source of energy, the need for renewable source of energy production multiplies manifolds. The time has arrived when there is a grave need to understand consequences of using non-renewable sources of energy leading to climate change and inevitable for countries to relish their energy deposits to meet energy needs [1]. The rising standards and quality of life being proportional to the dependence on power and technology has led to the over utilization of resources on this planet Earth. The balance of the earth's environment is in jeopardy due to disturbances in the current energy consumption pattern and excessive release of hazardous obnoxious gaseous pollutants in the air worldwide [1]. To mitigate the impact of developing human activities on climate, the dependability on non-renewable sources of energy shall be minimised and the level of energy intensity shall be limited. There is an urgent requirement to shift the resource pool to a more sustainable and eco-friendly resource base available in plenty and utilised inadequately. This shall gradually help in substituting the conventional sources of energy.

Renewable energy is that source of energy which encourages zero carbon dioxide (CO_2) emissions during the operation and are constantly replaceable in the human timescale [2]. The growing demand of energy can be met using green energy. The use of renewable sources of energy can be considered to be a crucial step towards 'sustainable development'. The various sources of renewable energy such as sunlight, rain, tides, wind, geothermal heat and oceanic waves can be tapped efficiently using effective modern techniques capable of identifying the resource potential, both spatially and temporally.

Techniques like Remote Sensing (RS) and Geographic Information Systems (GIS) can be helpful in mapping the resource zones and justify its exploitation amply in accordance to the site locations [3]. The demand for renewable energy increases and brings with this growth, the need for quantification of such resources available in scarcity or in plenty on/around earth. Such database can be useful in enriching the information bank related to decision making, technological improvement, and improvisation in technical/managerial manufacturing/development capacity.

2 Non-renewable Energy Sources

The need for renewable energy sources arises from the fact that the demand is increasing at a steady rate. The entire world is competing towards energy security and to bridge the gap between demand and supply, yet maintaining the environment's sanctity. Non-renewable and fossil fuel consumption mostly in the form of coal started way back in 4000 BC (Golas and Needham 1999). But, Industrial Revolution marked the beginning of large scale usage in all commercial and industrial sectors around the world around 1800 BC followed by crude oil and then natural gas consumption. Gradually, the world was sensitized with the ill-effects and threats posed by the burning reserves leading to negative impacts to the globe due to air pollution and greenhouse gas (GHGs) emissions. At present, the natural gas shares 28% while coal and crude oil share 33 and 39% of the total non-renewable energy respectively (https://ourworldindata.org/fossil-fuels). The global fossil fuel consumption grew from 5972.23 TWh in 1900 to 31011.14 TWh in 1960 and 132051.53 in 2016 (Smil 2017) (Fig. 1). Also, the fossil fuel production in India varied from 341.63–3947.92 TWh in 2014 (http://www.tsp-data-portal.org/) (Fig. 2).

While United States remains the lead in possessing coal reserves; Russia, China, Australia and India are also rich in coal reserves. Countries like Middle East, Russia, United States and China have very high stocks of Crude Oil. Also, Middle East (Iran with highest natural gas reserve), Russia and Turkmenistan followed by United States, Venezuala and Saudi Arabia have very high natural gas reserves. As per reserves to

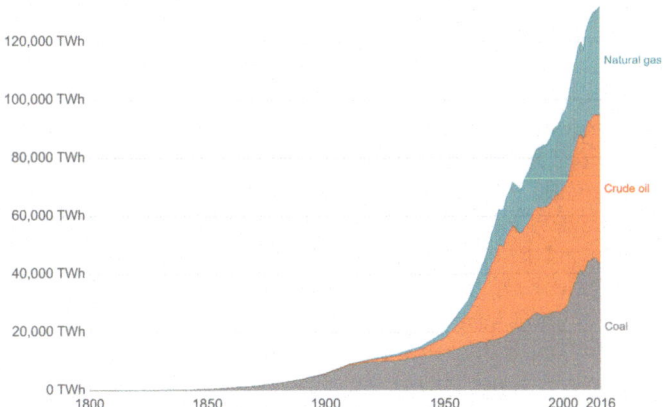

Fig. 1 Fossil fuel consumption (global). *Source* Vaclav Smil (2017) Energy Transitions: Global and National Perspectives and BP Statistical Review of World Energy Report

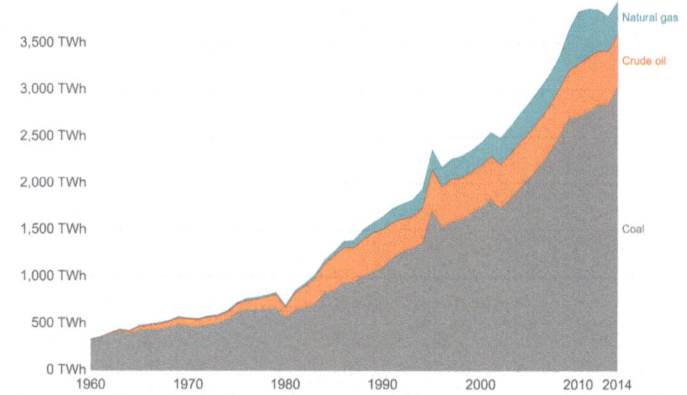

Fig. 2 Fossil fuel consumption in India. *Source* Coal Production—The SHIFT Project, Oil production—Etemad and Luciana, Gas production—Etemad and Luciana (http://www.tsp-data-portal.org/)

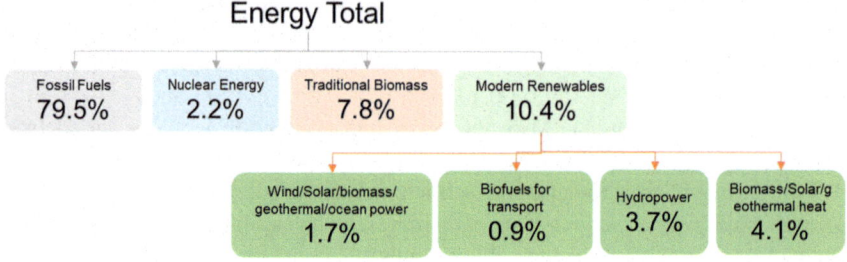

Fig. 3 Estimated renewable share of total final energy consumption in the world, 2016. *Source* [5]

production ratio (R/P) that measures the number of years left for production of coal, oil and natural gas, based on the availability and production level of reserves, 114 years of coal reserves, 52.8 years of Natural Gas reserve and 50.7 years of Oil reserve is left [4].

Although, traditional sources of fuel system account for a large portion (nearly 70–80%) of the total energy use, the reliance on variety of renewable energy sources available in the world is gaining momentum. In the last decade (2006–2016), the growth rate of modern renewables has grown by 5.4% [5] (Fig. 3). With upcoming policies globally and nationally in the past few decades, this growth only seems too modest or insufficient. Of the total renewable energy available in the world, nearly 27% contributes towards heating requirements, 3% is used for boosting the transport sector, while 25% share is used for power consumption/generation (as per 2015 statistics presented by Hales 2018).

3 Renewable Energy Sources in India

The mainstream renewable energy sources available in the world are wind, hydro, solar, geothermal, ocean thermal and bio energy. The location and geographic setup of India, provides advantage to the country for tapping almost all types of energy sources present on the planet (Fig. 4). In the world, India is one of the most prominent countries enjoying very large solar energy potential, potential of energy from wind and biomass energy [6]. However, India's per capita consumption of energy is very low (nearly 639 kWh) in the world, nearly one-third of the world average in 2015–16 [7]. In India, of the total installed capacity i.e. 344 GW, renewable energy sector share is 20% with 69.02 GW renewable power capacity installed [8] (Fig. 5).

3.1 Wind Energy

Energy from wind resource is considered to be one of the most favorable renewable energies. As per National Institute of Wind Energy, NIWE, India has a potential of 300 GW at a mast height of 100 m from the ground and 102 GW at a hub height of 80 m [8]. After hydro energy, wind energy is the most dependent energy source in India with more than 70% of installed potential. Most of the states sharing the

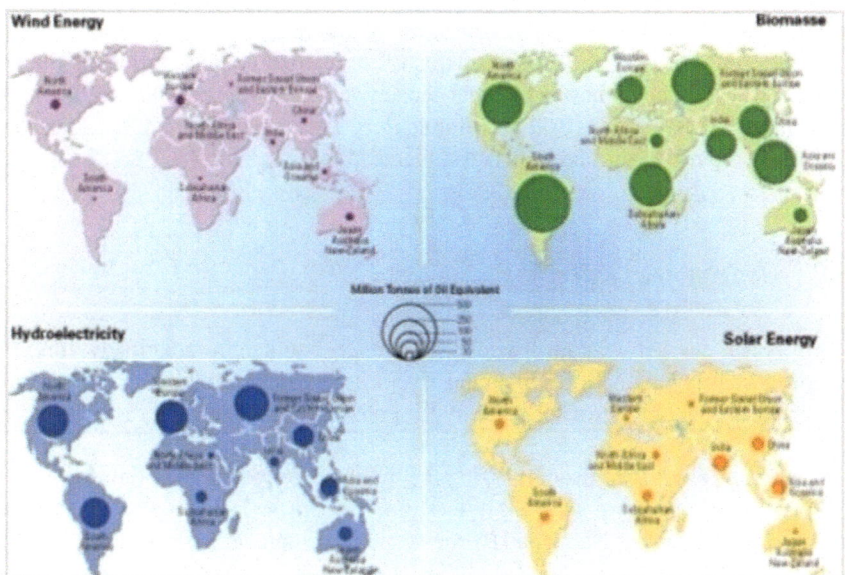

Fig. 4 Potential of renewable energy sources (wind, biomass, hydroelectricity and solar energy) in the world. *Source* Benjamin Dessus, Energies renouvelables, Global Chance, 2005

Fig. 5 Sources of energy and their installed capacity. *Source* MNRE, 2018

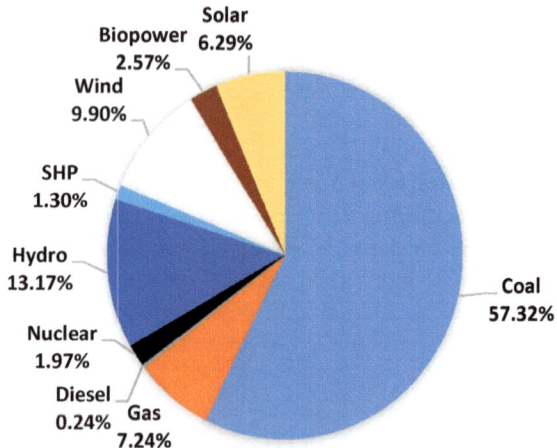

sea/ocean boundary have a very high wind potential. States like Gujarat, Karnataka, Kerala, Andhra Pradesh, Madhya Pradesh, Maharashtra, Orissa, Rajasthan and Tamil Nadu collectively (nine states) possess a potential of more than 48,561 MW, but the installed capacity is 10,891 MW [8, 9].

Within the "Wind Resource Assessment (WRA) Program" under "National Institute of Wind Energy (NIWE)" located in Chennai has an installed capacity of 836 monitoring stations for collecting the wind speed and direction data in 29 states and 3 union territories. NIWE uses advanced micro and meso-scale numerical and flow models validated using the measurements of the meteorological stations at 500 m spatial interval. The highest installable capacity (10,609 MW at 50 m, 35,071 MW at 80 m and 55,857 MW at 100 m) is found to be at Gujarat followed by Karnataka, Maharashtra and Andhra Pradesh (Figs. 6 and 7).

3.2 Hydropower Energy

Hydropower energy is the most widely used energy source in India today. It refers to the water flowing into streams and rivers used to convert kinetic energy into mechanical energy. Out of the total hydro potential available, India is using 17% [10].

India ranks fifth in terms of utilizing the hydro power potential in the world. The estimated hydropower potential of the various basins sum up to a total potential of 148,701 MW as per Narmada Hydroelectric Development Corporation [10]. But the potential of Small Hydro Power (SHP) projects i.e. projects having installed capacity of less than 25 MW, in India, is about 19,749 MW. The installed capacity of hydropower plants was nearly 3746.75 MW in 2016 [11] and is 4418.155 MW

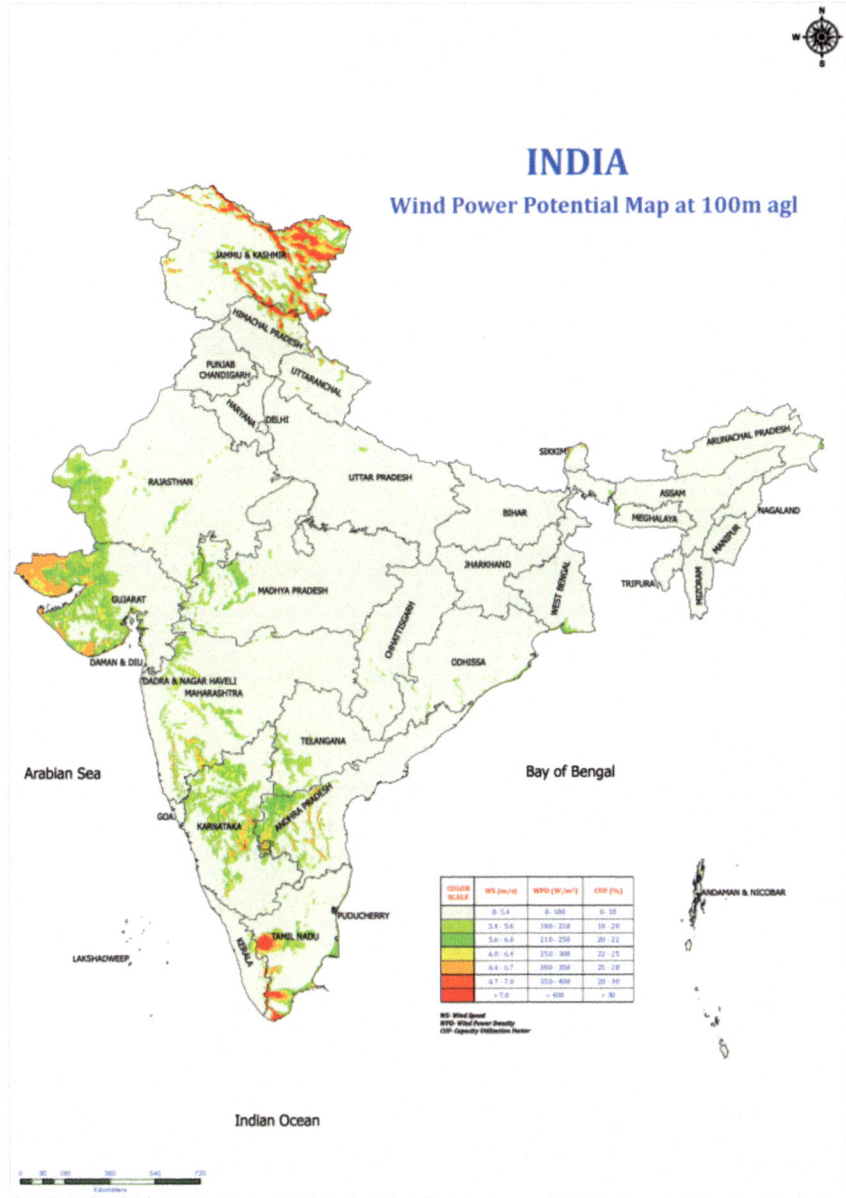

Fig. 6 Wind power density map at 100 m above ground level. *Source* Wind Resource Assessment Unit, Centre for Wind Energy Technology C-WET), Chennai (http://www.cwet.tn.nic.in/html/departments_wpdmap.html#)

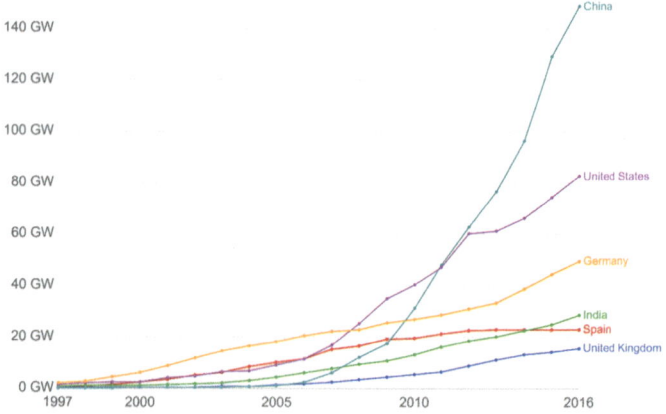

Fig. 7 Cumulative installed wind energy capacity (wind energy capacity including both onshore and offshore wind sources in GW). *Source* BP Statistical Review of Global Energy

with 1089 SHP in 2017 [8]. India has an estimated potential of installing SHP with 21 GW potential from 7,133 locational sites.

3.3   Geothermal Energy

Geothermal energy is contributing nearly 13,000 MW of installed capacity in the world with majority share from United States, Italy, Iceland and Japan [4]. Geological Survey of India identified locations of hot springs (340 in number) and may be utilized in the future.

3.4   Bio Energy

Bioenergy is primarily available through animal and human activities releasing carbon. It can be obtained through wood and related wastes, agricultural residual waste, aquatic waste and sewage/sludge/paper/industrial waste. The world produces around 9,500 TWh of biofuel and represents nearly 15% of the world energy consumption [10]. The total potential estimated for India is about 18 GW using the agricultural and residual waste, 7 GW using bagasse cogeneration (in sugar mills) making a total of 25 GW. India installed 8,414 MW capacity by December 2017 with highest capacity in Maharashtra followed by Uttar Pradesh and Karnataka [8].

3.5 Solar Energy

Solar energy is the most abundant form of energy available on the surface of the Earth. The three components viz. direct (beam), diffused and reflected components are available in plenty. Annual radiation falling on the surface of earth is approximately 3,400,000 EJ which is almost 7,500 times larger than the world's energy consumption from all sources of energy. India, with 300 sunny days, consumes nearly 12 TWh with an installed capacity of 9.01 GW in 2016 [4]. India receives nearly 5,000 trillion kWh/year if captured effectively can suffice the growing energy demand in the country [8].

4 Energy Status and Demand

As per BP Statistical Report on World Energy, India consumed 753.7 Mtoe of fuel using primary energy sources including renewables. The growth rate for energy consumption was observed to be 5.7% during 2006–16 with only 2.89% being contributed by renewable energy. This is a significant growth rate for a developing nation like India. If the world statistics were considered, India is one of the countries with least per capita consumption of energy (630 kWh/person/year), nearly one third of the world's average consumption. Currently, with an average Human Development Index, low GDP per capita and very low per capita energy consumption, the position of India is still safe. But, the total amount of energy being consumed by the country is not self-generated. Plenty fuel is being imported from other nations. Of the total fuel import, 86% crude oil, 37% natural gas and 22% coal is imported from Saudi Arabia, Middle East, and America. This is a matter of concern for the nation.

India is 31.2% urbanised today with more than one-thirds of the population without access to electricity. With an annual growth rate of 1.59%, India's population is also growing at a very fast pace. India is considered to be one amongst the fastest growing economies, only second after China with 9% GDP growth rate in 2007–08, 6.5–7% growth rates during 1990–2016, reviving growth rates due to economic changes in the last few years in the country and expected growth rates during 2016–40. It is also shocking to realise the carbon dioxide emissions due to conventional fuel consumption. The recorded carbon dioxide emissions in 2017 was 2,344.2 million tonnes with a growth rate of 6% in the last decade (2006–16). Statistics reveal that this is one of the highest growth rates in carbon emissions in the world. This clearly indicates that the nation could not explore the potential of abundant renewable energy sources present in the country.

BP Energy Outlook, 2018 reveals that the demand of energy is expected to increase manifolds in developed and developing nations like Africa, India, China and in 34 OECD (Organisation for Economic Co-operation and Development) countries, it shows a steadily declining trend after 2020. Two-thirds growth in energy consumption can be witnessed in India, China and other Asian countries (Fig. 8).

Fig. 8 Share of primary
energy in India (1990–2040
projected). *Source* BP
Energy Outlook, 2018

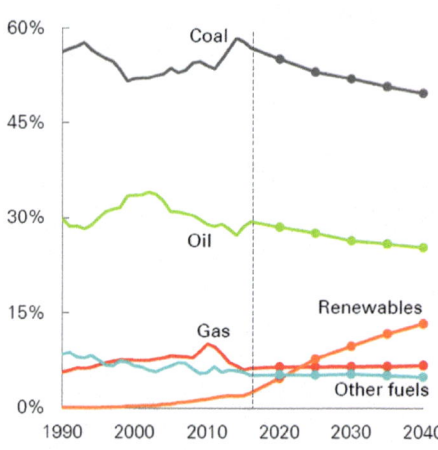

5 Solar Energy Potential in India

The rising trends of energy demand in the country poses a huge requirement for development of energy forms dispensed at affordable rates. *Solar Energy* is the most effective clean energy alternative available in the world today [12]. India lies between the Tropic of Cancer and the Equator (between 40°S and 40°N). The energy produced by tapping solar insolation over 1 km^2 of area is equivalent to burning 0.5 L of kerosene or 1 kg of coal. Moreover, it is a sustainable, energy efficient and clean source of energy. With an area of 3.28 million km^2, India receives 5,000 trillion kWh/year of solar energy equivalent. The average solar irradiance ranges from 4 to 7 kWh/m^2. Predominantly, India is considered to be nation with a very high solar insolation potential with 300 sunny days and approximately 2,300–3,200 sunshine hours per year. About 87.5% of the total land area of the country is used for various land cover/use, while preserving only 12.5% (0.413 million km^2) of the total area for solar power tapping.

Based on availability of land, the total assessed potential of solar power in the country is 750 GWp and the installed capacity today is 17052.37 MW as on December, 2017 [8]. National Institute of Solar Energy (NISE) assessed the potential of all states in India and calculated the state wise solar energy potential of the country (Table 1). NISE estimated the potential for all states in 2014–15 using wasteland data from Wastelands Atlas of India 2010, *Department of Land Resources, Ministry of Rural Development and National Remote Sensing Centre, ISRO*. The average area considered for installation of PV is 3% considering panel efficiency of 15%. Apart from wastelands, rooftop potential is calculated using data of urban area rooftops from Census of India 2011, Ministry of Home Affairs.

Table 1 Estimated solar energy potential in India (state-wise)

S. No.	State/UT	Solar potential (GWp)
1	Andhra Pradesh	38.44
2	Arunachal Pradesh	8.65
3	Assam	13.76
4	Bihar	11.20
5	Chhattisgarh	18.27
6	Delhi	2.05
7	Goa	0.88
8	Gujarat	35.77
9	Haryana	4.56
10	Himachal Pradesh	33.84
11	Jammu & Kashmir	111.05
12	Jharkhand	18.18
13	Karnataka	24.70
14	Kerala	6.11
15	Madhya Pradesh	61.66
16	Maharashtra	64.32
17	Manipur	10.63
18	Meghalaya	5.86
19	Mizoram	9.09
20	Nagaland	7.29
21	Odisha	25.78
22	Punjab	2.81
23	Rajasthan	142.31
24	Sikkim	4.94
25	Tamil Nadu	17.67
26	Telangana	20.41
27	Tripura	2.08
28	Uttar Pradesh	22.83
29	Uttarakhand	16.80
30	West Bengal	6.26
31	Other UTs	0.79

Source National Institute of Solar Energy, MNRE Annual Report 2017–18

6   Government Initiatives for Utilising Solar Energy Potential

Ministry of New and Renewable Energy (MNRE) in support with five institutions viz. Indian Renewable Energy Development Agency (IREDA) and Solar Energy Corporation of India (SECI) and three autonomous bodies- National Institute of Solar Energy (NISE), National Institute of Wind Energy (NIWE) and National Institute of Bio Energy (NIBE) implement various programs and activities through State Nodal Agencies (SNA), Public Sector undertakings, Central and State Electricity Authority, Institutions of Research and other government departments.

6.1   Jawaharlal Nehru National Solar Mission (JNNSM)

After the National Action Plan on Climate Change (NAPCC was announced in June 2008, Jawaharlal Nehru National Solar Mission (NSM) was launched on January 11, 2010 for implementing the development of 20 GW solar power by the year 2022. Later the target of 20 GW was revised to 100 GW. The mission was initially thought to be a three phase approach in which 2009–13 as Phase-1, 2013–17 as Phase II and 2017–22 as Phase III. It aims at generation capacity addition from solar technology (solar photovoltaic and thermal) in both grid connected and off grid connected modes.

6.2   Development of Solar Cities

MNRE launched the program in February, 2008. The purpose of Solar City Program aims at enabling the Urban Local Bodies under the Government to address energy challenges at city level by reducing the demand of conventional energy by 10%. A total of 60 cities/towns with population varying between 5 and 50 lakhs were proposed for the development. As of now, 46 cities were finalized, 39 cities have approved master Plan and 30 cities have solar city cell established (Table 2). 8 cities are to be developed as 'model solar cities' while 15 cities are to be develooped as 'pilot solar cities' [7, 13]. The cities with solar city master plan are mentioned in Table 2.

Under the same program Green Campus development is proposed for 13 campuses in the country namely [14]:

- Silver Jubilee Campus of Pondicherry University;
- Auroville Campus (Puducherry)
- Dayalbagh Nagar Panchayat
- School of Planning & Architecture (New Delhi)
- Malkapur Nagar Panchayat
- KIIT University

Table 2 Cities with sanctioned solar city master plan under solar cities program

State	City	Solar city cell
Andhra Pradesh	Vijayawada	Yes
Assam	Guwahati	No
	Jorhat	Yes
Arunachal Pradesh	Itanagar	Yes
Chandigarh	Chandigarh	Yes
Chhattisgarh	Bilaspur	Yes
	Raipur	Yes
Gujarat	Rajkot	Yes
	Gandhinagar	Yes
	Surat	Yes
Haryana	Gurgaon	No
	Faridabad	Yes
Himachal Pradesh	Shimla	Yes
	Hamirpur	Yes
Karnataka	Mysore	Yes
	Hubli-Dharwad	No
Maharashtra	Nagpur	Yes
	Thane	Yes
	Kalyan-Dombiwali	Yes
	Aurangabad	No
	Shirdi	Yes
Madhya Pradesh	Gwalior	Yes
	Rewa	Yes
Manipur	Imphal	Yes
Mizoram	Aizawl	Yes
Nagaland	Kohima	Yes
	Dimapur	No
Odisha	Bhubaneswar	No
Punjab	Amritsar	No
	Ludhiana	Yes
Rajasthan	Jodhpur	No
Tamil Nadu	Coimbatore	No
Tripura	Agartala	Yes
Uttarakhand	Dehradun	Yes

(continued)

Table 2 (continued)

State	City	Solar city cell
	Haridwar and Rishikesh	Yes
	Chamoli-Gopeshwar	Yes
Uttar Pradesh	Agra	No
	Moradabad	Yes
West Bengal	New Town Kolkata	Yes

Source Background paper, National Meet on Solar Cities

- Tezpur University
- Indian Institute of Engineering, Science & Technology (BESU)
- Jadhavpur University
- Writers Building (Kolkata)
- Madan Mohan Malaviya University of Technology (Gorakhpur)
- Orissa University of Agriculture and Technology (Bhubaneswar) and
- National Institute of Technology (Hamirpur)

6.3 "Solar Parks" and "Ultra Mega Solar Power Projects"

This scheme came into existence in December, 2014 as an initiative floated by MNRE. The scheme envisages to support the states in setting up the solar parks with sufficient communication system, road connectivity, transmission system, etc. as a part for setting the Solar Power Project. Initially, 25 Solar Parks and Ultra Mega Solar Power Projects were planned with 20,000 MW solar potential installed capacity. Later in 2017, the potential was revised to 40,000 MW from 50 solar parks. 35 solar parks In 21 states/UT have been granted approval with a capacity of 20,514 MW.

6.4 Other Schemes and Initiatives

- *300 MW of Grid Connected Solar PV Power Projects by Defence Establishments Under National Solar Mission (NSM)*: This scheme was an initiative for establishing solar power projects over rooftops and vacant lands in cantonment and Military Stations bearing a potential of 500 MW and Ordnance Factory Boards with 950 MW potential in January, 2015.
- *Grid Connected Solar PV Plants on Canal Banks and Canal Tops*: "Pilot-cum-Demonstration Project for Development of Grid Connected Solar PV Power Plants

on Canal Banks and Canal Tops" is planned under NSM with a target capacity of 50 MW each on canal bank and canal top.

- *1,000 MW of Grid Connected Solar PV Power Project by CPSUs and Govt. Organizations*: Launched in January 2015 for 1,000 MW capacity. MNRE has allocated 963 MW capacity to different CPSUs.
- *1,000 MW Capacity Grid Connected Solar Power Projects Implemented Through NVVN Under NSM*: 500 MW capacity each of Solar Thermal (ST) and Solar Photovoltaic (SPV) technology to be implemented.
- *Viability Gap Funding Scheme (VGF):* Under this scheme, Grid connected solar power projects of capacity 750, 2,000 and 5,000 MW are taken up by Solar Energy Corporation of India (SECI). Under 750 MW project, 680 MW capacity plants are commissioned in 7 states including Rajasthan, Gujarat, Maharashtra, Madhya Pradesh, Karnataka, Tamil Nadu and Odisha).
- *Solar Generation Based Incentive (GBI)*: The scheme was formulated for 25 MW capacity for demonstration purpose in 2008–09. Later in 2010, 100 MW scheme was allocated.
- *Grid Connected Rooftop and Small Power Plants Programme*: This program provides subsidy up to 30% for all states except North-eastern states with 70% subsidy for installing grid connected rooftop solar power plants in residential, public and institutional land use buildings. The estimated target for installation is 4,200 MW by the year 2019–20. Online calculator and solar rooftop calculator has also been developed for grid connected rooftops for entire country.
- *Mobile App for Solar Rooftop Systems*: Mobile app called **Atal Rooftop Solar User Navigator (ARUN)** was launched on January 24, 2017. It helps in understanding the about solar rooftop installation methods based on capacity and budget.
- *Initiatives by State Government*: Many states in the country (early 20 including Andhra Pradesh, Chhattisgarh, Delhi, etc. have solar policy for grid connected rooftop systems. States like Chattisgarh, Haryana, Chandigarh and Uttar Pradesh have mandated the installation of PV panels on rooftops in certain category of buildings.
- *Off Grid Solar Photovoltaics (SPV)*: Under the Decentralized programme, Ministry is deploying financial assistance called Central Financial Assistance (CFA) for lighting systems, solar street lights (under Atal Jyoti Yojana), solar pumps, etc.

7 Role of Geospatial Technology in Solar Potential Assessment

Solar energy is the most sustainable form of energy source due to its abundance in nature. Solar energy is clean and safe and hence its demand for sustainable urban development is increasing manifolds. Over the years, the technology using the PV (photovoltaic) effect for the production of electrical power has progressed immensely. Private investors and local authorities in the Government are showing keen interest in capturing the solar potential. Solar Energy is said to provide 2,850 times potential than the current global energy needs. Rooftop PV (RTPV) system leaders in the world were Japan, USA and Germany and recent growth is witnessed in countries like Italy, Australia and China [2].

Interestingly, the solar radiation incident on a particular surface can be estimated using remote sensing (RS) and geographical information system (GIS) by a number of methods. Several studies have tried to use remote sensing technology in collaboration of GIS for estimating the direct normal irradiance (DNI) or solar insolation [3, 15–21].

7.1 Solar Energy Potential Feasibility in India

Solar rooftop PV in India is not developed much but is maturing at a steady rate over the past few years. As per the 2011 census statistics, India has around 330 million houses of which 150 million houses have proper roofs i.e. are made up of cement concrete finishing. Apart from such buildings various institutional buildings can also be considered while installing solar panels. With the help of such rooftops an average of 1–2 KWp of solar PV system can be accommodated. Even accounting for average 1 kWp installation at around 30% households i.e. 150 million, the potential can reach up to a range of 45,000 MW. Besides such kind of buildings, commercial buildings, shopping complexes and offices can also accommodate large solar PV capacity. This can be helpful in increasing the potential manifolds. It can be made mandatory in development Control regulations (DCR) and Building Bye Laws that all households and Government/institutional buildings shall install rooftop panels. The rooftop installation has many benefits including reduction in additional usage of space, reduction in cost, value addition to otherwise unusable roof area, reduced dependency on grid power, etc.

Apart from rooftops, our country also possesses great network of canals which includes open main canal and its other tributary sub branches. Utilizing this particular network by instantly covering it by solar panels. It has immense potential and can provide GW scale potential in the country throughout. It has another advantage as no other area is further required for solar panel installation and limits evaporation loss of water from canals and hence tackling with twin challenges simultaneously providing security of water and providing energy. Besides all this, a huge quantum of wastelands is available in the Indian deserts. These wastelands offer maximum renewable potential supporting infrastructure ultimately fulfilling the demand of many. Considering the total potential of the country three areas of importance have been identified to harness huge renewable potential viz,

- Utilization of roofs/available open space with individual houses in form of distributed generation
- Canal top solar PV which additionally helps in saving of water from evaporation
- Wasteland utilization of various areas within the country like deserts.

7.2 Solar Energy: Components

Energy from the sun is received by the earth in half of the original intensity due to scattering, absorption and reflection in various atmospheric layers. The total flux is a combination of longwave and shortwave radiation. Only the radiant energy flux at a given place and time can be measured readily. There is heterogeneity in the solar radiation components due to factors like earth's inclination, surface orientation, topography and slope. Primarily, if the sky conditions are clear (clear sky), irradiation from sky is uniform (Fig. 9).

The total shortwave radiation emitted by the sun passing through the atmosphere (scattered and diffused) over a unit area in the horizontal and unit time is referred as *global solar radiation*. The total global solar radiation is divided further into two components called direct (beam) or diffuse. Conversely, the mathematical sum of the direct and diffused component is referred as the global radiation. *Direct radiation* or beam radiation is the component of radiation that travels through the atmosphere and space to the surface. *Diffused radiation* is that part of the global radiation which is scattered by the atmospheric constituents like clouds, aerosols and water vapor [3, 22, 23].

Another important component of the solar radiation is the reflected component. Approximately, 30% of the total radiation is reflected back to space and other remaining 20% is absorbed by clouds, dust, gases, water vapor and ozone and the remaining is received by the surface of the earth. The relationship between sunshine (N) with solar elevation <5° and daily total global radiation (G) is given by:

$$G/G_0 = a + (1 - a)\left(\frac{n}{N}\right) \tag{1}$$

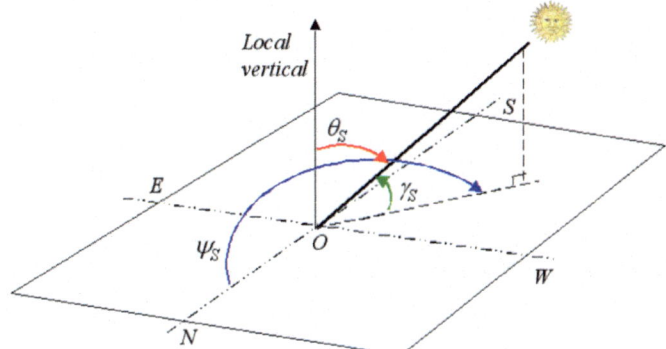

Fig. 9 Sky viewed from an observer O. θ_S—solar zenith angle, γ_S—solar elevation angle and Ψ_S—the azimuth. N, E, S, and W are the directions

where, G_0 is the cloud free daily global solar radiation, a is mean proportion of radiation received on a complete overcast day. G_0 is also replaced by "extra-terrestrial radiation (ETR)" In $kWh/m^2/day$ on a horizontal surface. In such a scenario:

$$G/ETR = a + b\left(\frac{n}{N}\right) \tag{2}$$

Also, diffused radiation (D) can be determined as:

$$D/G = c + d\left(\frac{G}{ETR}\right) \tag{3}$$

where, c and d are regression constants. Direct solar radiation can also be estimated by:

$$G_H = G - D \tag{4}$$

All the components of the solar radiation can be estimated using remote sensing using various empirical and physical models.

7.3 Factors Affecting Solar Insolation

Estimating energy received by a surface is essential for further estimating the potential in the area of concern. Radiation datasets based on satellite measurements have high spatial and temporal frequency compared to on site surface measurements using pyranometers and pyrheliometers. The sun-earth distance or the solar constant is a constant value i.e. 1,366 W/m^2 ±0.3% of which only half reaches the surface due to variations in intermediate interactive surfaces between the surface and atmosphere profile [3]. Roughly, the distance from sun, relief, terrain, aspect, rotation, tilt, etc. can be few considerable factors affecting the insolation. Accurate insolation data is required for many surface and landscape processes including climate, energy balance, snowmelt, runoff and moisture studies. It is hence, important to understand factors affecting the radiation received.

- **Shadow**: Factors like surface orientation, land surface gradient and natural topography has impact on the solar insolation due to varying elevation values and subsequent shadow effects. For installing PV panels in the northern hemisphere, a roof exposure onwards the south is optimal and in the southern hemisphere, north orientation is preferred. Sites with more than 2% slope can lead to shadow effects from panels themselves, whereas, 5% graded slope can lead to shadow effects. Hence, topography should be accounted for while installing PV panels and multiple rows of panels to avoid shade effects and maximize the solar exposure thereby increasing the potential [24, 25]. Trees, topographic and structural shading should be avoided while considering the installation of PV systems [26].

- **Tilt**: According to the geographic location and time period, the efficiency of panel can be improved using optimal tilt angle with the horizontal. The seasonal variation in tilt angle is advised based on latitude angle (φ) $\pm$ 15°, while the tilt angle for the entire year is equal to approximately (0.9 $\times$ φ) [27–30].
- **System performance factors**: Factors like transmission losses, distance of sub-stations to the PV installation site, snow cover, overcast and rains, dust and dust storms, irregular cleaning of PV panels, etc. are some of the major factors affecting the performance and results [24, 31].
- **Locational factor**: The location of the panels should be chosen in such a way so as to minimize the effect of water pollution risk and ample water availability for temperature cooling and cleaning.

Hence, site selection can b attributed to various factors like slope, aspect, direct Normal Irradiance (DNI), proximity to roads minimizing cost of construction and maintenance, distance to transmission line, dust potential zones, access to water source, environmental sensitivity, tilt, land cover/land use (LCLU), weather, season, rainfall, temperature, distance from river, etc.

8 Estimation of Solar Insolation

The solar potential of an area or the solar radiation and its components (direct, diffused and reflected) can be estimated using various techniques like:

- Using in situ direct measurements from dispersed weather stations measuring solar radiation
- Data using specially designing statistical and physical models using satellite derived parameters.

8.1 Direct Measurements

The solar radiation can be measured directly at weather stations using accurate radiometer instruments like *pyranometers and pyrheliometers*. Pyranometer is an instrument that is used for measuring hemispherical solar radiation. When measuring in the horizontal plane, the radiation is known as global horizontal irradiance (GHI). Reference GHI is calculated from reference direct normal irradiance (DNI) and reference diffuse horizontal irradiance (DHI). The receiving end of the pyranometer senses the heat energy converted by the incident solar radiation energy from the whole sky. Pyrheliometer is used for measuring the direct or beam component of the total global radiation from the sun. For correct measurement of direct solar radiation, the receiving surface is arranged normal to the direction of radiation by mounting it on a sun tracking device called *equatorial mount*.

In India, 45 solar radiation stations are mounted as on date. But, the sparse representation of pyranometer network in the country is insufficient in providing complete coverage and solar radiation variability. Apart from the cost of installation and limited network constraint, the measurements are prone to certain radiometer measurement errors due to sensitivity to wavelength, temperature, elevation and azimuth, and field of view. Hence, the instruments measure the radiation and the values are assigned either in actual numbers or interpolated using various interpolation techniques. The interpolation technique does not work well if number of points are insufficient. Therefore, the direct measurement technique holds good when working on sparse representation or large spatial scales. The pyrometer network is better for validation purpose [12, 32].

8.2 Modeling Approach

The most important factor influencing the energy potential estimation is the incoming solar insolation. Solar radiation is dependent upon the earth's rotation and revolution around the sun [33–35]. It is estimated that data collected though weather stations having a spatial resolution of few kilometers can have an RMSE of up-to 25% [36]. Another mode or attempt for obtaining solar radiation can be through modeling solar radiation using models. Literature suggests that the modeling for solar radiation at a given point can be divided on the basis of nine classification systems proposed below by [32]:

- **Type of output data**: Generally, the three components of radiation are direct, diffused and global. Many models derive or model for global irradiance and use this input for modeling the direct and diffuse component. The global radiation data ca be made available using satellite measurements
- **Type of Input data**: The input datasets required for the models may vary from meteorological to climatological components. The data can be retrieved using airborne/space-borne sensors
- **Spatial resolution**: Such models predict over specific location where inputs have been made available as point information, while others provide gridded data as output. The output may have greater spatial resolution to greater spatial coverage.
- **Temporal resolution**: The radiation output can be of varying time intervals varying from every minute, hourly, daily, season wise or a specific year and series of years. If analyzing the solar irradiance data for setting up Solar concentrators, high resolution or hourly data is needed.
- **Spectral resolution**: The solar shortwave radiation is modeled by the models within the range of 300–4,000 nm (0.3–4 μm). Certain models focus on the parts of the wavelength spectrum i.e. ultraviolet or photosynthetic.
- **Methodology based**: The methodology for the modeling approach can be statistical or deterministic. The statistical model or the stochastic model is based on modeling solar radiation based o some statistical properties of statistical time series

measures of radiation like mean, standard deviation, frequency distribution, etc. A deterministic model, models the past, present or future scenario of irradiance. It is also possible to find a combination of models having both stochastic and deterministic approach.

- **Type of algorithm**: physical or semi-physical models are derived directly from physical principles for a specific location and period.
- **Surface geometry**: the modeling requirement can be over topographically complex terrains or over horizontal/tilted plains.
- **Type of sky**: Sky conditions can either be clear or cloudy sky. Hence, modeling takes into account the clear sky and the presence of clouds. Some of the clear sky models are given in ASHRAE, Bird, Cloud Layer Sunshine (CLS), CPCR2, ESRA 2, Iqbal parametrization models, Kasten, METSAT, Mc Master (MAC), Meteorological Model (MRM5), REST2, Santamouris, Yang, etc.

Certain combinations of factors can be made for accessing the desired results. Certain parameters like linke turbidity etc. can be added for better results into the model.

8.3 Satellite Observations

The capability of geostationary satellites (at 36,000 km altitude) to capture information at a very high frequency of time (every 30 min or so) make it suitable for measuring the radiation through onboard radiometers. These radiometers aboard the geostationary satellite is capable of measuring the solar radiation reflected by the earth's surface either in the presence of clouds or no cloud mostly in the visible spectrum at 3–5 km spatial resolution. Some of the satellites providing the data for entire earth (sometimes excluding poles) include METEOSAT/SEVIRI mission (European), Geostationary Operational Environmental Satellites (GOES, American), The Indian National Satellite (INSAT), Chinese Fengyen and the Japanese Geostationary Meteorological Satellite (HIMAWARI) [37, 38]. Polar orbital satellites like NOAA/AVHRR also deliver the data but better temporal resolution and coverage of geostationary satellites deliver solar radiation at enhanced accuracy than other satellite missions possible.

8.4 Solar Radiation Database

There are many projects that started for extracting solar radiation around the globe. Global insolation high resolution datasets are available in the world like NASA SSE Global insolation datasets derived using physical models based on radiative transfer including scattering and absorption features. It provides $1° \times 1°$ spatial resolution for the entire world [12]. Some other datasets include ENMETSol (Germany), ESRA

(France), Helioclim-2,3 (France), Meteonorm 6 (Switzerland), PVGIS (Europe), SATL-LIGHT (France), SOLEMI (Germany), SODA (Europe), etc. to name a few.

Ramachandra et al. (2011) studied the database for India from NASA SSE product on more than 900 grids. A bilinear interpolation technique was used to extrude the monthly average Global insolation maps detailed with isohels having same radiation or equal radiation value. The study concluded that the Gangetic plains, plateau regions, Western dry areas, Ghats and coast and hilly regions receive annual global insolation of more than 5 kWh/m^2/day. These regions include states like Karnataka, Gujarat, Andhra Pradesh, Maharashtra, Madhya Pradesh, Rajasthan, Tamil Nadu, Haryana, Punjab, Kerala, Bihar, Uttar Pradesh and Chattisgarh and hilly terrains of few regions in Ladakh, Himachal Pradesh and Sikkim. Other states like Arunachal Pradesh, Nagaland and Assam also receive 4 kWh/m^2/day of insolation approximately.

8.4.1 Solar Resource Maps for India

For enhanced solar resource information in the country, some of the initiatives taken up by the Solar Energy Centre (SEC) are SEC-NREL collaborative project on "Solar Resource Assessment". Based on a bilateral partnership between India and US, National Renewable Energy Laboratory (NREL) has developed solar maps for India of last 15 years (2000–2014). The 10 × 10 km hourly solar resource map was developed using METEOSAT (weather satellite) measurements using solar modeling approach. In first phase, the maps were generated for 2002–2007 and were later updated in 2014. The dataset also contains enhanced aerosol (dust, smoke, haze, particulates) information for improved results. The products delivered by NREL available in two formats—maps and GIS database are:

- **Global horizontal irradiance (GHI)** or the total amount of shortwave radiation received from above by a surface horizontal to the ground. This value is of particular interest to photovoltaic installations and includes both Direct Normal Irradiance (DNI) and Diffuse Horizontal Irradiance (DHI)—annual average and monthly average (Fig. 10).
- **Direct normal irradiance (DNI)** or the amount of solar radiation received per unit area by a surface that is always held perpendicular (or normal) to the rays that come in a straight line from the direction of the sun at its current position in the sky—annual average and monthly average (Fig. 11).

8.4.2 The National Solar Radiation Database (NSRDB)

NSRDB is a database made by collection of hourly and half-hourly radiation datasets of global, direct and diffused horizontal irradiance and other meteorological products. This dataset is useful for regional scale representation of climate. The number of locations at which this data is made available varies for India. This dataset can

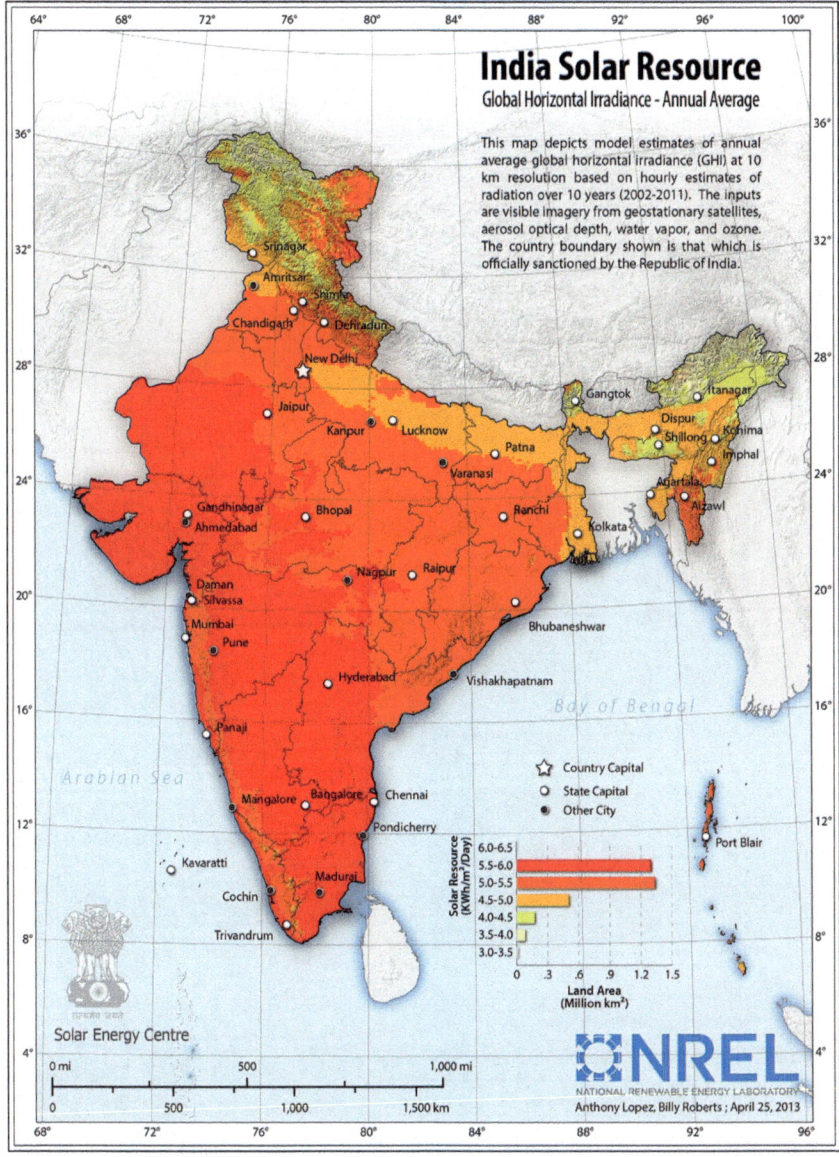

Fig. 10 Indian solar resource: global horizontal irradiance, annual average. *Source* NREL

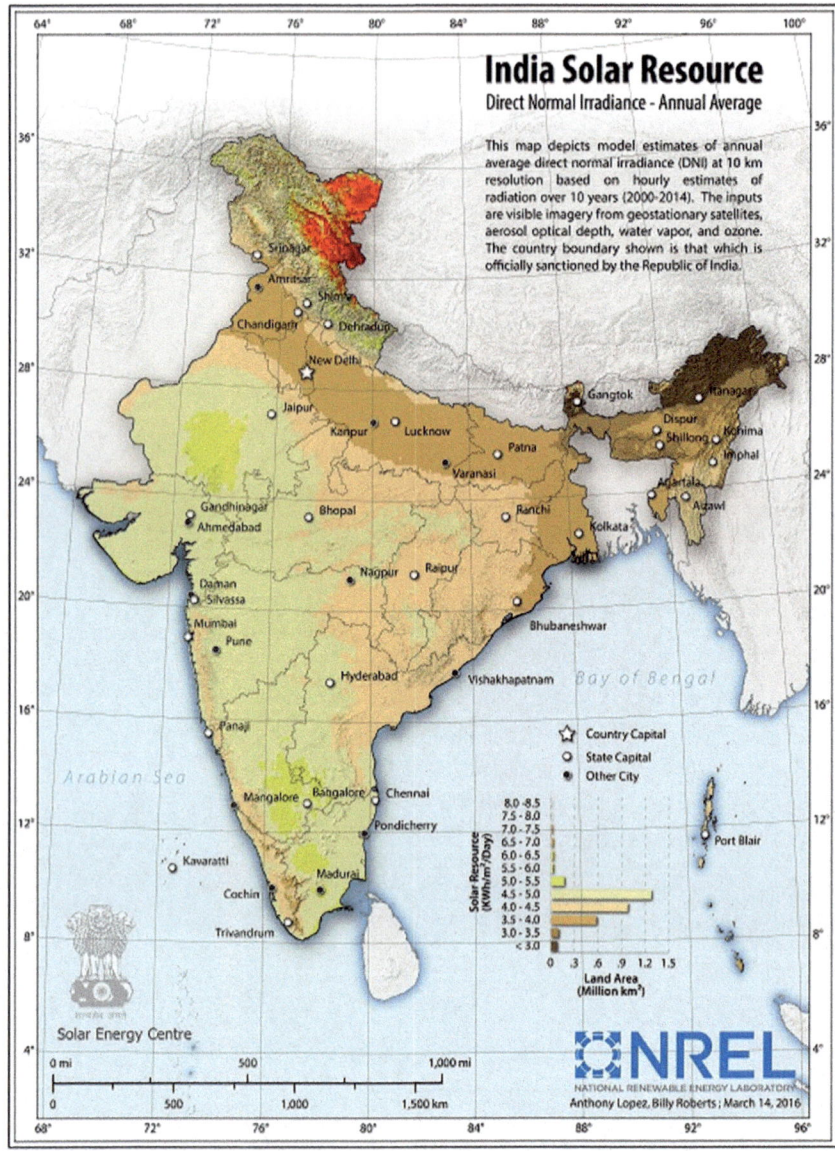

Fig. 11 Indian solar resource: direct normal irradiance, annual average. *Source* NREL

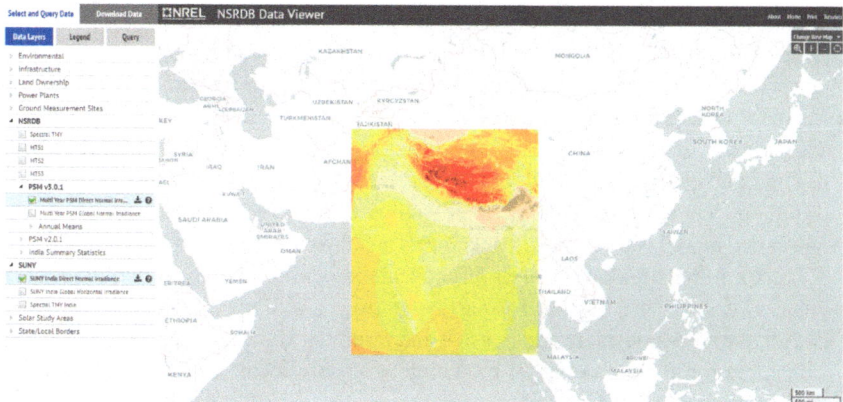

Fig. 12 NSRDB viewer: web-based GIS application for visualising and downloading solar resource data. *Source* www.nrel.gov

readily be used by numerical prediction models for predicting the solar radiation of the future in a desired location based on certain past conditions available (Fig. 12).

8.5 Solar Radiation Models in GIS Platform

Several empirical models are made using the geographical information system platform tools. GIS platforms help in faster processing, integration of all ancillary important considerations. Some of the most important GIS models available for use today are:

- **Solar Flux**: It is one of the most original and early models developed on ArcINFO platform by Hofierka and Suri. This model provided the platform for incorporating various surface parameters for calculating the solar radiation values through time like elevation, spatial coordinates, time, atmospheric parameters, etc. [39].
- **GIS Genasys**: Solar radiation algorithm developed commercially using AML script [34].
- **Solei in GIS IDRISI**: The model was developed and linked to the IDRISI software platform which used the elevation details through DEM [40].
- **Solar Analyst in ArcView GIS**: Solar Analyst is an extension to the Arc GIS module. It generates the hemispherical viewshed based on which the solar potential is estimated through the model. Solar Analyst is capable of detailed studies while incorporating detailed topographical parameters viz. DEM, sky size, coordinates of the area under consideration, time of the day and month of the year, transmittivity, etc. using area and point radiation tools [41].

- **SRAD**: The model was designed for mesoscale processes using a variety of landscape factors into consideration. It is used for shortwave and longwave solar radiation on the ground and the atmosphere from one day to a year [42, 43].
- **r.sun in GRASS GIS**: Suri and Hofierka [35] in 2004 developed the GIS model called r.sun. It is a model that runs in the open source platform of GRASS GIS. R.sun enables users to calculate direct, diffused and reflected component using a standard clear sky model unlike Solar Analyst tool that works on the hemispherical viewshed model and calculated only direct and diffused radiation. It helps modelers to perform the simulations at different scales (local to global) in two modes viz. rasters of solar irradiance at a particular time and daily raster maps which is a cumulative value of irradiances. It incorporates the shadow and facades into consideration for the radiation modeling at a more local scale with LiDAR data inclusion also [40].

Some of the models were applicable over smaller areas like Solar Flux, Solei and Genasys while others were applicable over larger areas and incorporated additional topographical and elevation parameters into consideration. Some open source models were also built in the Web-GIS tool format by The Energy Research Institute (TERI) for estimation of solar rooftop potential of Chandigarh area. The initiative was also collaborated with Shakti Sustainable Energy Foundation (SSEF). The model helped in estimating the roof-wise radiation considering the pre-modeled inputs saved by the modeler.

9   Solar Potential Estimation in Gandhinagar

Gujarat has always remained the leaders of using the cutting edge technology. They have remained in the forefront to use renewable energy source in the form of Gujarat Solar Park administered through Gujarat Power Corporation Limited (GPCL) in an attempt to use clean energy and reduce the climate change impact. 590 MW capacity of Solar Park is commissioned at District Patan in 5,384 acres of unused land. It also has a capacity to generate 100 MW of wind power. Additionally 1 MW canal top Solar Power Project was also implemented on he Sardar Sarovar Project's Sanand Branch Canal. The solar panel over dams is a revolutionary idea that helps in reducing water evaporation due to radiation and helps in generating 1.6 million units of "Clean and Green Energy".

9.1   Gandhinagar: Profile

Located at the bank of River Sabarmati, Gandhinagar was planned in 1960 as a new town and capital of Gujarat State bearing an area of 205 km^2. Gandhinagar is only 25 km away from Ahmedabad and has been approved as a city by the Ministry

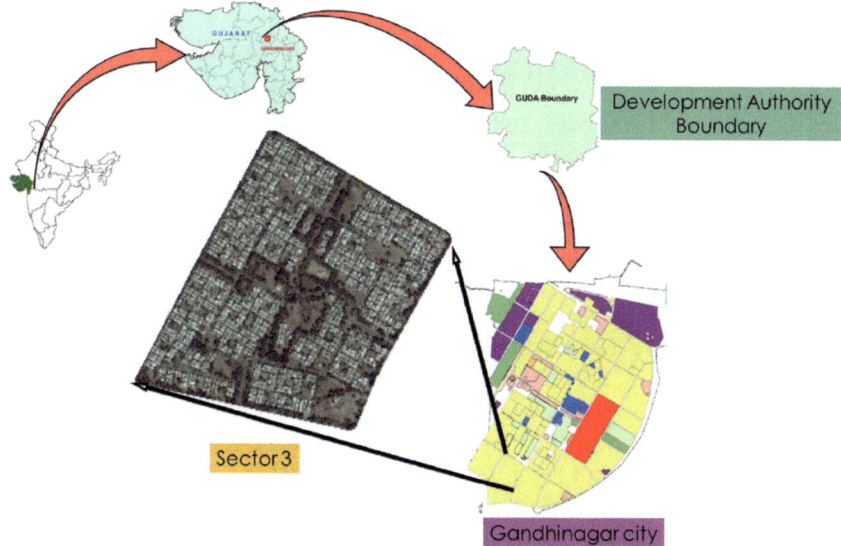

Fig. 13 Study area (India: Gujarat: GUDA: Gandhinagar city: sector 3)

of New and Renewable Energy (MNRE) for solar city development. Gandhinagar is a planned development and can be regarded to be a well-integrated structured development. The city has been divided into 30 sectors or grids. Each sector has its own facilities and showcases an extensive green planning. Standing at an average elevation of 265 ft, the city receives ample solar radiation, is collected to be one of the Smart City, and incorporates green areas (recreational land use as 25%) within the planned city framework [44]. The Gandhinagar area is just 57 km^2 with 43 km^2 as planned sectors (Fig. 13).

The first Model Solar City master Plan was prepared for Gandhinagar under the Solar City Mission by MNRE, Govt. of India. The program's objective was to substantially reduce (10%) the projected energy demand from conventional resources and switch to number of energy efficiency measures including non-renewable energy use. A Solar City cell is also created by the Gujarat Energy Development Agency. The preparation of Master Plan for Solar City in Gandhinagar was a five staged process:

- Preparation of base line for energy 2008
- Demand forecasting for the year 2018/2013 (five and ten years period)
- Strategies for different sectors using techno-economic feasibility survey.
- Developing and Action plan
- Financial Outlay and sharing of fund.

9.2 Gandhinagar: Energy Status

The 400 kV Gandhinagar service station, two 220 Service stations and one 132 kV service station located in the city is sufficient for sufficing the need of power requirement of the city. Additionally, there is also a 66 kV s/s across the district. If the potential of rooftop solar power is tapped, the energy generated by the stations can be routed to the main grid and other un-electrified areas can be given power supply especially using potential from the building rooftops.

Through the main sources of electricity viz. petrol, diesel, LPG, CNG and kerosene, the electricity distribution is zone wise where the city is divided into three zones (Table 3). Government and Industrial sector have the maximum share in energy consumption i.e. 67%. It is also pertinent to note that the residential area consume lesser electricity per plot but the area available with the low rise low density residential areas can provide a platform for usage of its available roof area for generating energy (Fig. 14). As per Census of India, 2001 census, the average electricity consumption per household is 2.68 kWH/day and annual consumption may go to 192 million units.

Table 3 Electricity consumption land use wise

	Electricity consumption (MU)			
Year	2007–08	2008–09	2009–10	2010–11
Residential	40.94	43.24	45.31	49.91
Commercial	12.46	13.61	13.79	15.19
Industrial	62.3	65.8	68.95	75.95
Government	62.3	65.8	68.95	75.95
Total	178	188	197	217

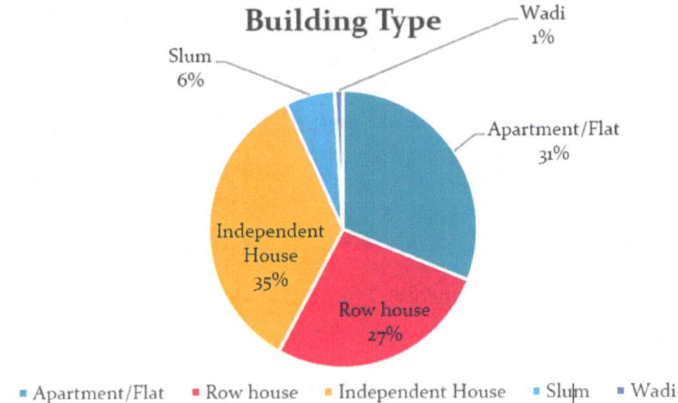

Fig. 14 Building type in Gandhinagar

9.3 Methodology: Rooftop Potential Estimation

The rooftop solar potential was to be assessed for Gandhinagar Sector 3 area in detail. The requirement was to understand the effect of height and shadow on the insolation change in an area. For the same a Worldview 2 stereo data with 0.46 m spatial resolution (at nadir) and 0.52 m spatial resolution (off nadir) panchromatic data was used. The 11 bit data is acquired at an orbital altitude of 770 km. The Digital Surface Model (DSM) was generated using the stereo pair data at 1.5 m spatial resolution. Footprints were extracted using two techniques: automated extraction and manual digitization. The accuracy of manual digitization was recorded better than automated extraction technique. After the DSM was generated in eATE module of LPS in Erdas imagine, the height of each pixel was assessed. The Solar Analyst tool embedded in Arc GIS platform was further utilised for calculating the year round solar potential in Gandhinagar using the dedicated parameters required for the tool (Fig. 15).

9.3.1 DSM

The generation of an accurate DSM is one of the most important steps towards preparing an accurate map for further analysis. The DSM was generated in the LPS module in Erdas Imagine software using 15 ground control points. 10 GCPs dis-

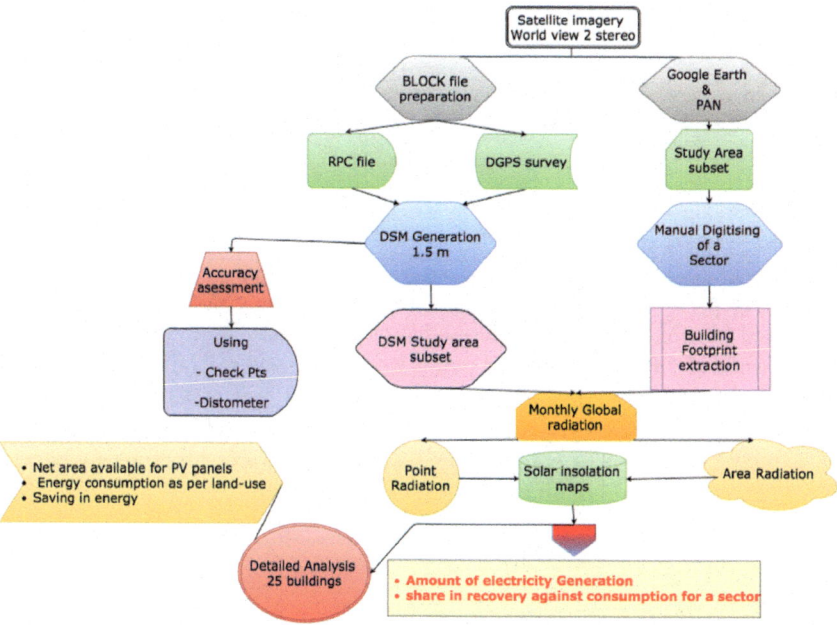

Fig. 15 Detailed methodology

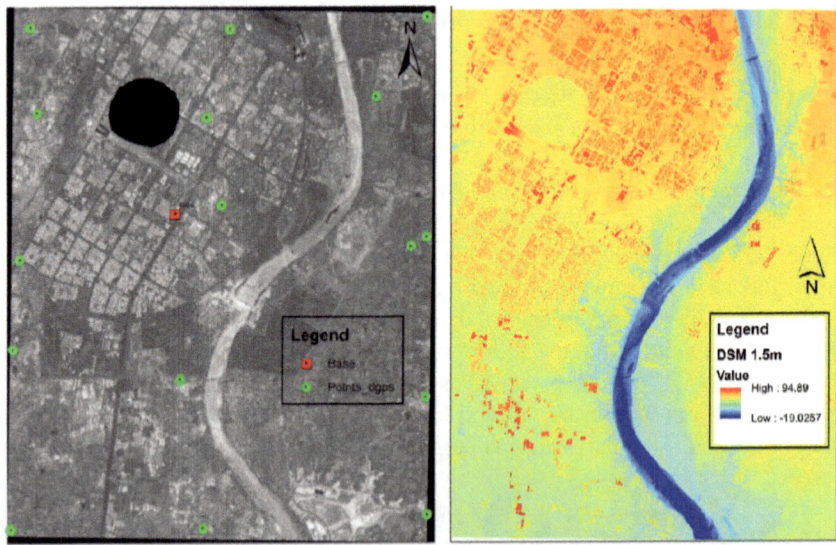

Fig. 16 Location of GCPs and generated DSM for Gandhinagar

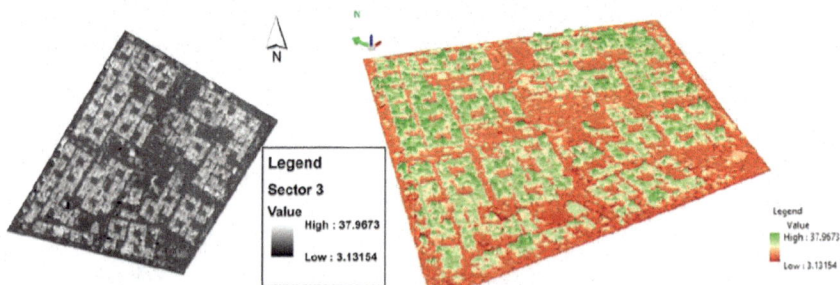

Fig. 17 DSM of study area

tributed evenly over the image for better image matching and better DSM result were used for while the others were used as check points. The GCPs were collected using Trimble DGPS survey and the points were processed in Trimble Business Centre bearing UTM projection, datum as WGS 1984 and zone as 43 N. As a standard procedure, block file was generated and stereo images were added. After adding the GCPs, check points and tie points, triangulation was performed following which the DSM was generated (Figs. 16 and 17).

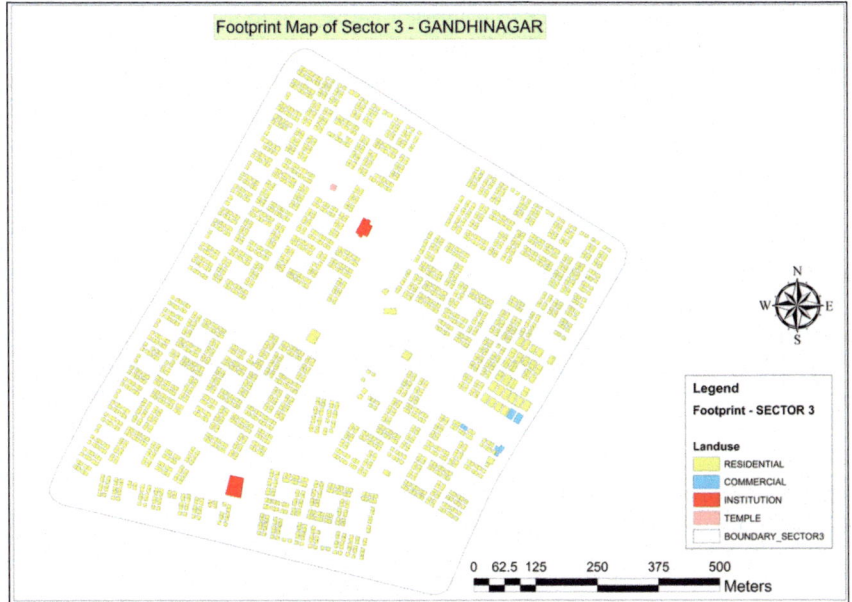

Fig. 18 Footprint map: sector 3: Gandhinagar

9.3.2 Footprint Extraction

The footprint was extracted using two modes i.e. manual digitization and automatic feature extraction using object based method image analysis (OBIA). Nearly 2,368 buildings were digitized manually using the Worldview 2 imagery and supporting data from Google Earth for the entire Sector 3 of Gandhinagar (Fig. 18).

9.4 Solar Analyst Tool: Point and Area Radiation

As discussed in Sect. 6.8.5, Solar Analyst tool is embedded in Arc toolbox. The point and area radiation tools are a part of the Solar Analyst tool. Area Radiation tool calculates the radiation over an entire area/landscape. The point radiation tool derives the incoming solar radiation on specific point locations in a point feature class. Some of the major input parameters required for the model are:

- **Input raster**: The spatial coordinates are fetched by the model with the help of Dem (in the study high resolution 1.5 m spatial resolution WV-2 DSM was used) from which the details of the location is picked. This is the input required in raster format.
- **Input point feature class**: It specifies location based points for further analysis purpose. Here the input is the shape file in vector format.

- **Latitude**: The spatial reference is already embedded in the input raster file and hence is subsequently picked up by the model.
- **Sky size**: Since the model works on hemispherical viewshed algorithm, sky size is the resolution of the viewshed. The hemispherical raster representations of the non-coordinate bearing the square grids of the sky.
- **Hemispherical viewsheds**: A viewshed explains the horizon angle or the angular distribution in sky obstruction calculated for each pixel depicting a region of interest.
- **Sun map**: From each sky direction, the sun map represents the amount of direct solar radiation based on sky sectors.
- **Time**: The time duration is annually or monthly.
- **Zenith and Azimuth**: The number of divisions representing the sky sectors.
- **Transmittivity**: The amount of radiation that can be transmitted through the atmosphere is referred as transmittivity. Value ranges from 0 to 1. Clear sky has default value 0.5.
- **Diffusion proportion**: It has inverse relation with transmittivity.
- **Hour interval**: No. of hour intervals in a day for calculating sky sectors.

9.5 Result and Discussion

The details of the results are discussed as follows:

9.5.1 Accuracy Assessment of DSM Generated

For accuracy assessment of space-based DSM, it is important to know how accurately the satellite image is georeferenced with the rational polynomial coefficients (RPCs) or using control points. RPC help in defining the positions but to a limited extent, hence control points were used for preparing the DSM. The accuracy of DSM is generally defined in horizontal and vertical terms and is improved using Ground Control points (GCPs) using the DGPS survey. Hence there are three methods adopted for accuracy assessment of DSM in both horizontal and vertical terms.

The statistical quantification of height can be done using root-mean-square error (RMSE) or "the square root of the average of the set of squared differences between height values of the DSM being evaluated and height values from GCPs with much higher accuracy". These independent GCPs are referred to as the reference data. The accuracy of a DSM is the closeness to reference data, and it is commonly referred to as "high" or "low" depending on the size of the RMSE value.

In the first process (Fig. 19), check point analysis approach is adopted. A check point compares the computed coordinates through photogrammetric measures on ground to the original values (Ground value). Check points are collected for both the study areas by DGPS survey. Through DGPS survey, total 15 GCPs are collected for Gandhinagar city. All 15 points are used as check point for accuracy assessment

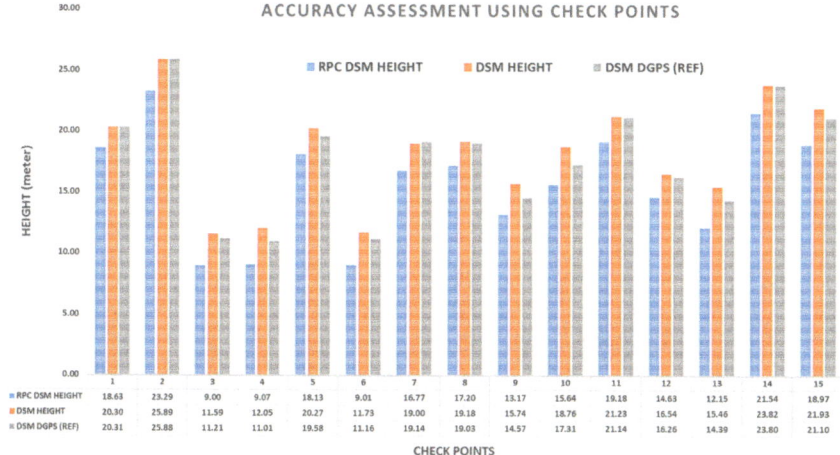

Fig. 19 Check points of DSM: Gandhi Nagar city (method 1)

because we have their z value. After collection of GCPs, processing is required for the removal of the ambiguity. Post processing has done with the help of the base station. After post processing, X, Y, Z of the check point compares with the respective pixel value for assessing the vertical accuracy.

In this method (Fig. 20), Laser distance meter was used for measuring the height of buildings. The obtained building height was taken as a reference data and then it was compared with the building height, computed through DSM. There were 4 building's height measured using Distometer. The building height accuracy was assessed for both GCP and only RPC DSM. Average of 0.80 m bias is coming for DSM made through GCP. Hence, it's also good accuracy for further analysis.

9.5.2   Global Solar Radiation

The selected area shows an increased solar insolation in kWh/day value for the month of May and June (Fig. 21). The lowest radiation is monitored for the month of November due to the presence of haze in the month of November and January. From November to February solar radiation is lowest as these months fall under winter season similarly estimation of radiance from April to August on Buildings roof top is higher. The direct component is 78% and diffuse radiation component is 22%. We can conclude that diffuse component also affects radiation. Reflected component is eliminated in solar radiation tool of ArcGIS because of too much low values.

	1	2	3	4
RPC DSM NET HEIGHT (TOP-GL)	6.97	10.43	11.91	7.10
DSM NET HEIGHT (TOP-GL)	7.97	12.32	10.76	9.74
DISTOMETER HEIGHT (REF)	9.50	12.50	9.50	9.50

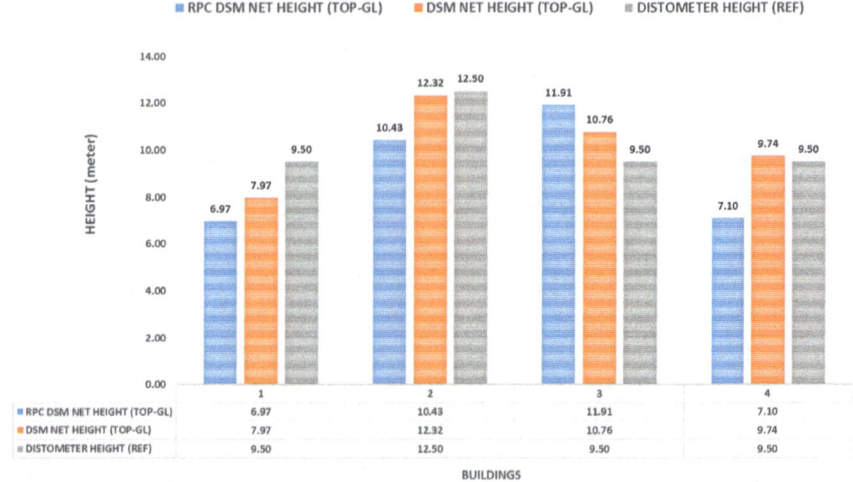

Fig. 20 Accuracy assessment for Gandhinagar DSM (method 2)

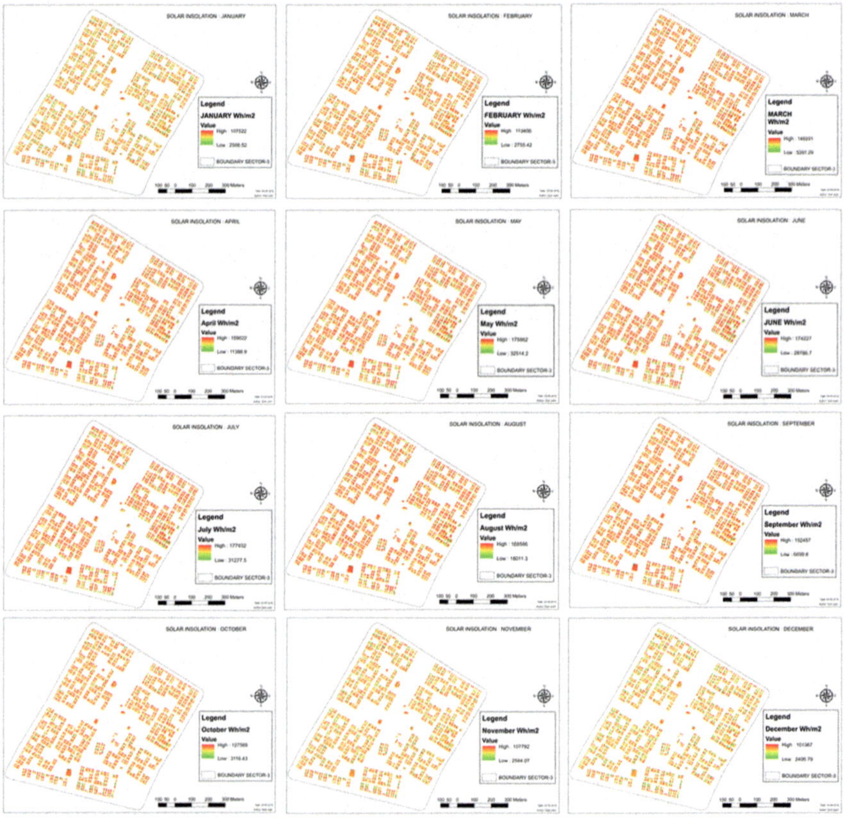

Fig. 21 Monthly global solar radiation maps for the study area (sector 3)

9.5.3   Global Solar Radiation

A relationship across the region was found between total roof area and generated electricity. Through the results, it can be concluded that roof top area is directly proportional to the electricity generation. This new understanding of roof area distribution and potential PV outputs will guide energy policy formulation in Gandhinagar city as city consist several high rise building so the roof area can be utilized in energy production.

To calculate electricity, generate by Solar PV panels following equation:

$$E = A * r * H * PR$$

E   Energy (kWh)
A   Total solar panel Area (m^2)
R   Solar panel yield (%)
H   Annual average irradiation on tilted panels (shadings not included) *
PR   Performance ratio, coefficient for losses (range between 0.9 and 0.5, default value = 0.75)

The sector having area of 0.725794 km^2 and having 2,368 buildings in it. Sector 3 having almost all residential use buildings so the electricity consumption is also taken equally all over sector for residential use only. As per solar city masterplan (2011) survey, the electricity consumption for residential buildings is 2.68 kWh/day. So we have taken 3 kWh per unit for the year 2015 for further analysis.

Rooftop of high rise buildings is generally occupied by miscellaneous structures like Stores, water tanks, solar water heaters etc. By taking this into consideration, only 30% area of available roof top is used for input calculation the energy consumption of sector 3 i.e. 2,368 buildings is 7,104 kWh/day by taking 3 kWh/day for single household. The solar rooftop energy generation is 18083.1256 kWh/day which is **254.54%** of required energy using only 30% of total plot area (Fig. 22). Exception is taken for some commercial buildings and other buildings in the sector (Figs. 23 and 24).

Fig. 22 Energy generated versus required providing a recovery by 254.54%

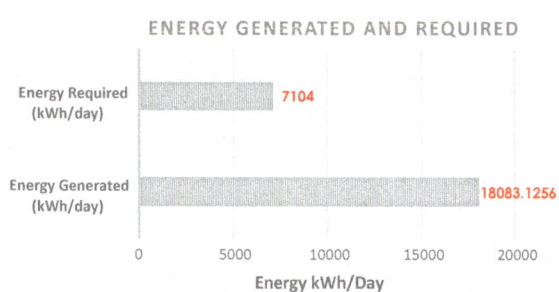

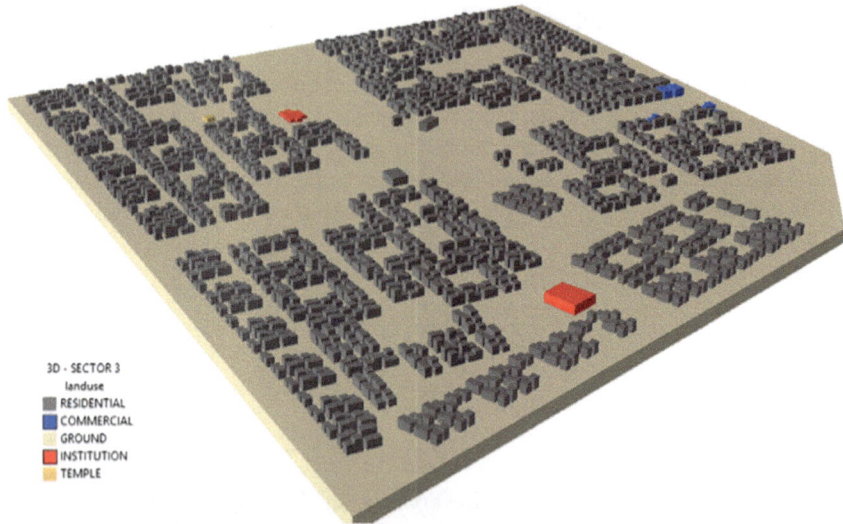

Fig. 23 3D representation of the sector-3

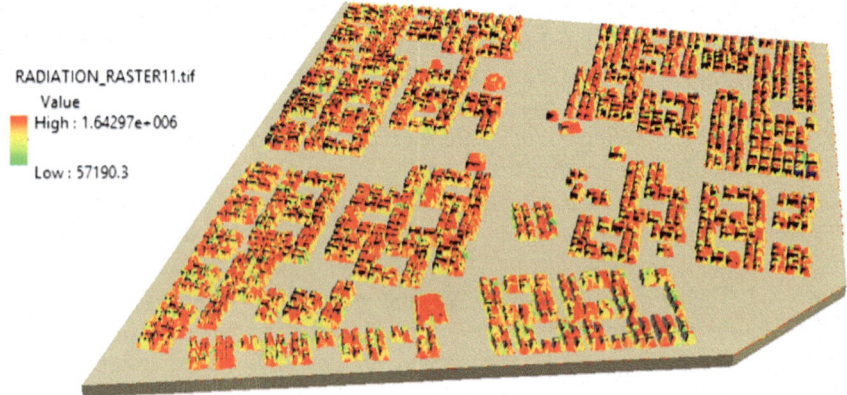

Fig. 24 Rooftop insolation on buildings in Arc SCENE in kWh/m^2/day

9.5.4 Global Solar Radiation: Validation and Inter-relationships

National renewable energy laboratory is a prime agency for renewable energy researches and data all over the world. This agency have all solar radiation data of India. Here, current research results were compared with NREL standard data. As shown in Fig. 25, we can see that the results were very similar for JUNE, JULY, AUGUST month and for other months slight variation is observed. The reasons behind this variation could be:

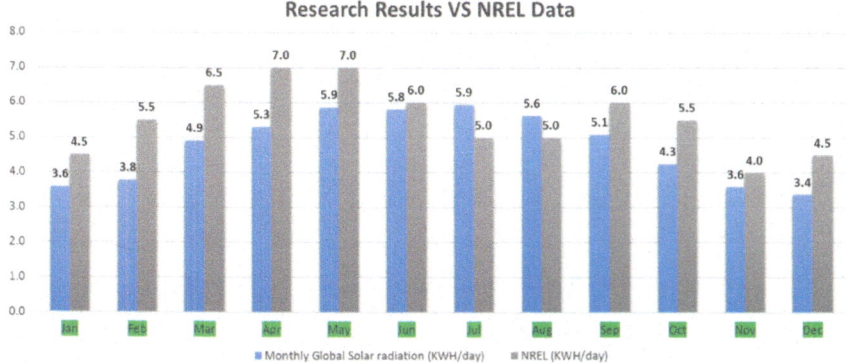

Fig. 25 Calculated global solar radiation validation with NREL data

- The NREL solar insolation is generated for the year 2010. In 5 years the seasonal variation can be witnessed.
- The NREL solar insolation is a 0.1° or a 10 km spatial resolution product representing an averaged value. The study uses the resolution of 1.5-m showing more detailed results different to that showed by NREL.
- The shadow effects were taken in this solar calculation process, so this results may be lower than NREL.

As a sample study, a detailed analysis of 25 buildings was also done to understand the relationship between land use and energy consumption pattern. 22 residential buildings, 1 commercial + residential, 1 commercial and 1 hospital (PSP) was chosen for the detailed sample study. The net area of rooftop for 25 buildings was calculated through digitizing only usable roof area of the buildings. 78% of total plot area was available.

Figure 26 clearly indicates that the energy consumption is more in commercial and public semi-public land use and least in residential land use. The results also indicate that the energy generated by residential areas is way higher than required by looking into the consumption pattern. Through analysis, it was also observed that Global Solar radiation has a strong positive correlation with the height of the building i.e. 0.5, which is not affected by shadow. Roof area has correlation with energy generated and it obvious that if area available for PV panel installation will be more then energy generation will be more. But there is little bit variations in energy output of the buildings having almost similar or small height difference. It can be due to shadow and other factors affecting solar radiation. The correlation of solar energy generation to surface rooftop area is 0.87.

Additionally, the solar radiation variation for buildings and surface shows that there is no difference for MAY, JUNE, JULY month and for other months there is a large difference (Fig. 27). The reason behind same could be that radiation for MAY, JUNE, JULY months the elevation angle will be higher (sun radiation perpendicular to ground at Gandhinagar location) and the tilt will be not there so there will be no

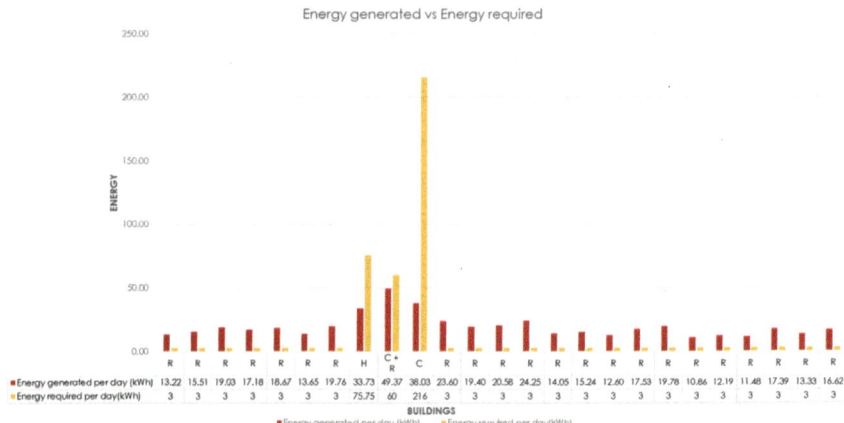

Fig. 26 Energy generation versus energy consumption for different building use

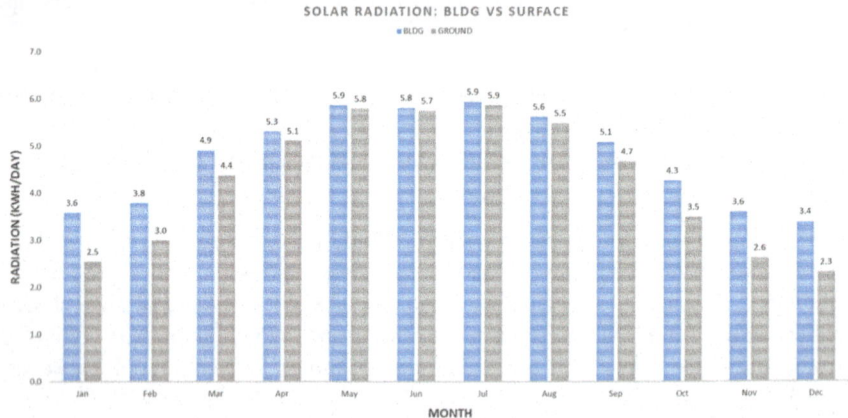

Fig. 27 Monthly solar radiation: building versus ground surface

shadow effect of buildings on surface and for other months the sun will be at some angle so the shadow effect will be witnessed and the buildings shadow will be on the ground ultimately affecting the radiation.

Analysis indicates that the height of the pixel is responsible for changing the global radiation dynamics of an area (Fig. 28).

10   Conclusion

The study was an attempt to understand the potential of solar roof tops at micro scale and find out the accurate solar radiation on each building footprint so as to maxi-

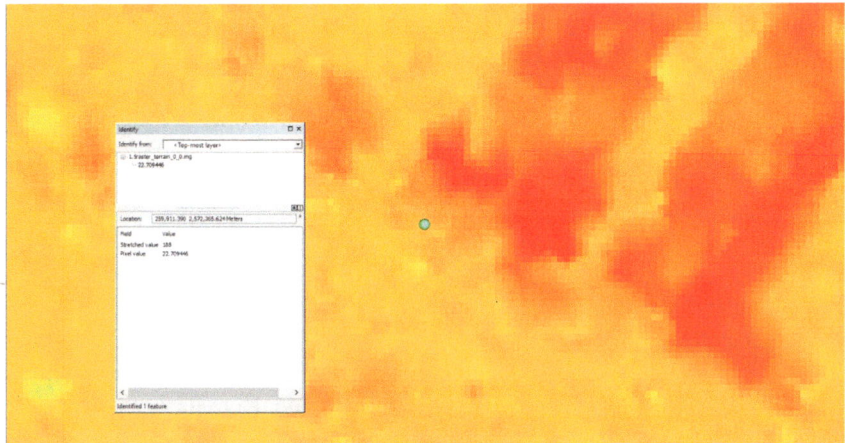

Fig. 28 The radiation is more on the building top than the surface

mize its usage for power generation. The study revolves around the need to utilize renewable energy resource as an alternate to ever increasing use of non-renewable energy resources causing a great damage to the potentially available mother Earth. Gandhinagar was chosen as case study area due to several factors explained earlier.

DSM was generated using high resolution Worldview 2 data for the study area Gandhinagar using LPS extension of Erdas Imagine. High quality Digital Surface Model was produced both using Empirical RF Method (using RPC files) and also by using Physical Model (using 10 GCPs). Since more robustness, consistency and confidence can be found with physical models, it is preferable to use them. The RMSE error reported at the time of block triangulation using RPC files for Gandhinagar DSM was 0.15998 m and the RMSE error using Physical model (considering 10 evenly spread and planmetrically distributed GCPs) is 0.26676 m. DSM was obtained having a considerably lower error value and a spatial resolution of 1.5 m. The resultant spatial resolution of a DSM is a function of the spatial resolution of the image and the b/h ratio and other sensor factors. Literature suggests that the DSM can be 3 times of the GSD of the image used for producing DSM. However, result was analyzed at various spatial resolutions like 2, 2.5 and 5 m. 1.5 m gave the best results in terms of visual interpretation and DSM quality (smooth and with less spikes). DSM created using ground control points which are collected through DGPS survey have better accuracy of height then DSM created through only RPC file. They have average bias for ground and buildings **0.5259** and **0.80 m** respectively.

The solar radiation insolation maps created for each month shows global solar radiation for each building of sector, for better visualization and interpretation of results the 3D extraction of buildings with solar insolation layer on it. But it can be used for only visual, not for calculations. For calculations of solar radiation for electricity generation, point solar radiation tool used which calculates for net rooftop area available for PV panel installation. The energy consumption of sector 3 i.e. 2,368 buildings is 7,104 kWh/day by taking 3 kWh/day for single household. The solar rooftop energy generation is 18083.1256 kWh/day which is **254.54%** of required energy using only 30% of total plot area. While on-site execution and installation of PV panels, losses may be witnessed, but energy required may be obtained using only 20–30% of the surface area available. The government is also giving 30% subsidy in cost and as per current scenario of Gujarat, 10,000 INR subsidy is given per kW energy plant installation.

We can produce much more energy for residential use by using solar rooftop energy. The generation of energy is almost 550% of current energy consumption. For commercial building use energy consumption will be 216 kWh/day and energy can be generated is only 38 kWh/day. So we can recover only **18%** of the current energy consumption. For mixed building use 60 kWh/day energy is required and 49.37 kWh/day can be generated through solar rooftop. Here **82%** energy can be recovered using solar energy. For hospital building use the energy can be generated using solar rooftop is 33.73 kWh/day and required is 75.75 kWh/day which can recover **45%** of current energy consumption. Automated footprint extraction through OBIA can be useful for larger area but it has very less accuracy compared to manual digitized footprints, that's why here manual digitizing method used for better accuracy.

We can conclude that by using solar rooftop energy as a smart energy alternative, we can make cities smart using RS and GIS we can make estimation of solar rooftop energy potential at large level or macro level. The MNRE (Ministry of New and Renewable energy) supporting solar concept at a maximum acceleration by giving subsidy and other arrangements. Government of Gujarat is also supporting solar rooftop program by giving 10,000 INR subsidy per kW production.

This study was undertaken to assist urban planners to estimate the potential of solar rooftop energy to make cities smart. Climate change and global warming is recent problem we are facing due to pollution in atmosphere so there is need to use renewable and clean energy solar energy. Solar rooftop energy is best option because it doesn't use the land as a resource neither you have to invest in land resource to use solar energy, anyone can tap solar energy potential on their rooftop and if production is more, then it can be used for public use by connecting it to grid. We can conclude that by using solar rooftop energy as a 'smart energy' initiative, cities can turn smart. Using RS and GIS we can make estimation of solar rooftop energy potential at macro/micro level precisely.

References

1. Mittal S (2010) Tapping the untapped: renewing the nation. Jaipur (2010)
2. Iglesias R (2013) Application of remote sensing-GIS to renewable energy resource assessment (2013)
3. Ramachandra TV (2007) Solar energy potential assessment using GIS. Energy Educ Sci Technol 18(2):101–114
4. BP (2018) BP Statistical Review of World Energy 2018

5. REN21 (2018) REN21: Renewables 2018 global status report
6. (Geni) Meisen P, (Geni) Quéneudec E, (Geni) Avinash HN, (Geni) Timbadiya P (2010) Overview of sustainable renewable energy potential of India," no. October 2006, pp 1–26
7. Sargsyan G, Bhatia M, Banerjee SG, Raghunathan K, Soni R (2011) Unleashing the potential of renewable energy in India, p 57
8. MNRE (2018) Annual Report 2017–18
9. Meisen P, Quéneudec E (2006) Overview of renewable energy potential of India, no. October, pp 1–20
10. Kumar A, Kumar K, Kaushik N, Sharma S, Mishra S (2010) Renewable energy in India: current status and future potentials. Renew Sustain Energy Rev 14(8):2434–2442
11. Tomar AKS, Gautam KK (2018) Renewable energy in India: current status and future prospects. Int J Eng Sci Invent 7(6):86–91
12. Ramachandra TV, Jain R, Krishnadas G (2011) Hotspots of solar potential in India. Renew Sustain Energy Rev 15(6):3178–3186
13. MNRE (2014) National meet on solar cities 'The Shivalik View Hotel', Chandigarh Assistance to Urban Local Governments
14. MNRE (2015) Development of solar city programme: status note on solar cities
15. Ramachandra TV, Krishnadas G, Jain R (2012) Solar potential in the Himalayan landscape. ISRN Renew Energy 2012:1–13
16. Aguayo P (2013) Solar energy potential analysis at building scale using LiDAR and satellite data, p 148
17. Agugiaro G, Nex F, De Remondino F, Filippi R, Droghetti S, Furlanello C (2012) Solar radiation estimation on building roofs and web-based solar cadastre. ISPRS Ann Photogramm Remote Sens Spat Inf Sci 1(2):177–182
18. Carl C (2014) Calculating solar photovoltaic potential on residential rooftops in Kailua Kona, Hawaii, no. May, pp 1–83
19. Bryan H, Rallapalli H, Rasmussen P, Fowles G (2010) Methodology for estimating the rooftop solar feasibility on an urban scale. Am Sol Energy Soc
20. Bassett T, Lannon SC, Waldron D, Jones PJ (2012) Calculating the solar potential of the urban fabric with SketchUp and HTB2. Sol Build Ski, Bressanone
21. Borfecchia F et al (2014) Remote sensing and GIS in planning photovoltaic potential of urban areas. Eur J Remote Sens 47(1):195–216
22. Wald L (2018) Basics in solar radiation at earth to cite this version: HAL Id: hal-01676634. Lect. Notes, p 57
23. Ramachandra TV, Jha RK, Krishna SV (2005) Solar energy decision support system. Sol Energy Decis Support Syst 24(4):207–224
24. Li D (2013) Using GIS and remote sensing techniques for solar panel installation site selection. Master Sci Diss Dep Geogr Univ Waterloo 1(1):1–140
25. Domínguez Bravo J, García Casals X, Pinedo Pascua I (2007) GIS approach to the definition of capacity and generation ceilings of renewable energy technologies. Energy Policy 35(10):4879–4892
26. Voegtle T, Steinle E, Tovari D (2005) Airborne laserscanning data for determination of suitable areas for photovoltaics. Int Arch 36:215–220
27. Cheng CL, Sanchez Jimenez CS, Lee MC (2009) Research of BIPV optimal tilted angle, use of latitude concept for south orientated plans. Renew Energy 34(6):1644–1650
28. Calabrò E (2009) Determining optimum tilt angles of photovoltaic panels at typical north-tropical latitudes. J Renew Sustain Energy 1(3):033104
29. Alves EDL, Lopes A (2017) The urban heat island effect and the role of vegetation to address the negative impacts of local climate changes in a small Brazilian City. Atmosphere (Basel) 8(2):18
30. Chen YM, Lee CH, Wu HC (2005) Calculation of the optimum installation angle for fixed solar-cell panels based on the genetic algorithm and the simulated-annealing method. IEEE Trans Energy Convers 20(2):467–473

31. Arán Carrión J, Espín Estrella A, Aznar Dols F, Zamorano Toro M, Rodríguez M, Ramos Ridao A (2008) Environmental decision-support systems for evaluating the carrying capacity of land areas: optimal site selection for grid-connected photovoltaic power plants. Renew Sustain Energy Rev 12(9):2358–2380

32. Gueymard CA, Myers DR (2008) Validation and ranking methodologies for solar radiation models. Model Sol Radiat Earth's Surf Recent Adv 479–509

33. Fu P, Rich PMP (1999) Design and implementation of the solar analyst: an ArcView extension for modeling solar radiation at landscape scales. In: 19th annual ESRI user conference, pp 1–24

34. Hofierka J, Suri M (2002) The solar radiation model for open source GIS: implementation and applications. In: International GRASS users conference in Trento, Italy, Sept 2002

35. Šúri M, Hofierka J (2004) A new GIS-based solar radiation model and its application to photovoltaic assessments. Trans GIS 8(2):175–190

36. Perez R, Seals R, Zelenka A (1997) Comparing satellite remote sensing and ground network measurements for the production of site/time specific irradiance data. Sol Energy 60(2):89–96

37. Bojanowski JS (2013) Quantifying solar radiation at the earth surface with meteorological and satellite data

38. Thies B, Bendix J (2011) Review-Satellite based remote sensing of weather and climate: recent achievements. Meteorol Appl 18:262–295

39. Dubayah R, Rich PM (1995) Topographic solar radiation models for GIS. Int J Geogr Inf Syst 9(4):405–419

40. Omitaomu OA, Singh N, Bhaduri BL (2015) Mapping suitability areas for concentrated solar power plants using remote sensing data. J Appl Remote Sens 9(1):097697

41. Hofierka J, Suri M (2002) The solar radiation model for Open source GIS: implementation and applications. In: International GRASS users conference in Trento, Italy, September 2002, no. October 2002

42. Wilson JP, Gallant JC (2000) Digital terrain analysis, pp 1–28

43. Mckenney DW (2010) Calibration and sensitivity analysis of a spatially—distributed solar radiation model 8816

44. MNRE Development of Gandhinagar solar city master plan

Part V
Kozhikode

Smart Water Management for Smart Kozhikode Metropolitan Area

T. M. Vinod Kumar, C. Mohammed Firoz, Puthuvayi Bimal, P. S. Harikumar and Praveen Sankaran

Abstract Smart Water is an essential component of the smart environment since life, sustenance, growth and death of the environment depends on water. The concept of smart water is derived from the concept of smart cities having one to one relationship in their building blocks. Smart Water Management is a very high responsive, intelligent digital system operated by IOTs and ICTs, clouds, related computer-based models along with humans to identify water-related issues and even automatically using artificial intelligence solving it in real time without human interventions. The chapter presents an attempt to develop a Smart Water Management for Kozhikode Metropolitan area. Water in the study area is studied in conjunction with the spatial distribution of community in a watershed. The issues arising out of the present and potential usage pattern for households for community wellbeing and economic development is the basis of water resources management in Kozhikode Metropolitan Area (KMA). KMA enjoys substantial precipitation a year and is bound by two major rivers on both sides and are lined by many streams within its jurisdiction. However, KMA faces many water-related problems. A SWOT analysis was performed to identify and consolidate the capabilities of various watersheds and its communities

T. M. Vinod Kumar
School of Planning and Architecture, New Delhi, India
e-mail: tmvinod@gmail.com

C. Mohammed Firoz · P. Bimal (✉)
Department of Architecture and Planning, National Institute of Technology Calicut, Kozhikode, Kerala, India
e-mail: bimalp@nitc.ac.in

C. Mohammed Firoz
e-mail: firoz@nitc.ac.in

P. S. Harikumar
Water Quality Division, Centre for Water Resources Development and Management, Kunnamangalam, Kerala, India
e-mail: hps@cwrdm.org

P. Sankaran
Department of Electronics and Communication Engineering, National Institute of Technology Calicut, Kozhikode, Kerala, India
e-mail: psankaran@nitc.ac.in

© Springer Nature Singapore Pte Ltd. 2020
T. M. Vinod Kumar (ed.), *Smart Environment for Smart Cities*, Advances in 21st Century Human Settlements, https://doi.org/10.1007/978-981-13-6822-6_7

241

in the study area. The issues faced by various communities were grouped into few categories and solutions were proposed for each community. Integrated Smart Water Management is proposed as a solution to the problems faced by the communities. It is footed on the principles of Water democracy, which is implemented through a system of ICT, IoT and decision support systems. The Integrated Smart Water Management System enables the community to be aware of the issues well in advance and find and implement solutions proactively. A spatial decision support system (SDSS) is proposed to help the community to take decisions related to water management. The SDSS takes many of the decisions as per the set procedure and alerts the community only cases where a systematic solution is not available. The integrated water management system implements the decisions taken by the SDSS, through its automatic sensors and actuators managing the water resources.

Keywords Integrated smart water management · SDSS · Smart water communities · E-water democracy · Kozhikode metropolitan area

1 Introduction

Smart Water is an important component of smart environment since life, sustenance, growth and death of environment depends on water. Water Resources in a Metropolis is a key Environmental Resource that links to households, community, land and land use with far-reaching impacts. Water in Metropolitan area is utilized for basic needs of human, animal and vegetation and urban non-agricultural activities. Environmental Water Resource Management here involves participating in scientific efforts to conserve the overall quantity and quality of Water in relation to the various typology of lands, and land uses including forests, parks, lakes, wetland, rivers, mountain ranges, and other areas of natural splendour. With the advent adverse effect of climatic change, water has assumed an important element of the environment to be developed and managed.

Smart water can exist better in a smart city because of the smart infrastructure endowment used for many needs of the inhabitants including water.

2 The Concept of Smart Water and Smart Cities

The concept of smart water is derived from the concept of smart cities having one to one relationship in their building blocks. A Smart City System comprises of six key building blocks: (i) smart people, (ii) smart city economy, (iii) smart mobility, (iv) smart environment, (v) smart living, and (vi) smart governance. These six building blocks are closely interlinked and contribute to the 'Smart City System', as illustrated in Fig. 1 Some authors treat the six elements of a Smart City System equally [1]. However, following Vinod Kumar [2], we give prominence to 'smart people' because

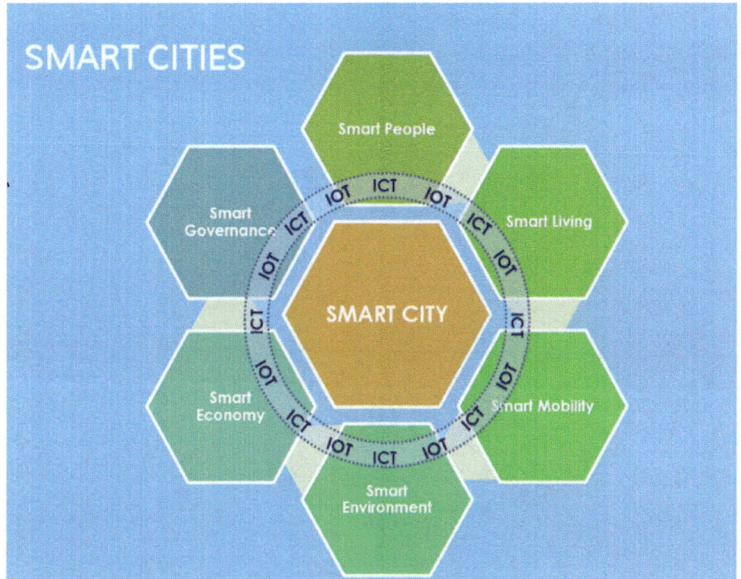

Fig. 1 Smart city system

without their active participation and involvement a Smart City System would not function in the first place. These six systems have strong linkages with IOTs and ICTs in a smart city performing designated functions in different latitude and longitude as per requirements. Figure 1 diagrammatically represents the smart city system.

Just like Smart city is considered holistically whether it is a metropolitan city or mega city. Water Resources in KMA is holistically considered and presented in Fig. 2.

As per Indian constitution, water resource is common property, and it is not owned by any single person or institutions. Common property is generally managed under capitalism, socialism and anthological system as shown in Fig. 3.

Private ownership of water resources has shown in the past adverse environment and livelihood results as for example experienced by the villagers around Pachamama in Kerala India, and local agitation of affected households and legal cases eventually evicted the Coca-Cola factory. There are several such experiences across India and the World. The socialist model of common property management involves nationalisation, and inefficient and often corrupt bureaucracy is not providing successful examples. However, the community system of management has been robust, result oriented and accepted for centuries by the local community and practised. This community model of common water property which is made digitally smart is accepted in this chapter. This community is built on smart people, and the characteristics of smart people are given below.

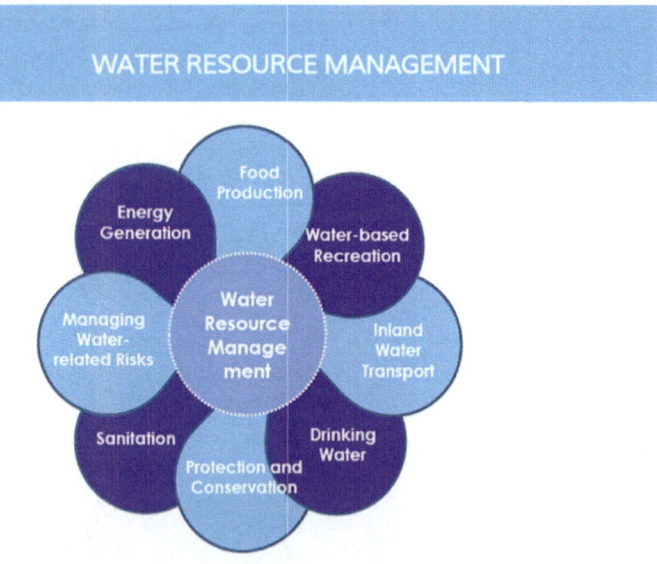

Fig. 2 Holistic view of water resources management

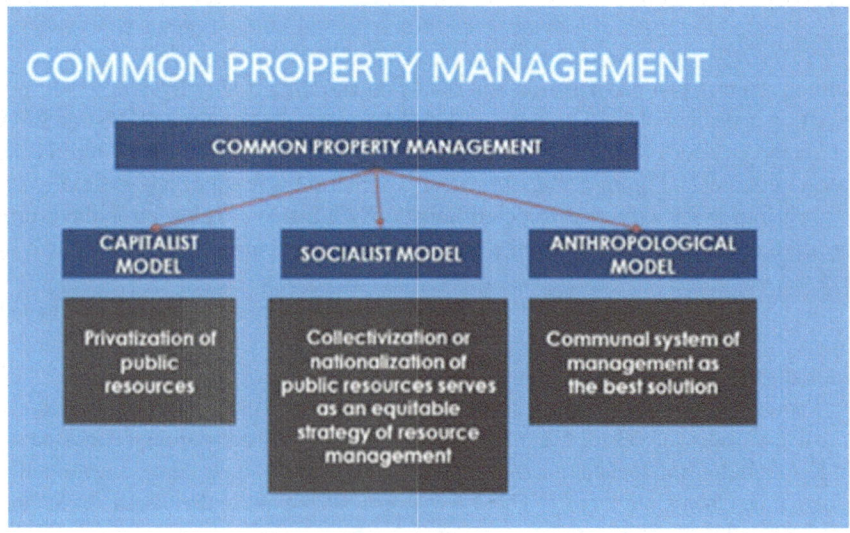

Fig. 3 Common property management

2.1 Smart People

'Smart People', the fundamental building block of a Smart City System, require many crucial attributes as given below.

(1) Smart people excel at what they do professionally.
(2) Smart people have a high Human Development Index [3].
(3) A smart city integrates its universities and colleges into all aspects of city life.
(4) It attracts high human capital, for example knowledge workers.
(5) A smart city maintains high Graduate Enrolment Ratio and has people with high level of qualifications and expertise.
(6) Its inhabitants opt for lifelong learning and use e-learning models.
(7) People in a smart city are highly flexible and resilient to the changing circumstances.
(8) Smart city inhabitants excel in creativity and find unique solutions to challenging issues.
(9) Smart people are cosmopolitan, are open-minded, and hold a multicultural perspective.
(10) Smart people maintain a healthy lifestyle.
(11) Smart people are actively involved in their city's sustainable development, its efficient and smooth functioning, its upkeep and management, and making it more liveable.

Smart people activate smart economy.

2.2 Smart Economy

'Smart City Economy', the second building block, requires the following attributes.

(1) A smart city understands its economic DNA.
(2) A smart city is driven by innovation and supported by universities that focus on cutting-edge research, not only for science, industry, and business but also for cultural heritage, architecture, planning, development, and the like.
(3) A smart city highly values creativity and welcomes new ideas.
(4) A smart city has enlightened entrepreneurial leadership.
(5) A smart city offers its citizens diverse economic opportunities.
(6) A smart city knows that all economics works at the local level.
(7) A smart city is prepared for the challenges posed by and opportunities of economic globalization.
(8) A smart city experiment, supports, and promotes sharing economy.
(9) A smart city thinks locally, acts regionally, and competes globally.
(10) A smart city makes strategic investments in its strategic assets.
(11) A smart city develops and supports compelling national brand/s.

(12) A smart city insists on balanced and sustainable economic development (growth).
(13) A smart city is a destination that people want to visit (tourism).
(14) A smart city is nationally competitive on selected and significant factors.
(15) A smart city is resourceful, making the most of its assets while finding solutions to problems.
(16) A smart city excels in productivity.
(17) A smart city has the high flexibility of labour market.
(18) A smart city welcomes human resources that enhance its wealth.
(19) A smart city's inhabitants strive for sustainable natural resource management and understand that without this its economy will not function indefinitely.

Smart people and the smart economy are most concerned about smart environment.

2.3 Smart Environment

'Smart Environment', the fourth building block, has the following attributes.

(1) A smart city lives with and protects the nature.
(2) A smart city is attractive and has a strong sense of place that is rooted in its natural setting.
(3) A smart city values its natural heritage, unique natural resources, biodiversity, and environment.
(4) A smart city conserves and preserves the ecological system in the city region.
(5) A smart city embraces and sustains biodiversity in the city region.
(6) A smart city efficiently and effectively manages its natural resource base.
(7) A smart city has recreational opportunities for people of all ages.
(8) A smart city is a green city.
(9) A smart city is a clean city.
(10) A smart city has adequate and accessible public green spaces.
(11) A smart city has an outdoor living room. Unlike the indoor living room in houses where we meet others, outdoor living rooms are aesthetically designed intimate, active, and dynamic urban realms where people meet face to face for a culturally and recreationally rich and enjoyable contact as part of living and work.
(12) A smart city has distinctive and vibrant neighbourhoods that encourage neighbourliness and a spirit of community.
(13) A smart city values and capitalizes on scenic resources without harming the ecological system, natural resources, and biodiversity.
(14) A smart city has an integrated system to manage its water resources, water supply system, wastewater, natural drainage, floods and inundation, especially in the watersheds where it is located, especially in view of the (impending) climate change.

(15) A smart city focuses on water conservation and minimizes the unnecessary consumption of water for residential, institutional, commercial, and industrial use, especially in the arid and semi-arid areas.

(16) A smart city has an efficient management system for the treatment and disposal of wastewater, and reuse of treated wastewater, particularly in the arid and semi-arid areas.

(17) A smart city has an efficient management system for the collection, treatment, and disposal of industrial wastewater.

(18) A smart city has an integrated and efficient management system for the collection, transfer, transportation, treatment, recycling, reuse, and disposal of municipal, hospital, industrial, and hazardous solid waste.

(19) A smart city has an efficient system to control air pollution and maintain clear air, especially in the air sheds where it is located.

(20) A smart city has an efficient and effective system for disaster risk reduction, response, recovery, and management.

(21) A smart city has and continually upgrades its urban resilience to the impacts of climate change.

(22) A smart city can create a low-carbon environment with a focus on energy efficiency, renewable energy, and the like.

Other three components have been discussed elsewhere [3].

3 Smart Water Communities

It is possible to visualise smart water analogous to smart city system given above.

Here smart people of the smart city are smart water communities. Smart mobility is smart mobility of water engineered and digitised that moves using IoT and ICT community decided water in quality and quantity to households most efficiently. Smart Economy transforms to smart water economy based on modified water usage pattern of households identified by the smart community. Smart Governance is smart water governance based on all water use legislation of Central and State Government. Smart Environment is Smart Water environment based on ecological, conservation and cultural design. Smart living becomes smart water living centred around the practice of urban design centred around public realm and tourism activities such as beach tourism, wetland tourism and so on.

4 Smart Water Management

Water management is about decision-making, specifically with the process of decision-making in relation to the use of natural resources namely water, the pollution of habitats and the modification of ecosystems. Water security is the goal of

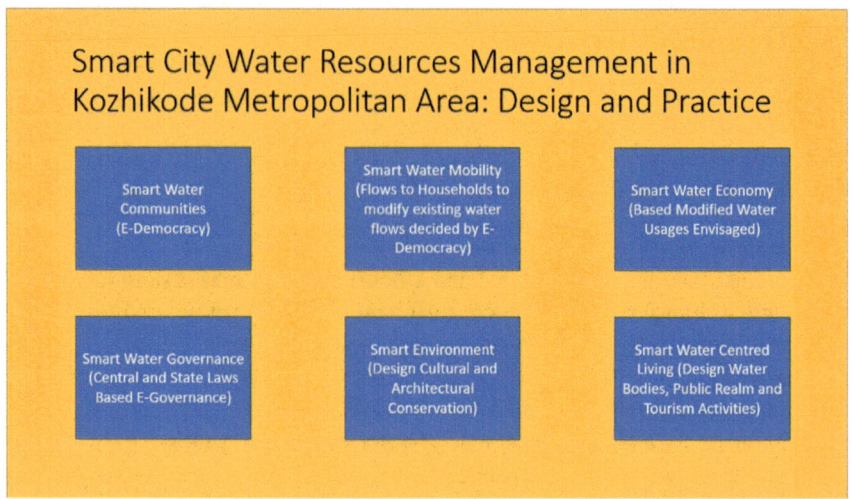

Fig. 4 Smart water system

water resources management. There is no single path to water security. Capacity building, adaptability, and resilience are essential for the future planning and management of water resources. Water Resources Management (WRM) is the process of planning, developing, and managing water resources, in terms of both water quantity and quality, across all water uses [4, 5]. Water Resource Management is a process which promotes the coordinated development and management of water, land and related resources in order to maximise the resultant economic and social welfare in an equitable manner, without compromising the sustainability of vital ecosystems [6]. Hence it is also called integrated water management.

Smart Water Management is not used widely before the advent of smart cities; hence it needs some explanation. Smart Water Management is a very high responsive intelligent digital system operated by IOTs and ICTs, clouds, related computer-based models along with humans to identify water-related issues and even automatically using artificial intelligence solving it in real time without human interventions (Fig. 4). This involves, controlling of devices remotely by means of wireless communication or powerline communication devices. Device communication, using middleware, and wireless communication to form a picture of connected environments and Information acquisition, dissemination from sensor networks, enhanced services by intelligent devices with predictive and decision-making capabilities. This can be better achieved by smart communities if these communities are exposed to continuing education by nearby institutions of higher studies. Smart water communities are strategic, purposeful, and resourceful. They are driven by long-term commitments to safeguard their natural resources and economic opportunities for future generations and preserving the beauty, vitality, and equity of the region [7].

5 The Study Area—Kozhikode Metropolitan Area

Water in the study area is studied in conjunction with the spatial distribution of community in a watershed. The issues arising out of the present and potential usage pattern for households for community wellbeing and economic development is the basis of water resources management in Kozhikode Metropolitan Area (KMA). Figure 5 shows the KMA area.

The Kozhikode Metropolitan Region has been delineated by aggregating the area of census 2011 urban agglomeration using the following set of maps of local self-

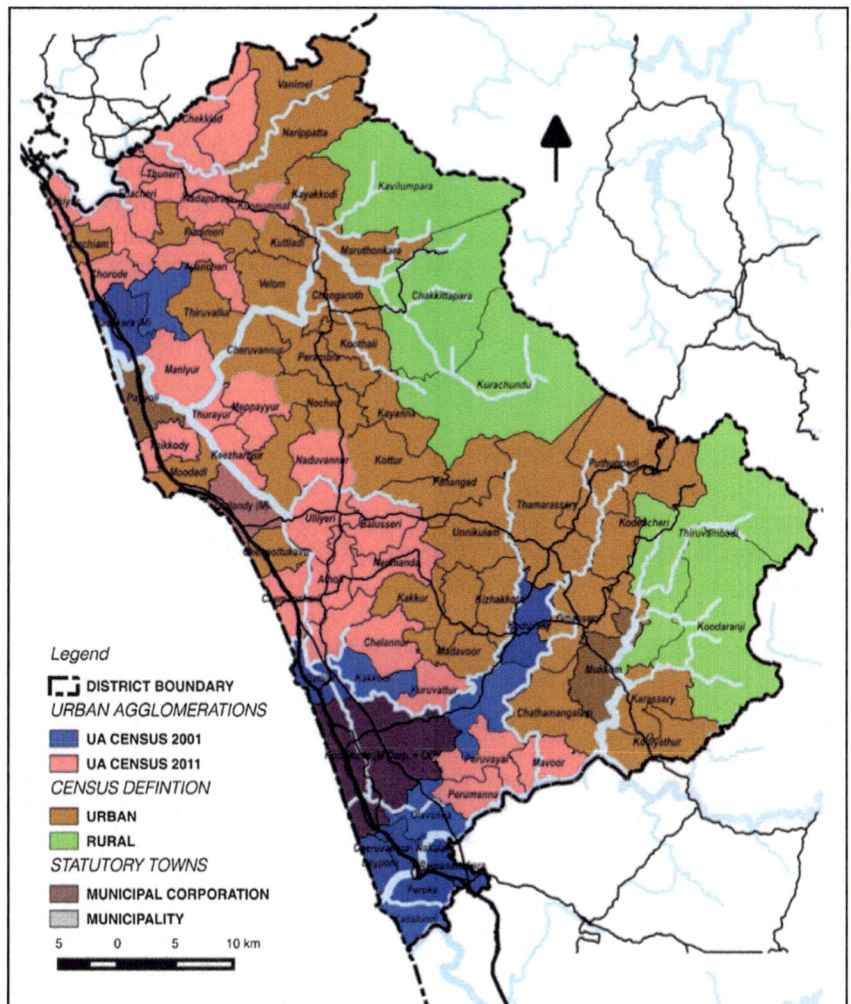

Fig. 5 Kozhikode metropolitan area 2011

government bodies satisfying some criteria. (i) Census towns, (ii) Census Urban areas. Local bodies that are surrounded by urban areas on all sides are also included even if those are rural areas. The final KMR is limited to the administrative boundary of the district. According to 2011 Census, Kozhikode Urban Agglomeration occupies a major proportion of Kozhikode district and has 75 local bodies and 8 statutory towns with a Metropolitan Area of 1803 km^2 and total population 2.3 Million (2,344,996) according to 2011 census. Vadakara, Quilandy, Payyoli, Feroke, Ramanattukara and Beypore are the major statutory urban bodies in the KMA based on population size. The Sex ratio of KMA varies from 981 to 1208 amongst the local bodies with an average sex ratio of 1106 which is higher than the state average of 1084 as per 2011 census. Major Employment is in the tertiary sector with high decrease in primary sector employment since 1991. Total working population in the region is 28% with the male work participation rate in the planning region is 50.76% whereas the female work participation rate is only 7.82%.

This Chapter will concentrate on the water in all its manifestations in KMA and design of a smart community resources management system using ICT, IoT, E-Democracy and E-Governance and work out a protocol for practice in 24 h 7-day framework. This study is based on several research works conducted in the past as well as limited primary surveys in selected watersheds having unique characteristics to identify local issues and solutions proposed by a local inhabitant.

5.1 Climate Topography and Drainage

The study region reports a maximum temperature of 36.5 °C and low temperature of 17.5 °C. The months of the April normally records the maximum temperature and minimum in January. Average precipitation in the region is 3130 mm with 126 days of rainy days in a year. The South West along with North East monsoons contribute the majority of the rainfall in the area with 82.77% of the rainfall. The months of January, February, March and April receives the remaining rain through some summer showers. When the North East monsoon fails, the region faces acute water shortage. Maximum rainfall is received during South West Monsoon from June to August.

The range of relative humidity varies from 74 to 92% during morning hours and from 64 to 89% in evening hours with the rainy season in the monsoon recording the highest humidity. The region has mostly alluvial soil along the low lands coastal plain and valley and paddy is predominantly cultivated. However, lateritic soil is seen mostly in the midlands and highlands of the region [8]. Therefore, it can be understood that the study area is blessed with a good tropical climate with abundant rainfall for at least 6 months with water scarcity reported only for less than 2 months a year.

The region is mainly divided into three elevation zones called, lowland,[1] mid-land,[2] and highland[3] regions. Topography of the region has affected the settlement pattern. Highly urbanized and developed areas fall in the coastal belt. Mid lands are moderately urbanized whereas the High lands are still rural in nature [5]. Because of this topography, whatever rains received in highlands eventually flow to the low-lands and ultimately reaches the Arabian sea. The terrain shows a dendritic drainage Pattern. Hence there is abundant scope for water conservation and retention so that effective water management practices shall be employed.

5.2 Transport and Linkages

The region is well connected with excellent transport network of rail, road and waterways. There are 2 national highways, and several state highways, major district roads and other important roads passing through the metropolitan region. The region is also well connected to other state of India not only by road, but also by rail transport. This mode of rail system not only helps for passenger transportation, but also for goods movement. Currently, waterways are being promoted as a mode of transport of goods (as well as tourism). This system has got Immense potential for development (was popular till early 80s). The proposed state waterways pass through the study region and the proposed master plan for 2035 envisages this concept of water transport. The river Kallai which was once a main transport corridor for transporting timber through water has a high scope for this mode of transport.

5.3 Land Use

The study area land use is divided into 10 categories. It can be seen that residential-agricultural mix dominate the total land use with 64%. This is because, residential developments in larger plots are also seen in Kerala. When the family grows, each of the siblings set up their own houses by subdividing the large plots with internal roads [10]. However, the plot continues to be used as an agricultural plot (Mainly coconut, Aracunut, Rubber etc.) with residences enclosed with in the agricultural land use. Table 1 gives the details of the land use, The land use classified as 'other land use' dominates with 12.33% followed by Forest dominates (8.6%) and pure agriculture land use (5.5%).

---

[1]The low land regions are gently sloping or level lands, with altitude below sea level to 7.5 m [9].

[2]The midland region is having slightly undulating topography. It has an altitude ranging from 7.5 to 75 m above mean sea level [9].

[3]The highland region is mainly covered with forest lands and plantation crops and has an elevation ranging from 75 to 750 m above mean sea level [9].

Table 1 Land use distribution of the study region

S. no	Land type	Area (km^2)	Percentage
1	Forest	154.85	8.59
2	Water bodies[a]	46.31	2.57
3	Marshy land	3.24	0.18
4	Residential	36.05	2.00
5	Agriculture	98.58	5.47
6	Plantation	3.6	0.2
7	Res/AgriMix[b]	1158.53	64.28
8	Other built-up[c]	219.36	12.17
9	Others[d]	60.00	3.33
10	Commercial	21.80	1.21
Total		1802.32	100

[a]Water bodies land use include perennial, reservoir/canal, reservoir bed/river bed/river island, sands/riverine/flood plain, water bodies/back waters and dams
[b]Res/AgriMix land use consists of land use mixed with residential units and catagories of agricultural plants like areca nut, banana, banana and tapioca, coconut/coconut and areca nut/coconut and tapioca, coconut dominant mixed crop, current fallow, mixed crop, rubber, tapioca etc.
[c]Other Built up includes Harbor/Port, Industrial/Industrial Park, Mixed Built-up/Mixed Built-up converted from paddy, Airport, Playground, Educational Institutions
[d]Other includes categories like Barren Rocky/Stone waste/sheet rock, Barren Rocky/Stone waste/sheet rock (RF), Coastal Sand, Land with scrub, Beaches, Mining/Industrial waste
Source [11]

5.4 Water Resource Inventory

The study region is drained by 5 major rivers namely the rivers of Kuttiadi (361.07 km^2), Korapuzha (646.41 km^2), Kallayi (96.0 km^2), Chaliyar (417.44 km^2) and Kadalundi (5.26 km^2) Mahi (284.8 km^2) [8, 12]. All the rivers are west flowing and drain into Arabian sea. Majority of the Kozhikode metropolitan region (KMA) area is covered by Korapuzha basin. Figure 6 shows the major river basins in the study area. Table 2 summaries the water inventory of the study region.

Among the rivers only Chaliyar and Kuttiadi rivers had hydroelectric power generations and irrigation projects. The only port in the region, Beypore is also located at the mouth of the Chaliyarriver. Tourism is widely active along the Kadalundi, Kuttiadi rivers with the famous Malabar river festival happening in Iruvazhinji river (tributary of Chaliyar) The Korapuzha 'Jalotsav'[4] is also a major water festival of Kerala state, India. The Kallai river is an important center in the world for timber business which is connected to Chaliyar river by Kallai canal and Korapuzha river via Canoli canal, both being an engineering marvel of the 18th century. Aquaculture and pen culture practice are also associated with some of these rivers. The region

---

[4]Jalotsav means water festival.

Table 2 Major water inventory in the study region

Name of the river	Origin and elevation (m)	Length (km)	Tributaries	Basin area (km²)	Avg. annual rainfall (mm)	Avg. annual stream flow (mm³)	Navigable length (km)	Irrigation projects	H.E projects
Kadalundi	CherakombanMala and 1160	130	Olipuzha, Veliar	1122	3400	1137	43.2	Nil	Nil
Chaliyar	Ilambalaari and 2066	169	Punnapuzha, Kanjira-puzha, Karimpuzha, Iruvan-jipuzha, Cherupuzha	2535	3800	5902	68.4	Chalipuzha Olipuzha Baipura-puzha	Pandiyar Punnam-puzha
Kallai	Cherikkulathur and 45	40	Cherikkulathur	96	3800	NA	9.6	Nil	Nil
Korapuzha	Arikkankunni and 610	40	Akalapuzha, Punoorpuzha	624	3800	222	24.8	Nil	Nil
Kuttiyadi	Narikotta and 1320	74	Onipuzha, Thottil-palampuzha, Kadiyangad-puzha, Thevan-nathilpuzha, Madappal-lipuzha	583	4500	1273	9.6	Kuttiyadi	Kuttiyadi

Source [8]

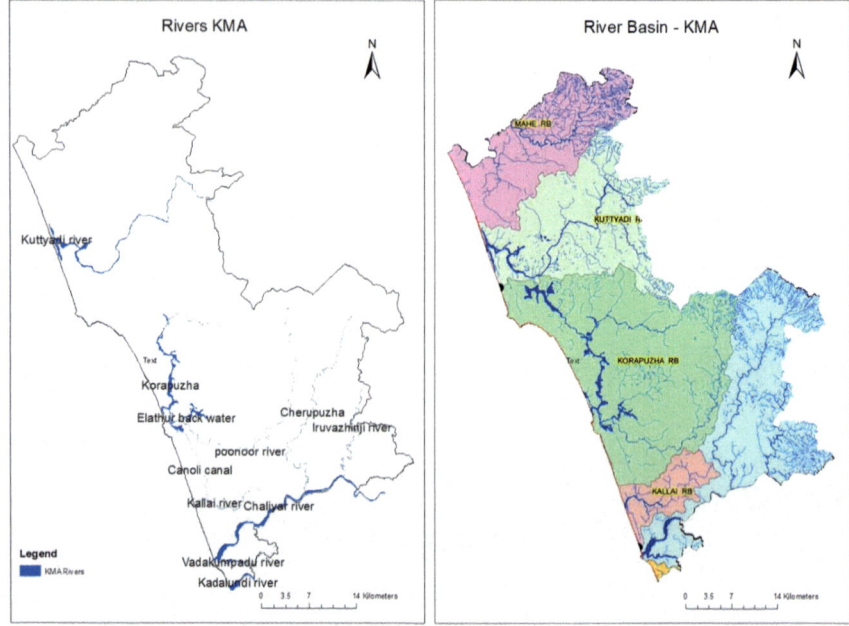

Fig. 6 Water resource inventory

also has certain back waters formed by the confluence of Akalapuzha and Poonoor puzha The backwaters serves 42 gramapanchayats,[5] 3 municipalities and forms a part of Kozhikode Corporation.

5.5   Watersheds in the Study Region

Watershed is defined as 'An area of land that drains all the streams and rainfall to a common outlet such as the outflow of a reservoir, mouth of a bay, or any point along a stream channel' [10]. The watershed of the study area consists of surface water sources like lakes, streams, reservoirs, and wetlands as well as the underground ground water sources. The watershed management therefore includes optimal management of precipitation, Infiltration, the soil characteristics, land cover and slopes, evaporation, transpiration etc.

---

[5]Grama panchayats are the lower most units of Governance in a 3 tire administrative unit in India.

5.5.1 Derivation of the Study Region Watershed

The watershed map of the study region is derived using the Arc GIS software. From the Digital elevation model of the study region Kozhikode metropolitan region (KMA), the flow direction and flow accumulation is made. From this, the drainage pattern of all streams that flowing is made. Finally, different pour points are selected on the stream channels and the watersheds are generated. A final number of 56 mini watersheds with population ranging from 5000 persons to 100 thousand populations are finally arrived. Figure 7 shows the derivation of the watershed map and Fig. 8 shows the final watershed map generated.

Referring to Table 3, It can be seen that except 2 blocks namely Kunnamangalam and Balussery categorized as semi critical, majority of the remaining 10 blocks in the region have critical shortage with respect to water availability. This is explained in Table 3. The table also shows the net annual ground water availability, existing ground water draft for irrigation, existing ground water draft for domestic and industrial water supply etc. It can be generally assumed that majority of the areas in the region do not have water scarcity and there is vast scope for irrigation using groundwater (Fig. 9).

5.6 Major Economic Resources of the Region

One of the major economic resources of the region as well as Kerala state is mainly from foreign remittance from the middle east along with economy primarily from service sector [9, 14, 15]. However, the local economic development and employment is also rooted in tourism, fisheries, Agriculture, Horticulture and small hydro resources. Some of the most prominent economic resources related to water management is listed below.

5.6.1 Tourism Development

The study region is blessed with all types of tourist destinations like ecotourism, beach tourism, heritage and cultural tourism etc. Though the district is rich with many diverse enthralling tourism spots, many of the destinations are not developed completely and there are potential sites that has to be explored. The district remains unexplored with rich culture, unexplored wildlife, soothing back waters, cool hill stations and attractions of pristine beaches. Despite its inherent advantages, including easy access with different modes of transport, Kozhikode has not been able to attract domestic and foreign tourists [9]. Table 4 explains the different types of tourism and places in the study region.

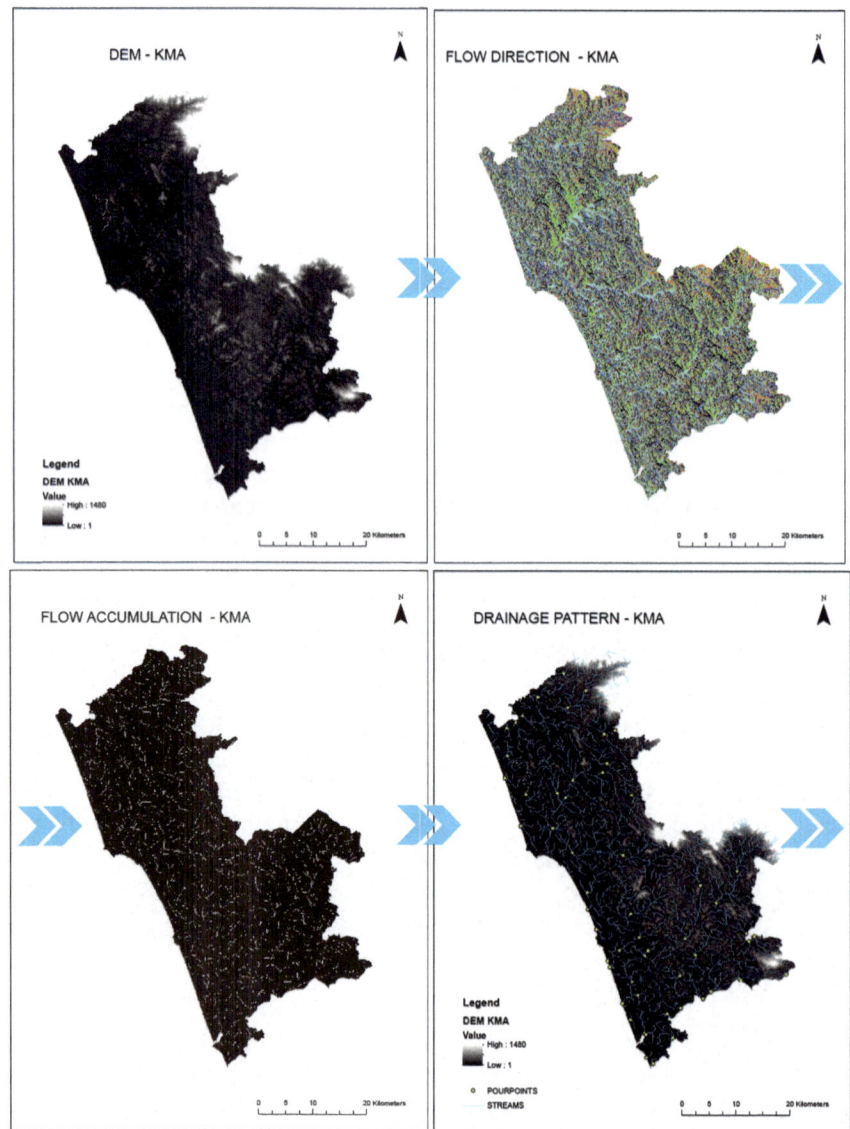

Fig. 7 Figure showing generation of watershed. *Source* [8]

Table 3 Block level ground water details of the study region

S. no	Block	Net annual GW availability	Existing gross GW draft for irrigation	Existing gross GW draft domestic and industrial water supply	Existing gross GW draft for all uses	Stage of GW development	Categorization of block
1	Badagara	16.03	1.46	8.22	9.68	60.37	Safe
2	Balussery	26.10	8.59	12.43	20.92	80.11	Semi critical
3	Kozhikode	37.7	5.03	26.09	31.12	82.53	Safe
4	Chevayoor	24.51	4.00	11.08	15.08	61.53	Safe
5	Koduvally	47.59	5.81	13.92	19.73	41.45	Safe
6	Kunnummel	26.37	3.67	10.09	13.76	52.19	Safe
7	Kunnamangalam	31.17	8.82	17.62	26.44	82.82	Semi critical
8	Meladi	31.25	3.14	6.7	9.85	31.51	Safe
9	Panthalayani	36.38	2.60	9.23	11.83	32.51	Safe
10	Perambra	36.00	4.49	8.47	12.96	35.99	Safe
11	Thodannur	16.79	1.47	6.78	8.24	49.09	Safe
12	Thooneri	17.44	2.92	7.19	10.11	57.97	Safe
	Total	347.38	52	137.71	189.72	54.61	

Source [8, 13]

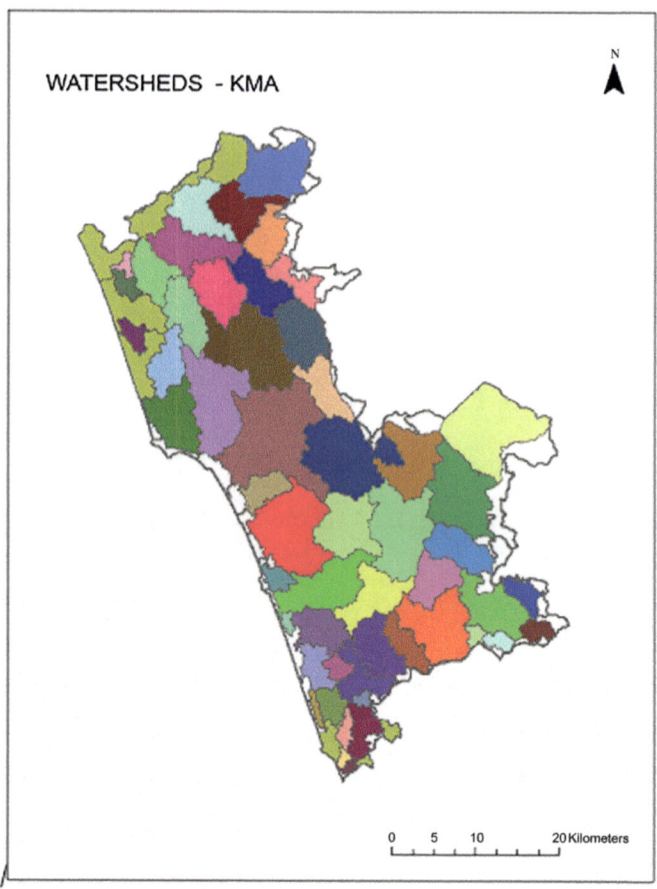

Fig. 8 Final watershed map derived for the study region. *Source* [8]

5.6.2   Inland Fisheries

The study region has got 9 minor inland fishing centers. According to [14], Kerala
inland fisheries. There are active fisherman populations of 1931 people with predominantly male population constituting of 669 families in the study region. Pen culture[6]
is also one of the most popular economic activity in the region. The fisheries projects
gained its momentum through the centrally sponsored schemes like Rashtriya Krishi
Vikas Yojana (RKVY), Matsya Samridhi project under Fish Farmers Development
Agency (FFDA), Matsya Keralam Project of Government of Kerala etc. where in
which community level local economic development took place. The Coastal Aqua-

---

[6]Pen culture and sometimes called as Cage culture is a fish farming method by which the fish or any
other aqua organisms are hold captive within an enclosed space whilst maintaining a free exchange
of water.

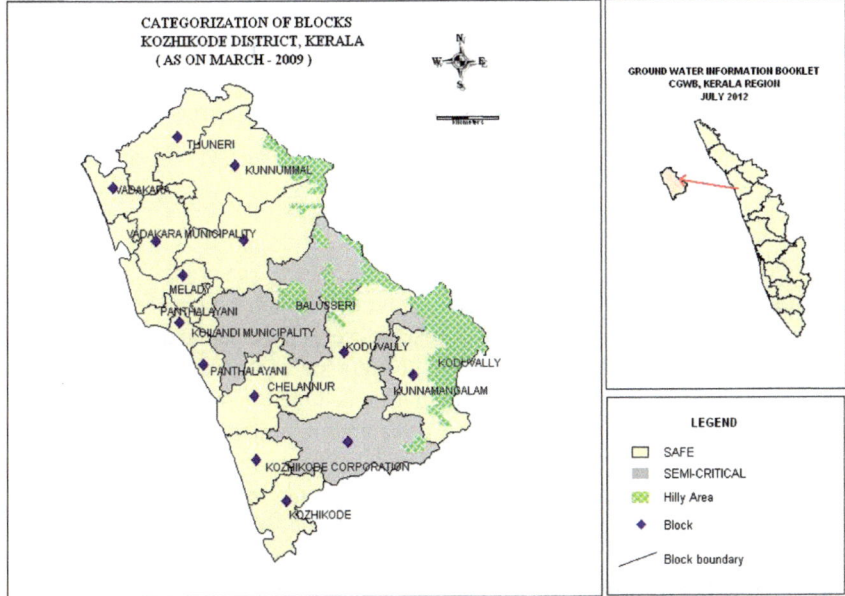

Fig. 9 Critical water shortage areas of the study region. *Source* [8, 13]

Table 4 Tourist resources of the study region

S. no	Type of tourism	List of places
1	Leisure tourism	Kozhikode City beach, Kolavipalam turtle beach, Beypore, Thikkodi Beach, Sand Banks in Badagara, Kappad Beach, Mananchira Square, Sarovaram Biopark
2	Ecotourism	Kakkayam, Kadalundi, Thusharagiri, Janakikad, Vanaparvam
3	Cultural and heritage tourism	Temple festivals, Thira, Kalaripayattu, SM Street, Malabar Cuisines, Ponmeri Shiva Temple, Thali temple, Misqual Mosque, Lokanarkavu temple, Sargaalaya Craft village, Pazhassi Raja Museum
4	Cultural tourism	Varakkal Devi temple, Pisharikavu temple, MisqualMosque, Thali temple
5	Medical tourism	
6	Fairs	Malabar River Festival, KorapuzhaJalotsavam

Source [4]

culture Authority Act, 2005 promotes the farming of shrimp, prawn, fish or any other aquatic life under controlled conditions in ponds, pens, enclosures or any other brackish water bodies and gives the appropriate legal backing for this economic activity [16]. The involvement of women is minimal in most of the fisheries related activities and needs to be promoted.

5.6.3 Agriculture and Horticulture

The study region though not very active in agriculture related economic activities, traces of protected cultivation at the household level is prevalent and is often gaining popularity. This helps households/farmers to cultivate vegetables and by depending on hi-tech green houses thereby using sophisticated environmental controlled situations. Felid investigations shows that this method of agriculture is successful with the farmers getting an 25–30% additional yield compared to open field cultivation with the average income amongst households drastically increased.

5.6.4 Small Hydro Projects

Small hydro power projects are further classified as micro hydro (up to 100 kW), Mini hydro (101–2000 kW) and small hydro (2000–2500 kW). The study region primarily depends on 4 power stations namely the Pathankayam small hydroelectric power project at the KodencheryPanchayat with a capacity of 8 MW, Vilangad small hydro electric project at the VanimelandNarippattapanchayat with an installed capacity of 7.5 MW, Poozhithode small hydroelectric project at the Maruthonkara panchayat with an installed capacity of 4.8 MW, and the Chembukadavu- Small Hydro Electric Project at the Thamarassery panchayat with a capacity of 2.7 MW are the major power supply sources. An agency for Non-conventional Energy and Rural Technology (ANERT) has been the key organization in implementing the non conventional energy resources in the region. They had also identified 3 additional microhydro projects namely at places namely Narippatta (Watershed 2, 3, 4), Puthuppady (Watershed 22, 27) and Kodencheri (Watershed 22, 27).

5.6.5 Ports

The study region is blessed with the presence of the Beypore port, which was one of the prime ports of India once. Though this port is not highly effective now a days on account of it smaller size and difficulty in handling large vessels, this port has a high potential for development as a major port of India. Currently the Beypore port is being used for the export of copra, coir, fiber, fish, timber, cement, iron, steel machineries, food grains etc. This port also serves the main transport mode place to Lakshadweep islands of the Indian territory. The study region also has 4 other harbors namely at the Vellayil (watershed 43), Chombala (watershed 7) Koyilandy (watershed 24) and at Puthiyappa (watershed 33). Currently a high-speed hydrofoil ferry is being operational for tourism purpose which takes a travel time of just 3 h from Kozhikode to Kochi.

6 SWOT Analysis

To understand the strength, weakness, opportunities and threats of the study region, a SWOT analysis (Table 5) is undertaken. Accordingly, the study regions are subjected to analysis under the heads of People, Economy, Mobility, Environment, Living and Governance. These heads are the essential components of consideration of a smart city.

7 Watershed Issues in the Study Region

Based on the SWOT analysis and a field survey, certain specific problems are identified for the study region. Accordingly, the whole study region is analysed, and the major problems of the regions are classified under the following heads

Typology 1—Water Scarcity
Typology 2—Water Pollution
Typology 3—Ground Water Depletion
Typology 4—Salinity

The analysis of the regions is done based on the delineated boundary of a watershed. The following section describes the analysis.

7.1 Water Scarcity

One of the major problems in the KMA study region is the scarcity of water specially during summer seasons. The scarcity varies from a few days in certain watersheds to 2–3 months in others. The Calicut city is one of the most affected areas with a minimal requirement of 200 mld against a supply of 84 mld per day [4] Water scarcity is severely affected in the highland regions specially during the peak summer. The current water supply system by the Kerala Water Authority cannot meet the demand. Based on our survey, it is found that 26% of the areas has scarcity of water with the following region namely Kayakkodi, moodadi, Arikulam, Chathamangalam, Kadalundi and Kuruvattur being severely affected. Figure 10 shows the regions which are reported to be having water scarcity.

From the field survey, it can be found that the watershed Number 37 is the one having the highest scarcity (Fig. 11). Hence, a detailed analysis of the same

Table 5 SWOT analysis

	Strength	Weakness
People	Active community level organizations for the protection of water bodies with good participation of volunteers. Several local drinking water schemes present	Political influence of the schemes and non-availability of protected drinking water especially from government agencies such as Kerala Water Authority
Economy	Location of industries near to water bodies. Tourist potential of available rivers and wetlands	Absence proper and effective effluent treatment system which is polluting the rivers Lack of good tourist infrastructure and services
Mobility	Established transportation infrastructure like Canoli canal and port connectivity from Beypore	Under developed and unutilized existing water lanes
Environment	High water table with a terrain having a natural slope. Presence of mangroves and ecologically sensitive areas which are not highly damaged	Coastal areas with ground water salinity intrusion and point and non point sources of pollution at several areas. Lack of technological innovations to safeguard water bodies from human interventions
Living	Educated population with high social awareness to conserve water and environment	Adaptation time for new technological intervention with political and social intervention possibilities from different groups
Governance	Increased awareness in community level meetings. Wide usage of technological aids such as internets and smart phones	Lack of time and interest in participation among certain groups

(continued)

Table 5 (continued)

	Opportunities	Threats
People	Local technology for water management and educated skilled people	Entire population is not very much aware about the schemes
Economy	Unexplored possibilities of water related economic activities including potential cultivable land utilization	Environmental problems out of related economic activities like pesticides, over utilization and over exploitation of resource
Mobility	Proposal for inland water transportation connecting various regions of Kerala	Water logging and flood, Environmental degradation by overuse of resources
Environment	Provisions for rain water harvesting and ground water recharging, Wetland management to conserve biodiversity	Ground water depletion at a high rate, increase in contamination, reclamation of wetlands
Living	Possibilities of water harvesting and recharge systems at the household level – ICT/IOT monitoring systems	Out migration of the skilled workforce
Governance	Implementation of ICT/IOT systems for online public participation and monitoring	Misuse of transparency and privacy concerns

Source [4]

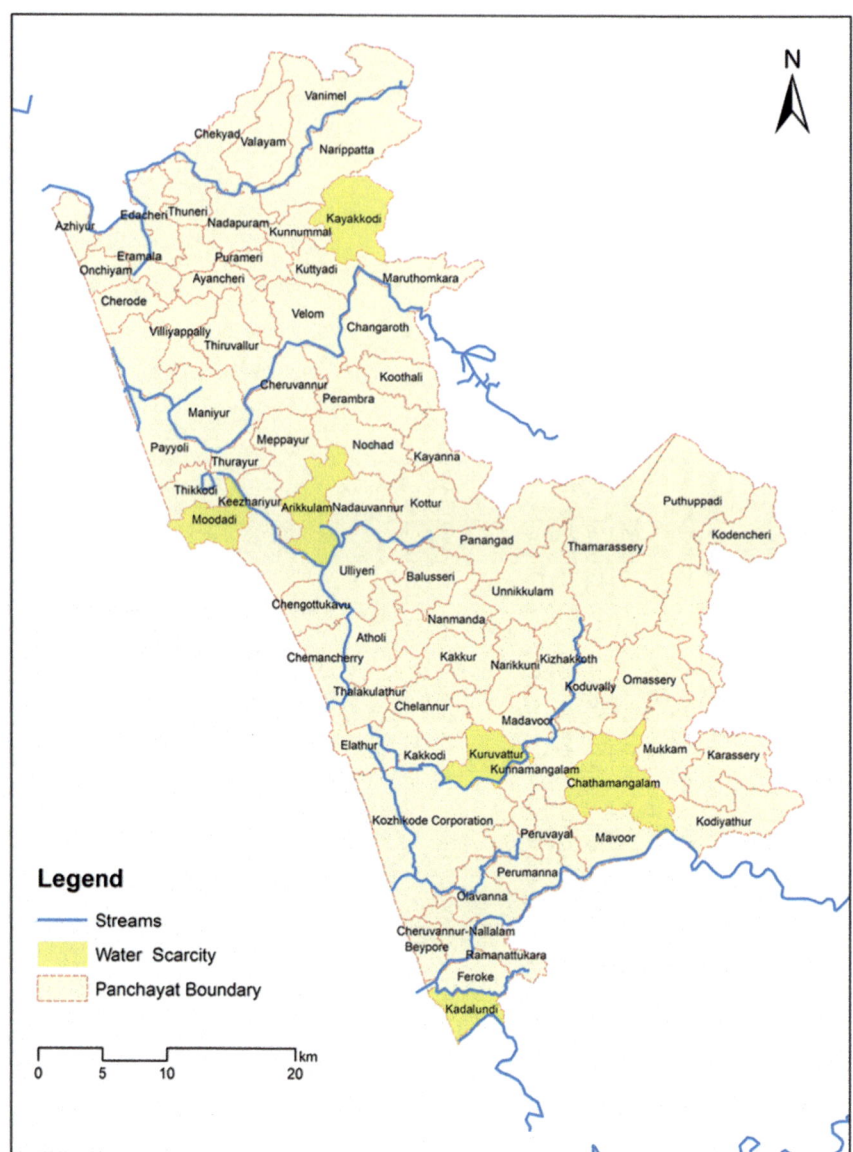

Fig. 10 Water scarcity affected areas of the study region. *Source* [8]

Fig. 11 Locational details of watershed 37 in the study region. *Source* [8]

region is undertaken. The watershed is distributed in 4 local bodies namely Kuruvattur, Madavoor, Kunnamangalam panchayats and Kozhikode Corporation. The study watershed area constitutes of an area of 18.18 km^2 with a population of 44,736 persons per square kilometer and 10,654 households. The region elevation ranges from 10 to 150 m and the water scarce areas are mainly located at the hilly areas.

A household survey is undertaken for the region and the results are as follows. Majority of the people in the watershed are depending on the open well water (65%), followed by 20% households depending on public water supply. The remaining depends on community wells, bore wells etc. The water scarcity extends to even 7 months in a year in certain areas with at least 40% of the households having water shortage up to 4 months of severe water scarcity. Little or no community water management efforts are seen in the region. The primary survey also reveals that average spending of water per day during the extreme seasons range up to Rs. 100 per day and in some extreme cases, up to Rs. 1000/day per family and during such period, they primarily depend on tankers. The survey results identify the excessive dependence on bore wells which directly results in groundwater depletion. It necessitates the significance of community level water management practices and advocating rain water harvesting practices.

7.2   Water Pollution

Water pollution is one of the major issue of the KMA region. This pollution can be categorized into 2 types. Point sources namely pollution from household, industries, hospitals, commercial establishments and non-point source like agriculture, urban land use etc. The rivers mainly Iruvazhinji, Chaliyar, Kallayi, Ponnur and the Canoli canal are highly polluted [8]. Almost all water bodies including the smaller streams etc. are severely polluted in the major urban areas. Njeliyamparamba region the land fill region of Calicut city is the most polluted region with the pH values ranging from 4.58 to 7 during pre-monsoon and 4.45–6.84 during post monsoon [8].

The electrical conductivity of the samples varied from 127.3 to 2680 μS/cm during pre-monsoon and 118.50–2050 during monsoon and 131.20–2670 during post-monsoon. Source: [4, 17]. The other issues are changing colour of open well water, presence of coli form bacteria etc. Figure 12 shows the water polluted region of KMA.

On analyzing the groundwater samples, it is found that, high level of bacteriological contamination is observed specially during the monsoon season with high count of total coli forms which ranges between less than 3 MPN/100 ml and greater than 2400 MPN/100 ml [17]. About 78% of the monsoon water samples show a total coliform count higher than the permissible limit with contamination domestic, municipal and hospital wastes. About 86% of the monsoon water samples show a faecal coli form count higher than the permissible limit in the study area [17]. The spatial distribution of total coliform count is shown in Fig. 13.

A detailed survey analysis of the watershed no 37 and 51 is further undertaken. The survey reveals that more than 50% of households depend on open well, and the others on public tap. The survey also revealed that approximately, 38% of the population has reported poor quality water specially during rainy seasons in terms of spatial distribution of total coli form (Fig. 13), total hardness (Fig. 14) and high turbidity with reddish and yellowish colour (Fig. 15).

Waste dumping into water bodies and percolation of leachate containing pesticides from agriculture field are causing severe water pollution. The prime reason for

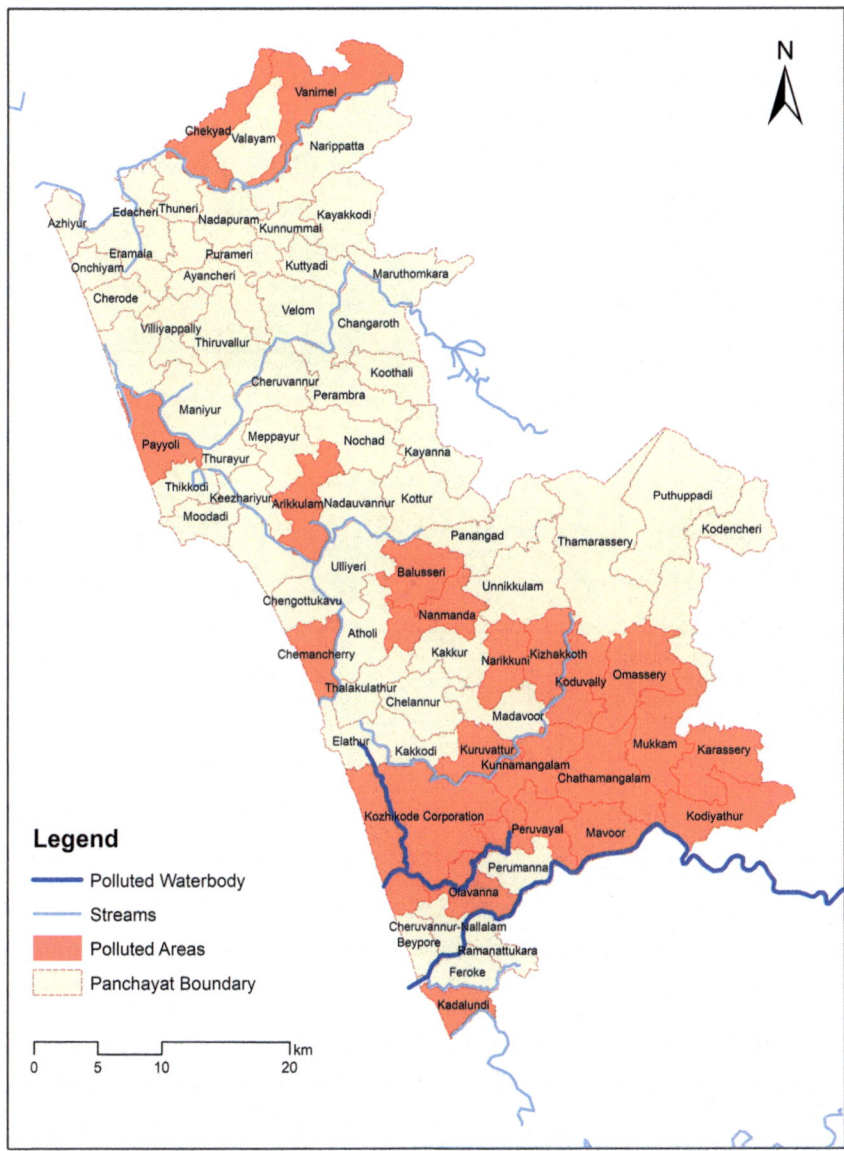

Fig. 12 Water polluted areas of the study region

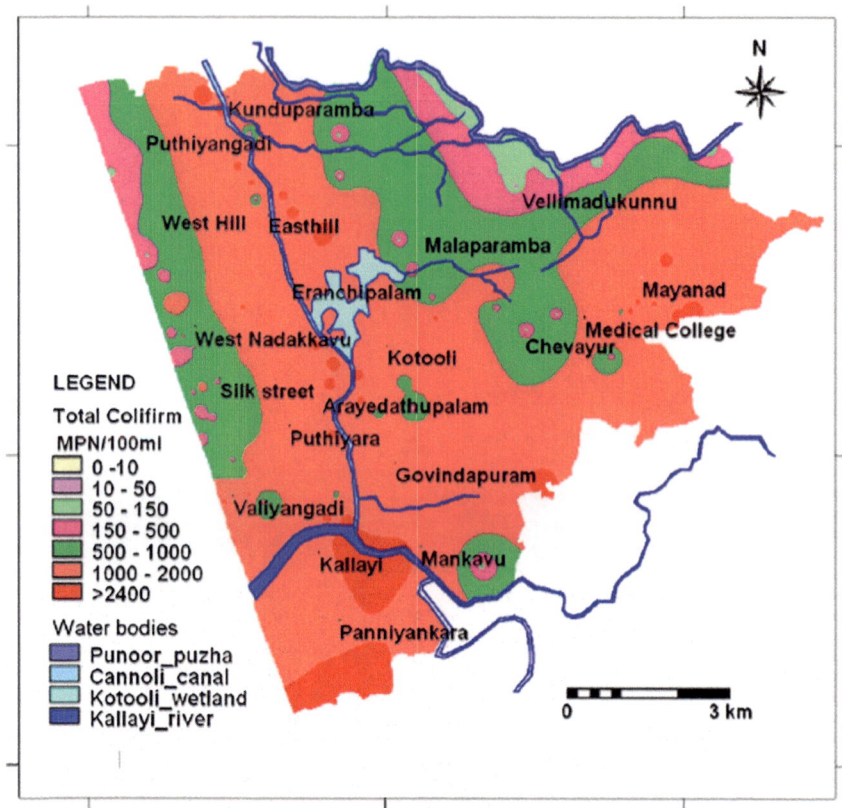

Fig. 13 Spatial distribution of coli form count in the study region. *Source* [17]

presence of e-coli bacteria in the drinking water sources water is due to the nearness of the wells to the leach pits/tanks due to the sprawl settlement system and with moderate to high density population distribution. The Njeliyamparamba landfill area (which is the solid waste dumping yard for Kozhikode), lying in the study region is one of the major point source of pollution. The water quality is so severe that the water is blackish in colour in most of the wells nearby and many wells far away from it. It also pollutes ponds and other sources of water and thus creates not only environmental, but social problems as well [8]. Another major polluting source is the vicinity of Canoly canal area where many of the households discharge their waste and leach pits to the canal along with discharge from the timber yards of the wood industries located nearby causing water pollution to other areas.

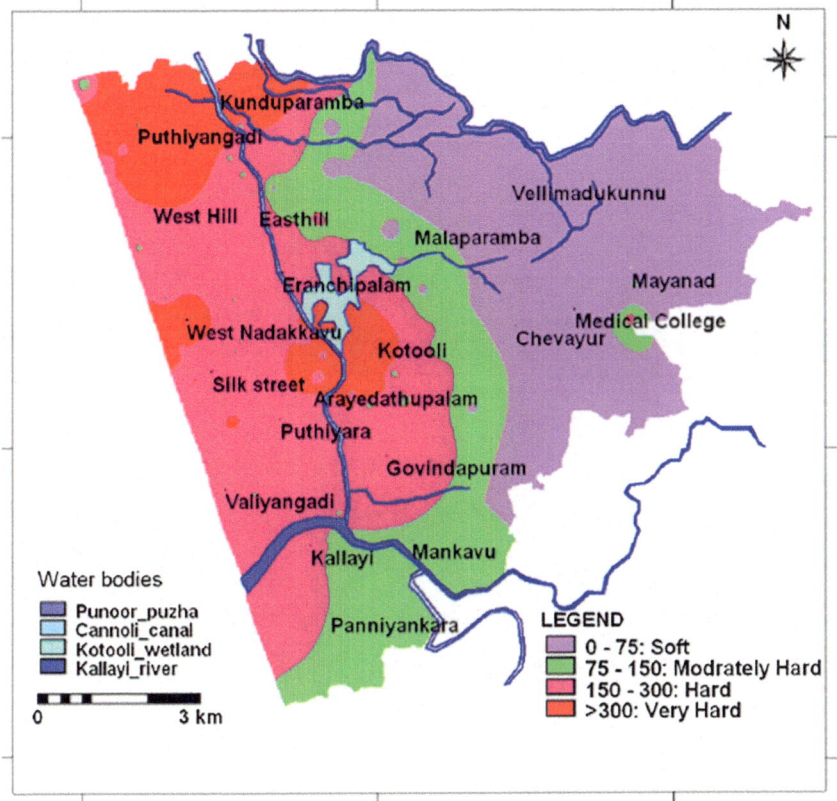

Fig. 14 Spatial distribution of total hardness distribution in the study region. *Source* [17]

7.3 Ground Water Depletion

Ground water depletion is mainly due to excessive exploitation, no recharge of water back to soil, and excessive wastage of water. The increase in number of bore wells gives the exact reason for this. Ground water exploitation issues are mainly in Kadalundi, Unnikulam, Chekkiad and Kunnamangalam regions. 4 out of the 54 watersheds are having serious ground water issues. Figure 16 shows the map explaining water issue areas of the study region. A recent survey by the groundwater Department with a sample size of 54 wells shows a steep depletion of ground water ranging from 1 m to 20 cm at different locations of the region. The reasons attribute are Indiscriminate consumption, growing rate of urbanization, and resultant factors such the levelling of paddy fields have brought down groundwater recharge capacity [8].

The places like Nallalam, Cheruvannur, Perumanna and Palayad are the worst hit areas where in some areas like palayad, the water depletion is about 1.29 m in a year, possibly one amongst the highest in Kerala [18].

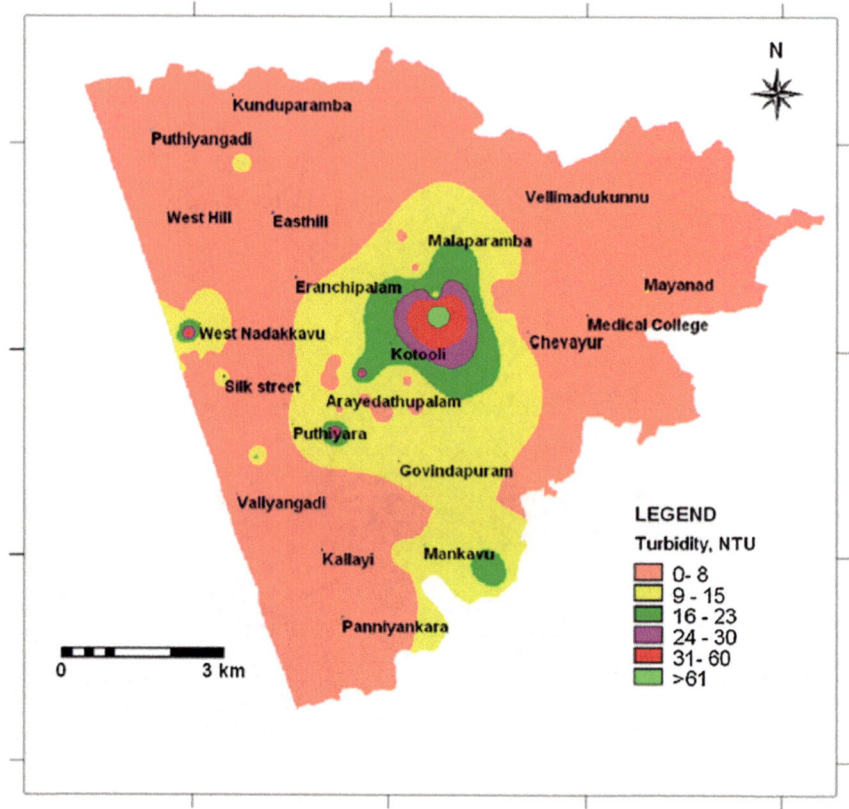

Fig. 15 Spatial distribution of turbidity in the study region. *Source* [17]

7.4 Salinity Issues

Salinity is mainly due to the intrusion of salt water from areas near to the sea. Most of the coastal areas has got issues with respect to water salinity. The areas namely Feroke, Kadalundi, Nallalam, Beypore, Chorode, Thikkodi, Elathur, Kappad are the major areas affected by salt water. Figure 17 shows the regions affected by water salinity.

Severe salinity intrusion is seen along the areas near to sea. The depletion in groundwater level has resulted in seepage of salt water from sea and rivers into wells. In addition to the imminent drought owing to deficit rainfall, salinity ingress is also posing a threat to the people [19]. With the decrease in water table and increase in consumption, saline water from backwaters and sea is intruding to wells and water bodies [16]. According to the field survey analysis, the households near the Canoli canal region are most affected.

Fig. 16 Ground water issue areas of the study region. *Source* [8]

Fig. 17 Salinity affected areas of the study region. *Source* [8]

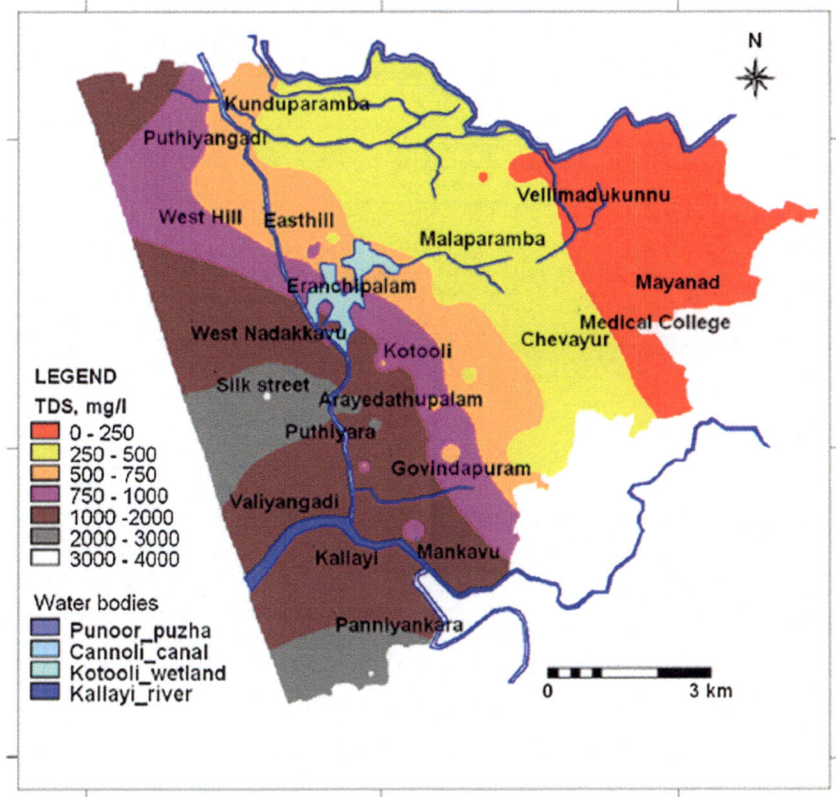

Fig. 18 Salinity in borewells of the study region. *Source* [17]

Alternatively, the water quality status of groundwater as obtained from the borewells as samples were analysed and the results shows high concentration of TDS, chloride, sodium and total hardness. Arayedathupalam and places near Civil station Calicut shows high concentration of salinity in bore well water. Very minimal bacteriological contamination is observed in the bore well water samples analysed. The high concentration of salinity in bore well water in these places may be due to the salts present in the aquifer system or because of the influence of seawater [17]. Figure 18 shows the distribution of salinity in bore wells in the study region.

8 Methodology

The methodology is given below which is self-explanatory (Fig. 19).

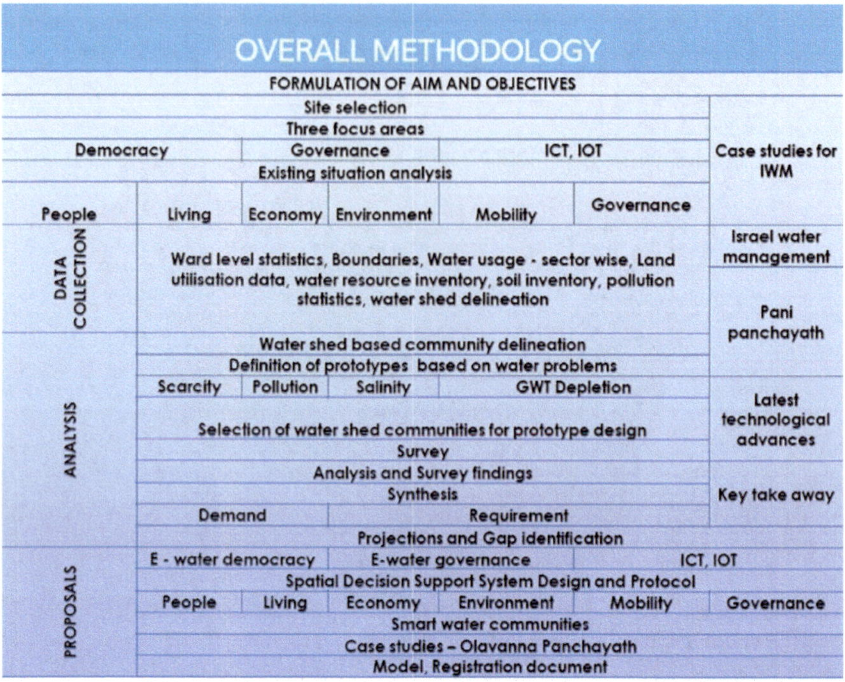

OVERALL METHODOLOGY

			FORMULATION OF AIM AND OBJECTIVES				
			Site selection				
			Three focus areas				
	Democracy		Governance		ICT, IOT		Case studies for IWM
			Existing situation analysis				
	People	Living	Economy	Environment	Mobility	Governance	
DATA COLLECTION		Ward level statistics, Boundaries, Water usage - sector wise, Land utilisation data, water resource inventory, soil inventory, pollution statistics, water shed delineation					Israel water management
		Water shed based community delineation					Pani panchayath
ANALYSIS		Definition of prototypes based on water problems					
	Scarcity	Pollution	Salinity	GWT Depletion			Latest technological advances
		Selection of water shed communities for prototype design					
		Survey					
		Analysis and Survey findings					
		Synthesis					Key take away
	Demand		Requirement				
		Projections and Gap identification					
PROPOSALS	E - water democracy		E-water governance			ICT, IOT	
	Spatial Decision Support System Design and Protocol						
	People	Living	Economy	Environment	Mobility	Governance	
		Smart water communities					
		Case studies – Olavanna Panchayath					
		Model, Registration document					

Fig. 19 Integrated water management system in KMA

9 Proposals

The watersheds in the study area were analysed for the major issues they face and their potentials. Each of these watersheds become unique with their own set of problems and potentials. A common solution cannot be applied to all these watersheds. Smart water communities are proposed to take care of the problems and solutions for each watershed. However, the smartness can be achieved only if the community is enabled to continuously monitor the resources, predict the demand, identify and rectify problems at the earliest. An Integrated Smart Water Management System is proposed for continuous monitoring. This also enables the water communities to effectively interact with each other and find solutions to the problems together.

The research attempts to achieve the principles of water democracy [20] with the help of the IoT and ICT framework of the smart region. Various IoT/ICT solutions are identified for each of the issues identified for watersheds of KMA. These solutions were integrated to form a Spatial Decision Support System (SDSS) [21].

10 Integrated Smart Water Management in KMA

The major attributes of the Integrated Smart Water Management (ISWM) are illustrated in Fig. 20. The ISWM forms a broader picture of a connected environment through the interconnection of various devices taking part in water management. The ISWM need to be capable of acquiring various information related to water management through a system of sensors measuring various parameters related to water resources, distribution, and consumption. It will be capable of assimilating various information and making predictions and appropriate decisions for various predicted scenarios. With the help of the predictive capability and availability of all possible information, the system should be able to implement the decisions through a system of intelligent devices, which control the water supply and management system. The system also disseminates useful and actionable information for each stakeholder through a variety of methods, which include, web interface, mobile platforms and voice enabled interface.

Integrated Smart Water Management (ISWM) in KMA is diagrammatically shown in Fig. 21. It integrates six components of smart water to ICT and IoT design, E-Water Democracy Practice and E-Water Governance Practice. E-Water Democracy is the support and enhancement of democracy, democratic institutions and democratic process by means of ICT. E-Water Governance refers to the use of government agencies of information technologies that have the ability to transform relations with citizen business and other arms of Government. These Technologies can serve a variety of different ends: better delivery of Government services to citizens, improved interaction with business and industry, citizen empowerment through access to information or more efficient government management. The resulting benefits are less corruption, increased transparency, greater convenience, revenue growth and cost reduction [2].

10.1 IoT and ICT Design for ISWM

The key difference of smart water communities from other water communities is their dependence on continuous real-time data about the state of the water sources, consumption, demand and the distribution process. This is achieved through the deployment of an array of IoT sensors throughout the water network.

The IoT and ICT framework consists of sensor networks, their communication infrastructure, data centres, data analysis centres and information dissemination systems. Figure 22 presents the ICT backbone arrangement proposed for KMA.

IoT sensors are standalone devices with their own source of power and means of communication to the internet. The sensors are designed to measure a specified quantity at a regular interval and communicate the measurements to the gateway through a suitable communication protocol. There are a number of communication techniques being used for IoT communication. Figure 23 compares some of the most used transport layer protocols used for the IoT communication. Figure 24 graphically

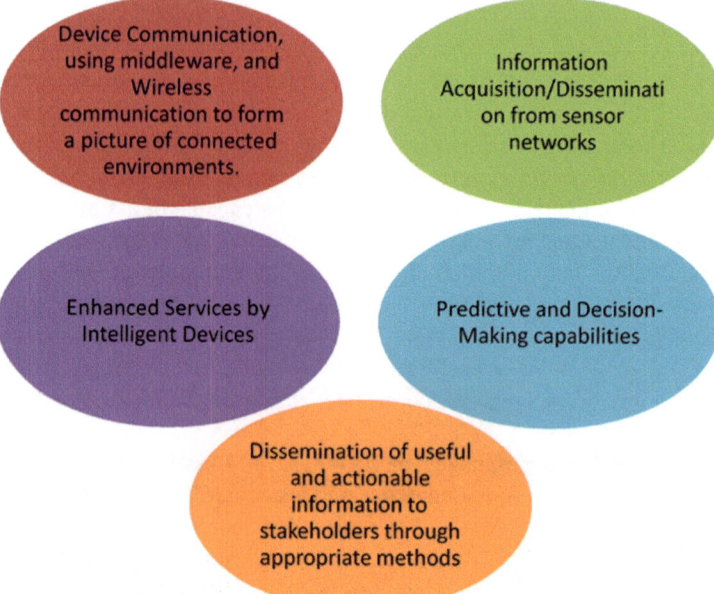

Fig. 20 Major attributes of integrated smart water management system

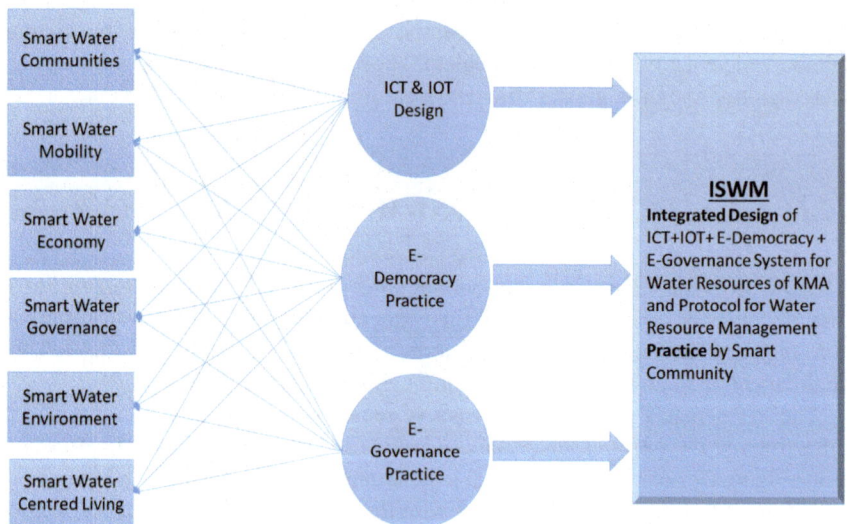

Fig. 21 Components of ISWM

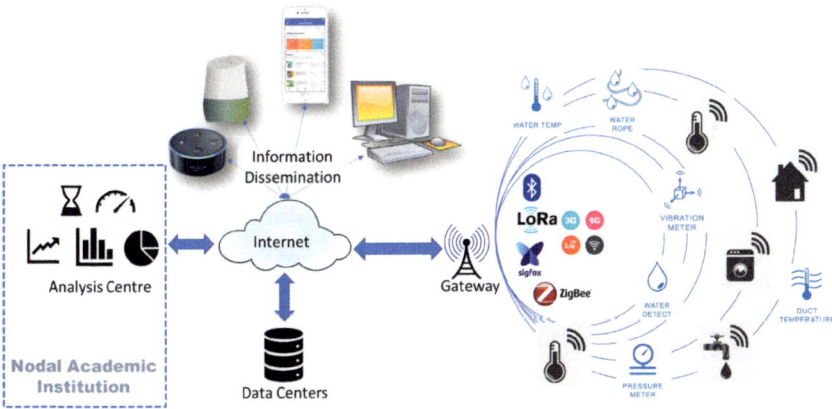

Fig. 22 ICT backbone for KMA

Technology	Frequency	Data Rate	Range	Power Usage	Cost
2G/3G	Cellular Bands	10 Mbps	Several Miles	High	High
Bluetooth/BLE	2.4Ghz	1, 2, 3 Mbps	~300 feet	Low	Low
802.15.4	subGhz, 2.4GHz	40, 250 kbps	> 100 square miles	Low	Low
LoRa	subGhz	< 50 kbps	1-3 miles	Low	Medium
LTE Cat 0/1	Cellular Bands	1-10 Mbps	Several Miles	Medium	High
NB-IoT	Cellular Bands	0.1-1 Mbps	Several Miles	Medium	High
SigFox	subGhz	< 1 kbps	Several Miles	Low	Medium
Weightless	subGhz	0.1-24 Mbps	Several Miles	Low	Low
Wi-Fi	subGhz, 2.4Ghz, 5Ghz	0.1-54 Mbps	< 300 feet	Medium	Low
WirelessHART	2.4Ghz	250 kbps	~300 feet	Medium	Medium
ZigBee	2.4Ghz	250 kbps	~300 feet	Low	Medium
Z-Wave	subGhz	40 kbps	~100 feet	Low	Medium

Source: https://blog.helium.com/802-15-4-wireless-for-internet-of-things-developers-1948fc313b2e

Fig. 23 Comparison of transport layer protocols

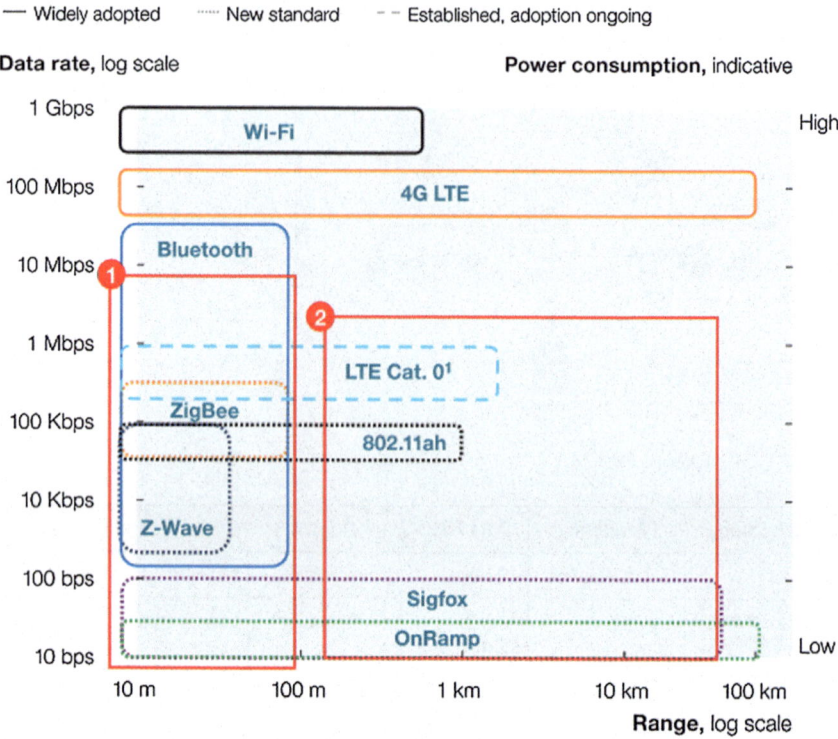

Fig. 24 Comparison of transport layer protocols based on data rate and range. *Source* https://blog. helium.com/802-15-4-wireless-for-internet-of-thingsdevelopers-1948fc313b2e

compares the data rate and range of these communication protocols. It is evident that each protocol has it's own peculiarity. 3G/4G LTE networks are suitable for long range and high data rate requirements, but they are expensive and consumes lot power. However, most of the IoT data transmissions are very tiny and needs only less than 1 Kbps data rate. The power consumption of IoT devices is a critical factor, as it is difficult to get the continuous power supply to all the devices located at various parts of the water network. The solution is to use low power devices with a built-in battery, which will last for several years. SigFox is a suitable protocol with these properties.

Gateways communicate the measurements through the internet infrastructure to the data centres. The data gets accumulated at the data centres. The data centres can be anywhere on the globe and can be accessed via the internet.

Figure 25 shows the GSM (Global System for Mobile) coverage which includes 2G, 3G and 4G versions. The eastern side has poor connectivity owing low population density and highly undulating terrain of Western Ghat. However, many of the rivers originate from this Western-Ghats range and are crucial for managing water sources. However, most of the houses in the study area including the Western-Ghats range

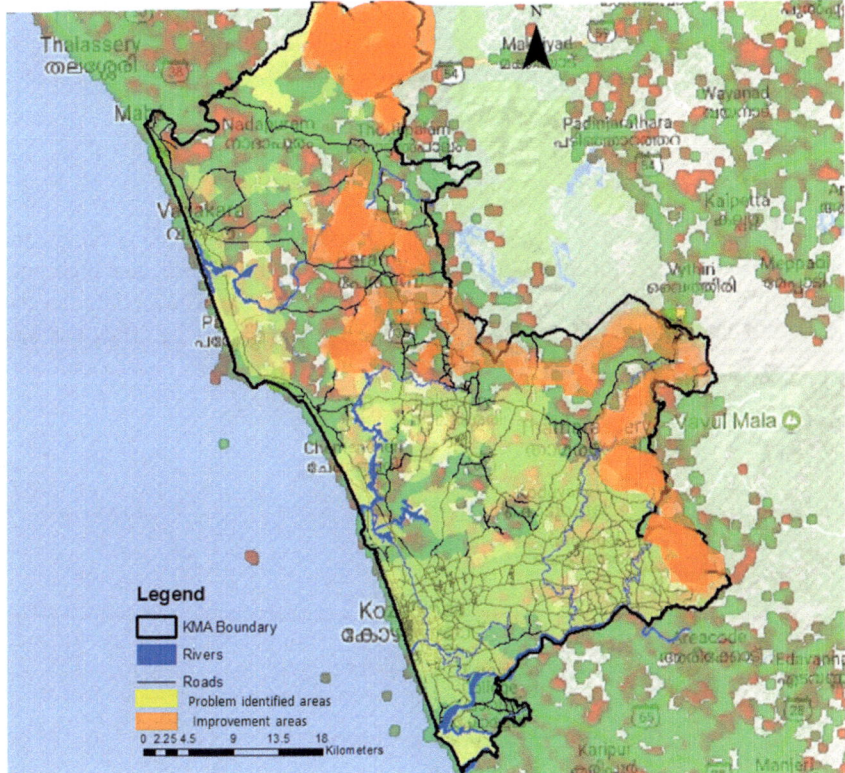

Fig. 25 GSM coverage in KMA

are having optical fibre cables laid for the cable TV network. Though the cables also provide high-speed data connectivity, they are hardly used in the present scenario. The gateways can be connected together using this optical fibre network to form a Metropolitan Area Network (MAN) [22]. This will help in securing the devises and data from external access. However, there are multiple Internet Service Providers (ISP) available, and it is possible to use multiple ISPs for different gateways. In such a case a virtual metropolitan area network (VMAN) may be used to bind together the components securely.

Each smart water community works along with a Nodal Academic Institution (NAI) in the region, which also serves as the Sync Stations for the uploaded data. As per the Triple Helix Concept [23], the role of the nodal institutions are much broader. They are responsible for providing indigenous solutions for the technological problems faced by the Smart Water Community. The NAI utilises the expertise of their faculty and students to manage the research and innovation requirements of the Smart Water.

Community, at zero marginal cost. Four NDIs were identified in the KMA as listed below.

1. National Institute of Technology, Calicut
2. Indian Institute of Management, Kozhikode
3. College of Engineering, Vadakara (Affiliated to CUSA T)
4. Model Polytechnic College, Vadakara.

These NDIs are already connected with high speed dedicated optical cables, and they have enough spare bandwidth for the use by SWCs. These dedicated highspeed connections serve as the ICT backbone for the SWC (Fig. 26). These NDIs can also act as the data centres too, provided they have the spare capacity. They can also

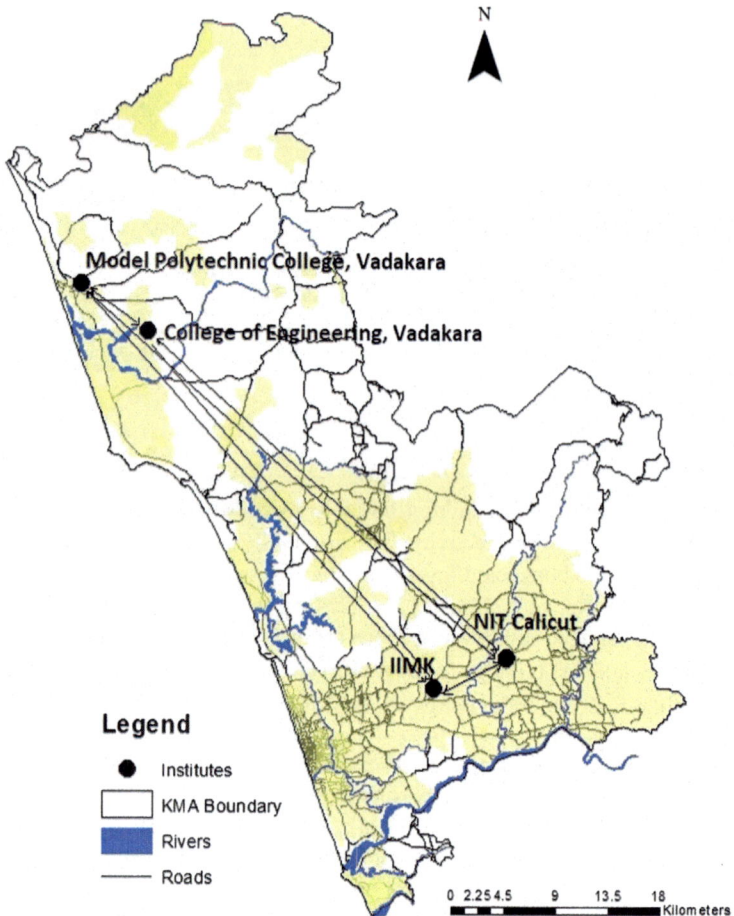

Fig. 26 Nodal academic institutions

synchronize the data each other to prevent any potential data loss due to technical failure of any one among them.

11 E-Water Democracy

The concept of water democracy is that water is a gift of nature and a right for any living being on the planet. This right is to use the water but not to own it. Shiva [24] emphasises that the right for water is not accorded by the state, but that evolve out of the ecological context of human existence. Hence the management of surface and groundwater has to be done with utmost accountability and oversight. The concept of water democracy points out the water conservation, protection and equitable distribution of water for the benefit of every living being of nature.

Vandana Shiva in her famous book "Water wars" [24] had laid out nine principles of water democracy, as given below.

1. **Water is nature's gift**: We owe it to nature to use this gift in accordance with our sustenance needs, to keep it clean and in adequate quantity. Diversions that create arid or waterlogged regions violate the principles of ecological democracy.
2. **Water is essential to life**: Water is the source of life for all species: All species and ecosystems have a right to their share of water on the planet.
3. **Life is interconnected through water**: Water connects all beings and all parts of the planet through the water cycle. We all have a duty to ensure that our actions do not cause harm to other species and other people.
4. **Water must be free for sustenance needs**: Since nature gives water to us free of cost, buying and selling it for profit violates our inherent right to nature's gift and denies the poor of their human rights.
5. **Water is limited and can be exhausted**: Water is limited and exhaustible if used non-sustainably. Non-sustainable use includes extracting more water from eco-systems than nature can recharge (ecological non-sustainability) and consuming more than one's legitimate share, given the rights of others to a fair share (social non-sustainability).
6. **Water must be conserved**: Everyone has a duty to conserve water and use water sustainably, within ecological and just limits.
7. **Water is a commons**: Water is not a human invention. It cannot be bound and has no boundaries. It is by nature a commons. It cannot be Owned as private property and sold as a commodity.
8. **No one holds right to destroy**: No one has a right to overuse, abuse, waste, or pollute water systems. Tradable-pollution permits violate the principle of sustainable and just use.
9. **Water cannot be substituted**: Water is intrinsically different from other resources and products. It cannot be treated as a commodity.

E-Water democracy point to the application of electronic communication technology for ensuring highest level of efficiency and accountability in the management of water and its resources. This chapter emphasises particularly on those solutions that have become more efficient by the application of ICT related techniques.

11.1 Case Studies of Water Democracy

Few case studies were conducted to understand various implementations of water democracy principles.

11.1.1 'Pani Panchayat' Maharashtra

Pani Panchayat is a movement of farmers of Naigaon village motivated by Mr. Vilasrao Salunke to meet the water requirement of the village of the drought-prone Purandhar taluka of Maharashtra [25]. It is a voluntary activity of a group of farmers engaged in collective management (harvesting and distribution) of surface water and groundwater. The government's inability to deal with the drought situation lead to the movement. They took a 40 acre land on lease from the village temple trust and developed a recharge pond, a dug well in the discharge zone, and a lift irrigation system (Fig. 27). It was later scaled up to meet the demands of the farmers. They formed Gram Gaurav Pratishthan (GGP) who manage both groundwater and surface water. This was with a vision of equitable distribution of water to all of its people in the village through sustainable development of the watershed to improve the quality of life of its inhabitants and participating communities through education, training and active participation in sustainable production activities [25].
Key lessons:

- Assured water for the population who are depending on land for livelihood.
- Natural resource protection and conservation especially water.
- Protecting water rights for weaker section of the society.
- Safeguarding agricultural income or provide alternate livelihood support to the people.
- Practice organic farming for sustainable development.
- Promote harmony among the inhabitant and the environment.
- To develop a watershed model based on water equity that would transfer philosophy and technology nationally and internationally in agricultural, environmental and conservation of water sector in an integrated way.
- Water user group for each lift irrigation scheme which in turn had a representation in the village level Pani Panchayat.
- Decentralized decision making with public participation.

11.1.2   Community Action for Self Sufficiency in Drinking Water Supply Olavanna, Kerala

Olavanna Gram Panchayat in Kozhikode district had a drinking water scarcity problem. Olavanna presented a classic case of 'water, water everywhere, but not a drop to drink'. The 3 rivers, including the big Chaliyar, flowing through the Panchayath are saline. Other non-saline surface water bodies in the Panchayath go dry in February.

Small groups of villagers of Olavanna Gram Panchayat in Kozhikode district have been organizing themselves into groups to find solutions to their water needs. They collected money and did setup micro piped water supply projects and manage that themselves without waiting for the government intervention. The villagers realized that local needs required local solutions. A registered co-operative society was formed to provide drinking water for their own needs [26]. The society is organized as a general body consisting of all beneficiaries which counts to average 50, an Executive Committee of 7–11 members and four elected members for the posts of President, Vice-president, Secretary and Treasurer. Table 6 compares the private micro water supply schemes to the grama panchayat and Kerala Water Authority (KWA) schemes. It is evident that the private schemes perform better than the other by providing water 24 h against two to three hours of supply in other schemes.

After the success of the model, the Government of Kerala had formed a scheme for micro water supply projects which is known as Jalanidhi with the help of world bank funding. Jalanidhi helps such self-help groups to organize themselves, gives guidelines, and helps to arrange technical know-how [27, 28].

Key lessons:

Fig. 27 Water conservation work—village Kumbarwalan. *Source* https://www.indiawaterportal. org

Table 6 Comparative assessment of rural water supply schemes in Olavanna

	Private scheme	Gram panchayat	KWA
Per family share in capital cost	Rs. 4500 (full recovery)	Rs. 7000 (25% recovery)	Rs. 7000 (no recovery)
Average capital cost	Rs. 2.5 lakhs	Rs. 3.75 lakhs	Rs. 16.8 lakhs
Per family share in O&M cost (per month)	Rs. 25–50 (full recovery)	Rs. 10–20 (75% recovery)	Rs. 17 (25% recovery)
Average number of house connections	54	52	240
Number of public stand posts	–	20	45 (25 in use)
Supply hours	24 h (10 h in Apr–may)	2–3 h (1 and a half hrs in Apr–May	Uncertain and poor service
Supply months	12	12	9-Aug
Management responsibility	Society	GP	KWA
Number of schemes	26 (6 under construction)	18 (12 under construction)	3

- Willingness to pay—capital cost, O&M cost
- Private schemes are more cost-effective
- User management has led to user satisfaction
- Local expertise.

11.2 Planning for E-Water Democracy

Each of the micro-watersheds in KMA was identified in Sect. 5.5. Water communities were identified based on the watershed they belong. Each of these watersheds was analysed to identify their water-related problems. The communities were categorised based on the kind of problems they face. Prototype communities were identified to study the issues in detail. Household surveys were conducted, and detailed data collection and analysis were performed to understand the issues and identify solutions. Detailed proposals were prepared for each of these prototype communities, which may be adoptable by similar communities.

As discussed earlier in Sect. 5.1, KMA region enjoys a heavy rainfall for approximately six months in a year. Water retention and preventing contamination are the key challenges to be tackled for KMA. The rainwater available on the rooftop of a residence itself is sufficient to meet the demand of a household for a year as evident from the calculation given below.

Average annual rainfall in Calicut (R)   3200 mm/year

The roof area of a typical house (A)	100 m^2
Runoff coefficient (C)	0.85
Annual water harvesting potential of a household	$100 \times 3.2 \times 0.85 = 272 \text{ m}^3$
	$= \textbf{2,72,000 L}$
Annual water requirement	Household size $\times$ Per-capita consumption per day $\times$ 365
	$= 4.5 \times 135 \times 365 = \textbf{2,21,737.5 L}$

Form the above calculation it is evident that, the potential is not only enough to meet the requirement, but also capable of providing a surplus of 50,000 L per annum.

11.3 Proposals for E-Water Democracy

Proposals for the e-water democracy are discussed based on the prototype watersheds identified with the set of problems they face.

11.3.1 Ground Water Recharge

In the areas where there is a problem of groundwater depletion efficient groundwater recharge options can be adopted. Paddy fields are considered to areas of natural recharge. In watersheds where paddy field is present, the stormwater surface runoff can be directed to paddy fields. Watershed 37 has reported the problem of groundwater depletion. As the watershed contains a considerable area of paddy fields as shown in Fig. 28. it can be used as recharge areas. Drains can be provided along the natural drainage which will collect rainwater from each housing plots to the nearby paddy field. The particulars of the watershed 37 is presented in Table 7.

The residents are encouraged to adopt any one or more of the proposed schemes given in Table 8 to harvest water.

Table 7 General particulars of watershed 37

Watershed number	37
Local body	Unnikulam
Area	19.2 km^2
Population 2011	46,354
Number of households	10,990

11.3.2   Agriculture

The KMA is having tremendous potential in agriculture as it got a large area of paddy and fallow land. The watershed 51 has got a considerable area of fallow land which can be used for agriculture and horticulture as given in Fig. 29 and Table 9. However, the water scarcity in the summer months is deterrent for vegetable cultivation. The solution is to utilize every drop of water efficiently and avoid wastages. New techniques such as drip irrigation and sprinkler irrigation can be effectively applied here with effective water management.

IoT for Drip Irrigation

In drip irrigation, water is delivered to plant roots through a series of pipes, tubes, and valves. It is controlled by emitters and pumps, which allows water to be focused in an area. Also, reduce evaporation and runoff and contribute to water conservation [30]. Cost of installation for a drip system is around Rs. 60,000–75,000 per acre for growing vegetables and around 35,000 per acre for fruits.

The drip irrigation can be made smart by incorporating sensors and actuators to control the soil moisture level. This helps to reduce overall water requirement even further, but at the same time ensures the required level of water moisture all the time. The IoT based sensors can relay the soil moisture, and nutrients in the soil, which may be used for arriving at the optimum amount of irrigation and timing of irrigation.

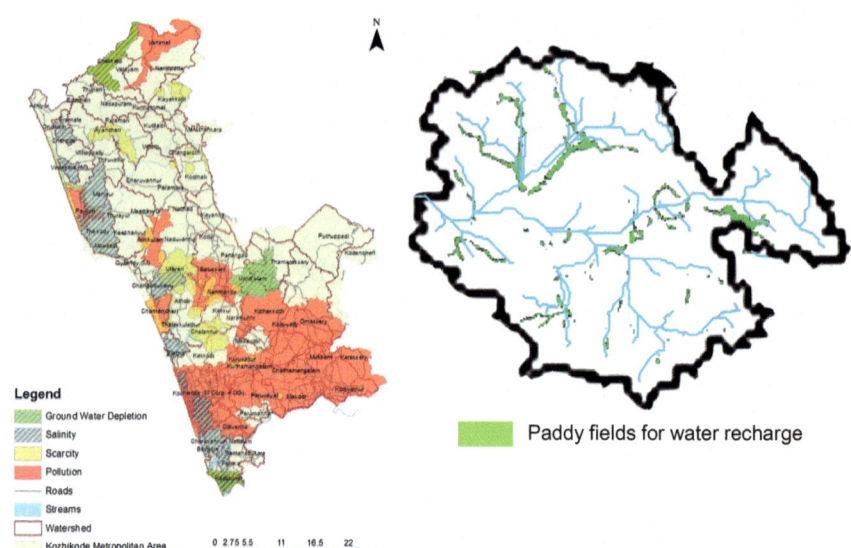

Fig. 28 Paddy fields in watershed 37

Table 8 Schemes for water recharge and harvesting

Sl no.	Method	Description	Technical feasibility	Economic feasibility
1	'Mazhapolima' (Govt. of Kerala promoted scheme) [29]	Harvest the rooftop water by filtering and redirecting it to the already existing open well	Feasible in almost all the houses. Low technical expertise required	Low cost. One-time expenditure for the plumbing and filter. Feasible for any economic class
2	'Jalanidhi' (Govt. of Kerala promoted scheme)	Harvest the rooftop water by filtering and storing in ferro-cement tanks	Feasible in almost all the houses. Ferrocement tank making and plumbing need expertise	Low cost. One-time expenditure for the plumbing, filter and tank. Tank costs about Rs.1.7/L. Feasible to any economic class as government subsidies are available for the tank
3	Community ponds	Harvest the rooftop as well as surface runoff water by directing them to community ponds	The terrain needs to be analysed to identify the feasible location for the community ponds. Suitable land pooling techniques may be employed to avail the land if it belongs to the private party	Land purchases are expensive and hence may require a larger number of beneficiaries to make the individual contribution within the feasibility limit
4	Recharge through paddy fields	Recharge groundwater by redirecting the runoff water to paddy fields and allowing water to be stagnant there for a longer time	It is feasible only to regions, where there are paddy fields existing in the path of natural drainage	Feasible to all classes as there is no major intervention is required, apart from keeping the paddy fields safe from encroachments

It is also possible to analyses this knowledge with weather predictions and optimizes the water usage accordingly [30].

The smart features can be controlled through an ordinary smartphone, from literally anywhere. It also enables the farmers to keep a watch on their crop all the time. Another important aspect is to compare the growth of plants from similar farmers, straight away on the mobile screen, and keep learned about fertilisation regime they follow for better growth. Figure 31 shows the mobile interface for the smart irrigation control system (Fig. 30).

Micro Hydro Projects

The eastern side of KMA has highly undulating terrain, and plenty of streams originate from the sides of the Western Ghats. These streams may be tapped to generate power. Micro hydro is a type of hydroelectric power that typically produces from 5 to 100 kW of electricity using the natural flow of water. A schematic of the micro hydro project is presented in Fig. 33. These installations can provide power to an isolated home or small community or are sometimes connected to electric power networks. Micro hydro systems complement solar PV power systems when water flows and solar energy is at a minimum. Micro hydro is frequently accomplished with a Pelton wheel for high head, low flow water supply. As there are potential areas in KMA

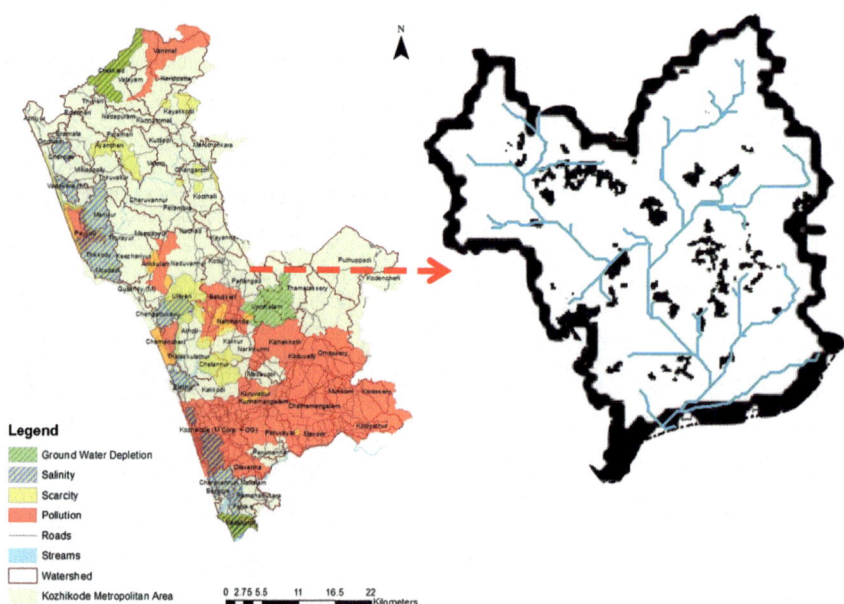

Fig. 29 Fallow land in watershed 51

Table 9 General particulars of watershed 51

Watershed number	51
Area	11.89 km^2
Population 2011	66,941
Number of households	13,384

community can set up micro-hydro systems which can produce electricity necessary to run small-scale business. Excess electricity if produced can also be transferred to the power grid which will fetch them income. Potential locations for installation of micro-hydro projects were identified in KMA region as given in Fig. 32.

11.4 Tourism

The tourism potential of KMA is not utilised extensively. Kerala has its own footprint in tourism. Many watersheds of KMA has the potential for tourism development due to the presence of backwaters and beaches.

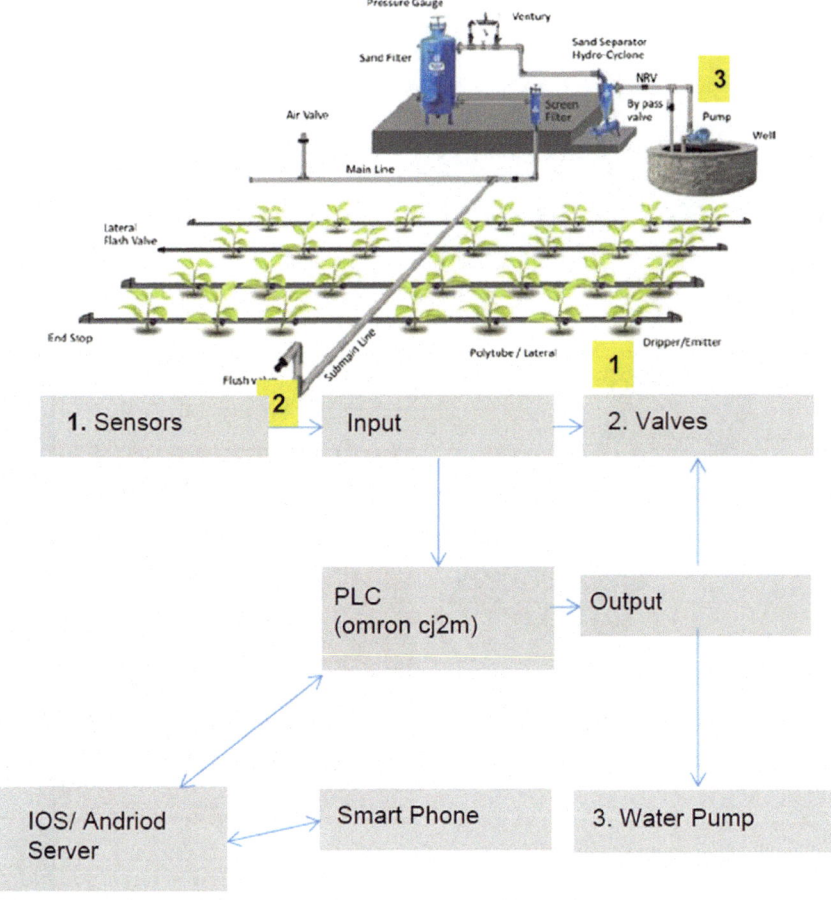

Fig. 30 IoT arrangement for drip irrigation

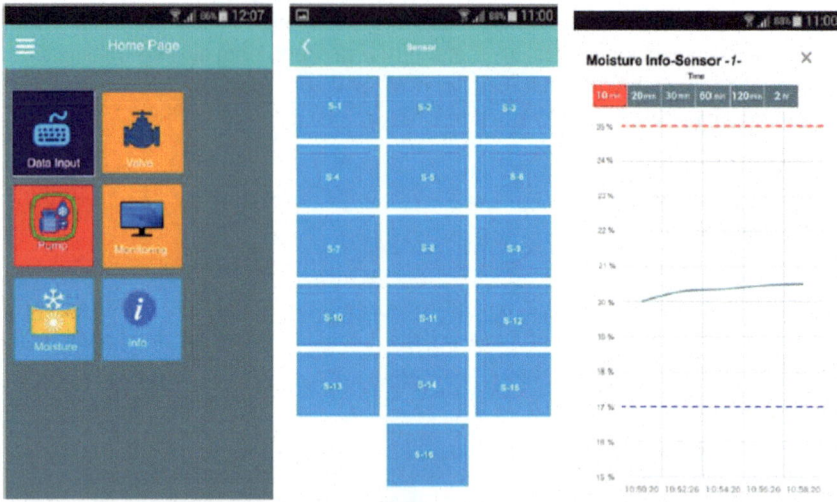

Fig. 31 Interface for smart drip irrigation

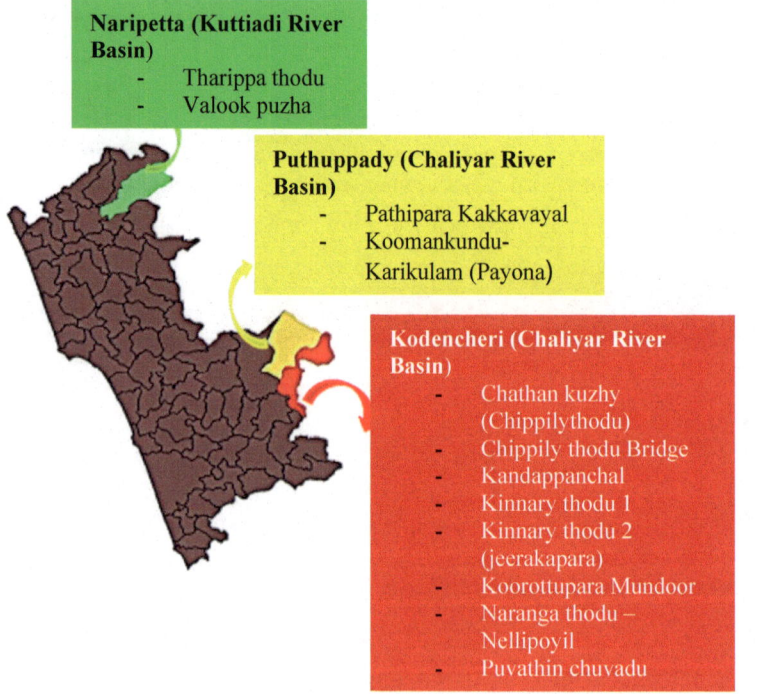

Fig. 32 Micro-hydro potential areas in KMA

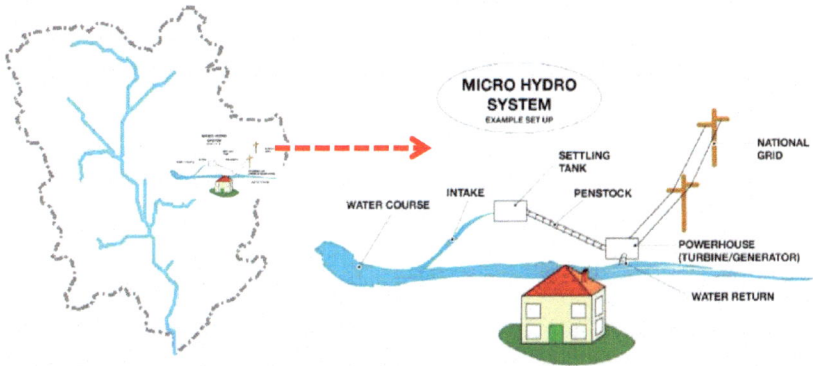

Fig. 33 Schematics of micro hydro power

11.4.1 Back Water Tourism

Even though KMA got a backwater, the potential is not yet tapped to the extent possible. Figure 34 shows the watersheds having tourism potential due to the presence of backwaters. The proposed backwater network will be connecting the Elathur backwater and the Kallai river through Canoli canal with a navigable length of 20 km.

Elathur backwater is in the watershed 30 which has an area of 68.9 km², population according to 2011 census as 27,361 and a total of 6810 households. The other watersheds in which the backwater network is passing through are 30, 35, 33, 38, 43 and 48. The tour packages will be of one day, two days etc. and it includes the facility for staying, Fishing and cooking.

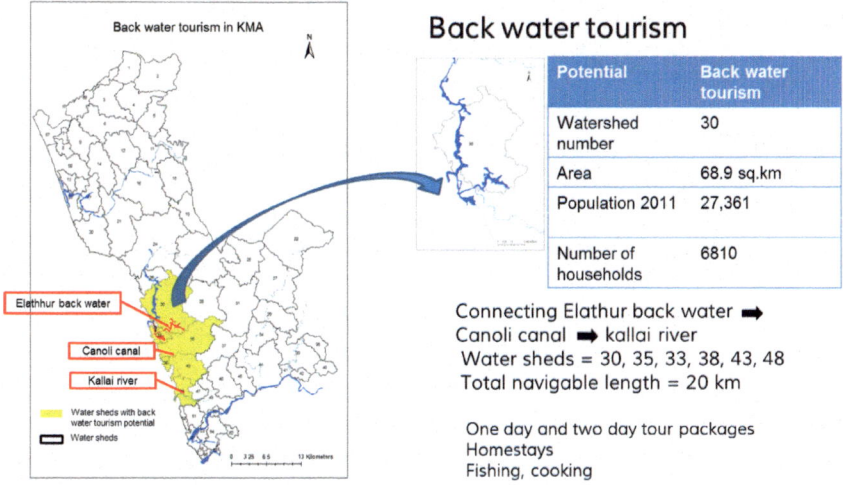

Fig. 34 Back water tourism route in KMA

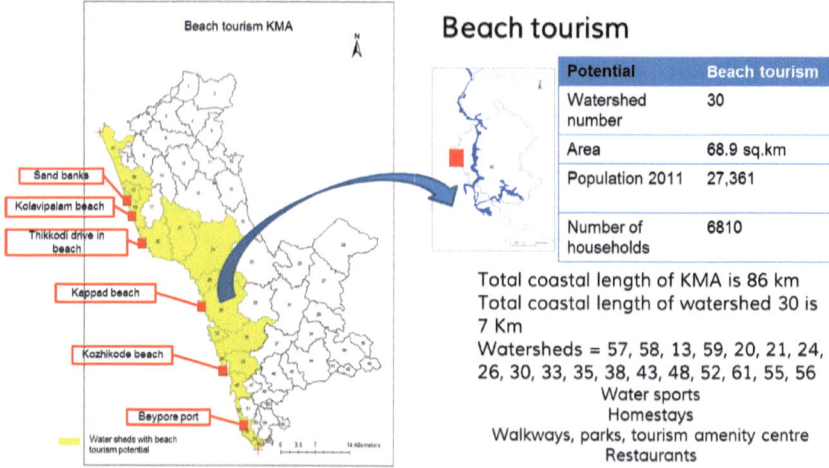

Fig. 35 Watersheds in KMA with beach tourism potential

Fig. 36 Kappad beach. *Source* https://travellifejourneys.blogspot.com/2016/04/kappad-beach-kerala-india.html

11.4.2 Beach Tourism

Total coastal length of KMA is of 86 km in which only a few beaches like Kozhikode and Kappad (Fig. 36) are attracting people. The other places identified for developing beach tourism facilities are Beypore port, Thikkodi drive in beach, Kolavipalam, beach, and sandbanks etc. There are 18 watersheds in KMA which has untapped beach tourism potentials. Figure 35 shows the watershed with potential for beach tourism.

11.4.3 Ecotourism

Potential ecotourism destinations identified in the KMA are Janakikad, Vanaparvam, Tusharagiri waterfalls in the eastern side and Kadalundi bird sanctuary in the west. However, these destinations are least used as they are not connected to each other. These can attract more tourists if there is an identified tourism circuit.

11.4.4 Tourism Circuits

Tourism circuits are the one day or two-day tour packages with a fixed destination. Two tourism circuits are proposed in the KMA region which are,

(1) Kadalundi—Beypore port—Kozhikode beach—Kappad beach
(2) Tusharagiri—Vanaparvam—Janakikad.

The first circuit is a mix of ecotourism and beach tourism in the western part of KMA which will helps to tap the potential of beaches in the KMA. The second one completely constitutes ecotourism destinations which includes a waterfall and two forests. The circuit is located on the eastern side of KMA. Tourism circuits of KMA are shown in the figure.

11.4.5 ICT and IoT Proposals for Tourism

(1) **Participatory Sensing**: Here the visitors or the local community can upload digital photos or videos about the destination that they have visited which can be used for mapping and tracking flora and fauna activity and habitats.
(2) **Carrying Capacity**: Here the sensors will be placed in each of the tourism destinations which will constantly monitor the number of visitors in the destinations. Once the destinations reach its carrying capacity the same will be shown on the website to avoid the further entry of visitors. CCTV footages may be subjected to image sensing techniques to assess the number of visitors.

11.5 Fisheries

For effective monitoring of the water quality level which is important for the promotion of fish health, a set of smart sensors are proposed. The data from these sensors make it easy to do monitoring of fish behaviour, fault diagnosis, automatic control, information management and equipment monitoring.

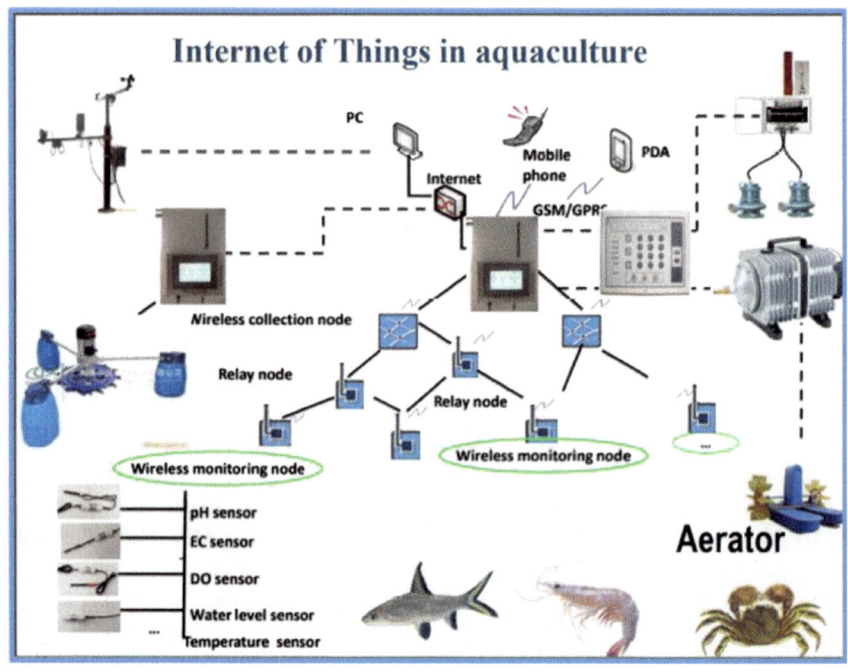

Fig. 37 IoT for aquaculture. *Source* [31]

11.5.1 Internet of Things in Aquaculture

Under the IoT based fish farm management system, sensors are installed on each fish tank to measure water temperature, quality and oxygen level etc. The water quality will be constantly monitored and uploaded to the cloud (Fig. 37). If any abnormality is found, even in the middle of the night, the management system identifies suitable remedies and implants that using actuators connected to equipment like aerator, shades etc. It alerts the farmers when their interventions are required and thereby ensuring stable and efficient farm management.

12 Water Governance

Water governance is highly contextual, the water policies need to be tailored to different water resources and places, and that governance responses must adapt to changing circumstances. Water governance cycle is illustrated in Fig. 38. Basic points in the concept of water governance are,

- Water diplomacy
- Conflict resolution
- Public participation
- Strategic planning
- Finance administration
- Public policy and law.

The concept of water diplomacy is defined by various academics and organizations in a different manner. Water diplomacy includes all measures by state and non-state actors that can be undertaken to prevent or peacefully resolve (emerging) conflicts and facilitate cooperation related to water availability, allocation or use between and within states and public and private stakeholders [32]. Practice shows that water-related conflict prevention and resolution is largely the outcome of processes of research and fact-finding, negotiation, mediation and conciliation that are rooted in an in-depth understanding of the social/cultural/economic/environmental conditions and the political context. This should be supported by a sound assessment and integrated analysis of the water system [33]. There are few main governance gaps hindering water policy design and implementation.

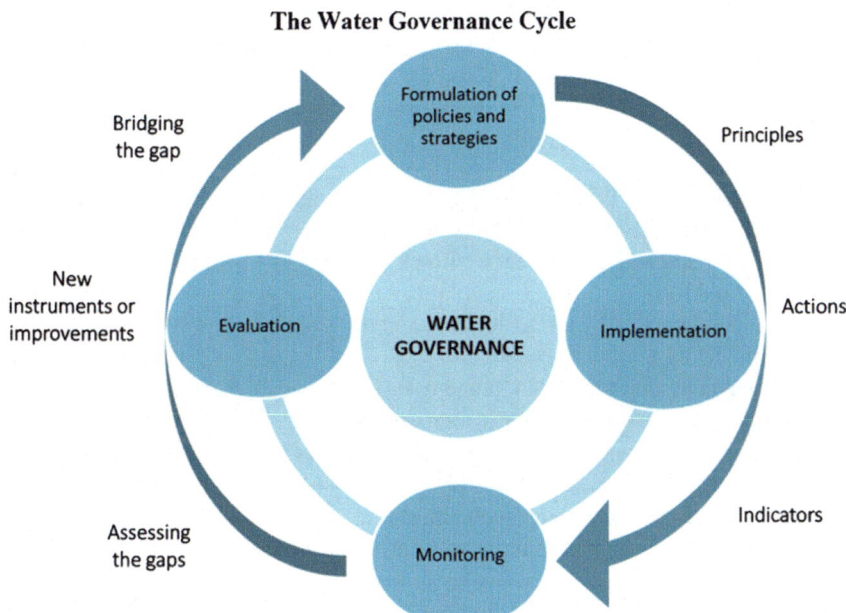

Fig. 38 Water governance cycle. *Source* Forthcoming, OECD working paper, 2015, Water Governance Indicators

12.1   Functions of Different Authorities

As per the constitution of India, the responsibility of providing wholesome water is vested in the 3 levels of local self-government in case of rural area and in municipalities/corporations in case of the urban area. The duties related to public water supply vested in these governing bodies are extracted and given below.

Grama Panchayat

(1) Maintenance of traditional drinking water sources.
(2) Preservation of ponds and other water tanks.
(3) Maintenance of waterways and canals under the control of Village Panchayats.
(4) Collection and disposal of solid waste and regulation of liquid waste disposal.
(5) Stormwater drainage.
(6) Management of water supply schemes within a village panchayat.
(7) Setting up of water supply schemes within a village panchayat.
(8) Examining the complaints against the Public Distribution System and find out and implement remedial measures.

Block Panchayat

(1) Provide technical assistance to Village Panchayats.
(2) Prepare schemes taking into consideration the schemes of village panchayats in order to avoid duplication and to provide backward, forward linkage.
(3) Implementation and maintenance of all Lift Irrigation Schemes and Minor Irrigation.
(4) Schemes, covering more than one village panchayat.

District Panchayat

(1) The mobilisation of the technical expertise available from Government-non-Government institutions.
(2) Provide technical assistance to Block Panchayats, Village Panchayats and Municipalities.
(3) Prepare schemes after taking into account the schemes of the Village Panchayat and the Block.
(4) Panchayat to avoid duplication and to provide forward-backwards linkage.
(5) Development of groundwater resources.
(6) Construction and maintenance of minor irrigation schemes covering more than one Block.
(7) Panchayat Command area development.
(8) Implementation of water supply schemes covering more than one Village Panchayat.
(9) Taking over of water supply schemes covering more than one Village Panchayat.

Municipality

(1) Conservation of traditional drinking water sources.
(2) Preservation of ponds and other water tanks.
(3) Maintenance of waterways and canals under the control of the Municipality.
(4) Stream water drainage.
(5) Implementation and maintenance of all minor and lift irrigation projects within the Municipal areas.
(6) Implementation and Maintenance of all micro-irrigation projects.
(7) Carry out conservation of water.
(8) Implementation of groundwater resources development.
(9) Maintain water supply schemes within the respective Municipal area.
(10) Arrange water supply schemes within the respective municipalities.
(11) Examine complaints against public distribution system and to find out and implement remedial measures.

12.2 Proposals

The general problems identified in KMA are presented in Table 10. A spatial decision support system (SDSS) is proposed to help in implementing the governance system for water. SDSS is an interactive, computer-based system designed to assist in decision making when spatial aspects are involved the problem or its solution. It is designed to assist the decision maker with guidance. A system which models decisions could be used to help identify the most effective decision path. The SDSS for water governance was designed through four steps.

(1) Issue identification
(2) Data collection
(3) Alternate potential solutions

Table 10 General problems Identified in KMA

Problem	Aim	Objectives
Water pollution	To resolve the water pollution problems in KMA	• To identify the pollution levels of water bodies • To monitor and regulate water polluting activities
Water scarcity	To ensure enough water supply equitably	• To regulate the withdrawal and usage
Salinity	To treat the issue of salinity	• To control and monitor activities resulting in salinity
Groundwater depletion	To monitor ground water depleting	• To regulate water tapping activities • To monitor land use violations

(4) Selection of optimal solution.

Variety of data are being collected by the sensors on all parameters of water supply and management. This information is linked to GIS to generate spatial representation and analyse the problems spatially. Alternate solutions are prepared for each category of problem, and a final solution is identified by analysing the data for each scenario separately.

12.2.1 Existing Laws and Regulations

There are six existing laws to deal with water resource management and distribution among that two are central and four are state laws. These rules were studied to identify the legal provisions to deal with several violations that may arise in water management. A summary of the act, potential violation, and its legal provisions are given in Table 11.

12.3 SDSS Implementation of Selected Solutions

Some of the solutions implemented through SDSS for prototype water communities are presented below.

12.3.1 Determination of Source of Pollution Outlet Along Rivers

According to survey findings, illicit discharge of waste was into water bodies such as Chaliyar and Canoli canal was one of the major problems. According to the water (prevention and control of pollution) rules 1975, to permit any poisonous, noxious or polluting matter into any stream or well or sewer or on land based on standards by state pollution control board is punishable by law. The penalty is 1.5 years of imprisonment which can extend up to 6 years with fine. Polluting activities are rampant which can be monitored with the help of SDSS as shown in Fig. 39.

12.3.2 Monitoring Zonal Violation

A zonal violation can be monitored with the help of a monitoring framework implemented with the help of SDSS and ICT. It was found that most of the violations are intentional while few are out of ignorance about the existing land use control. The intentional violations happen on holidays to avoid immediate action by authorities, and often get regularised as there was no record of when the violations took place. This requires an automated system to constantly monitor the land use changes and record the violations as soon as it takes place.

Table 11 Summary of the act, potential violation, and its legal provisions

Law/act	Year	Rules	Objective	Penalty
Water (prevention and control of pollution) rules—Ministry of Environment and Forests	1975	Anybody who do or permit any poisonous, noxious or polluting matter into any stream or well or sewer or on land shall be prosecuted and imprisoned for 1.5 to 6 years	Provide for the prevention and control of water pollution, and for the maintaining or restoring of wholesomeness of water in the country	1.5 year of imprisonment extend up to 6 years with fine
		Industry likely to discharge sewage to river cannot be established without permission		2 years to 6 years of imprisonment with fine
Water (prevention and control of pollution) rules—Ministry of Environment and Forests	1978	Every person carrying on any specified industry; and every local authority, consuming water shall pay cess	An act to provide for the levy and collection of a cess on water consumed by industries and local authorities	1–6 months of imprisonment, fine of Rs. 1000 or both
Kerala water supply and sewerage act	1986	• Water supply for domestic purposes not to be used for non-domestic purpose • Prohibition of wastage of water • Unlawfully obstruct the flow of or flush, draw off or divert or take water from, any water works • Any actions to pollute water	To provide for the establishment of an autonomous authority for the development and regulation of water supply and wastewater collection and disposal	Fine of Rs.1000 to an extension of Rs. 25,000

(continued)

Table 11 (continued)

Law/act	Year	Rules	Objective	Penalty
The Kerala groundwater (control and regulation) act	2002	• In a notified area person shall submit an application to dig or convert well to bore well • Cannot dig borewell within 30 m of a public water source	To provide for the conservation of groundwater and for the regulation and control of its extraction and use in the State of Kerala	Fine up to Rs. 500 Second time—fine Rs. 1000
		Unauthorised digging, construction or use of well		Rs. 2000 Second time—6 months imprisonment or fine of Rs. 6000
Kerala irrigation and water conservation act 2003	2003	• Cannot abstract water from a water course by installing mechanical equipment with 5HP • Not allowed to make river diversion works weir or any other permanent structure in or across any water course	To consolidate and amend the laws related to construction, maintenance and regulation of irrigation works and involve farmers in water utilisation system	1 year of imprisonment/Rs. 5000 or both
		• To divert any river or interlink two or more rivers or effect inter-basin transfer of water from such rivers		3 years or fine extend up to 1 lakh or both
		Quarry sand in any area in a water course within a distance of 500 m from any dam, check dam, reservoir or any other structure or construction on or across such watercourse are prohibited		2 years or Rs. 25,000

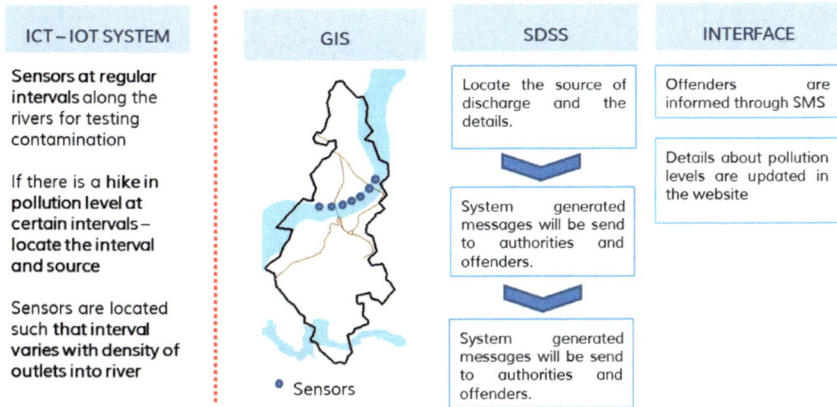

Fig. 39 SDSS for monitoring polluting activities

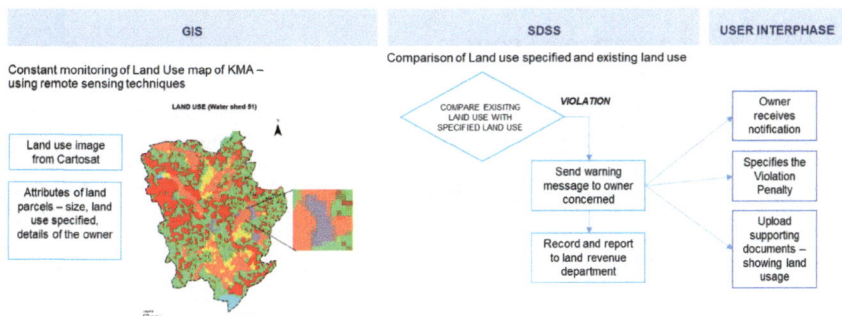

Fig. 40 SDSS for monitoring zonal violation

Change in land use pattern can be monitored with the help of remote sensing satellite imageries and cross-checking with the zonal plan of the area. Any violation such as the absence of a buffer zone around industries, illicit conversion of wetlands and paddy fields can be spotted, and the owner of the land parcel attributed by the cadastral map can be informed regarding the same. At the same time, an alert can be sent to the development authorities to inform the same. The process is explained in Fig. 40.

12.3.3 Monitoring of Water Supply

Continuous monitoring and regulation of the water supply network are possible with the help of SDSS and ICT. The entire water supply network is updated on the GIS with each segment attributed to an identified plumber who is given the responsibility

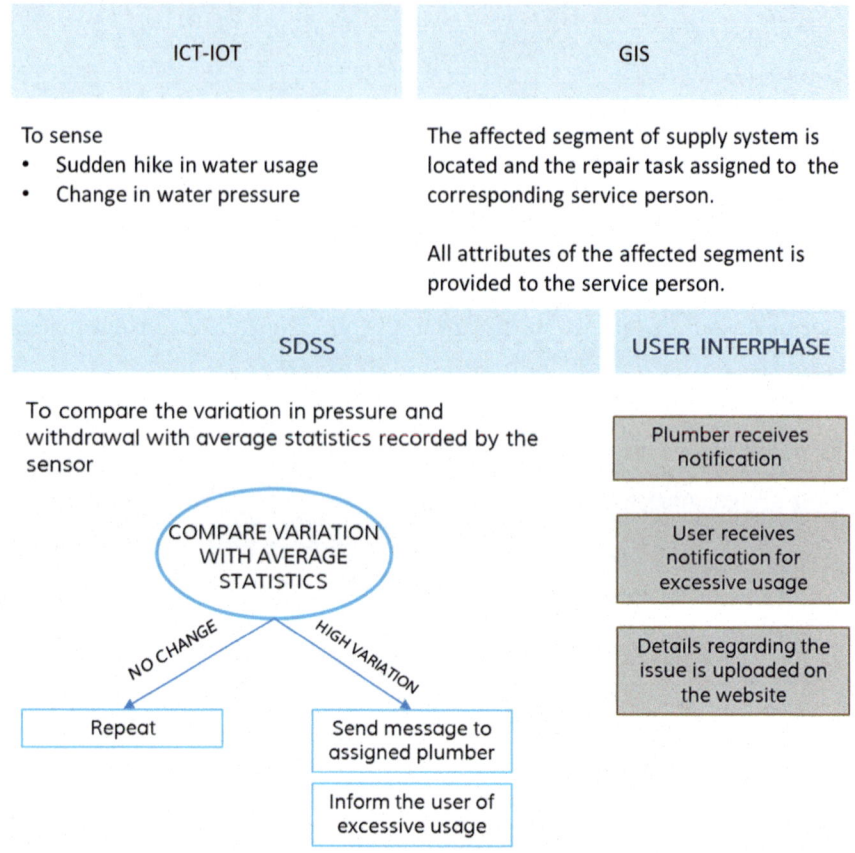

Fig. 41 SDSS for monitoring water supply

to maintain that segment of the network. He or she will be informed in the case of any leakage or maintenance work.

Sensors are deployed along the network to detect any anomalies such as a sudden change in pressure, abnormal vibration or hike in water usage. A notification is sent to the plumber in charge with the help of SDSS who inspects the water supply system and carries out the maintenance work. If the situation is recognised as pipe breakage, which is characterised by large vibrations and sudden drop in pressure, the valve on the upper side can be closed automatically through the IoT enabled valve. This will prevent the loss of water (Fig. 41).

12.3.4 Proposal for Monitoring River Sand Mining

River Sand Mining is an issue in many areas in KMA. This leads to the deepening of water bodies and the destruction of riverbanks. According to Kerala Protection of River Banks and Regulation of Removal of Sand Act, 2001 the penalty for illegal sand mining is 2 years of imprisonment and a fine of Rs. 25,000. This should be monitored and regulated using a spatial decision support system for KMA.

The temporal data can be obtained with the help of remote sensing, and the change in riverbed and shoreline can be identified. Cartosat is a satellite that monitors mining activities with image interpretation using change in tone, texture, colour etc.

In GIS, which is having the database of all information of KMA in a spatial format, the location of the issue is marked. The database is checked for the status of ownership of the land and whether it is exceeding the leased boundaries for mining if it is an approved site for sand mining. Illegal mining sites also can be identified in the same manner. The landowner is identified, and the Spatial Decision Support System sends notifications to the owner or leaser regarding the penalty to be paid. The violator should pay the fine through the website created for the Smart Water Management in KMA. The flow chart of the process is shown in Fig. 42.

13 Conclusions

Efficient and equitable management of drinking water is one of the greatest challenges the humanity faces currently. Right for fare use of water is directly linked to the ecological context of human existence. Private ownership of water resources and its management had resulted in an adverse environment and livelihood results. However, the public ownership and socialistic management of water resources appear to be ideal but had proven to be inefficient and leads to corruption. This chapter identifies certain local community practices, which are practised for centuries, and proven to be efficient and equitable. The chapter further looks, how such practices can be implemented to Kozhikode Metropolitan Region, to take care of its diverse issues related to water resource management. Smartness imparted to the region through the development of ICT framework, and other components of the smart city may be utilised to enhance the water resource management. The research attempts to achieve the principles of water democracy [20] with the help of the IoT and ICT framework of the smart region.

The region was mapped extensively to take stock of its resources, and geographic peculiarities. The region was divided into a number of watersheds. Each of these watersheds was analysed to match its water requirements and resources. Various issues faced by the people in these watersheds were enumerated, and four major issues namely scarcity, contamination, groundwater depletion, and salinity were identified.

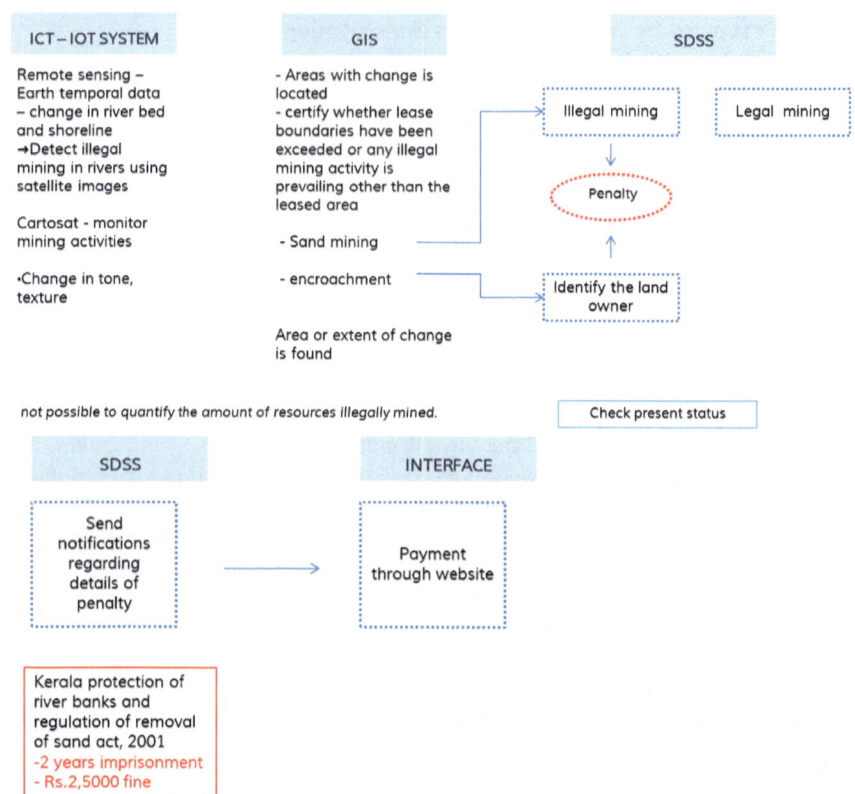

Fig. 42 SDSS for monitoring river sand mining

A number of case studies were done to get the state-of-the-art solutions to the problems faced by these communities. Various smart solutions were identified for these watersheds to forecast, detect and monitor the water resource utilisation. An IoT, ICT framework is proposed to integrate various solutions activities required for the management of the water resources. Smart water communities were proposed for the management of the water resources, which is envisaged as an agile body to monitor the resources and its equitable usage constantly. Three nodal institutions in the region were identified to help these communities to evolve solutions for their unique problems and impart training to them.

An SDSS framework is proposed to help these communities to function efficiently. The SDSS automates many of the managing tasks and accomplishes them mostly without human intervention. The SDSS was designed by looking at various legal provisions and responsibilities of various governing bodies.

Acknowledgements The work explained in this chapter is based on one semester work of the second semester students of M.Plan programme at The Department of Architecture and Planning of NIT Calicut during 2016–17. The authors would like to gratefully acknowledge the efforts of our students Ms. Anagha K. J., Mr. Arjun P., Ms. Bigly Sathyapal, Ms. Fathima Zehba M. P., Ms. Gopika C. B., Mr. Jose B. Thomas, Ms. Kesowe-U Wetsah, Mr. Kuruva Manoj Kumar, Mr. Lekkala Venkatakrishna Reddy, Ms. Mekhana S. S., Mr. R. S. Vishnu, and Ms. Salimah Hasnah. We also acknowledge the support extended by all our colleagues in the department for this work.

References

1. Batty M et al (2012) Smart cities of the future. Eur Phys J Spec Top 214(1):481–518
2. Vinod Kumar TM (2015) E-governance for smart cities. Springer, Singapore
3. Vinod Kumar TM (ed) (2016) Smart economy in smart cities. Springer, Singapore
4. Cardwell HE, Cole RA, Cartwright LA, Martin LA (2009) Integrated water resources management: definitions and conceptual musings. J Contemp Water Res Educ 135(1):8–18
5. World Bank (2018) Water resources management overview. [Online]. Available: https://www.worldbank.org/en/topic/waterresourcesmanagement. Accessed: 09 Dec 2018
6. GWP (2000) Integrated water resources management global water partnership Technical Advisory Committee (TAC) background paper no. 4
7. The University of Tennessee (2018) What is a smart community? [Online]. Available: https://servicelearning.utk.edu/smart-communities-initiative/what-is-a-smart-community/. Accessed: 10 Dec 2018
8. Studio Report (2017) Smart water management for Kozhikode metropolitan region
9. Firoz M (2015) Reclassification of the typology and pattern of composite settlement systems: a case of Kerala, India
10. Firoz M, Banerji H, Sen J (2014) A methodology to define the typology of rural urban continuum settlements in Kerala. J Reg Dev Plan 3(1):49–60
11. Land Use Board (2014) Landuse data for Kerala
12. Ministry of Water Resources, Ganga Rejuvenation Board, and Central Ground Water Board (2015) Ground water year book of Kerala (2014–2015)
13. Ravi A (2013) Ground water information booklet of Kozhikode district, Kerala State
14. Firoz M (2006) Spatial planing in rural urban interface in Kerala. Inst T Planners India 3(3):1–6
15. Firoz M, Banerji H, Sen J (2014) An enquiry into the quality of life and infrastructure delivery in the rural urban continuum—a case study of Kerala settlements India. In: International conference on quality of life
16. Director of Fisheries (2013) Kerala inland fisheries statistics 2013
17. Purakkat DV (2009) Groundwater information system for Calicut corporation on a GIS platform. PhD thesis, University of Calicut
18. S. A. Indian Express (2017) Salinity ingress posing threat to wells in Kozhikode
19. Maya KP (2017) Tourism and recreation planning for Kozhikode. NIT, Calicut
20. Godina E (2005) Water wars: privatization, pollution, and profit. By Vandana Shiva. J Biosoc Sci 37(03):381–382
21. Giupponi C, Sgobbi A (2013) Decision support systems for water resources management in developing countries: learning from experiences in Africa. Water 5(2):798–818
22. Wikipedia Contributors (2018) Metropolitan area network—Wikipedia, the free encyclopedia. [Online]. Available: https://en.wikipedia.org/w/index.php?title=Metropolitan_area_network&oldid=863400766. Accessed: 01 Oct 2018
23. Smith L, Der Panne V (2012) The triple helix concept. no 1993, pp 2010–2011
24. Shiva V (2002) Water wars: privatization, pollution and profit. Pluto Press, London
25. ACWADAM (2010) Pani panchayat: a model of groundwater management—a presentation by ACWADAM. [Online]. Available: http://www.indiawaterportal.org/articles/pani-panchayat-model-groundwater-management-presentation-acwadam. Accessed: 27 Oct 2018
26. Dhanuraj D (2005) Drinking water utilisation—Olavanna shows the way. Cochin

27. World Bank (2013) Environmental assessment and environmental management framework. Saint Lucia

28. Priya K, Ajay A, Nayar SK (2016) Sustainability in rural water supply: a case study of Jalanidhi, Kerala. Int J Innov Res Sci 5:266–271

29. Jacob N, Gopalan R, Lala S (2008) Solution exchange for the water community—participatory well recharge programme—Mazhapolima—experiences compiled. India Water Portal

30. Jha B, Mali SS, Naik SK, Kumar A, Singh AK (2015) Optimal planting geometry and growth stage based fertigation in vegetable crops. Technical bulletin no R-56/Ranchi-25. Research Centre Ranchi, ICAR-Research Complex for Eastern Region, Patna, India

31. Li D (2012) Internet of things in aquaculture. Beijing Engineering Research Center for Internet of Things in Agriculture, China Agricultural University, Beijing

32. Huntjens P, Yasuda Y, Swain A, De Man R, Magsig B, Islam S (2016) The multi-track water diplomacy framework: a legal and political economy analysis for advancing cooperation over shared waters. Hague, Netherlands

33. Pohl B et al (2014) The rise of hydro-diplomacy strengthening foreign policy for transboundary waters. Berlin

34. USGS (2016) United States ground water survey

35. Salaj SS, Ramesh D, Suresh Babu DS (2018) Assessment of coastal change impact on seawater intrusion vulnerability in assessment of coastal change impact on seawater intrusion vulnerability in Kozhikode coastal stretch, South India using geospatial technique. J Coast Sci 5(1):27–41

Part VI
New Delhi

Visualizing Environmental Impact of Smart New Delhi

Shovan K. Saha, Mahendra Sethi and Achintya Kumar Sen Gupta

Abstract The character of the parcel of land in the Aravalli-Raisina Hill region selected for building New Delhi was almost dramatically transformed by Edwin Lutyens from its original rocky, semi-arid landscape to that of a pleasant garden city consisting of a low rise—low density habitat set amidst evergreen tree lined avenues, large open spaces and an expansive green lawn with shallow water bodies in the middle of the new capital city. From being the capital of a nation of 279 million souls in 1931, New Delhi presently serves as the capital city of the most populous democracy of the world having a population of over 1.2 billion. The idea of transforming New Delhi under the Smart City Mission of the Government of India launched in 2015 was possibly aimed at achieving two goals: (i) to be counted among the renowned smart capital cities of the world and (ii) to ensure ushering in of a new lease of life for New Delhi, far into the 21st century. Significantly increased dependence on solar energy, software driven management of city services and generally improved efficiency of traffic flows, security and other aspects of citizens' daily life represent the perceived dimensions of smart New Delhi. Visualizing the environmental cost of such transformation is the concern of this chapter. In the process it is discovered, that many of the implemented and ongoing improvements are also in the list of Smart City project of New Delhi. Several of them have been already attempted as the city confronted a variety of challenges from time to time. The aspect of environmental impact resulting from addressing the challenges such as erecting the he LIC building at the periphery of Connaught Place (Rajeev Chowk), DMRC node in the central park of Connaught Place (Rajeev Chowk) and redensifying the single storeyed residential neighbourhoods by four storeyed apartments were hardly visualized and much less addressed. After a close review of the New Delhi Smart City Plan, against a mul-

---

S. K. Saha (✉)
Sharda University, Greater Noida, UP, India
e-mail: shovanksaha@gmail.com

M. Sethi
Dr. APJ Abdul Kalam Technical University, Lucknow, India
e-mail: mahendrasethi@hotmail.com

A. K. Sen Gupta
Institution for Hygiene and Environmental Sanitation, Delhi, India
e-mail: ak.sengupta48@gmail.com

© Springer Nature Singapore Pte Ltd. 2020
T. M. Vinod Kumar (ed.), *Smart Environment for Smart Cities*, Advances in 21st Century Human Settlements, https://doi.org/10.1007/978-981-13-6822-6_8

titude crucial parameters like population density, landuse, traffic, water, sanitation, power and waste management, the present chapter conducts an impact analysis and proposes a comprehensive environmental management plan.

Keywords New Delhi · Environment · Smart city · Land use · Environmental impact · Environmental management plan

Abbreviations

C&D	Construction and Demolition
CPHEEO	Central Public Health and Environmental Engineering Organization
CGWB	Central Ground Water Board
DDA	Delhi Development Authority
DJB	Delhi Jal Board
DUAC	Delhi Urban Art Commission
EIA	Environmental Impact Assessments
EMP	Environmental Management Plans
GoI	Government of India
HH	Household
ICT	Information, Communication and Technology
JNNURM	Jawaharlal Nehru National Urban Renewal Mission
LBZ	Lutyens' Bungalow Zone
LPCD	Litres per Capita per Day
MBBR	Moving Bed Biofilm Reactor
MLD	Million litres per day
MoUD	Ministry of Urban Development
MTD	Million Tonnes per Day
NCTD	National Capital Territory of Delhi
NDMC	New Delhi Municipal Corporation
O&M	Operation and Management
PPA	Power Purchase Agreement
SBR	Sequencing Batch Reactor
ULB	Urban Local Bodies
WtE	Waste to Energy
ZDP	Zonal Development Plan

1 Introduction

Nature and civilization have competed through centuries to dominate planet Earth by adopting newer, smarter means as if they are rivals sharing a habitat, keen to express supremacy over one another. Many of the new cities built in India through the millennia—including some new capital cities—proved to be smart enough to continue to flourish in spite of multiple changes in their contexts and conditions that resulted from inevitable evolution of nature and society, while many others went into oblivion [24]. By building New Delhi (1911–31), Edwin Lutyens transformed the rocky, semi-arid character of the selected pocket of the Raisina Hill region of the Aravalli Ranges to that of a pleasant, carefully landscaped 'garden city' [4] consisting of a low rise—low density habitat set amidst evergreen tree lined avenues and large open spaces. In the middle of New Delhi—also known as Lutyens' Delhi or Imperial Delhi—Lutyens designed the Kingsway or the Central Vista (presently Rajpath), an expansive linear space, linking President's House (Viceroy's Palace) on top of the Raisina Hill on the West to the gently sloping bank of River Yamuna on the East proposed to be used as the grand processional route. Rows of evergreen Black Plum trees (Eugenia Jambolana/Syzygium Cuminit, locally known as Jamun) combined with a pair of shallow water bodies on both sides of Rajpath further enhanced its axial visibility and grandeur. Lutynes' Delhi enjoys the distinction of being one of the most impressive planned capital cities of the world not only through its urban form but also as an example of architectural creativity that captured some typical characteristics of classical Indian and European architecture [12] (Fig. 1).

A description of the building of New Delhi by Lutyens depicts the magnitude of the project: "The stone yard… employed over 3500 men who dressed over 3,50,000

Fig. 1 Coronation memorial plaque announcing transfer of capital of British India from Calcutta (Kolkata) to Delhi ([36] open source)

cubic feet of marble. To the south of the city, 700 million bricks were made out of 27 kilns…there were 84 miles [134 km] of electric distribution cables and 130 miles [208 km] of street lighting, 50 miles [80 km] of roads and 30 miles [48 km] of service roads…" [24].

New Delhi was separated by a 'sanitation green' from the preceding Imperial Delhi, developed in 1648 by the Mughul Emperor Shahjahan [16]. That imperial city commonly referred to as Old Delhi or Shahjahanabad, had a population of a little over 400,000 in 1911. In terms of area, New Delhi (43.7 km²) was nearly eight times larger than Old Delhi (5.4 km²). Prior to building New Delhi, on the north of Old Delhi the temporary but new capital city of British India was built immediately after the announcement of December 1911 by King George V to shift the capital city from Calcutta (present Kolkata) to Delhi (Fig. 2). On the south-west of Old Delhi, the grand Imperial Delhi was built from 1912 to 1931.

Though built 283 years apart, in terms of location, Old Delhi and New Delhi were close enough to achieve not only a visual link but also encouraged the citizens of both cities to develop and maintain an intense relationship of functional interdependence that were at many instances, complimentary. The Lady Hardinge Hospital and College located within two kilometres of Connaught Place was inaugurated in 1916 served as the first modern women's hospital of Delhi and offered to Indian women, also for the first time, an opportunity to serve the society as qualified medical doctors [36]. Irwin Hospital (presently Lok Nayak Hospital) built at the north-east corner of New Delhi enabled residents of Old Delhi to easily access the new and modern healthcare facility. The Railway Station of Old Delhi built in 1867 served New Delhi too, providing the much required and heavily utilized rail link to the rest of India. The relationship between the old and the new imperial cities of Delhi grew stronger with the passage of time, as both cities grew in population size and activities. A map of Lutyens' projected "Imperial Delhi," from Encyclopedia Britannica Eleventh Edition, December 1911© This image (or other media file) is in the public domain because its copyright has expired.

During the last 87 years (1931–2018) New Delhi not only retained her fundamental identity as the capital city of India, but also grew by many folds in several dimensions. During this period, population of New Delhi, and India grew by over four times and that of Delhi by over three times (Table 1) due to several forces. Primary among them were regional politico-cultural adjustments in south Asia leading to possibly one of the highest international migration and high fertility rate especially during the first half of this period. Cumulatively, they led to steady increase of (i) average density, (ii) number of government offices and buildings accommodating them, (iii) built up area, (iv) diversity and intensity of activities and (v) volume of vehicular traffic on the avenues of New Delhi to name only a few dimensions.

Initially, a majority of residents in New Delhi were white collared employees of the Imperial administration who moved in with their jobs from Kolkata (then Calcutta) in 1912 onwards. Over time, the socio-cultural composition of the government employees transformed towards a mosaic that somewhat represented the diversity of India. Till date, New Delhi continues to be a sanitised island of Delhi (refer Fig. 3) and the NCR inhabited by the most powerful decision makers of the Government of

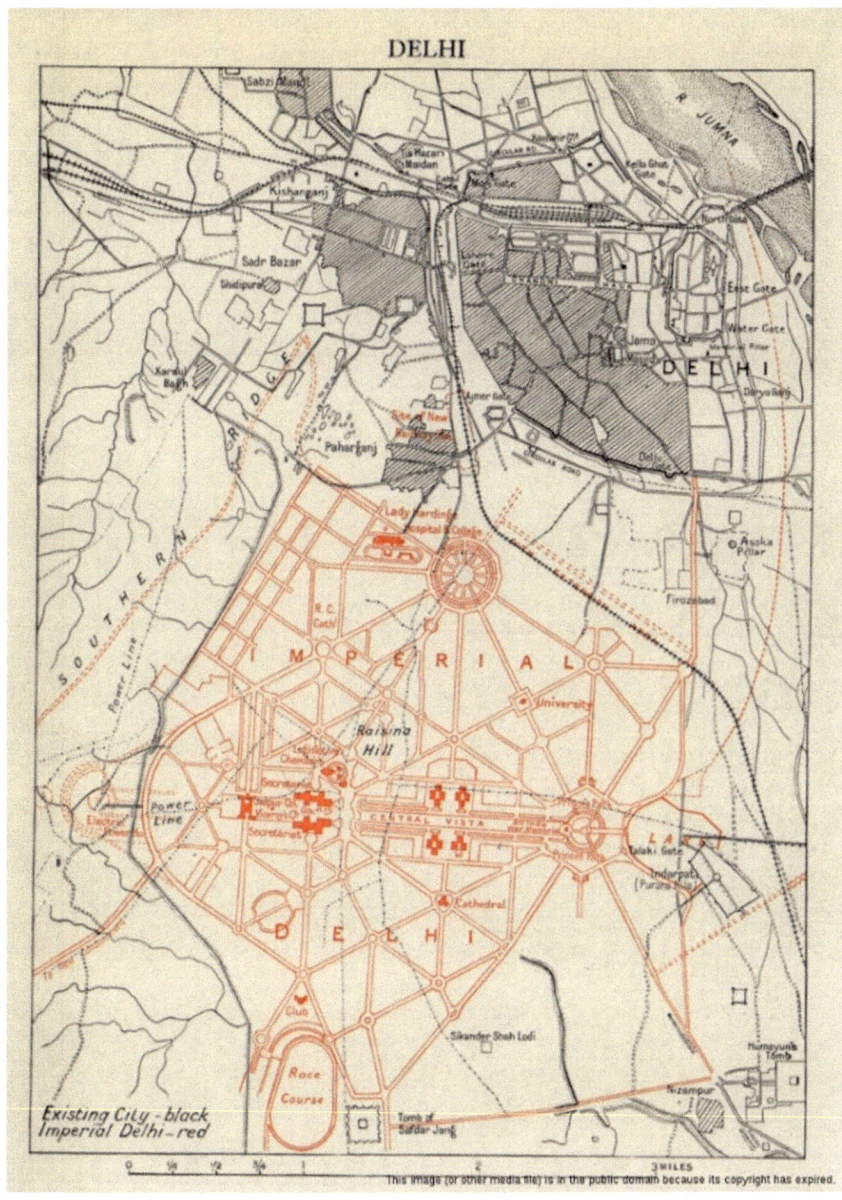

Fig. 2 Plan of Old Delhi (Shahjahanabad) and Imperial Delhi (New Delhi or Lutyens Delhi) inaugurated in 1648 and in 1931 respectively. *Source* Encyclopedia (2011)

Table 1 Population and area of New Delhi, Delhi, NCR and India 1921, 1931, 1951, 2001 and 2018

S. No.	Area of reference	Population density in persons per sq km						Remarks
		1921	1931	1951	2001	2011	2018[c]	
1.	**New Delhi** 47.2 km²[b]	31,000[a]	65,000[a]/ 73,653 1377	94,000 (1941) 1991	1,79,112 3794	1,33,713 2832	2,94,000 6228	
2.	**Zone D** (including New Delhi) 68.5 km²				5,87,000		8,13,000 (2021)	
3.	**Delhi** 1483 km²	4,88,452	6,36,246 429	1.74 m 1173	13.8 m 9305	16.7 m 11,260	19.4 m 13,081	Includes all the seven historic cities of Delhi built since 10th century
4.	**National Capital Region (NCR)** nearly 55,000 km²				37.1 m	46.0 m	50 m	Established in 1985 through NCRPB Act 1985
5.	**India** 3.28 million km² (decadal growth of urban population in percent)	251 m	279 m (19)	361 m (41)	1028 m (31.5)	1210 m (31.8)	1358 m	

Source [a]Census of India [6], [b]NDMC Act [27], [c]Estimated figures

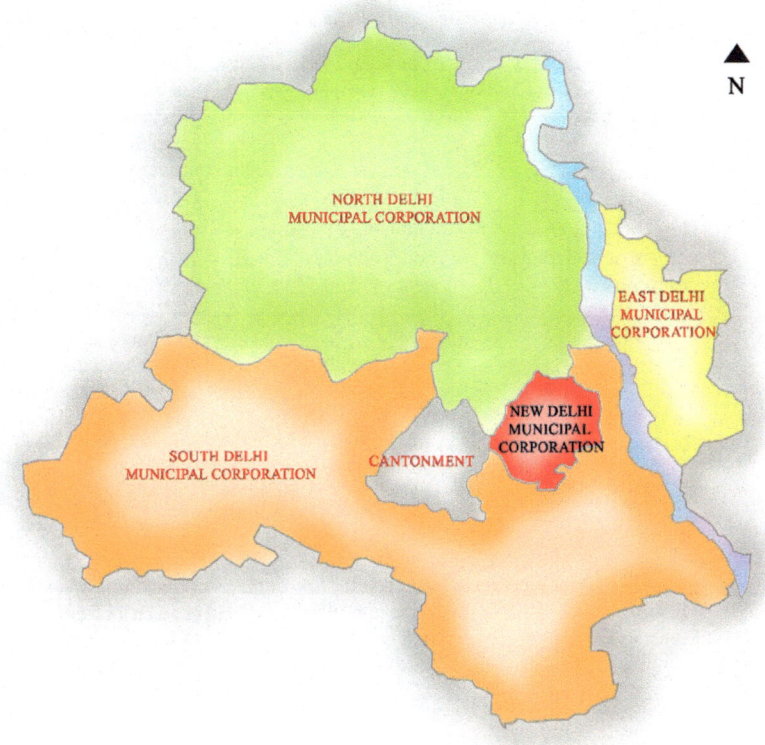

Fig. 3 MAP—showing N. Delhi within Delhi. *Source* Authors

India and others associated with them directly and indirectly. Being the capital city of one of the most healthy democracies of the world, with the passage of time, New Delhi received more and more citizens not necessarily as government employees or business visitors to the Ministries but as curious tourists from the neighbouring regions and also from far flung areas of India.

Densities New Delhi with an area of 47.2 km^2, was resided by about 3,00,000 persons or 1.7% of Delhi's population of 18 million in 2016. In terms of space, it accounts for nearly 2.9% of the Delhi's area of 1483 km^2 (NCT). Thus, the average density of Delhi is nearly 1.5 times higher than that of New Delhi, 9294 p/km^2 and 6660 p/km^2 respectively. By 2021, New Delhi's population is projected at 0.419 million, raising New Delhi's average density to 9805 p/km^2. Major hospitals, cultural facilities in New Delhi serve not only Delhi but also the NCR and beyond [18]. With the starting of the metro rail in 2002 and its subsequent rapid growth radiating from Delhi to longer distances in the hinterland, accessibility to New Delhi has grown at an increasing pace [36]. The major development initiatives in New Delhi during the 87 years are captured in Table 2.

Table 2 Major development initiatives for New Delhi 1930s till the initiation of smart city scheme

Year/period	Development initiative	Location	Impact on New Delhi
1930s to 1940s	Four movie theatres were built Regal, Rivoli, Plaza, Odeon	Connaught Place	Connaught Place assumed the role of providing urbane recreation and entertainment of New Delhi
1940s to 1960s	New buildings built to accommodate ministry offices, other govt. offices built	Within New Delhi, near capitol complex, Central Vista	Since soon after attaining independence from colonial rule in 1947, the number of ministries, workers, built up area grew rapidly; except Krishi Bhavan and Udyog Bhavan, aesthetic quality of other buildings mismatched with New Delhi's original architectural character
1967	New Delhi Railway Station started functioning	Just outside New Delhi on the north; Railway track cut through New Delhi	Direct mass access to New Delhi from the rest of India; volume of passengers and goods increased as Delhi and NCR grew
1968	Removal of the statue of King George V	Park at the eastern end of Central Vista	Weakened the iconic image of New Delhi
1970s	State Emporia and Palika Bazaar started functioning	Baba Kharak Singh Marg (Irwin Road) New Delhi	Attracted domestic and international visitors from far and wide locations
	Bank of Baroda, State Bank of India, American Library and other multifloored buildings constructed	Sansad Marg, (Parliament Street) Barakhamba Road, Kasturba Gandhi Marg (Curzon Road)	
	New Delhi District Centre by Ar Kuldeep Singh and Raj Rewal through a national competition	Sansad Marg (Parliament Street)	This 23-storey building was one of the first major change in the skyline of new Delhi
1972	NDRAC commissioned schemes from architects to rejuvenate Rajiv Chowk (CP) while conserving its original character designed by Ar. Robert Tor Russel	Mainly for Rajiv Chowk (Connaught Place) and also for other selected areas within New Delhi	The proposals received by NDRAC were probably too advanced for the then decision makers to accept or approve and they never got implemented on ground

(continued)

Table 2 (continued)

Year/period	Development initiative	Location	Impact on New Delhi
1973	DUAC created	Delhi	TOR "preserve and maintain aesthetic quality of Delhi"
1983	The plans for most of the 12 subzones covering New Delhi prepared earlier were approved by 1983	New Delhi	Overall intensification of activities as proposed in plans
1985	IGNCA designed through international design competition won by Ar Ralph/Lerner, Princeton, USA	Intersection of Rajpath and Janpath, the two cardinal axes of New Delhi	As an architectural statement, the design conformed well with that of New Delhi
1986	LIC building designed by Charles Correa, India's celebrated Architect	Rajiv Chowk (CP) at the intersection of Sansad Marg and outer circle	Changed the low profile sky line of New Delhi
1988	LBZ boundary delineated to be exempted from development control regulations for the rest of the zone	A linear zone extending from Rashtrapati Bhavan to India Gate Park on both sides of Central Vista	LBZ excluded the plots along the main roads demarcating the zone
2002	Introduction of mass transit—Delhi Metro	Immediately outside New Delhi, at interstate bus terminal	The commercial core of New Delhi, Connaught Place received its own metro station, namely Rajiv Chowk, in 2005
2003	Re-delineated LBZ after the first delineation in 1988	As above	The latter boundary included the first row of plots along the main roads demarcating the zone in order to ensure aesthetic harmony of on both sides
2005	Multi-level metro station hub (DMRC) constructed below central park of CP	New Delhi, in particular, CP	Dramatically increased the accessibility of CP and New Delhi from all over NCR
2010	XIX Commonwealth Games, Delhi 2010. Several projects were initiated in CP aimed at improvement of services and its image	Projects aimed at functional and aesthetic improvement of New Delhi were attempted	The agencies involved in governance of Delhi including NDMC actively participated in readying the city to host this event

(continued)

Table 2 (continued)

Year/period	Development initiative	Location	Impact on New Delhi
2015	Included in Round 1 list of Smart City initiative of Govt of India	New Delhi, Zone D of Master Plan of Delhi 2021	Presented in this chapter

Sources Authors
CP Connaught Place; *DMRC* Delhi Metro Rail Corporation; *DUAC* Delhi Urban Art Commission; *LIC* Life Insurance Corporation; *LBZ* Lutyens Bungalow Zone; *NDRAC* New Delhi Redevelopment Advisory Committee; *IGNCA* Indira Gandhi National Centre for Arts; *NCR* National Capital Region

In addition, the city of New Delhi has been an anchor for several national and international mega events, conferences, games and sporting events, including Asian Games, Commonwealth Games, etc. that triggered physical development from time to time.

New Delhi responded to the array of challenges by introducing new laws, master plans, development control regulations, amendments to existing laws, policies and programmes, accommodated in new offices and more. In the process, the quality of life of the residents of New Delhi changed at varying paces through this period for the ordinary citizens, bureaucrats and political masters of India. New activities and nodes emerged in the city some of which replaced exiting activities. Needless to say, these changes of activities that took place across New Delhi induced transformation in the land use pattern of the city both qualitatively and quantitatively.

Land use

Conceptually, New Delhi was designed as a quiet administrative township rather than a holistic city that would continue to evolve with her inhabitants, far into the future. Therefore, an overall ambience of tranquillity interspersed with carefully crafted buildings set amidst landscaped greens was created where the highest level of decisions related to governance of India would be taken in nearly complete seclusion from the hustle and bustle of the city.

Land use based long term city planning was not yet a universally accepted practice during early 20th century though Patrick Geddes was already working on Indore and Chennai (Madras) and many other cities of India (1915–23). With the introduction of the land use based first Master Plan of Delhi 1962–81 New Delhi was included in Zone D having a total area of 6855 ha. Based on the second and third Master Plan of Delhi 1981–2001 and 2001–21 respectively, the Zonal Plan for Zone D proposed a few modifications in the land use distribution as shown in Table 3. Land under Transportation was increased by 12.5 ha and that under Public & Semi-public use by 59.9 ha, while land under Recreational was reduced by 77.9 ha. Some of these land use redistributions affected New Delhi directly. Nevertheless, they would act as catalysts of further intensification of the already intense the activity pattern of New Delhi.

Lutyens' Bungalow Zone (LBZ) was evidently perceived as the core area of New Delhi, to be resided by high level bureaucrats and elected representatives who would

Table 3 Land use distribution of Zone D of Delhi as per MPD 2021 and as modified up to 21-11-2016

Land use	Area as per MPD 2021		Area as modified up to 21-11-2016		Changes (in Ha)	
	Ha	Percent	Ha	Percent	Increase	Decrease
Residential	2443.6	35.7	2453.4	35.8	9.8	
Commercial	216.1	3.1	217.7	3.2	1.6	
Manufacturing	28.6	0.4	28.6	0.4		
Recreational	2291.4	33.5	2213.5	32.3		77.9
Transportation	780.8	11.4	793.3	11.6	12.5	
Utility	109.7	1.6	109.7	1.6		
Governmental	483.4	7.0	477.5	6.9		5.9
Publi and semi-public facilities	501.4	7.3	561.3	8.2	59.9	
Total	6855.0	100	6855.0	100	83.8	83.8

Source Developed by Author based on Zonal Development Plan for Zone D, New Delhi

author policies and laws for British India. Particularly during the post-independence era, discussions and debates about the character of development of LBZ has been undertaken from time to time by several Committees etc. appointed by the Government including architects, city planners, urban designers and bureaucrats. Such exercises generally opined that the LBZ must have customized development norms that should ensure conservation of its pristine ambience compared to that of the rest of New Delhi and Delhi.

As reported by the DUAC in its study [12] the LBZ occupies 23.6 km^2 or a little more than 55% of New Delhi. The development controls recommended by DUAC [12] ensure that a very low density of 37 p/ha in LBZ compared to 3375 p/ha for the rest of Delhi [2]. As such, a wide range of low-rise built forms with virtually no restriction on the aesthetic aspects such as façade control through application of specific materials, colour, shape or size of apertures etc. are proposed. Enforcement of detailed norms related to built-form and imageability would ensure that LBZ represent an urban design statement in harmony with the aesthetics of New Delhi.

"While doing a diagnostic survey of Lutyens Bungalow Zone it became apparent that the entire area does not have a uniform architectural character. An analysis of the characteristics of land within the zone reveals 6 typologies, as shown in Fig. 4".

1. Land which is public green and forest.
2. Land in use for public institutions, including government and other public offices, cultural institutions, hotels, and protected historic monument.
3. Land under government residences (bungalows and other types).
4. Land leased for private residences.
5. Land for markets.
6. Land under barracks/hutments of temporary construction.

(a)

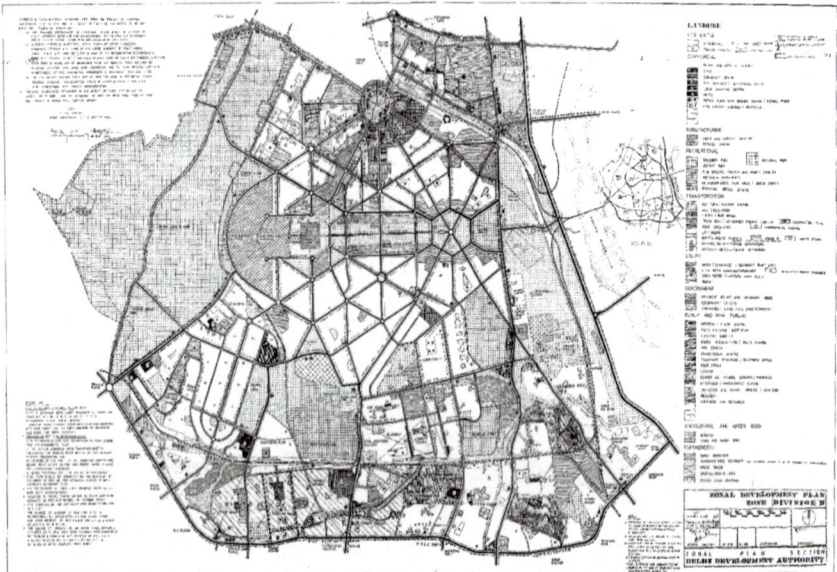

(b)

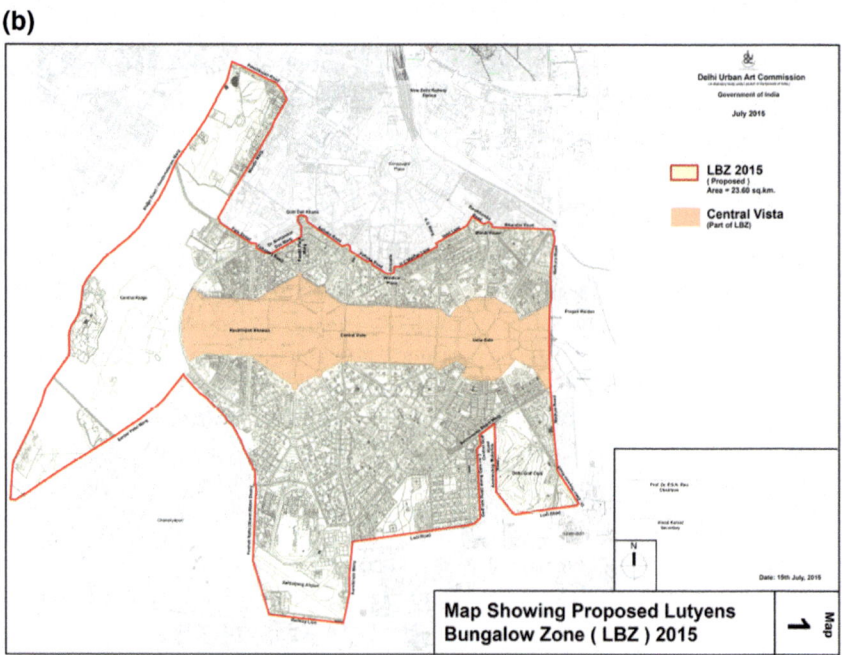

Fig. 4 **a** Zonal development plan for zond D—New Delhi. (*Source* DDA [9]). **b** Proposed Lutyen's Bungalow zone for the year 2015. (*Source* DUAC [12])

Certain areas outside but contiguous with the LBZ boundary are also considered. These are the area north of Ashoka Road and Feroz Shah Road up to and including Connaught Place, the area East of Mathura Road up to the river Yamuna and including the Purana Qila. Pragati Maidan and the zoological park; and the area south of Lodi Road.... [14].

Recent initiatives: The official concern for the wellbeing of towns and cities in India were expressed since the days of British rule through initiatives such as establishment of municipalities and enactment of town planning acts [14]. With the advent of the post-independence era on 15th August 1947, such initiatives have been in terms of creation of Development Authorities for the largest and some other cities, preparation of master plans for them as well as appointment of various committees, commissions and task forces by the national government from time to time. Enactment of the Delhi Development Act in 1957 led to establishment of the Delhi Development Authority (DDA) with the specific task of preparing the first Master Plan of Delhi 1961–81 (launched on 2nd September 1962) and building Delhi according to the Master Plan. Unfortunately, the appropriate financial support and follow-up actions required to translate the city master plans and well-intended recommendations of expert groups into reality were hardly there and therefore not been effective. Two recent initiatives, the Jawaharlal Nehru National Urban Renewal Mission (JNNURM) in 2005 and the Smart Cities Mission launched on 25th June 2015 respectively were exceptions. The first was allocated INR500 trillion or USD11 trillion approx. (@INR46 per USD) (England, 2005). Delhi with a population of 16.75 million (2001) and an area of 1483 km^2 headed the list of 63 (later 65) selected cities to be 'renewed' under JNNURM scheme. The second was allocated INR980 trillion or USD15 billion approx. (@INR64 per USD) (England, 2015). New Delhi or Lutyens' Delhi with a population of little over 0.3 million (2011) and an area of 43 km^2, was the first among the list of 100 selected cities under the Smart City Mission.

2 Smart City Mission for New Delhi

Smart cities is a term used globally since its evolution during the last two decades (Fig. 5). The word smart highlights the importance of Information and communication technologies (ICT). The literature interprets smart cities to describe a city's ability to respond to the needs of its citizens. In fact, the term 'Smart City' was not the original term that scholars defined. Initially the research was on the terms Digital City and ICT [35]. It is further found that the earliest evidence, seen as a 'Digital City' practice was in Amsterdam on 15 January 1994 where for the first time, internet access was made available to a large group of users. This was followed by literary evidence on Digital city and Virtual City in 1997. The start of local cyber based communities came to the fore. However, virtual cities mentioned the digital representations and manifestation of cities. In 1998, smart city as a term was perhaps first practiced in Dubai, followed by evidences of eco city in 2005. This marked the beginning of smart cities from technology to urban planning perspective, while UN

(a)   **(b)**

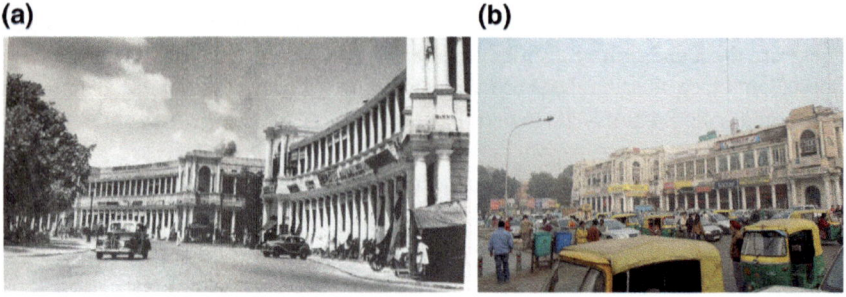

Fig. 5 **a** Connaught Place, **b** Rajiv Chowk (1995). *Sources* Wikipedia [36]

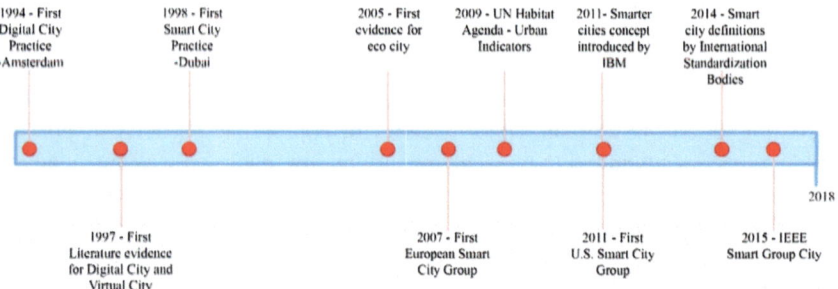

Fig. 6 Smart city timeline. *Source* Upadhyay [35]

Habitat Agenda in 2009 mentioned the urban indicators regarding the ICT based smart cities, thereby globalizing the discourse. On 4th November 2011, IBM coined the term 'smarter cities' and registered it as a company's trademark, taking smart city to a new level that led to the inclination of cities towards technology and digital education of its citizens. The first US smart city group was established in 2011, which then succeeded with the definitions of smart city by international standardization bodies. The current world sees a large number of cities transforming to smart cities where ICT is used as a tool to run the urban structure (Fig. 6).

The Smart Cities Mission in India was launched on 25 June 2015 with an objective to develop infrastructure and services with added smart solutions that fulfil the needs of its citizens. Reviewing the relevance of the smart concept at the national level, it is argued that Smart Cities pose both immense challenges and possibilities to attain not just sustainable urbanisation in India [29], but also to bring qualitative improvement in urban living of scathes of population [19]. The 41 projects proposed for New Delhi under Smart City Mission include a wide range of sectors such as solar energy, automated parking, telecommunications, strengthening of water supply and sanitation network and beautification of New Delhi including conserving the original image of Connaught Place. Many of these were part of the initiatives taken prior to the XIXth Commonwealth Games 2010 hosted by India and held in Delhi. As a result, visibility of the progress of work of the Smart City projects on sites within New Delhi has been limited.

Table 4 indicates the typology and budget allocations of the Smart City projects proposed by NDMC. As reflected by the amounts allocated, nearly two thirds of the total budget was for Pan City Projects, that is, projects aimed at improving the quality of life for all citizens of New Delhi by installation of software driven monitoring and management of infrastructure and services. Of these, three-fourths or Rs. 12,860 m (USD 201 m) was allocated for smart grid and energy management projects and 80% of the budget for Smart Health projects was proposed to be used for integration through application of software. Solar power projects were allocated over 22% or Rs. 4300 m (USD 67 m).

As evident from Table 4, the priority of the Smart City initiatives for New Delhi was towards mechanisation and automation of services and infrastructure including parking, health services, e-learning and e-governance, driven by software. Among pan-city projects, high priority is given to provision of energy, particularly through smart grid and renewable sources like solar energy. A detailed discussion is included later under the section 8.6 & 8.7. NDMC's total budget for the years 2018–19 was 30% more than the total budget for the listed smart city projects (Final Budget Speech of NDMC, January 2018). The types of projects proposed to be taken up by NDMC and those proposed under the Smart City Mission were largely common, enhancing the probability of their smooth implementation. The budget speech announced a few projects in addition to those on the smart city list and indicated NDMC's intention to establish link with the city of Leuven, Belgium through a Twin City Agreement signed in November 2017. As part of greening New Delhi, the Zonal Development Plan (ZDP) of Zone D as per MPD 2021 also recommended nurturing of trees of New Delhi as some of them lived beyond their known life expectancy while many others are approaching that stage [9]. Yet, heritage and landscape projects were not included in the list of Smart City projects for New Delhi.

3   Water Supply, Sanitation and SWM

Around 90% of Delhi's water demand is met from surface water sources, of which more than 60% are drawn from external sources mainly dependent on river Yamuna, Ganga and Bhakra Beas system (Sl. No. 3 to 9). River Yamuna, Western Yamuna Canal and the Upper Ganga Canal, are surface water sources for Delhi. Around 446 tube wells drilled in Yamuna bed and areas within city to meet the water requirement. Delhi Jal board is equipped to treat 790 MGD of water inclusive of about 100 MGD of Ground water abstraction. This over dependence on external resources is highly vulnerable—not only economically but also environmentally. Moreover, erratic rainfall, decline in river flow, uncertainty of dam based resource augmentation, declining groundwater output, drying rivers and aquifers, shrinking glaciers and increasing demand and friction with riparian states [25] has made these external sources most vulnerable.

Table 4 List of projects proposed for NDMC area under smart city mission

S. No.	Type of projects	Number of projects	Budget in million Rs/USD[a] (%)	Remarks
1	Urban mobility and smart parking	10	2560/40 (8.6)	Nearly 3/4th of the budget is allocated to building multi-level automated parking at three locations in and around Connaught Place
2	Area Based Development (ABD)	9	4645/73 (15.7)	Nearly 1/3rd of the budget is for Sensor based Common Service Utility Duct followed by Rooftop solar panels (22%)
3	Municipal SWM	5	406/6 (1.3)	Nearly 3/4th of the budget is for mechanization and automation of cleaning activities
4	ABD	7	205/3 (0.7)	Four projects are aimed at improving and strengthening the presence of New Delhi among the world cities
5	Pan City	4	19,260/302 (64.7)	These projects are of highest priority as evident from the budget allocated Over 3/4th of the budget is allocated to smart grid and energy management projects
6	Smart water and waste water management	1	1904/30 (6.4)	This is a pan city project
7	Smart education	2	450/7 (1.5)	The emphasis is on eLearning with over 3/4th of the budget
8	Smart health	3	248/4 (0.8)	Over 80% of the budget is allocated for integration of public health facilities through software application
	Total all projects	41	29,768/465 (100)	

Source MoUD [22]

[a] @INR64 per USD

Table 5 Internationally, average per capita water supply in some developed countries

Country	Average per capita water supply daily (l)
Netherlands	116
Germany	122
UK	144
France	151
Spain	154
Italy	188

Source Raphaël Demouliere [11]

The details shown in Fig. 7 are representative and not to scale. Most of the water sources are located between 100 and 400 Km away from Delhi (Table 5).

Even if we consider institutional and commercial demand separately, New Delhi with around 0.3 million populations gets 127 million litres per day (MLD) of treated water with an average per capita supply of more than 400 L, which is 2.5 times the standards prescribed by Central Public Health and Environmental Engineering Organization (CPHEEO). This is more so when some other parts of Delhi are getting hardly 20–30 lpcd. Moreover, it is not practicable to consume such high quantum of water domestically unless the unaccounted for water loss consisting of large scale gardening, car washing, un-accounted water connection and wastage/leakage is taking place. While the skewed pattern of consumption even within Delhi has been increasingly pronounced with the haphazard growth of the metropolis demographically and spatially (Fig. 8).

As per tariff structure, treated water supply per month up to 20,000 litres/connection in domestic establishment is provided free of cost. This comes to around 135 lpcd for a family size of 5, which is the norm fixed by CPHEEO for urban household. Earlier studies have shown that when people pay for a provision, they take care of it and also use it prudently. Moreover, the management staff may resort to slackness which may increase the losses as the consumers getting free water supply are not in a position to ask for a better system. Finally, there is a huge economic, social and environmental loss to the system as water is brought from hundreds of kilometres, treated and then provided on charity. People will not understand true value of this costly water. With passage of time each drop of water will count and sustainability of such a system is questionable.

To reduce the per capita consumption of fresh water brought from far off places it is absolutely necessary to develop 'Smart Designs' for reuse and recycle of wastewater (grey water) from domestic, commercial and institutional establishments to be conveniently used for irrigation, flushing of toilets, car washing etc. There need to be a coherent policy in this regard. To reduce the use of fresh water further, even there is a need to the adopt technologies like water less urinals in all community toilets, commercial and institutional establishments throughout the New Delhi. Of course, this should be followed with proper monitoring of the systems to assess their function and environmental impact.

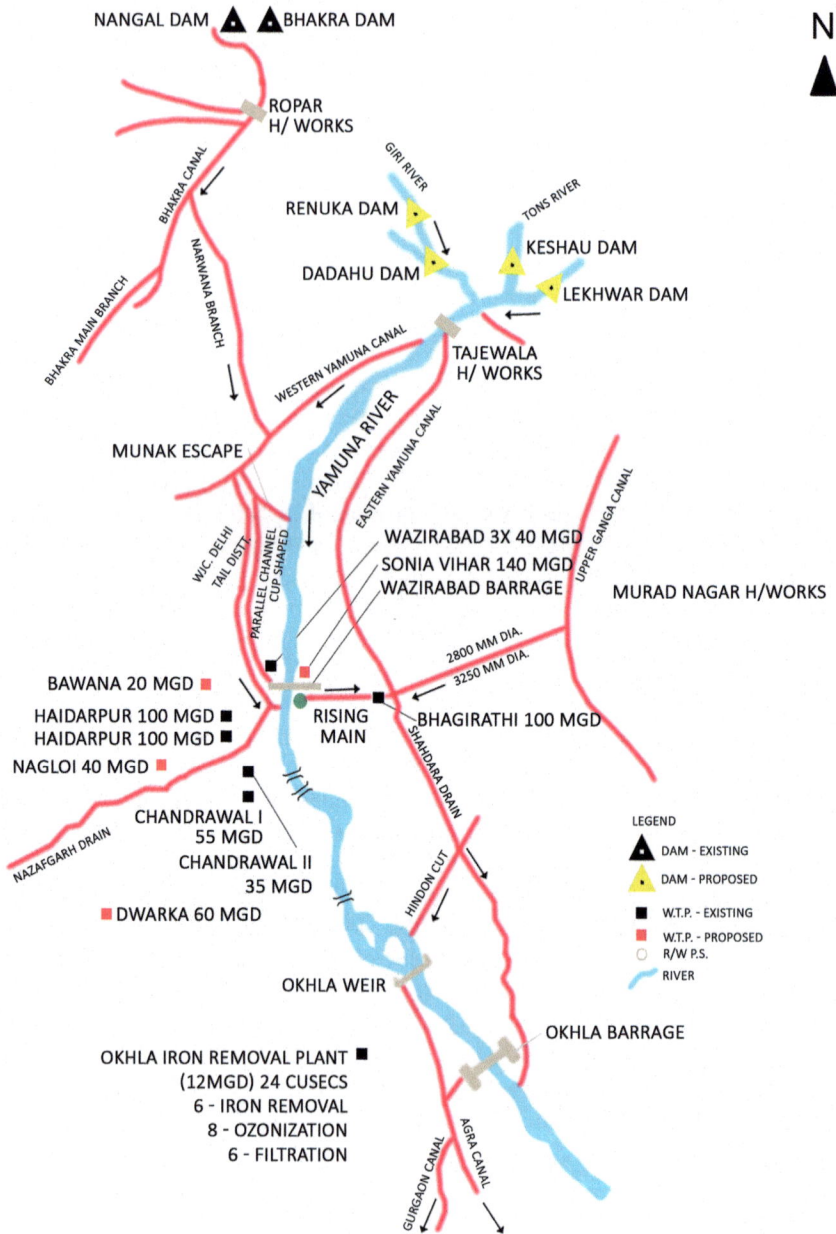

Fig. 7 Water treatment plants in Delhi and sources of raw water. *Map Source* Authors, based on information from IL&FS Ecosmart [18]

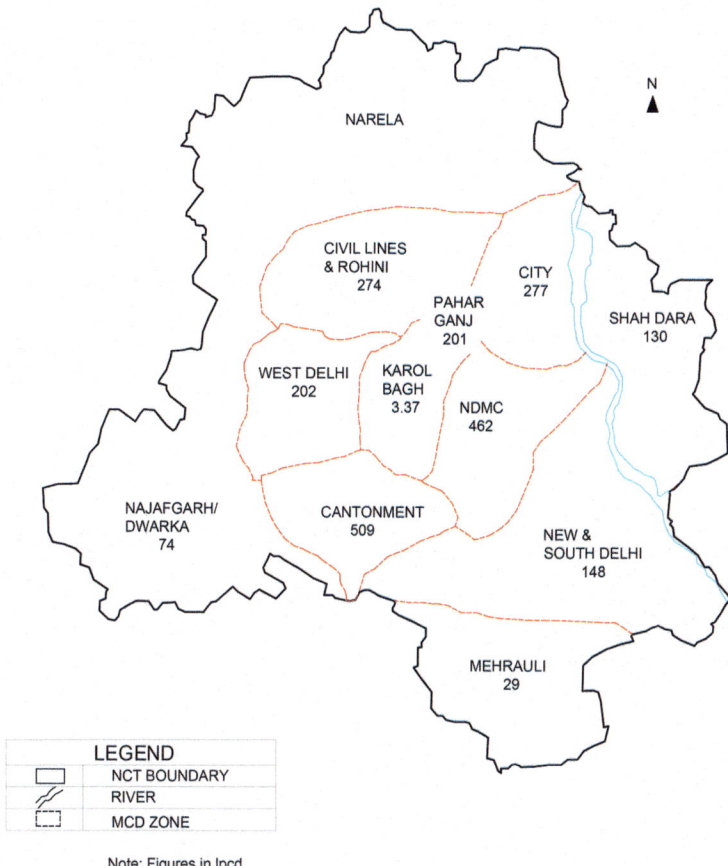

Fig. 8 In-equality of water supply in different parts of the city. *Map Source* Authors, based on information from Delhi Jal Board

Another important factor that the Fig. 5 reveals is the inequitable distribution of water within Delhi. Whereas some parts of Delhi are getting 400 lpcd, some other parts are having around 20 lpcd and thus have to depend on tanker supply, where both quality and quantity of water is compromised. The mere fact that such a large fleet of water tankers (more than 450) is maintained by DJB to provide drinking water supply to many colonies of the city regularly only shows its management deficiency. So, the whole water distribution system network within Delhi needs 'Smart' technical design and revamping. It would invariably consider an ideal tariff structure for a balance between the economic, environmental and social demands on water resources and supplies.

Studies have shown that there is significant water loss within the water supply systems (Approx. 16% leakage is in transmission main and another 24% in distribution system) within Delhi [3]. It is necessary to strengthen leak detection processes by having electronic instruments like pressure and acoustic sensors. Unless distribution losses are restricted less than 10%, 24 × 7 water supplies will add to the revenue losses. Since, there are newer 'Smart technologies' are presently available for locating and correcting distribution line losses there is a need to adapt the same. Apart from the issues mentioned earlier, there is a need for application of policies, which will include restriction and bans on certain type of water use, standards for fixtures and appliances etc. and technology advances to use less water for the same function. Meanwhile, rainwater harvesting in small communities, apartment complexes, institutions and even individual houses is crucial to meet the irrigation and flushing needs. This will significantly reduce the burden on fresh water resources for Delhi.

Safe water supplies for all populations can only be guaranteed when access, equity and sustainability are assured. Variation of water availability occurs within Delhi at a large scale. Due to subsidised water tariffs, water rich areas misuse the costly resource for non-essential purposes such as washing cars, watering lawns and filling tanks. In poorer areas, thousands of people are only having access to standpipes, which are having intermittent supply. Or even worse, they are buying water of doubtful quality from private vendors at prices that may be 10 to 20 times higher than inner city water tariffs. The health consequences resulting from this deprivation caused by this inequity are sometimes considerable, as evident by differentiated incidental rates of water borne diseases amongst rich and poor.

There has been a worldwide retreat of glaciers, and a similar trend has been reported for a few Himalayan glaciers as well [28]. A study by the Geological Survey of India has revealed that glaciers are receding at an alarming rate: for example, Gangotri is receding by 17.5 m/year, Dkriani by 17.5 m/yr, Milam by 13.3 m/yr, Pindari by 23.5 m/yr and Zemu by 13.2 m/yr [34]. In future, the run-off of Himalayan Rivers is expected to be highly vulnerable to climate change because warmer climate increases the melting of snow and ice earlier and faster. These changes can affect the availability of freshwater for natural system and human use. Melting of glaciers will have direct impact on water resources affecting drinking water supply, irrigation, hydropower generation and other uses. Thus, the water sources for Delhi are depleting and quite vulnerable, the National Capital is very high in consumption with an insatiable demand for water. It is imperative to make New Delhi smart towards this vital natural resource.

4   Sanitation and Sewerage System in Delhi

Delhi Jal Board (DJB) has provided sewage facilities in all the approved colonies, all 44 re-settlement colonies and 126 urban villages, 541 unauthorized-regularized colonies and 100 unauthorized colonies. There are 36 Sewage Treatment plants at 21 locations [10].

According to study conducted by Central Pollution Control Board (CPCB) for monitoring the performance evaluation of STPs located in Delhi in 2012, the city generated 3,800 MLD of wastewater whereas the installed treatment capacity was 2,603.7 MLD and the actual utilisation was only 1,575.8 MLD. While, in terms of total capacity, the existing STPs can treat most of Delhi's sewage, they are unable to do so because the infrastructure to carry the sewage to the plants is yet to be laid. In some other cases, operational problems to run the sewage treatment plants do persist, which means that either the sewage is just not treated or it is not treated to optimal environmental standards [33].

In case of New Delhi, though it generates a substantial quantum of wastewater, the treatment process is handled outside the district boundary. Recently, the following technologies such as Sequencing Batch Reactor (SBR) and Moving Bed Biofilm Reactor (MBBR)/Fluidized Aerobic Bioreactor have been approved under JNNURM projects due to their advantages such as less requirement of land, high effluent quality etc. [7]. There are a number of newer treatment technologies that have come into practice in recent times and they do merit attention in their own way. If Delhi goes for decentralized wastewater treatment processes, there will be smaller sewerage network, which will be cost effective. Moreover, if faecal matter and urine streams are separated at the source both become useful resources [25] (Fig. 9).

As per the findings of a study conducted by NIUA, the sanitation services hardly generate any significant revenue, though NDMC may be an exception. It is suggested that in the initial years, tariffs should be set to ensure that they recover at least the O&M costs of the sewerage system. Once operational efficiency is demonstrated with the infusion of private sector participation, the user acceptability of a tariff increase by ULBs would also improve. The tariffs can be increased to recover the capital costs also in addition to the O&M.

For improving sanitation services suggestions are:

Since having sewerage network along with sewage treatment units and finally disposing it away are all high cost ventures, it is necessary for urban areas to have closed and smart resource loops with decentralized treatment process to avoid high transportation cost.

There is a need for a policy change and awareness campaign for advocating the efficient use of treated wastewater for irrigation, toilet flushing and car washing activities. Community has to be involved in this campaign and necessary design changes in the system need to be incorporated. The smart approach is to see sewage as resource and wastewater to be recycled and reused. The urban society needs to be set up incrementally increasing targets on recycle and reuse of wastewater and implementing agencies need to provide all support to do so. Even STPs can have a smart and integrated command area approach to recycle and reuse the treated sewage [25]. With 24 × 7 water supply and water efficiency rating system being planned for Delhi, per capita water consumption is expected to go down quite appreciably. Less water consumption will certainly lead to less sewage generation and thereby lower power consumption and land footprint.

Distribution of vector-borne diseases increases substantially with the deterioration of the environment. As majority of mosquito and fly related diseases is associated with environmental conditions, including land and water or wastewater manage-

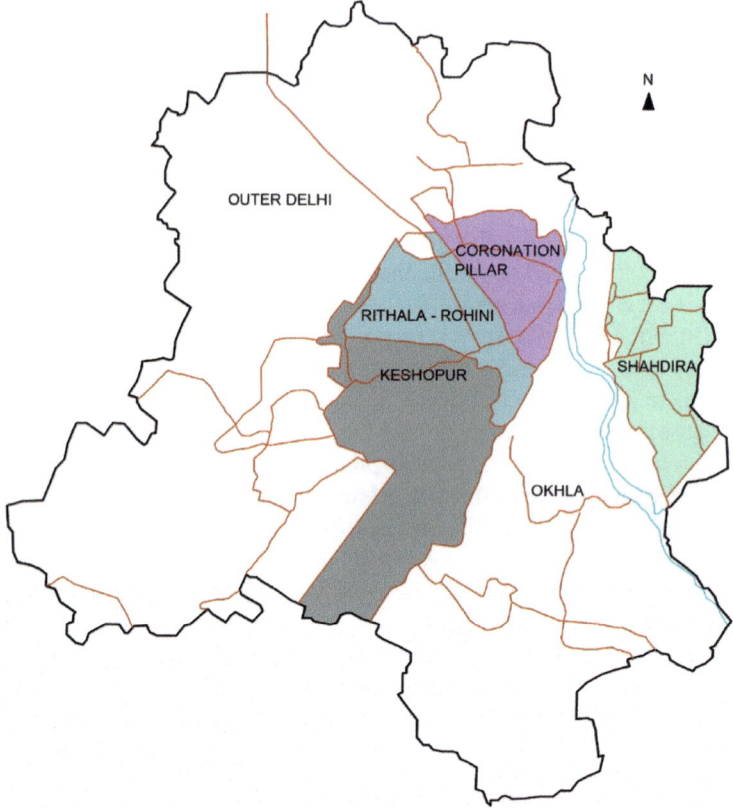

Fig. 9 Six drainage zones of Delhi. *Map Source* Authors, based on information obtained from Delhi Jal Board

ment, a very high global burden of disease is attributable to environmental factors. Environmental quality is thus an important direct and indirect determinant of human health. Deteriorating environmental conditions are a major contributory factor to poor health and poor quality of life and hinder sustainable development. Major challenges to sustainable and smart development are posed by mismanagement of natural resources, excessive waste production and associated environmental conditions that effect health.

5 Solid Waste Management

NDMC area generates around 300MT of municipal solid waste daily. An efficient, house to house solid waste collection system with reporting mechanism has been developed almost throughout the NDMC area. But segregation of waste at HH level and the final disposal of waste in an environmental friendly manner are yet to be achieved. Apart from this, on an average 50MT of Construction and Demolition (C&D) waste and 50MT of Drain silt are generated on daily basis within NDMC area. The per capita waste generation is around 556 g/day in NDMC area which is more or less same for Delhi [26]. Presently, mixed waste is usually collected from households and sent to the dump-sites/Waste to Energy plants/compost plants. Hardly, 2% of the solid waste is getting segregated at the source. Efforts are being made in the NDMC area to develop effective system of waste segregation at source, collection, transportation, processing and disposal. In NDMC area, private parties have been engaged to carry out door to door collection and segregation of the waste. Road sweeping is also carried out by mechanical road sweepers on some selected roads of NDMC area.

A smart city invariably demands several measures. Presently, there is lack of public awareness about sanitation and cleanliness, particularly on dumping of waste on roadside or other vacant plots. Some aggressive campaign need to be launched by the local authorities to encourage people to keep their area and roads, lanes and by-lanes clean. This will help in behavioural changes amongst the public. In addition, rag pickers/informal waste collectors form one of the most important roles of solid waste management system. So far, their roles have been recognized informally. Local authorities may like to integrate their roles in the smart Solid Waste Management system by registration and assigning them roles and responsibilities.

All the three existing dumpsites for solid waste of Delhi have exceeded their capacities years ago. The dumping sites do not have any methanisation or gasifiers to control the methane gas being produced naturally. There is no fire protection system and these are potential health hazards. The leachate carrying heavy metals from these dump sites, especially at Bhalswa one of the sites, are draining into the major city level drain connecting to water bodies as well as adjacent areas. A study need to be conducted to access the ground water condition around these dumping ground areas as possibilities of ground water pollution around these areas are very high. As all the dumping sites are located outside NDMC area, the residents near these dumping sites are facing health hazards.

Most of studies conducted to assess composition of Indian municipal solid waste shows that 50–60% consist of compostable materials. In May 2017, a survey conducted by Shriram Institute for Industrial Research, Delhi for samples collected from South Delhi Municipal Corporation area shows biodegradable part of waste was between 55 and 60%. The same report shows that the mixed waste has a calorific value in the range of 1274.25–1324 kcal/kg. As per Solid Waste Management Rules, 2016, minimum waste calorific value for incineration is 1500 kcal/kg. Meanwhile, Waste to Energy (WtE) Guidelines, 2017 suggest a calorific value of over 1600 kcal/kg to run the WtE plant without use of any auxiliary fuel. Hence, as per the report of the Shriram Institute the waste composition for the city is not conducive to incineration-based technologies. According to Planning Commission report, 2014 and National Green Tribunal order OA 199 of 2014 dated 20 March, 2015, only the non-recyclable high calorific value waste should be used for WtE projects.

Table 6 Impacts of removal of fractions from WtE plant that can be processed, recycled

Fraction removed	Prime impacts of removal on remaining waste
Glass, metals, ash, minerals from construction and demolition waste	Increased calorific value Decreased quantity of slag and recoverable metals
Paper, cardboard and plastic	Calorific value decreases Chlorine loads (e.g. from PNC) in emissions decreases
Organic waste from kitchen and garden	Decreased moisture loads Increased calorific value
Hazardous waste (e.g. electronics, batteries)	Reduced efforts to remove toxic volatile heavy metals from air emissions (e.g. mercury) Reduced concentration of toxic pollutants in slag and fly ash (e.g. cadmium, lead, zinc)

Source Mutz and Hengevoss [23]

As composition of waste should be the basic deciding factor for the selection of waste disposal technologies, it is absolutely necessary to study the composition of waste properly before a final decision. The study of municipal solid waste of Delhi shows that on an average about 10% of the waste is expected to continue to be recovered for recycling by the informal sector, 65% could be treated by using either biological or thermal treatment technologies and remaining 25% is inert material. Of the 65% treatable material, 60–70% can be processed through composting or bio-methanation technologies, while 18–20% (textile, cloth, rubber, LPV) can be thermally treated through incineration based technologies. Presently, everything (mixed waste) including recyclables is encouraged by Delhi Municipal Committees to be put in the WtE plants. A study on Waste to Energy Options in Solid Waste Management, 2017 indicates prior separation of recyclables influences the characteristics of the remaining waste to a great extent. The suggestions are Table 6.

Delhi has three WtE plants one at Okhla (2000 tonnes), Gazipur (1300 tonnes) and Narela Bawana (2000 tonne). Mixed recyclable and organic waste are being fed into these plants in order to meet the high calorific value of 1400 kcal/kg, which is in contradiction with direction of National Green Tribunal as well as SWM Rules. NDMC is also planning to install another 300 TPD treatment unit including waste to energy plant.

Two centralized composting plants at Okhla and Bawana with total capacity of 1200 TPD are functional to handle biodegradable waste. However, it is reported that the quality of compost obtained these plants is highly compromised upon. As these compost plants are getting un-segregated waste, the plants are showing poor efficiency and products are many a time remaining unsold (NCT, Recommendation for Long Term Action Plan for Solid Waste management in Delhi, 2 August 2017). In addition, the domestic hazardous waste like discarded paint drums, pesticide cans, CFL bulbs, tube lights, expired medicines, broken mercury thermometers, used batteries, used needles and syringes and contaminated gauges etc., generated at the

Table 7 Existing infrastructure for waste disposal/proposal

Composting sites	1. Okhla—200 tonnes 2. Bawana—1000 tonnes
Waste to energy plants	1. Okhla—2000 tonnes, 16 MW 2. Ghazipur—1300 tonnes, 14 MW 3. Narela-Bawana—2000 tonnes, 24 MW
Construction and demolition waste process	1. Shastri Park—500 tonnes 2. Burari—2000 tonnes
Landfills/dumpsites for waste disposal	1. Ghazipur—70 Acres, oversaturated 2. Okhla—56 Acres, oversaturated 3. Bhlasawa—40 Acres, oversaturated 4. Bawana—Integrated waste management plant in 100 acres

household level need to be stored separately in a bag and be given to waste collector for disposal separately. No such process has been initiated in any area so far. Moreover, the missing plans also include final disposal of bio-medical waste, hazardous waste including electronic waste units, which are adding the environmental degradation in a big way.

A Construction and Demolition Waste processing plant of capacity 2000 MTD has been installed by M/s IL & FS at Jahangirpuri and 500 MTD at Shastri Park, where processed construction and demolition material is used for making tiles/pavement blocks and also for ready mix concrete, aggregates etc. Around 50 MTD of C&D waste are generated in the NDMC area daily. These waste need to be carried to the processing units for final disposal (Table 7).

Plastic waste especially carry bags has been creating nuisance in Delhi despite over 12 years of massive awareness campaign "Say No to Plastic Bags". The Hon'ble High Court of Delhi had passed a judgment in August 2008 for imposing ban on plastic carry bags in main markets, local shopping centers, etc. subsequent to which Government of Delhi had issued a notification on 07.01.2009, but the situation continued to worsen even after. The Ministry of Environment, Forests and Climate Change, GOI has notified revised Plastic Waste Management Rules 2016 on 18 March 2016 that entrust the responsibility of plastic waste management with Urban Development Department and the urban local bodies. Overall there should be a holistic approach towards tackling the problem. This need to be taken care with a strong will to follow the rules and guidelines, community participation along with political will to solve the environmental issues.

6 Urban Energy

Evidence from 209 countries demonstrates that as developing countries urbanize, their cities' consumption of energy and contribution to carbon emissions start to

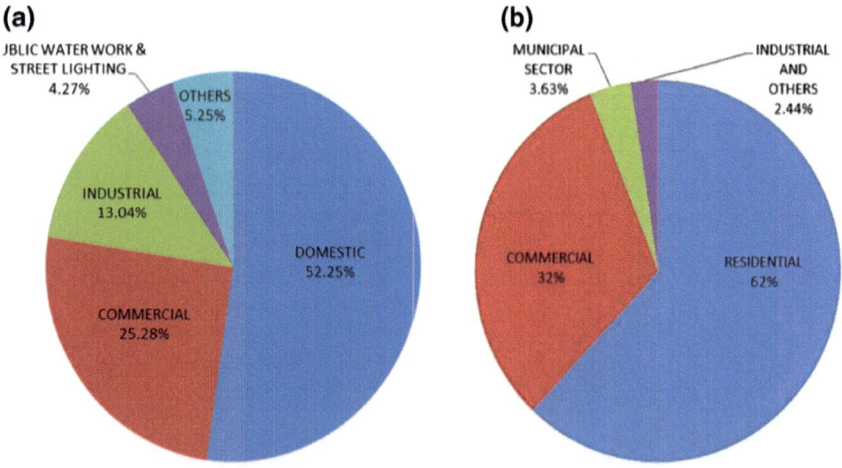

Fig. 10 **a** Energy scenario in Delhi and **b** consumption of energy across different sectors in NDMC area. *Source* CEA [5] and NDMC (2017)

become disproportionately high in comparison to their population and wealth [31, 32]. Indian cities prove to be a breeding ground for highly intensive economy, energy and emissions linkages and co-benefits at their intersection [30]. Delhi, being the national capital is one of the most energy consuming city in India that has also been selected to become a smart city. In spite of having no internationally accepted definition of a Smart City or a national urbanization policy in India, the overarching concept of Smart Cities nonetheless holds immense potential to achieve multiple benefits of sustainability, systems efficiency, economic growth, participatory governance and better quality of life [20]. Certain inroads are clearly visible in the area of electricity, public transport and waste to energy, as discussed henceforth.

As per the report of the Central Electricity Authority on Load Generation Balance Report 2015–2016, Delhi is consuming more electricity than the states of H.P, J&K, Uttarakhand, Chhattisgarh, Goa, Kerala, Bihar, etc. and all other states of Northeast. Already, in Delhi, the household electricity consumption per capita is about 43 units per month against a national average of 25 units. According to Delhi Electricity Regulatory Commission (2016), the electricity locally generated is 5941 MU, while that purchased from other states is 24,618 MU. As per Centre for Science and Environment's analysis of [5] data of energy use in Delhi, it has been identified that the domestic sector is the biggest consumer of energy (Fig. 10a). Delhi's average consumption is only about 181 Kwh and about 2/5th of the households consume less than 100 Kwh per month. Meanwhile, energy consumption scenario in NDMC area for which the Smart City is being implemented is (sector wise) Residential: 62% (277 MU), Commercial: 32% (1153 MU), Municipal sector: 3.63% followed by Industrial and other sectors: 2.44% (Fig. 10b).

Table 8 Sector wise number of consumers in NDMC (2017)

Particulars	Number
Domestic	43,664
Non domestic	22,114
Small industrial power	10
Street lighting	75
Other (including NDMC officials)	2465
J.J. clusters	1701
Total	70,029

Table 9 Technology type, capacity and location of power plants in Delhi

Name	Type	Capacity (MW)
1. Badarpur thermal power station	3 × 95 coal (Indirectly fired boilers) 2 × 210 coal (Directly fired boilers)	705
2. Pragati power station	2 × 104 GTs (open cycle) 1 × 122 STG (combined Cycle steam turbine)	330
3. Indraprastha power generation Co. Ltd.	30 X 6 GTs 3 × 30 STGs 2 × 67.5 coal (2 GT + 1 ST depend on liquid fuel)	405
4. Pragati-III CCGP project, Bawana	4 × 250 GTs 2 × 250 STGs	1500 (N)

The sector-wise consumers in these sectors are enlisted in Table 8, suggesting that the major sources of energy in NDMC are Electricity, Kerosene, LPG and Diesel/Petrol, with their consumption as 1441, 13, 12, 5 and 0.01 MU respectively.

Delhi's peak demand has doubled in the last 10 years. In 2015, Delhi's peak demand was 6,006 MW, in 2016, it crossed 6,300, which was higher than Mumbai, Kolkata and Chandigarh. It is estimated to be about 12,000 MW by 2021 (8–10% is the average annual increase in power consumption in Delhi). About 60% of Delhi's peak electricity in domestic sector is consumed by air conditioners. This shows that this high consumption of energy in Delhi can be reduced if city's design and planning become sensible to reduce the heat gain. Now, the summer heat gets trapped by the insensitive planning of Delhi and even in night, when it is cooler as compared to day, the trapped heat is not released effectively. For a very long time, Delhi's power supply has been dependent on its four power plants, namely Badarpur Thermal Power Station (705 MW), Pragati Power Station (330 MW), Indraprastha Power Generation (405 MW) and Pragati-III CCGP Project, Bawana (1500 N). The technology type, capacity and location of these plants is mentioned in Table 9.

7 Renewable Energy

As far as renewable energy is concerned, grid interactive renewable power as on
31.03.2014/15 data for solar power shows that 5.15/6.71 MW evolved to about
40 MW by Dec 2016. The state also commissioned 177 MWh (362.5 KWp) of
solar power projects till July 2016. This is quite miniscule to the actual solar potential of the city. Delhi has about 300 sunny days and available space for rooftop solar
PV is estimated to be 30 km^2. Depending on various factors and basis of calculation,
technical estimates for solar potential of Delhi range between 2050 MW by Ministry
of New and Renewable Energy [21] and 2500 MW by Energy Efficiency and Renewable Energy Management Centre [13] with annual generation approximately 3500
million KWh by 2025. The state considers that with the right solar policy and implementation, Delhi has the potential to attain 1 GW Solar Power capacity in 5 years
[13]. As per these estimates, a major portion of Delhi's potential clean energy generation from rooftop solar PVs is vested in residential buildings (49%), followed equally
by government buildings (26%) and commercial & industrial buildings (25%). The
target of NDMC with corresponding achievement and emission reduction against
both renewable energy and energy efficiency strategy are listed in Table 10.

While national level solar policy (National Solar Mission) aimed to establish large
scale solar plants (750 MW + capacity), it misses out opportunities in harnessing
rooftop solar. Thus, in 2016, the government of NCT Delhi came up with New Solar
Policy a five year programme from 2016 to 2021, which actually started implemen-

Table 10 Renewable Energy (RE) and Energy Efficiency (EE) strategy, target of NDMC with
corresponding achievement and emission reduction

RE and EE strategy for NDMC	Energy saving target (MU)	% of saving target to achieve	Emission reduction
RE for residential Sector	54.06	28	43,789
RE for commercial sector	16.22	8	13,189
RE for municipal sector	31.26	12	19,715
Total RE strategy	**102**	**48**	**76,693**
EE for residential sector	115.52	59	94,006
EE for commercial sector	14.77	7.52	14,529
EE for municipal sector	11.40	6	9437
Total EE strategy	**142**	**72**	**117,972**
Total RE and EE combined strategy	**244**	**120**	**194,664**

Table 11 Number of registered vehicles in Delhi up to 31 March, 2018

Type of vehicle	Number[a]	Contribution to air pollution in percent[b]	Remarks
Motor car/cab	3,240,489	10	
M. Cycle/Scooter	6,957,223	33	
Goods carrier	153,389	45	
Others	405,667	12	Includes goods carrier, tractor, ambulance, three wheeler etc
Total	10,756,768	100	

Source [a]GNCT Delhi [15], [b]IIT Kanpur [17]

tation in 2017. The policy is applicable on solar PV generation of 1 KWp or more at the local/rooftop level & aims at improving energy security along with cutting down of electricity expenses for Delhi, especially during peak demand. It will also reduce the need for new power purchase agreement (PPA). The agenda is "Achieve aggressive yet realistic rooftop solar growth in Delhi". This would be achieved through two targets of generating: (a) 1.0 GW solar energy by 2020 (14% of peak load, 4% of total energy), and (b) 2.0 GW solar energy by 2025 (21% of peak load, 7% of total energy). The annual target set under this agenda is shown in Table 10. The NSP envisages to create an additional generation capacity of 1945 MW by 2025. The smart city tends to follow and implement these targets on the ground.

Any strategy for clean energy in a megacity has to evaluate the environmental costs of public mobility. It is one of the major sectors that guzzles fossil fuel and contributes to GHGs, local air pollution and impacts on human health disorders. As per the Census of India [6] data, while population of the city increased from 9,879,000 (2000) to 16,750,000 (2011) and 18,686,902 (2017), the corresponding number of vehicles have increased from 3,456,579 (2000) to 6,011,731 (2010) and to 10,756,768 up to 31 March 2018. It is important to note that while trucks registered in Delhi, adhere to BS-III norms, those registered in other states but passing through the city daily follow the inferior BS-II norms. Similarly, 4 W (cars) follow BS-III and 2 W adhere to BS-IV norms, which are a notch above those 4 wheelers and 2 wheelers registered outside Delhi. As such, trucks contribute 20% of PM2.5 and 40% NOx [8]. Meanwhile, a diesel car emits 3 times the NOx and 7 times PM2.5 of a petrol car. The cumulative impacts of these multiple air pollutants play havoc on Delhi's population generating health hazards like respiratory disease, asthma, cardiovascular diseases, Alzheimer's and Parkinson's disease, etc. Fortunately, commercial trucks are not permitted to ply within NDMC area but diverted to ring roads and by-pass roads. Similarly, non-destined trucks are by-passed through the newly constructed peripheral roads.

As far as mass transit is concerned, it has been studied that while Delhi Metro has not promoted positive land cover change (increasing built-up area), either along rail lines or stations, on the contrary it has influenced land cover change in peripheral areas, mostly in the form of informal settlements [1]. This shows that Delhi is a victim of poor land-use planning due to which people are forced to travel long distances for work and other activities (Table 11).

8 Visualizing and Mitigating Environmental Impacts Under Smart Cities Mission

NDMC is committed to Smart Cities Mission through implementation of a Smart City Plan. NDMC has set a target of 10% reduction in conventional energy which is equivalent to 197 MU in electricity through solar power. Delhi has identified solar power as a key renewable energy source and accordingly set a target of installing 1000 MW of solar power capacity by 2020 and 2000 MW by 2025. For detailed targets and preliminary achievements refer Table 10. In addition, Delhi has a huge opportunity in becoming smart city by significantly recalibrating its energy portfolio. The strategies include shifting from conventional fuels in transportation to battery operated or electric vehicles, energy efficiency in design and construction and maintenance of buildings, and promotion of waste to energy through innovative, scientific municipal waste management. For instance, waste to energy facilities like the Timarpur—Okhla WTE Plant provide new and considerable opportunities to convert Delhi's municipal waste into electricity. This section discusses the potentials and challenges in actualizing these strategies and achieving these targets.

In New Delhi, Lutyens simply attempted to make a statement—and did so very successfully—that drew heavily from his knowledge of the profound interdependence of sustainable good environment of cities and nature. Lutyens envisioned with sufficient clarity that the semi-arid natural climate of Delhi region that was in great contrast with that of Calcutta region in Bengal had to be moistened and greened appropriately. Such a vision could be thought of and implemented primarily because New Delhi was essentially a 'new town' built on virgin land. Creation of a man-made city forest spread across the ridge on the west and south-west of Rashtrapati Bhavan (Viceroy's Palace) has been a permanent feature that contributed significantly to modify the micro-climate of New Delhi. Similarly, the large green open spaces, especially in the Lutyens' Bungalow Zone and the tree lined avenues were designed to create a sustainable, comfortable ambience and a pleasant visual experience throughout the length and breadth of New Delhi. This pristine green environment of New Delhi has been generally conserved in spite of many fold growth of population and activities inside the city.

As New Delhi grew its roots, her resident population, floating population; core, secondary and tertiary activities also grew at a matching pace hardly allowing NDMC and other associated agencies to sit back and visualize the consequences of such growth jeopardising the smooth functioning of the city services and image of the city. The emergent problematic issues were: far too many white collared workers in and around the capitol complex, traffic congestion around India Gate and the 'hexagon', lack of car parking space in Connaught Place and other locations, encroachments by vendors on the footpaths of Connaught Place, to name a few.

The demand on the city's infrastructure and vehicular circulation system to serve the increasing intensity of activities compelled introduction of new means and measures to cope with the pressure without altering the spirit, structure and geometry of New Delhi. Installation of traffic signals, construction of underground car parking facility were necessary for meeting the needs of the Delhiites as well as visitors to New Delhi. Construction of the multi-layered metro rail hub in the central park of Connaught Place has greatly facilitated accessibility of New Delhi in general and Connaught Place in particular not only from far flung neighbourhoods of Delhi but also those living in the NCR. As a direct impact of operationalizing the metro hub, there has been a dramatic increase in the day time population of the city, particularly in and around India Gate and Connaught Place. As per Authors' estimates, the current ratio of night-time and day-time population of Connaught Place is estimated to be 1:20.

Meanwhile, the access and coverage of smart city project in New Delhi area is concentrated in select areas. Thus the likelihood of significant changes either in the quality of life or the activity pattern in New Delhi particularly due to implementation of the 41 projects under the Smart City Mission appear to be limited. As per MPD 2021 and Zonal Development Plan of Zone D which is administered by NDMC, first twelve out of the 20 Subzones (D1 to D12) cover New Delhi. Though the proposals for these subzones were prepared in 1966, modified and approved by 1983 but they were finally approved by the DDA in 2017 [9].

Table 12 indicates the growth of population and activities that cumulatively occurred in New Delhi and their impact on the infrastructure and resources. In sub-zone D-1, the number of shops, restaurants and commercial activities grew heavily while the resident population eventually declined in recent years. But the day-time population grew dramatically in recent years—and continues to do so—with the establishment of the DMRC hub. In comparison, in subzones D-2 and D-5 as the single storeyed tenements were replaced by multi-floored apartments for government employees, the resident population grew by about six times and the associated commercial activities grew in proportion. It is evident that Sub-zones D-1, D-3, D-4, D-8, D-9 and D-11 hosted very high modifications in terms of day-time population and built up area and the consequent impacts in terms of consumption of water, food and electric power and produced correspondingly larger quantity of solid waste. Moreover, environmental impacts of such an intra-city growth pattern in terms of deepening of water table, increased air and noise pollution, disposal of municipal waste etc. must be assessed and addressed.

As the character of subzones vary significantly, the resultant environmental impact assessments (EIAs) and environmental management plans (EMPs) would also be diverse. Thus, a specific stage wise methodology for identifying the EIA is proposed as below.

Table 12 Subzone wise growth and development of New Delhi and their impact on city services and resources

Subzone No.	Name of locality(as per ZDP)	Growth, development and impacts							
		T	FF	RP	DP	W	F	SW	E
D-1	**Connaught Place and extension** Larger and growing number of shops and offices. 13 floor, 44 m high LIC building built 1986 DMRC hub constructed in Central Park 2005	HH	HH	L	HHH	H	HH	HH	HH
		Continued rise in: traffic, footfall, day-time population, consumption of water, food, solid waste, energy							
D-2	**Mata Sundari area** Single storeyed row houses replaced by 8 floored apartments	H	H	H	H	H	H	H	H
		Rise in: traffic, resident and day-time population, consumption of water, food, solid waste, energy							
D-3	**Kasturba Gandhi Marg** Multi floored buildings built 1970s onwards	H	HH	L	H	H	H	H	HH
D-4	**Sansad Marg** Multi floored buildings built 1970s onwards, New Delhi District Centre 1983	HH	HH	L	H	H	H	H	HH
D-5	**DIZ Area** Single storeyed row houses replaced by 8 floored apartments	H	H	H	H	H	H	H	H
D-6	**Upper ridge Area**								
D-7	**Purana Qila** Zoological Park developed, increased footfall	H	H		H	H	H	H	H
D-8	**India Gate** Increased footfall	HH	HH	L	HH	H	H	H	H
D-9	**Central Secretariat** Multi-fold Increase of white collared workers	HH	HH	H	HH	H	HH	HH	HH
D-10	**Buddha Jayanti Park** Increased footfall	h	h	h	H	h	h	h	h
D-11	**Khan Market** Increased footfall and vehicular traffic	H	H	h	H	H	H	HH	H
D-12	**Akbar Road** Increased vehicular traffic	H	H	h	h	H	H	H	H

T Traffic; *FF* Footfall; *RP* Resident population; *DP* Daytime population; *W* Water consumption; *F* Food consumption; *SW* Solid Waste generation; *E* Energy consumption; *h* Moderately high; *H* High; *HH* Very High; *L* Low; *Based on windscreen survey September-October 2018

Stage 1: Statement of existing/baseline status

This would include identification of status of activities in terms of both diversity and intensity. For this purpose, the starting point may be documentation of the pattern of existing land uses and tracing their evolution as well as trends since early days of inception of the city. Similarly, status of key dimensions of environment including the trends of their changes may be documented. An analysis of the above documentation may be undertaken to identify the cause and effect relationships between trends of activities and their impact on environment, subzone wise. The output of this first stage maybe in terms of a comprehensive data-base, that may be updatable and retrievable as and when needed. The second output would be identification of the cause and effect relationships from past till date.

Stage 2: Projected status of activities with reference to specific year

The developmental activities of New Delhi have been essentially undertaken through formal process of approvals and sanctions. Starting with population growth, such activities have to be projected for a specific year in the future, that must match with those of the approved development plan for the subzone and financial cycles. The status of environment as a result of impacts due to the projected activity scenario may be undertaken. The implications of the modified environment may then be documented for the society at large. The implications must be critically assessed form the points of view of conserving the character of New Delhi and sustainability of the projected activity patterns. Keeping in view the development scenarios, their environmental implications may be evolved for comparative analysis of environmental costs.

Stage 3: Selection of the most acceptable scenario

After detailed evaluation of the envisioned scenarios, the most feasible scenario may be selected for further development and implementation. Starting with a standard set of criteria and their benchmarks, for a given subzone, depending upon its character and role in the functioning of New Delhi, the relative weightages of the components of the set of criteria may be designated. The fine tuning of the weightages is necessary as a specific subzone evolved to accommodate functions and activities driven by not only those of its own but also those of neighbouring subzones. Therefore, consideration must be given to the functions it performed through the last 87 years (1931–2018), how it impacted the environment and the intended future status of a subzone.

Stage 4: Preparation of EMP

How best the adverse environmental impacts of the selected scenario may be eliminated or minimized is the objective of proposing an EMP for a subzone. The inclusiveness of the EMP would depend upon who are the residents and or beneficiaries of a given subzone. EMP units would need to be established at the subzone level who will be responsible for (a) Creating the data base and maintaining records, updating the same in a easily retrievable form; (b) Administering the EMP in the subzone and coordination with concerned actors, other subzones as well as financial management.

Monitoring the progress of implementing the EMP entails resolving bottlenecks and/or obtaining the appropriate advice from the EMP unit at the city level that is for New Delhi or for the entire Zone D. The EMP unit having a close view and sustained experience of the status of both the spatial and smart development and its environmental impact including management of the same would also be in a good position to visualise the situation over the next few years or more. Therefore, suggestive directions and prescriptions may be recommended.

Planned development being a cyclic process, the inputs offered by the EMP unit would be necessary for indicating limits of further growth of population and activities in a subzone in view of their potential adverse environmental impact. Secondly, the activity pattern of a subzone must be in continuity at least partly in order to maintain its original identity and role in New Delhi.

9 Conclusion

Smart Cities Mission launched in 2015 is a clear indication of the high priority given by Government of India to accelerate planned development of the urban settlements, so urgently required to carry forward the larger interests of a holistic national development. Inclusion of New Delhi among the first round of cities further symbolises the importance attached to the programme and also an intention to set an example among the competing cities.

First and foremost, New Delhi must be recognized and perceived as a city occupying an area of 47.27 km^2 designed and built essentially as the new capital city of British. Delhi or the National Capital Territory of Delhi (NCTD) has an area of 1483 km^2 including New Delhi located at about its centre. Assuming New Delhi's fundamental role and identity would remain as such, the agencies and actors responsible for its sustenance will have to be smart enough to cope with the changing scenario outside its boundary where a wide range of forces will exert unprecedented pressure on New Delhi. The challenge might well be creation of a system operated by the concerned agencies and actors that would be able to respond to the pressures from time to time.

Inclusion of New Delhi in the first list of cities to be transformed as a Smart City is a good opportunity to achieve that objective. Among the necessary preconditions to that end, the foremost is availability of data pertaining to various dimensions of New Delhi. As such, the Zone D data is misleading and confusing for evolving and monitoring smart city interventions for New Delhi. Secondly, in order that 'Smart New Delhi' is sustainable, all the users of New Delhi, citizens as well as visitors must participate appropriately to operate and manage the smart systems for functioning of services and facilities.

According to the UN World Cities Report 2016, India's urban population is estimated to grow from 377 million in 2011 to 677 million in 2030, that might account for about half of India's population. Also by then, there will be seven megacities (a city having a population of 10 million or more) in India Mumbai, New Delhi, Kolkata, Bengaluru, Chennai, Hyderabad and Ahmedabad instead of three in 2011. The task of smartening them through the Mission is a tough and complex task. New Delhi can hardly afford to adopt a lackadaisical attitude. The concerned agencies should therefore revisit the proposed development initiatives of all the twelve subzones of Zone D, representing New Delhi for better resilience and sustainability.

Notes

1. Four examples of ancient Indian cities still teeming with life are: Varanasi, Ujjain, Puri, Madurai.
2. The phrase 'garden city' merely implies here a city generously interspersed with gardens and opens spaces and not Ebenezer Howard's concept of Garden City (1898).
3. Sanitation Green refers to an open space created to ensure maintenance of a 'safe' distance between existing indigenous towns and planned settlements commonly referred to as Civil Lines built during the British rule in India.
4. Several acts were enacted during the period of British rule in India, such as City Improvement acts, including Bombay Improvement Act 1898, Mysore Improvement Act 1903, Calcutta Improvement Act 1911 and Town Planning acts like Bombay Town Planning Act 1915, Madras Town Planning Act 1920.

Acknowledgements The authors would like to acknowledge Indian Society for Applied Research & Development (ISARD), supported by Kyoto University, Kyushu University, Asia-Pacific Network for Global Climate Change Research (APN) for conducting Expert Workshop on Low Emission Development Strategies in Delhi on May 21, 2018 that was crucial in deliberating issues with the concerned stakeholders for New Delhi to become a smart city. Thanks is also due to Aesha Upadhyay and Prachi Gupta, both masters students of Dr. APJ Abdul Kalam Technical University, India for formatting the draft chapter, painstakingly compiling citations and list of references.

References

1. Ahmad S, Avtar R, Sethi M, Surjan A (2016) Delhi's land cover. In: Sethi M (2017) Climate change and urban settlements—a spatial perspective of carbon footprint and beyond (ISBN: 9781138226005). Taylor & Francis, Routledge, London, U.K
2. Balachandran M, Karnik M (2015) Lutyens' Delhi may be about to change for good. Retrieved from https://scroll.in/article/754724/lutyens-delhi-may-be-about-to-change-for-good
3. CAG (2012) Performance Audit of Jawaharlal Nehru National Urban Renewal Mission (JNNURM), Chapter 8. Ministry of Housing and Urban Poverty, Govt. of India, New Delhi
4. Cavale R (2017) Pattrick Geddes in India: anti colonial nationalism and the historical time of cities in evolution. Landscape Urban Plann 71–81
5. CEA (2017) Peak power supply position report (2016–2017). Central Electricity Authority, Ministry of Power, Government of India, New Delhi
6. Census of India (2011) Provisional population totals 2011, paper II, 2. Census of India, New Delhi
7. CPHEEO (2012) Recent trends in technologies in sewerage system. Ministry of Urban Development, New Delhi
8. CSE (2015) Delhi clean-air action plan. Centre for Science & Environment, New Delhi
9. DDA (2017) kZonal Development Plan, Zone-D (As per MPD 2021)

10. Delhi Jal Board website. Last updated 20 April, 2018
11. Demouliere R, Berger J (2012) Public water supply and sanitation services in France—economic, social and environmental data. BIPE
12. DUAC (2015) Report on lutyens bungalow zone (LBZ) boundary and development guidelines. Govt. of India, Delhi
13. EEREM (2016) Delhi Solar Policy, 2016 (Notification). New Delhi: Energy Efficiency and Renewable Energy Management Centre, Department of Power, Government of NCT Delhi. Change in post transit era. Cities 50:111–118
14. Ganju MNA (1999) Lutyens bungalow zone. Archit Des Indian J Archit 6(Nov–Dec):34
15. GNCT Delhi (2010) State of environment report for Delhi, 2010. Department of Environment and Forests, Government of NCT of Delhi, New Delhi. http://www.indiaenvironmentportal. org.in/files/SoEDelhi2010.pdf
16. Hutton JH (1933) Census India 1931: the population problem in Delhi. Retrieved from http:// indpaedia.com/ind/index.php/Census_India_1931:_The_Population_Problem_in_Delhi
17. IIT Kanpur Study (2016) Comprehensive study on air pollution and green house gases (GHGs) in Delhi (final report: air pollution component). Department of Civil Engineering, Indian Institute of Technology, Kanpur
18. IL&FS Ecosmart (2007) JNNURM city water supply system. Govt. of India, Delhi
19. Mittal S, Sethi M (2016) Are smart cities for real: will they bring qualitative improvement in urban living? In: SPANDREL 2015–16 special issue: making cities smart and competitive, issue 11, pp 1–12
20. Mittal S, Sethi M (2018) Smart and livable cities: opportunities to enhance quality of life and realize multiple co-benefits. In Sethi, Puppim de Oliveira (eds) Mainstreaming climate co-benefits in Indian cities. Springer Nature, Singapore
21. MNRE (2016) Annual report 2016–17. Ministry of New and Renewable Energy, Government of India, New Delhi
22. MoUD (2015) The smart city challenges, stage 2, smart city proposal. NDMC, New Delhi
23. Mutz D, Hengevoss D (2017) Waste-to-energy options in municipal solid waste management. Deutsche Gesellschaft für Internationale Zusammenarbeit GmbH, Eschborn
24. Nath A, Mehra A (2002) Dome over India: Rashtrapati Bhavan. India Book House, Mumbai
25. NCT (2016) Water policy for Delhi. Govt. of NCT of Delhi, New Delhi
26. NCT (2017) Recommendations for long term action plan for solid waste management in Delhi. Government of NCT Delhi, Delhi
27. NDMC (1994) The New Delhi municipal council act. Govt. of India, Delhi
28. Sengupta A (2012) Resources and infrastructure. Climate Change and Disease Dynamics in India, TERI
29. Sethi M (2015) Smart cities in India: challenges and possibilities to attain sustainable urbanisation. Nagarlok 47(3):20–37
30. Sethi M (2018) Co-benefits from the energy sector. In: Sethi, Puppim de Oliveira (eds) Mainstreaming climate co-benefits in Indian cities. Springer Nature, Singapore
31. Sethi M, de Oliveira Puppim (2015) From global 'north-south' to local 'urban-rural': a shifting paradigm in climate governance? Urban Clim 14(4):529–543
32. Sethi M (2017) Climate change and urban settlements—a spatial perspective of carbon footprint and beyond (ISBN: 9781138226005). Taylor & Francis, Routledge, London, U.K
33. Singh S (2017) Why untreated sewage continues to be dumped into the Yamuna. Retrieved from NDTV: https://swachhindia.ndtv.com/untreated-sewage-continues-dumped-yamuna-6622/
34. Tangri AK (2000) Integration of remote sensing data with conventional methodologies in snow melt run-off modelling in Bhagirathi river basin. Technical Remote

35. Upadhyay A (2018) Rethinking smart cities towards an integrated approach. M. Arch Dissertation Report. Faculty of Architecture, Dr. APJ Abdul Kalam Technical University, Lucknow
36. Wikipedia (2018) A plague at the coronation park. Retrieved from Wikipedia: https://upload.wikimedia.org/wikipedia/commons/6/69/A_Plaque_at_the_Coronation_Park.JPG

Part VII
Surat

Amidst the Governance Challenges in Environmental Management and Sustainable Urbanization in Surat

Bhasker Vijaykumar Bhatt, Shashikant Kumar
and Neerajkumar D. Sharma

Abstract The urbanization is becoming inevitable across the globe and for one of the fastest growing cities, Surat, the status is reported to be alarming. Faster growth of population and settlement with an increase in enterprise establishments generating opportunities are acting as counter magnets in the case of Surat. The city has a natural advantage by means of a perennial flowing river, Tapi, carrying fresh water to match the requirements posed by the citizens for domestic as well other usages. Keeping the environmental concerns at the central thought around urbanization in Surat, the chapter explores through the extent of development and resulting effects. Objectives for the study were set to understand different aspects and chalk out smarter proposals for governance of environmental resources through promoting sustainable practices. Keeping aware of the natural resources utilized or mobilized in terms of quantity and quality is considered a manner towards sustainable use. The chapter discusses the situation in Surat regarding the demographics, weather, natural drainage and topography, irrigation and agricultural activities, transport and linkages along with vehicle growth, air quality, water resources, vegetation cover and, industrial activities as drivers of the economy for the citizens. Further, an exploration of the state of essential service infrastructure of water supply, sewerage, solid waste and aquaculture is discussed. Keeping various activities pertaining to environmental concern, the role and challenges faced by various government agencies are deliberated. The concluding remarks are drawn identifying some of the critical concerns and alternative solutions towards smarter governance by adopting technological advances and reforming ongoing practice is recommended. The use of GNSS based solutions is emphasised for various purpose of monitoring and positioning considering the era

B. V. Bhatt (✉)
Bhaikaka Centre for Human Settlements (APIED), Nr. Bhaikaka Library, Vallabh Vidyanagar,
Anand 388120, Gujarat, India
e-mail: er.bhasker@gmail.com

S. Kumar
Bhaikaka Centre for Human Settlements (APIED), Vallabh Vidyanagar, Anand, Gujarat, India
e-mail: shashikant.kumar@apied.edu.in

N. D. Sharma
GIDC Degree Engineering College, Abrama, Navsari, Gujarat, India
e-mail: neerajrwh@gmail.com

© Springer Nature Singapore Pte Ltd. 2020
T. M. Vinod Kumar (ed.), *Smart Environment for Smart Cities*, Advances in 21st Century
Human Settlements, https://doi.org/10.1007/978-981-13-6822-6_9

when there is a maximum number of satellites available for fetching GNSS signals in the space above South-Asia. Employing and promoting the use of interoperability of various positioning satellite constellations through RTK solutions are briefly discussed.

Keywords Smart infrastructure · Sustainable environment · Services · Surat · Urbanization

List of abbreviations

AD	Anno Domini
ASICS	Annual Survey of India's City-Systems
BC	Before Christ
BRTS	Bus Rapid Transit System
CE	Common Era
cm	Centimetre
ESR	Elevated Storage Reservoir (for Water)
GEB	Gujarat Electricity Board
GEC	Gujarat Ecology Commission
GIDC	Gujarat Industrial Development Corporation
GIS	Geographic Information System
GPCB	Gujarat Pollution Control Board
INR	Indian National Rupees
IoT	Internet of Things
km	Kilometre
km^2	Square kilometre (Area)
m	Meter
ML	Million Liters
MLD	Million Liters per Day
MSL	Mean Sea Level
MT	Metric Tonne
NH	National Highway
NWRWSK	Narmada, Water Resources, Water Supply and Kalpasar Department
PPP	Public-Private Partnership
RDBMS	Relational Data Base Management System
RS	Remote Sensing
RWH	Rain Water Harvesting
SGCCI	The Southern Gujarat Chamber of Commerce and Industries, Surat
SH	State Highway
SIC	Surat Irrigation Circle, Surat
SOx	Sulfur Oxide composites
SMC	Surat Municipal Corporation, Surat
STP	Sewage Treatment Plant

SUDA   Surat Urban Development Authority, Surat
T.P.   Town Planning (as in T. P. Scheme)
UASB   Up-flow Anaerobic Sludge Blanket
UGSR   Underground Storage Reservoir (for Water)
ULB   Urban Local Body
WTP   Water Treatment Plant

1 Introduction

Environmental management is a crucial element in any urban settlement. If not managed properly, it can result in several instead of a variety of short-term and long-term adversities. A variety of waste results at the end of user consumption, and when it is about the urban user, there exists a significant diversification in terms of goods and forms of resources consumed. With the rise in the urbanization, not only the natural resources are depleted but also, the state of phenomenal actions is also modified. In general, the environmental management term includes for the ecology, earth science, atmospheric science, water science, climate change, society and culture, law, assessments, mitigation and monitoring, practices and sustainability aspects.

Urbanization is dynamic in nature with mostly, increasing populations generating diverse demands and creating pressure on natural resources available in the surrounding. With an urban spread, the Urban Local Bodies need to empower itself in terms of physical and intellectual capital to cater to the needs for managing and governing the utilization of resources and avoid exploitation. Figure 1 shows a conceptual relationship in terms of the effects of urbanization through urban environmental shifts. The governance for environmental management aspects is essential for a sustainable society to leverage natural resources for the next generations.

The current work presents exciting details of the existing situation in practices of managing several resources by the local government and inter-alia state government departments for the city of Surat—an emerging metropolitan in the Southern Gujarat State of India.

There are various natural elements existing on the earth. These are partially monitored, and to an extent, its use is regulated by several agencies. The term SMART is an acronym in the management, especially in the cases of setting up goals and objectives [1]. Further, the concept of SMART—as the thought of an abbreviated form of Specific, Measurable, Assignable, Relevant and, Time-based are explored in the context of the environmental management of the study area. For the city of Surat, the same is discussed in the subsequent section of the chapter.

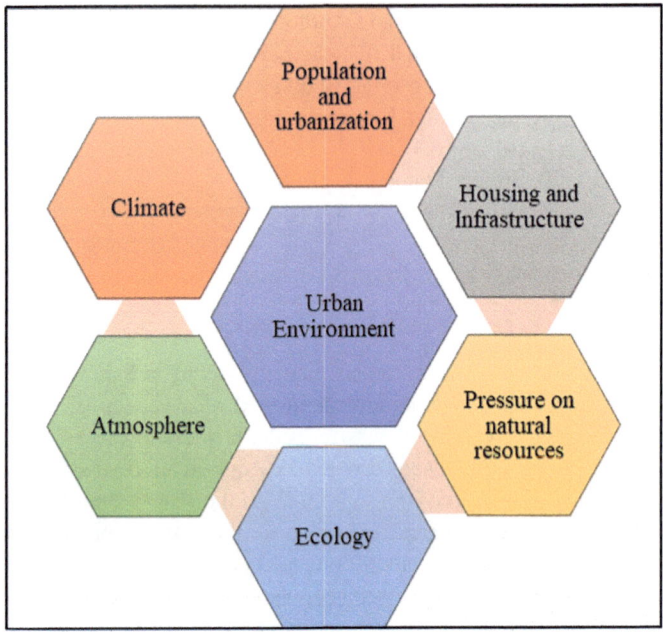

Fig. 1 The Urban environmental crux and effects

2   Objectives of the Study

The study is conducted to understand the extent of environmental resources and natural setup in the region. With the expansion of Surat city as an urban centre, it has affected the natural state. The urbanization process seldom preserves the environment. There are several government organizations that are looking after and performing various tasks pertaining to safeguarding the environment. The current work has the following objectives:

(1) To explore the regional set up in and around Surat city for various environmental parameters;
(2) To understand the extent and effect of infrastructure and services in Surat city;
(3) To identify and propose smarter options pertaining to environmental management in a sustainable manner in the Surat city with a particular focus on solid waste management.

3   About Study Area—Surat City

The history of the city dates to 300 BC yet no specific date of urban establishment is documented [2]. Following the principle of human settlements, the city emerged at the southern bank of the Tapi river as shown in Fig. 2, initially and expanded in all directions. Today, it can be called as a city residing on the banks of two rivers namely, Tapi at the centre and Mindhola at the South.

Economics of the city is governed by the industrial establishments (more than 50,000 units) that includes small, medium and large-scale enterprises [3]. The primary sector for economic involvement of working personnel is textiles and chemicals as well as diamond cutting and polishing. Broad-spectrum opportunities for earning have been an attractive magnet for citizens from across the nation to reach for and settle in Surat. On a positive cycle, increasing migration has resulted in increasing economic opportunities, and the trend is continued. The city is well known as a diamond city and Textile city. Traditionally, gold and silver thread work embedded in the cloth was a mastered art among local citizens.

It is the *eighth largest city* and *ninth largest urban agglomeration* [4] in the country of India. It became the *fourth fastest growing city* [5] by the year 2016. In the year 2013, it was awarded as *The Best City* by Annual Survey of India's City-Systems (ASICS), and in the year 2014, the city was awarded as Runners-up for 'Quality of life' and 'Quality of City-Systems' [6]. The city has reported making leading progress over several facets in terms of Smart cities movement initiated by the Government of India. The Ministry of Urban Development declared it to be the *third Cleanliest City* of India in the year 2010 [7]. The extent of urbanization can be visualized in Fig. 3 which is a clipped image of layer stacked satellite image bands composed in a False Color Composition Image. The image shows the extent of the Surat region under the administration of Urban Development Authority of Surat (SUDA).

The municipal administrative limits are extended to the extent of 326.515 km^2 within the urban agglomeration of 985 km^2 [9]. About two decades back, the city had a spatial spread over merely 8.18 km^2 [10]. The Census of India, 2011 revealed that the city was housing about 44.67 Lakh citizens [11] within the administrative boundary of Surat Municipal Corporation with the population of urban agglomeration of about 60 Lakh residing in an area of about 722 km^2. The literacy rate was reported to be higher than 86%. Males constitute about 53% and females about 47% depicting for mostly a gender-balanced society [12]. Table 1 shows the city limit expansion year wise. In 1961 area of the city was only 8.18 km^2, today that area is known as an old city. After that, the city limit was expanded gradually. The major expansion was made in July 2006 in which city limit was extended up to 326.515 km^2.

Figure 4 is a graphical representation of the study area, Surat city, with the growth in the regulatory limits over past decades. Before the latest extension, the S.M.C area on the northern side of the river was 19.63 km^2 while that towards the south was 92.65 km^2. After the extension in the year 2006, the area of the city on the northern side of the river is 79.1 km^2 while that on the southern side is 254.91 km^2. Figure 4 shows the extent of regulatory limit extensions.

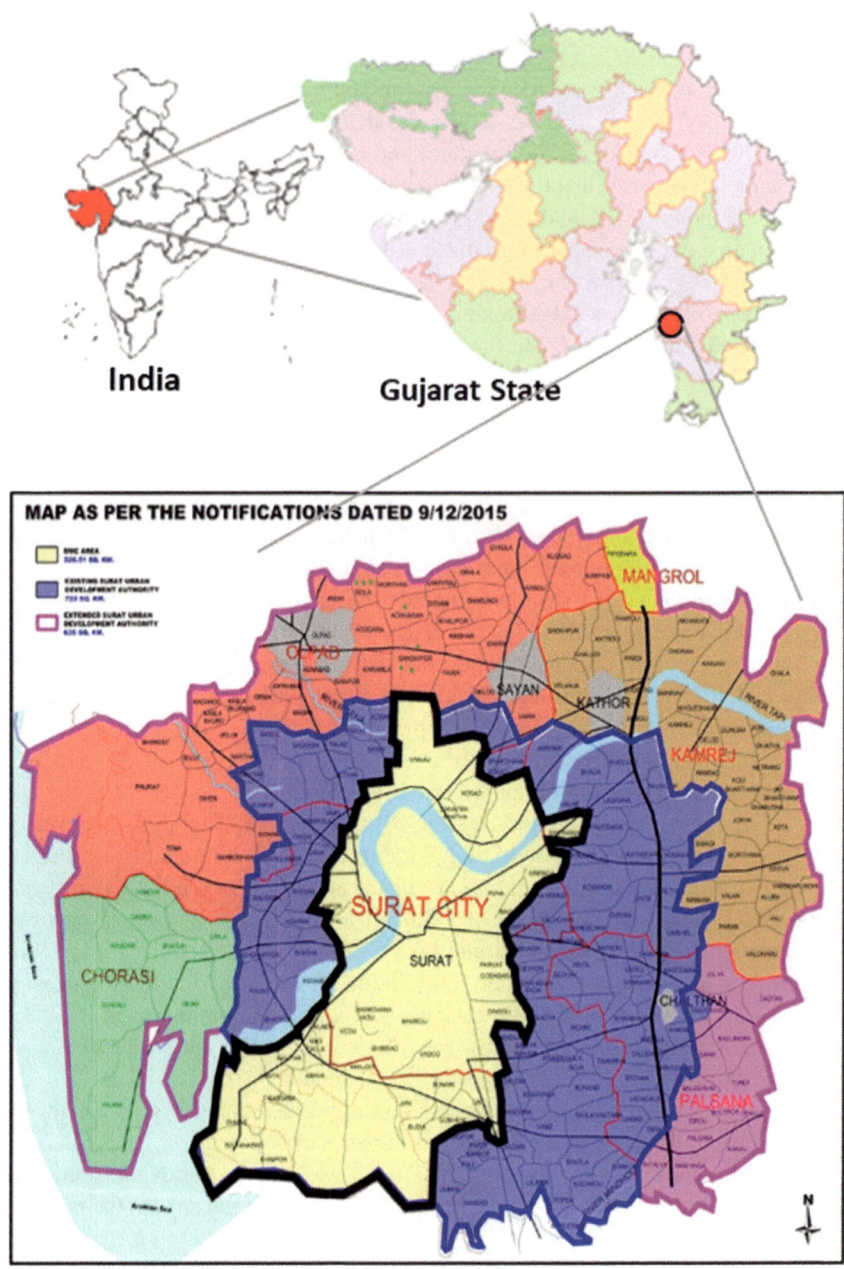

Fig. 2 Study area of Surat City

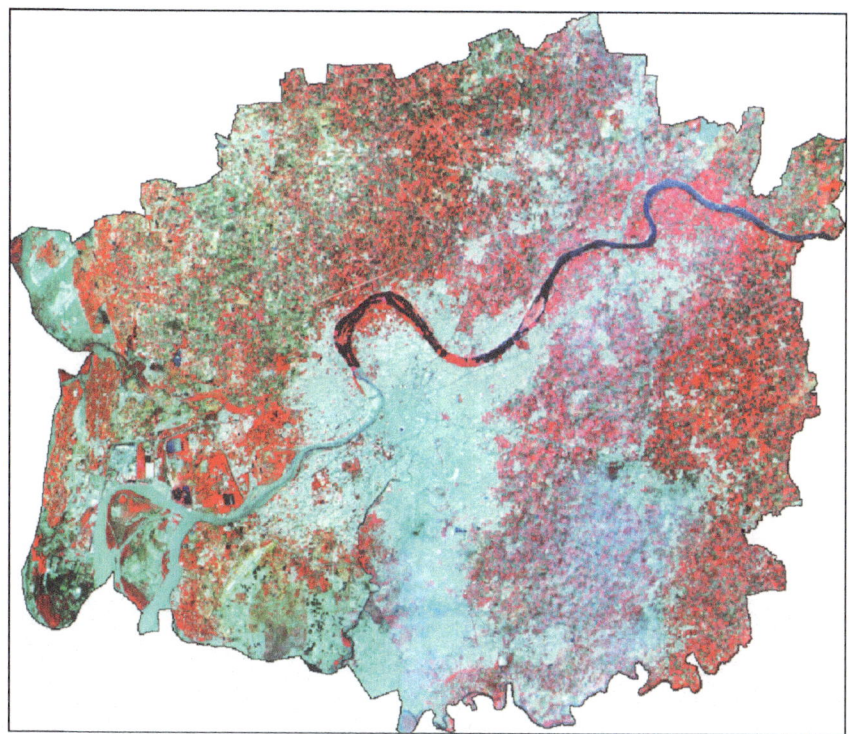

Fig. 3 Surat Region urbanization in FCC image. *Source* EarthExplorer, USGS, Landsat OLI 8 Images, January 2018 [8]

Table 1 SMC administrative limit extensions and census populations

Year	Area in (km^2)	Census year	Population	Decadal growth rate (%)
1961 (old city)	8.18	1931	98,936	–
1963	21.95	1941	171,443	73.29
1970	33.80	1951	223,182	30.18
1975	55.56	1961	288,026	29.05
1986	111.16	1971	471,656	63.75
1994	112.28	1981	776,583	64.65
2006 (February)	146.45	1991	1,498,817	93.00
2006 (July)	326.51	2001	2,433,835	62.38
2018 (December)	326.51	2011	4,466,826	83.53

Source SMC, 2018 [13]

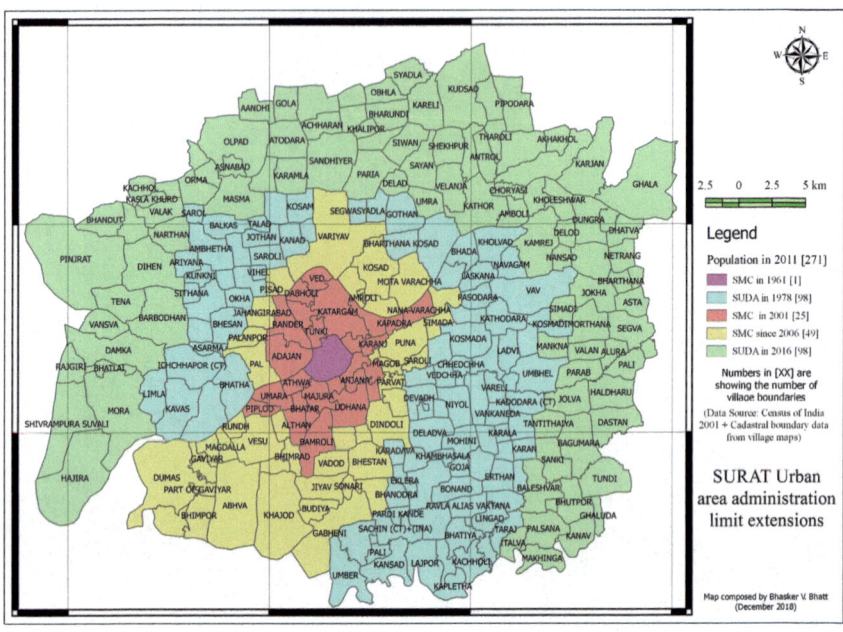

Fig. 4 Growth of Administrative limits of SMC. *Source* Authors, 2018

3.1 Demographics

In the earlier times, in the year 1981, the city population was reported and recorded as 776,583 persons housed in a spatial area of 55.56 km^2. The status of municipal administration was a *'Borough'* in those times [14]. It can be seen at the centre part of the Fig. 5. Till 1980s the population increase was identified to have a low rate of growth, however, since then after, the growth of the population increased. In the year 2001, the city reported having the highest growth rate in the entire nation to an extent crossing 88% in the then decade for the urban agglomeration whereas the city level growth rate was about 62%.

In the decade of the year 2001, the population recorded as 2,433,835 persons for the area of 112.28 km^2 of Surat Municipal Corporation rising from 1,498,817 persons recorded in the year 1991 within an area of 111.16 km^2. In the year 2006, there was a city limit extension by the State Government and the area under the administration of SMC reached 326.515 km^2. In the census year 2011, the population was reported as 4,466,826 persons within the city limits. The spread of population is visualized in Figs. 6 and 7 which represents the population as well as the density of persons residing in various administrative wards of the Surat region. The population density in the core city (old walled city) is identified as high as crossing 1000 ppha. In addition, the village in the fringe also reported having increased in the population during the Census of India, 2011.

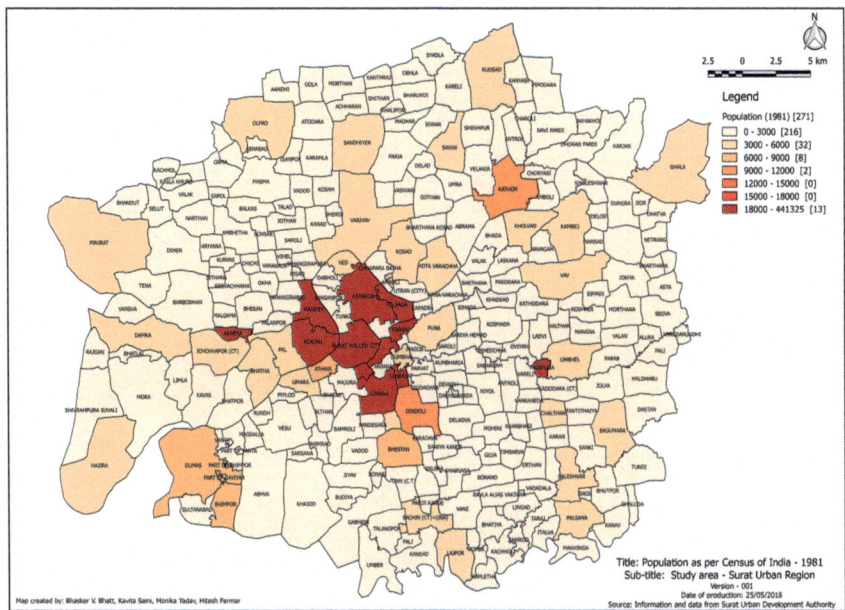

Fig. 5 Population in the region during the year 1961. *Data Source* Census of India

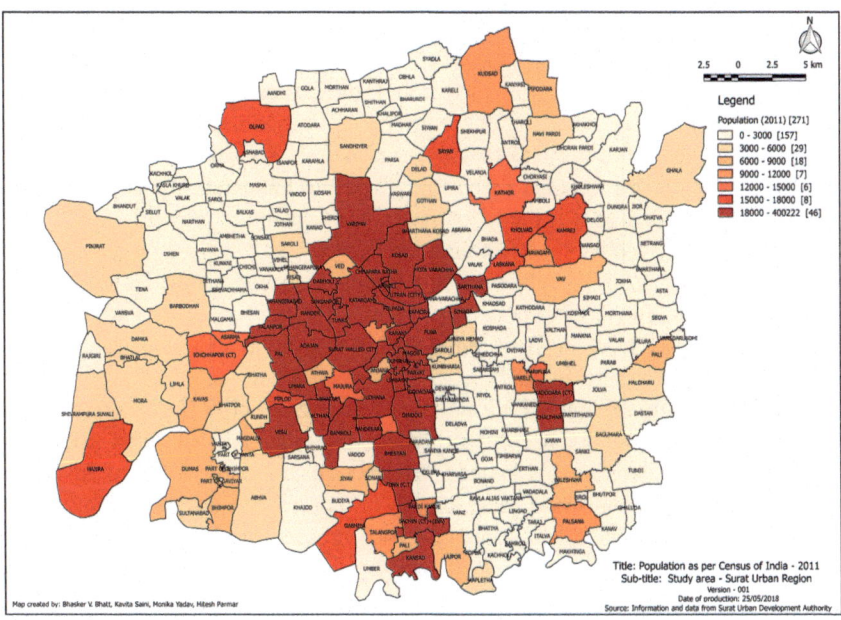

Fig. 6 Population in the region during the year 2011. *Data Source* Census of India

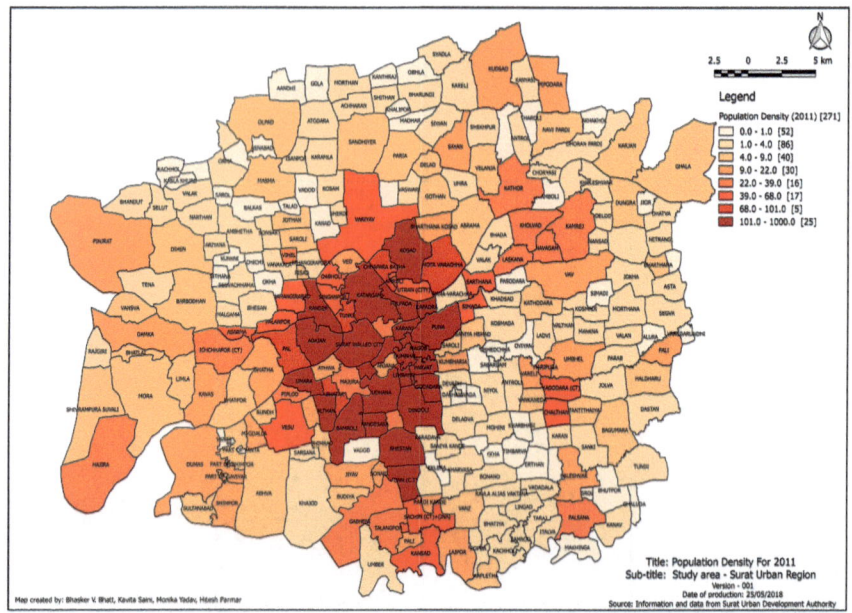

Fig. 7 Population density during the year 2011. *Data Source* Census of India

An analysis was conducted to identify the extent of the rise in the built-up spaces using the satellite images (downloaded from EarthExplorer of USGS) for the years of 2000, 2009 and 2018. The images used were for the months of January for specific years so as there are fewer cloud covers and consistent visibility. In about 18 years, it is found that the built-up spaces have increased to the extent of about 250 km². Broadly, the rate of rising in the built spaces per year in the region is found to be alarming as about 13 km² in and around Surat city.

Figure 8 shows a corresponding rise in the built-up spaces (area in Ha) in the Surat region (SUDA Administrative boundary) which includes for the area administered by SMC as well as the State Government. This built-up includes for residential as well as industrial and institutional spaces.

Figure 9 shows the extent of built-up spaces in the Surat city. The image extracted is in the form of False RGB composition for the city limits of Surat Municipal Corporation administration. Of the total 326 km² area, more than 50% of the spatial area is covered with built-up spaces. Results of a previous study based on satellite image analysis show that the built-up area has a subsequent rise in each decade and reached up to 31.59% by the year 2011 which was 17.57% in 1991 [16]. The more of the natural surface is getting paved. Subsequently, the area for agriculture and plantation is reducing.

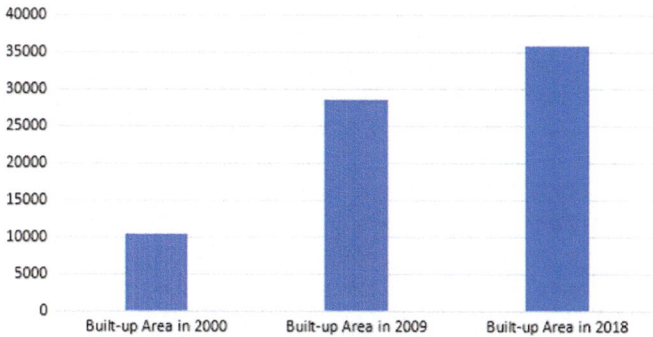

Fig. 8 Comparative built-up areas of Surat using spatial image analysis

Fig. 9 FCC image for SMC area. *Source* USGS, Landsat TM 4-5 band Images, January 2011 [15]

3.2   Weather, Drainage and Topography

The climate of the city is actively moderated by the Arabian Sea (to the West) and the Gulf of Cambay. The climate is classified to be tropical savannah having Summer to have a duration from March to June [17]. The monsoon begins with June–July, and average precipitation is about 1200 mm annually [18].

The city is prone to monsoon floods when high-tide in the Sea hinders the flow of city stormwater to the river. Figure 10 shows a drainage map of the study area. The area is supported by a system of about 12 drains merging either into Tapi River or Mindhola River. These drains are distinguished with a lower ridge line, in almost a flatter terrain. Most of these natural drains carry waters left unused in the canal tails during the off-monsoon. The drain basins are visualized in Fig. 11. These drains (also known as '*khadi*') carry a high volume of waters and often flooded during the monsoon, making the city vulnerable for river floods (significant occurrences in the years 1998, 2004 and 2006) as well as floods in these drains. Dhiman et al. identified based on a study that recent significant floods in these natural drains resulted in a considerable loss and disturbance in the city during the years 2004, 2005, 2007 and so on [19].

The region has a dense canal network dependent on the Tapi river water. The canals supply water from Ukai Dam and Kakrapar Weir located at about 85 and 60 km towards east from the city of Surat. The canal network was constructed in the duration of 1980s and are well maintained by the Surat Irrigation Circle, Government of

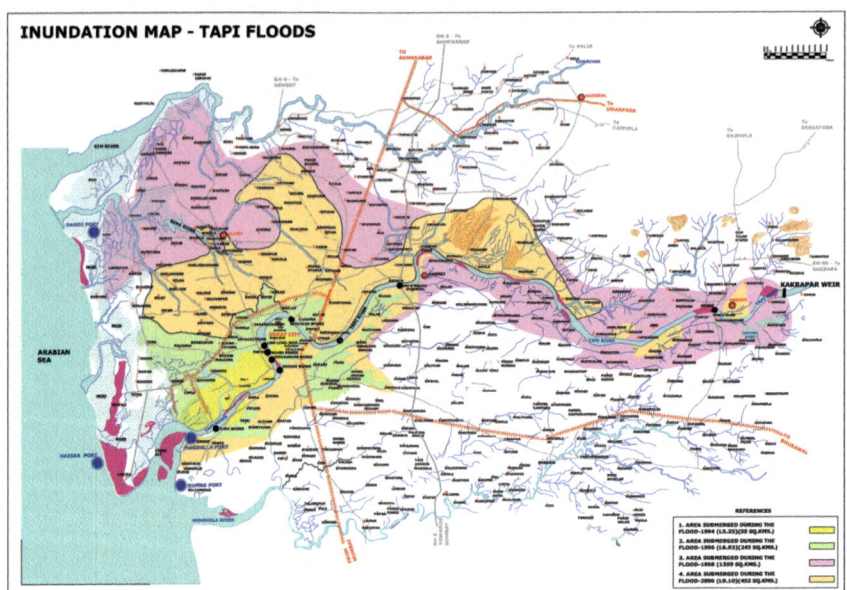

Fig. 10 Inundation map for floods in Tapi River. *Source* Surat Irrigation Circle (2011)

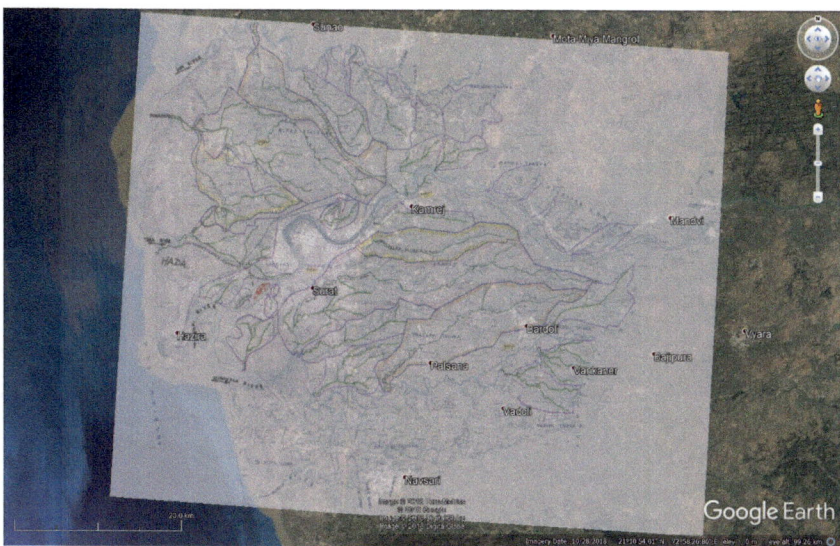

Fig. 11 Drainage map of Surat region overlay in Google Earth. *Source* Drainage map—Surat Irrigation Circle

Gujarat. The circle office works under the state ministry of Narmada Water Resources, Water Supply and Kalpsar Department. It establishes a 'Flood Cell' every year during the months of May to November, based at Surat and monitors as well as coordinates among agencies dealing with floods and water resources in the region.

The terrain of the city is mildly sloping towards the west, i.e. the bay of Cambay of the Arabian Sea. An average elevation of the city is said to be 18 m above the mean sea level as reported on the railway station of the city. Patel et al. [20] identified by analyzing DEM based submergence in the Surat city and concluded from a study that about 85% of the city area could get submerged during the floods like occurred in the year 2006.

3.3 Irrigation and Agricultural Activities

For the fecilitation of irrigated agriculture in the Tapi basin, Ukai-Kakrapar Scheme, the biggest irrigation scheme in Gujarat, was taken up by the Government of Gujarat and Completed in two stages, the First stage in the year 1957–58 and the second in the year 1971–72 (partially). The total service area (culturable command area-CCA) under the scheme in its final stage of development as planned (in the year 1966) was to be 379,940 ha (Revised area 356,480 ha. in the year 1978) with a total irrigation intensity of 109% under Kakrapar and 103.5% under Ukai. Irrigation network in 359,280 hectares (revised area 337,200 ha in the year 1978) has been

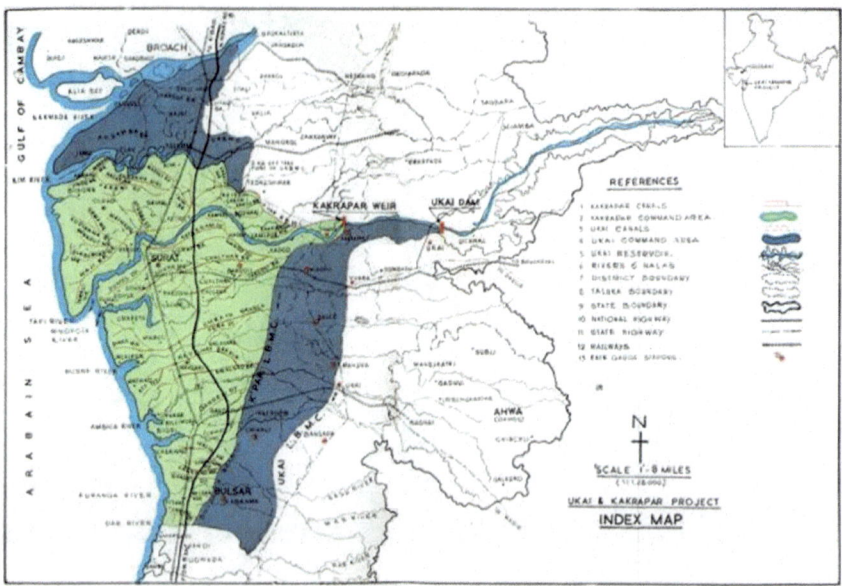

Fig. 12 Canal command area under Ukai Dam and Kakrapar weir. *Source* Surat Irrigation Department

completed. Agriculture development in the command area has been reasonably rapid since the year 1958 after the completion of Kakrapar system and has gathered greater momentum after the completion of the Ukai dam in the year 1971. Figure 12 shows the canal irrigation command area of Ukai dam and Kakrapar weir.

Fertile land, good rainfall, huge storage at Ukai dam, along with possible good marketing facilities, the rapid development of agro-industries, and progressive nature of farming community in the command area, is all that is nurturing requirement accelerating the pace of agriculture development. On the other hand, inadequate conveyance capacities of canal to meet future irrigation demand due to increased irrigation intensity with increasing coverage of cash crops such as sugarcane and other perennial crops, scarce drainage provisions and their resulting effects on cropping pattern, fewer control structures and constraints in ensuring efficient water management have imposed serve limitations in the planned development of the area.

In the year 1978, the project preparation and monitoring cell of Govt. of India and area development commissioner, Govt. of Gujarat jointly prepared a report wherein a revised cropping pattern (as a part of modernization of the system) was proposed for Canal systems in Surat region, based on experience and existing crop growing trends at that time. The then proposed cropping pattern was as under (which is applicable till recent times) (Table 2).

With the establishment of sugar factories and "Khandsari" units from the years 1959 to 1977, sugarcane cultivation got great momentum in the command area. With the yet increased number of sugar mills in the command area Ukai dam and Kakrapar weir, the farmers have shown a shift towards perennial crops especially sugarcane

Table 2 Proposed cropping pattern for K.R.B.M.C and U.R.B.M.C (in 1978)

S. No.	Particulars of crop	% of CCA	
Perennials			
1	Sugarcane	8	10
2	Banana	1	
3	Orchards	1	
Kharif			
4	Paddy (short duration)	7.5	33
5	Paddy (long duration)	7.5	
6	Sorghum (Juwar/Bajri)	10	
7	Groundnut	8	
Rabi			
8	Wheat	15	40
9	Sorghum (Juwar/Bajri)	15	
10	Pulses	10	
Two seasonal (Kharif and Rabi)			
11	Cotton	30	32
12	Vegetables	2	
Total		115	

Source Surat Irrigation Circle, 2015

and cotton. These crops are water intensive and cash crops that can be cultivated around the year. These crops have brought economic prosperity among people yet affected the groundwater and soil quality adversely.

3.4 Transport and Linkages

The region is well connected by means of having linkages of NH-48 (Mumbai–Delhi), NH-53 (towards Kolkata), railway (Mumbai–Delhi and central India locations), a domestic airport (in the process of conversion to an international status), and ports of Magdalla and Hazira in the West. The proposal of a coastal highway of Gujarat, expressway of Vadodara to Mumbai, Dedicated rail freight corridor, ring roads and other roads are at various stages of different proposals.

The region is well connected by means of waterways. There are four ports managed by the Gujarat Maritime Board (GMB), Government of Gujarat. The ports are under development. In addition, there is a proposal under development to increase the transit using the Ro-Ro Ferry services. At present, the service is available in a limited route, from Dahej (to about 40 km North to Surat) to Ghogha (Western part of Gujarat) for human as well as goods movement. It has resulted in a considerable decrease in travel time and savings of resources.

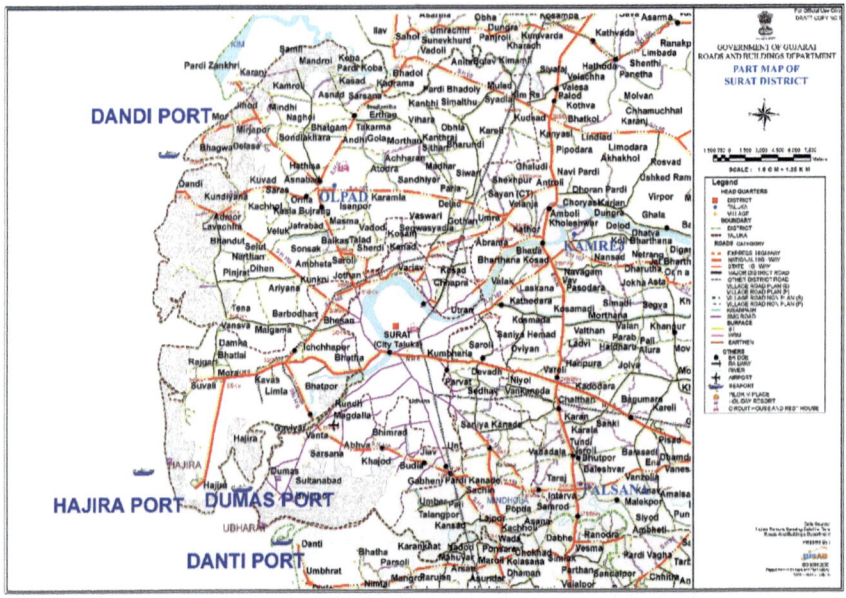

Fig. 13 Transportation network in the Surat Region. *Source* Roads and Building Department, Surat Circle

The Mumbai–Delhi railway line is being converted to 4-lane. The work is under progress. In addition, the central government of India has initiated land acquisition for a prestigious project of a bullet train with financial and technological support from Japan. In addition, the work is under progress for the Dedicated Freight Corridor (DFC) which is aligned through the study region. Mumbai–Vadodara Expressway proposal also is under progress. Alignment for the proposed road is finalized, and the project is under the land acquisition stage. Figure 13 show the regional level road and railway network and Fig. 14 shows the proposed alignment of the high-speed rail (bullet train) passing through the study area.

Such a dense road network and railway along with air and waterway connectivity provide ample opportunity to Surat for movement of people and goods. Within the city, the work for MRTS is also in progress in addition to BRTS with two phases in operation.

3.5 Water Resources

The regional topography has a mild slope towards the west (the Arabian Sea), and the city of Surat has spread of about 18 m above the Mean Sea Level. The region is rich with water resources. Many natural drains and tanks are existing with a major perineal river Tapi flowing from the centre of the city. Until the 1980s the city was highly prone to the flooding disasters during the monsoon duration. However, the risk was reduced

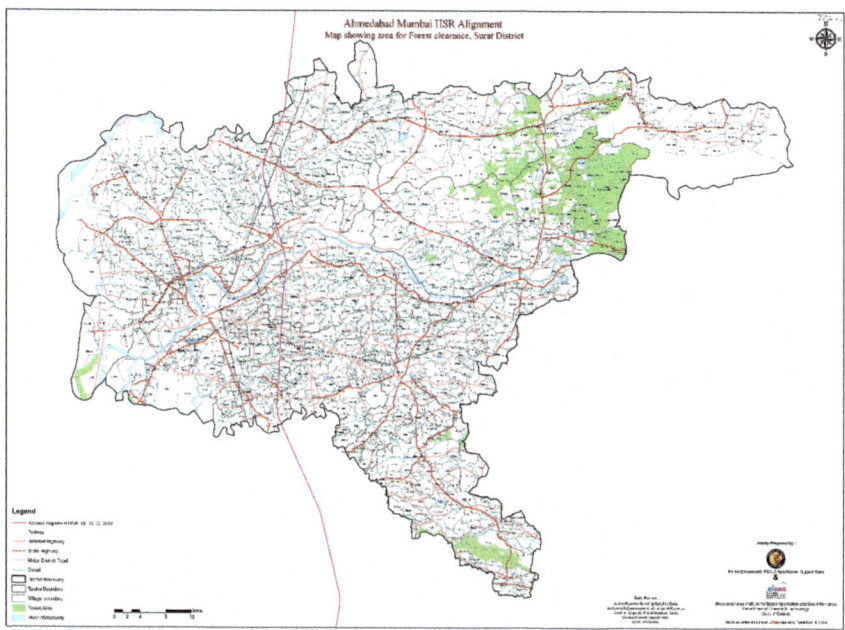

Fig. 14 HSR alignment passing through Surat Region. *Source* BISAG, Gujarat, February 2018

after the construction of Ukai dam (located at about 80 km to the East) and Kakrapar weir (about 15 km downstream to the dam). These two megastructures in the region provide the potable and irrigation waters to various places located in an area of about 200 km by 70 km in South Gujarat. Good rainfall and availability of water round the year resulted in the flourishing of the regional growth and urbanization, especially in Surat. The water resources helped the growth of the textile industries at large as it consumes a high amount of water during the processing and cloth manufacturing. In addition to it, the availability of water played a crucial role in the migration of people from the water scarce western part of Gujarat and the community helped in the growth of the diamond cutting, polishing industry. During the summer of 2017 and 2018, there was a substantial reduction in the available water. Public awareness drives and programs are initiated by the State and Local government advocating for water conservation.

On the other hand, with the ingress of saline water from the Arabian sea due to groundwater extraction, most of the habitable areas in the western part of the city are facing a shortage of potable groundwater. Broadly, the citizens of Surat are dependent on the water supplied by the ULB and private sources. The Development Control and Regulation set by the local authority, and state government are emphasizing the water conservation by imposing for the construction of groundwater recharge wells. However, the measure is limited to more significant buildings. The city is also lacking from such wells constructed in public places by any organizations except a few to locate.

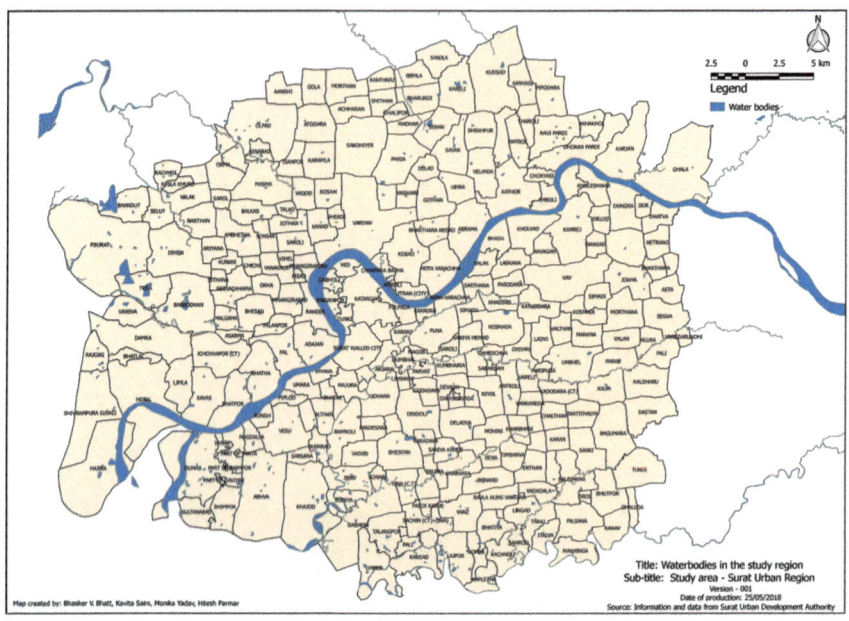

Fig. 15 Waterbodies in the Surat Region. *Source* Developed by authors using multiple sources

The agricultural activity in the region is predominated by cash crops (as cotton, sugarcane and alike) as the black cotton soil having a good quality of groundwater allows for. In addition to natural water resources, the state government has constructed a vast network of the canal to serve the purpose of irrigation from storage reservoir of Ukai (located at about 80 km to the East) so as to maintain the natural water resource balance to an extent possible. Figure 15 shows the existing water bodies in the Surat region. There are about 300 waterbodies (in the form of lakes, ponds) of various sizes are existing as per the revenue records.

3.6 Vegetation Cover

A study was carried out using the satellite images made available by the EarthExplorer, USGS. The Landsat Missions based images for Red band and Infra-red bands were used for calculation of the vegetation indices. The image resolution is 30 m sized pixel for both images. Change detection in the duration of the year 2000 and 2018 was attempted to understand the extent of vegetation cover and quality thereof.

The images used was captured in the months of January for the specific years. The season has post-monsoon cultivation in the agricultural fields. It was apparently identified that the NDVI values are reducing, not only in the bounding range of upper and minor limits as well as overall value as well. It is an indicator of degradation in the vegetation covers over the duration. The NDVI values for the year 2000 ranges

Fig. 16 NDVI for the year 2000

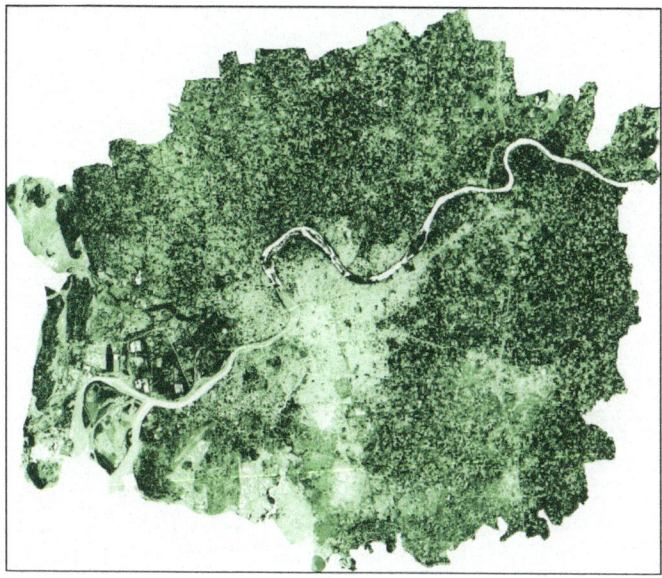

Fig. 17 NDVI for the year 2018

from −0.004353 to 0.246034 whereas for the year 2018, the value ranges from −0.000231 to 0.178733. The barren lands are increasing in the study area. Although, since the year 2000, thick vegetation or forests are not identified in the study region and the land surface seems to be shielded with sparse vegetation only.

The sparse vegetation is also an indicator for the crops as cotton, sugarcane and alike that are cultivated in the region. Figures 16 and 17 shows the NDVI extracted from the band images using QGIS tool. The increase in the built-up spaces also can be visualized in these images.

3.7 Industrial Activities and Economy Drivers

The city is booming with industrial establishments. It is among the key reasons for the prosperity in the region and considerable migration of people from across the nation in search of opportunities (Fig. 18).

The city itself has more than 50,000 working units as small to medium scale enterprises. Table 3 shows details of the number of units established and in operation in the region. Majority of these enterprises are dealing in the textile industries, chemicals and allied sectors that includes for plastic, mechanical, IT and service sector. These units are established across the city and functioning since long past. In the duration

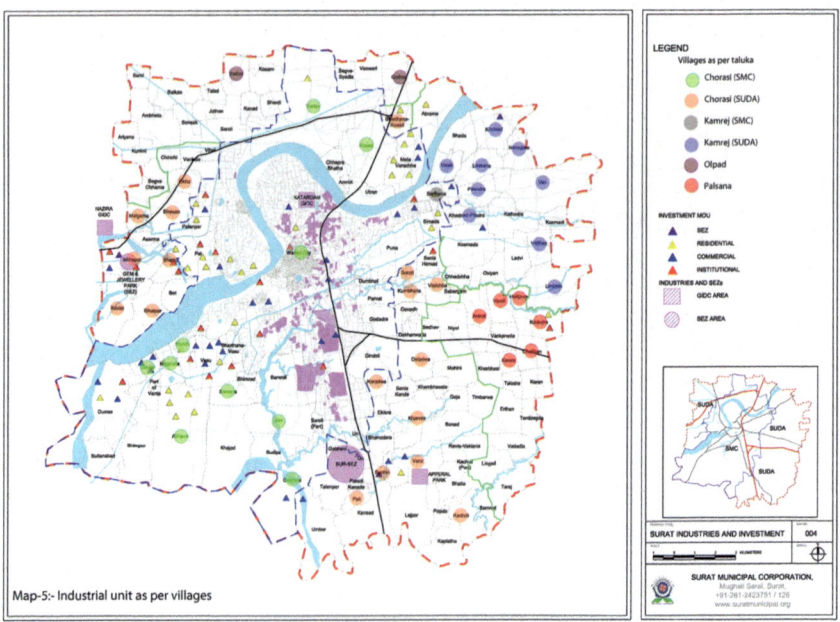

Fig. 18 Economic activities in the Surat Region. *Source* Surat Municipal Corporation, 2015

Table 3 Establishment of enterprises in Surat region

Enterprise establishments	Nos.
In SMC	49,437
In SUDA (Ex-SMC)	743
In Ex-SUDA Palsana	302
In Ex-SUDA Kamrej	98
In Ex-SUDA Chorasi	23
In Ex-SUDA Olpad	330
Total in SUDA (with SMC)	50,180
Total in peri-SUDA area	753

Source Compiled information from GIDC and DIC, Surat, 2014

of 1970, while these units were establishing, there was no land use plan nor any mechanism to regulate it. Hence, these units, with the passage of time, developed over places wherever land parcels were available. Today, units are functional at several places within the city. In the absence of zoning regulations, are difficult to trace in terms of their location and grouping. Within the city core, there are many houses where mixed use in a building can be observed. Such working places generate solid waste in a complicated mixture if the segregation is not carried out by individuals.

3.8 Vehicle Growth and Air Quality

In the last decade the rapid growth rate in vehicular population has been observed due to the amount and spread of activities concentrated in the city is quite worrying. The socio-economic development and inadequate mass transport system have led to the growth of the vehicle population in Surat are high. In the year 2003, the total vehicles were nearly 8.96 lakh rising to nearly 20.82 lakh by the year 2013. It shows an increase of 132.12% in ten years. The category wise data mentions that the number of motorcycles, motorcars, auto rickshaws were increased by 126.72, 217.17 and 116.94% respectively in the decade.

The two-wheelers have an average decadal share of 86 per cent in the total vehicles registered in the city. The cars have a share of a decadal average of about ten per cent. It depicts for predominating own mode of transportation in the city. These vehicles float on the city roads, emitting exhausts. It results in depletion of the air quality.

Recently, the SMC has established air quality monitoring stations at Varachha and Limbayat. There are many industrial establishments in the Limbayat area. The Varachha area is a mostly residential area with established industrial units of yarn spinning and twisting. The reports are generated and displayed publicly and disclosed through the website of SMC [21]. The website is supplied with real-time data by the stations, and Air Quality Index is calculated and shown with interactive graphical representations for 24-h records of PM-10, PM-2.5, CO_2, CO, NO, NO_2,

SO_2, temperature and humidity. The data on air quality through monitoring stations is maintained and made available through open data portal of the SMC.

4   State of Infrastructure and Environmental Concerns

This sub-section of the chapter discusses the details pertaining to physical and virtual infrastructure in the Surat city. The city has about 1.4 million houses that include the use for residential purpose in full or part as per records of Census, 2011. Citizens are provided with water supply, sewage collection connection, storm drain network, roads, streetlights, solid waste collection, fire and emergency services as well as social infrastructure. The sub-section discusses a few vital services only. The municipal corporation of Surat initiated computerization of its records back in the year 2001 onward. Over the period, the SMC is taking up initiatives to convert the manual systems to automation by means of centralized operation units with the deployment of SCADA. The scale of infrastructure is tremendous to be converted to smarter options demanding for funds yet; the SMC has a track record of zero-debt operations since the year 2001.

As per the report on Surat SWM project under JnNURM [22], it reveals the past related to infrastructure services and specifically about the impact of ill-management solid waste generating in the city. "The development of the city and its infrastructure could not keep pace with the increasing population leading to inadequate service provisions thereby resulting in a low quality of life in the city. The city doubled in size during the years 1981–1991 due to which many fringe areas remained un-served with basic amenities like water supply and sanitation. The rapid population growth in Surat caused several management problems for the ULB. The SMC is responsible for the provision and maintenance of the entire range of civic infrastructure services in the city (including water supply systems, sanitation and drainage facilities, solid waste collection, and disposal). During the early 90s, the basic infrastructure services were low with only 35% coverage of water supply (access to piped water supply) and 33% of the city being covered through the closed drainage system. The sewerage network covered less than 30% of the city's area. Municipal SWM was in a poor state with waste being disposed of in drains and water bodies. *The waste collection efficiency of the city was reported at only 40% in a study conducted in the year 1995.* This sporadic development and lack of essential services and infrastructure led to the outbreak of Plague in the year 1994 which claimed several lives. The primary cause of the outbreak was considered to be ineffective waste management, which led to the blockage of stormwater drains resulting in flooding of the fringe areas of the city."

Table 4 Water supply capacity of SMC

S. No.	Particulars	Year 2015–16
1.	Intake well capacity (MLD)	2033.00
2.	Water treatment plants (MLD)	1486.00
3.	UGSR Capacity (in ML)	699.10
4.	ESR Capacity (in ML)	120.70
5.	Pipelines (in km)	3350.00

Source Surat Municipal Corporation, 2016 [23]

4.1 Water Supply System

The city has almost flat terrain that leverages an advantage of providing water supply through a gravity system of operations. There is an interlinked network of Elevated Reservoirs, and the system is operated in a loop. The supply is intermittent to the users, and in few areas of the new north zone, the system is made 24×7 by deploying smart meters on a pilot basis. The raw water for supply is obtained from the Tapi river by means of radial wells. The water is treated to the quality norms as set by the state government pollution control board—The Gujarat Pollution Control Board (GPCB). With a pre-chlorination dosage, the water is lifted to the elevated reservoirs and then intermittent supply is maintained. Table 4 underneath shows a summary of the extent of water supplied by the SMC. The administrators are committed for 100% coverage through piped water supply in the city.

Recently, the SMC has also initiated for metered water supply in the industrial-residential areas of Pandesara and Khatodara (in the southern part of the city). The checking and payment of water charges are facilitated through web-based operations.

4.2 Sewerage Network

The piped water supply will return about 80% of water in the form of sewage. The SMC has laid down an underground sewer network of length greater than 1400 km. the terrain and geographic locations of the sewage treatment plants have led the network to have the support of pumping stations. The SMC has already placed about 34 sewage pumping stations across the city. About 150 km^2 of the city area is covered with piped sewers. The installed capacity of the sewage treatment plant is about 726.5 MLD which is almost at par with the water supply capacities as developed. The wastewater is treated and released in the natural drains (not in the river) as per the quality norms set by the GPCB and daily water samples of treated-discharged water are obtained and tested at the established laboratories. With the aid of the World Bank, in the year 2001, the SMC established a SCADA based sewage treatment plant at Anjana and the plant has the capability to generate green energy which is used by the plant primarily, and the surplus energy is merged in the national grid. Most of the Sewage Treatment Plants of providing a secondary level of treatment (conventional activated sludge) to sewage. One plant at Bamroli (located to the southern parts in the city) is established with tertiary treatment (UASB + Extended aeration) capacity also [24]. All these STPs are operated and managed by private agencies on contractual arrangements. The sewer system is maintained clean for clogging by means of contractual arrangements for cleaning and desilting on a regular basis.

4.3 Solid Waste Management

The solid waste management by the department of SMC is taking an approach of integration with various activities in the city. It is integrated with operations related to water supply, sewage, health and so on [25]. The city generates about 4000 MT of solid waste every day at an average of about 700 gm per person [26]. The waste is collected by means of the door-to-door garbage collection system, container lifting, hotel-kitchen waste management and night scrapping-brushing activities. It is brought at transfer stations from collection points. The waste is transferred to the treatment site by means of the secondary transfer system. A scientific landfill activity is performed at the disposal site located at Khajod (Southern part of the city) since the year 2006. There is a strict observance of no open burning of solid waste in the city. The biomedical waste is collected by NGO and treated separately by means of shredding and incineration in an autoclave, observing safe disposal [25].

The proposals include for the establishment of the waste-to-energy plant on BOOT base of PPP model; plastic waste collection system; strict observance of segregation at the source for dry and wet wastes; organic waste treatment facility; development of four more waste transfer station in addition to existing six stations and so on.

The SMC made remarkable efforts after the year 1995 for managing the solid waste generating in the city. Since after the outburst of Plague in the year 1994, the SMC undertook a massive clean-up drive with regular activities and transformed the city among one of the cleanest cities within a span of 15 months. The governance played a crucial role here along with the will and cooperation rendered by the citizens—some of the significant and noteworthy initiatives included for rearrangement of waste management zones and about 52 sanitary districts (wards); daily monitoring and reporting on the waste collection efficiency; posting of sweepers for round the clock duty ensuring cleaning of roads twice a day; assigning duties to contractors for night sweeping and scrapping activities; slum improvement mission with help of NGOs; revamping of the administration system and financial management of SMC; introduction of a complaint redressal system through policy decisions towards making the systems responsive [22].

In the later period, during the years of 2005–2010, under the JnNURM, the SMC initiated and implemented efficient door-to-door garbage collection system through PPP mechanism; increase in storage and transfer vehicles along with street sweeping equipment and machineries; treatment and conversion of waste at a total cost of the then amounting more than INR 50 crores.

Presently, the waste management is taken care of by the involvement of various agencies, as shown in the table below.

Waste management action	Responsible agency
Door-to-door collection	Private operator
Street sweeping	SMC staff
Drain cleaning	SMC staff
Primary collection from commercial waste generators and marketplaces	Private operator
Secondary collection	SMC (except one ward)
Waste transportation	Private operator
Waste treatment	Private operator
Waste disposal	Private operator

Source SMC, 2013 [22]

It is evident from the above status that by the involvement of private operator/agencies, the efficiency in the solid waste management could be observed. Since the pace of population increase and development of built-up spaces and upcoming settlements is high, the trend needs to be continued to maintain the efficiency. About ten years ago, the daily solid waste collection was averaging about 2000 MT which at present has crossed a daily average of 4000 MT.

4.4 Aqua-Culture Activity

Since past about two years, i.e. after the year 2016, the barren land parcels in the southern part of the city (on the bank of Mindhola River) are converted to aquaculture and prawn farming activities. The low-lying land served as a floodplain spreading the high discharges from the natural drainage system between two rivers. The extent of the spatial spread of the activity is more than 80 km^2 (Almost 25% of the municipal administrative area). These land parcels were kind of non-productive lands due to tidal water ingress at regular interval as per tidal cycle. The land became salty due to frequent seawater intrusion.

As the farming activity is taking place within earthen bunds, the sea water will lose its natural course and may affect adversely. Many of past research has identified that the ecosystems are ruined due to such activities. However, initially, such activities are bringing prosperity among citizens at the cost of environment. The natural mangrove cover along with existing ecology is disturbed due to the activity and need to be paid attention to. There is a scope for a detailed study to understand the impact of the activity on the ecosystems, flora and fauna as well as on living being in the surrounding (Fig. 19).

Fig. 19 Aqua-culture activities in the southern part of the City. *Image courtesy* Google Earth Pro 2018

5 The Smart Governance and Environmental Management

As the city expands and rate of urbanization increases, the concerns have been raised about the management of solid waste. The SMC is well responding to the aspect and has placed the door-to-door waste collection by applying the principle of segregation-at-source and daily night sweeping rounds in service to the citizens. The plastic waste objects are of a concern as it is a significant source of land pollution in the surrounding. As a resultant effect of National Sanitation Program (Since the year 2015 known as Swachh Bharat Mission), the city government has achieved various benchmarks by fulfilling the mission objectives effectively. Further, the case will be elaborated in the chapter.

The water obtained from the Tapi river as a raw source is treated by the SMC and supplied to the citizens and for public purposes. There exists an extensive network of water supply pipelines that are having mains in a loop and has the ability to serve in any and/or all part of the city. To catch the supplied water after use as a release in the form of domestic wastewater is collected through a gravity-based underground pipeline network and treated at various secondary level wastewater treatment plants. The treated water is disposed-off in the surrounding natural drains and the same at times can be a threat to aquatic life when the concentration of pollutants is high during the Summer season.

The transportation, specifically private mode is predominating in the city. The SMC has been recently providing an effective city-bus and BRTS connectivity and has further ambitious plans for e-bus as a public transportation option reducing the air pollution. Also, public bike sharing project is to be taken up as an initiative for local area connectivity and transit option. Such efforts may leave a positive impact in reducing the generation of SOx and NOx pollutants through the vehicles.

Several organizations are playing a role in managing natural resources and the urban environment. The next sub-section discusses the governance paradigm related to these organizations and challenges faced and opportunities by employing smarter alternatives for making a sustainable environment.

5.1 Surat Municipal Corporation and SUDA

Firstly, the SUDA is responsible, mainly for administering the development by regulation over zoning and building construction activities in the urban region. SUDA has, under the provisions of the Gujarat Town Planning and Urban Development Act, 1976 and amendments, imposed for development control regulations by means of building bye-laws. The recent announcement by the state government for observing and imposing current building bye-laws is implemented by the SUDA. Online building plan approval system is developed by the organization, and the same is progressively adopted by practising professionals. The features of the online system enabled for time and resources saving.

The Comprehensive General Development and Control Regulations (CGDCR) as imposed by the state government has provisions for compulsory rainwater harvesting, tree plantation, parking space provisions, building margin and open spaces related provisions. A prompt follow-up on these provisions shall ensure somewhat recovery on the lost potable state of groundwater as well as assist in maintaining the air quality. Yet, SUDA has not been enabled to keep a check and detect changes in the region for land use and land cover (such as farming, agriculture, mangrove cover, coastal and river stream bank alignments and so on), the organization may adopt the remote sensing techniques in routine to monitor the changes in remote places within the area under administration. The images of Landsat OLI 8 bands are processed and made available on the lower resolution by EarthExplorer within a span of a week. A higher resolution image also can be made available from IIRS-ISRO as captured through LISS 3 for Indian sub-continent.

In addition, the provisions for terrace farming, vertical gardens, solar roof-tops, use of smarter security devices, use of non-conventional energy sources in buildings and alike are not among included bye-laws. Such can be contained within the set of norms to activate a push-factor. Some citizens opt for such alternatives. However, an effective shift can only be observed if mechanisms are enforced with the support of the law.

The SMC, on the other hand, is maintaining a consistent improvement through the provision of web-based and mobile-based services. The SMS has enabled for the following services through its website.

- Property tax, Professional tax and tax invoice related tasks;
- Registration and certifications for shops and establishments, Nursing home, birth and death, contractor lodger certification;
- Bookings for halls, stalls, community halls, society civic facilities;
- Water meter related services, 24 × 7 water connections;
- Use of e-library, e-magazines, various online forms;
- Feedback, complaints and redressal for all kind of services that SMC is involved with.

In addition to the above, the citizen participation is involved through mobile apps—Citizen's connect, SAFAL, Heritage walk, Surat Sitilink (for mass transportation) and Surat Smart city. Data and records pertaining to deaths in road accidents, vehicles registration, weather parameters and so on are made public through open data portal of SMC. A Web-GIS based portal is also developed by the SMC which assists in locating specific information in an interactive manner. Various guidelines, norms and awareness related information are made available on the website.

The SMC and SUDA have well adopted the e-governance and m-governance principles. Yet there is a vast scope in terms of position-based actions and live reporting systems. On a trial basis, as discussed previously, the SMC has installed an online air quality monitoring and display system.

5.2 Surat Irrigation Circle

It is an organization of the State Government, which accounts for all sort of water resources in the region. It has separate offices established for the river, dam and weirs, canal system, flood monitoring, rainfall and runoff, groundwater resources, ponds and lakes, natural drains, tidal levels and so on. During the monsoon, the organization is in action for monitoring and managing the high-water levels and floods in Tapi river as well as other drains in coordination with the collector office, Disaster Management units, SMC and other organizations as per the guidelines of the state government. The primary task round the year is to maintain and provide water through a canal system constructed in the command area of Ukai dam and Kakrapar weir. It supplies water for agriculture, domestic and industrial usage.

The organization has scope for adopting real-time monitoring and management systems. The department makes use of accurate maps developed by BISAG to an extent. Yet, for online operations that include for web-based and mobile-based management, applications of sensors are some of the aspects that are largely lacking and has scope for inclusion.

5.3   Ground Water Resources Development Corporation Ltd.

The city is situated near the Arabian sea. Since about half a century, there has been a considerable extraction of groundwater for use in the developmental activity. The groundwater levels have drown-down with depletion in the quality as well. With lesser recharges of groundwater, the salinity ingress is increased and spread. The extent of saline water in the sub-surface resources has spread up to the walled city of Surat. The water has lost the quality to be regarded as 'potable'.

Since the year 1975, Ground Water Resources Development Corporation Ltd. is working on monitoring and tracking the groundwater resources through investigation, exploration, management and recharge works across the state. It is functioning under the state government. The organization is assigned to perform tasks to carry out research and maintaining records related to hydrological projects, lift irrigation, drip irrigation and artificial groundwater recharge related activities.

The department is functioning conventionally, and there is considerable scope for improvement by adopting smarter solutions, investment of funds in procuring technology and capacity building. The department at present seems to lack in keeping pace with the increase in urbanization as well as managing pressure on natural resource of groundwater. The organization has no presence in terms of e-governance and m-governance and online monitoring as well as management systems. If the aspects are explored and adopted appropriately, the quality of ground water in the study area can be maintained to potable level. The organization is more focused on activities in the tribal areas in the Gujarat state, and very few projects are undertaken in the urbanized places.

5.4   The Collectorate of Surat District

The collector is a state government representative of the revenue ministry. The primary function of the Collector office is to govern the use and manage the land parcels within the area of the district administration. Post spatial split in the year 2007, the area of Surat district is 4549 km^2. The major urban centre is Surat, and apart from that, there are other urban centres as Bardoli, Olpad, Kim, Kamrej, Mahuva and Mandvi. The population of Surat district was recorded at 6.08 million in the census year 2011 which was 4.27 million during previous census year of 2001 [27].

It was reported that about 79% population in the district is urban population. With urban agglomeration of Surat, the urban population resides on about 25% of a land parcel of the total district land. Rest of the land is used for other activities and natural features. All the conversion of land from agriculture to non-agriculture purpose is monitored by the District Collector office. The office is maintaining the records of a wide variety and has computerized most of the records. Yet, the dynamic understanding of granted permission for development is beyond that reflect as a grouped activity. Use of GIS-based inventory for land is an essential requirement.

The duties for a collector office also include for public security, revenue functions, implementation of land reforms, welfare of the agriculturists, enforcement of various acts, rehabilitation in case of disaster occurrence, maintain rural statistics, law and order and preventive actions against any incident that is disruptive in nature and to perform other government duties as assigned by the home ministry of the state government. The collector has supreme powers in the district. The office needs a comprehensive integration with all active government organizations in the district.

5.5 Gujarat Ecological Commission

The organization primarily functions to plan and work for the restoration of ecosystems that are degraded within the state of Gujarat [28]. The task is performed with the assistance from various NGOs in action at various regions. The organization also look after tasks pertaining to the socio-economic development of coastal communities, mangrove plantation, research, communication and capacity building [29]. It works under the Gujarat state forest department for monitoring and maintaining ecological balance.

The organization have obtained comprehensive data and records from diverse sources. These records are mostly in a computerized form. The GEC was the first government organization that implemented a Web-GIS based platform in an interactive way. The department needs to be active in terms of implementation of policies through powers delegated to the district collectors. The Surat district has a coastline towards the west, and the river delta of Tapi and Mindhola has mangrove covers. The interventions and understanding of ecological and environmental parameters as developed by the GEC in a comprehensive manner needs to be integrated into the planning tools such as Development Plan. These plans are prepared by the urban development authorities keeping the facilitation of urbanization in focus. The GEC is leading the coordination with several state agencies for the world bank funded integrated coastal zone management project at the Gulf of Kutch.

5.6 Gujarat Forests Department

The Gujarat state forest department maintains the records and publishes in the form of annual reports. The department functions not only protecting the forests but also for wildlife in national parks and sanctuaries and maintain the records of species, monitoring for permissions for visitors to forests, developing nature campsites and programs for nature education camps, the supply of a variety of woods, registration and monitoring of sawmills and alike.

In the Surat district, the department reports that there is 496.72 km^2 area under the forests which is about 11% of the total area of the district. As of the year 2017–18, out of the total forest area in Surat district, 489.08 km^2 is reserved forest land that is

Fig. 20 Forest areas in the Surat Region. *Source* Authors

restricted from access and human interventions. The forest in Surat district is comprising of mangroves that is moderately dense [30]. Such flora in the forest has no monetary value, but it is an ecological asset. For citizens who are not much concerned with the preservation of the environment, the mangrove land is considered as a wasteland, and permissions for developmental activities are sought. The concentration of the forest lands is seen in Fig. 20, at the western parts of the Surat region which is a coastal region showing the location of the mangrove forests.

5.7 Gujarat Industrial Development Corporation

The GIDC is an institution that was created to ensure the orderly establishment and organization of industries in the industrial areas and estates in the state. There are six GIDC industrial estates in Surat city. In the Gujarat state, there are more than 200 industrial estates. The GIDC procures the land parcels from the district collector, notifies an area and provides access and facilities to the industrial units for functioning. The organization is providing online support for applications for land allocation, water and drainage connection, plan approvals and plinth level checking, building completion level checking, right of use permission, subletting of allocated plots, amalgamation, sub-division of land parcels within an estate and so on.

The industrial units in GIDC estate are regulated for its establishment, however, upon entering into the operational phase, the industrial pollution and extraction of waste is not monitored by the GIDC. The degradation of water streams in surrounding or depletion of soil quality and such other aspects are not addressed by the GIDC. The pollution is monitored by another institute—Gujarat state Pollution Control Board (GPCB) which has laid down stringent norms regarding industrial extraction of polluting substances. Implementation of these norms by consistent monitoring is a significant issue that needs to be addressed by taking the support of smarter technological alternatives.

6 Concluding Remarks and Way Forward

Considering the discussions in the previous sub-sections of the chapter in the context of Surat city, few key points are identified and summarized as below:

- Surat is among fastest growing cities having a larger population and high growth rate of built-up spaces;
- The study area is losing its character for agricultural activities, and the farmlands are getting converted to habitable land parcels;
- The city is prone to floods through the river and natural drains;
- The continued rise in the number of vehicles within the city;
- Conversion of marshy lands to aquaculture activities;
- A large number of industrial establishments within the city;
- Larger the population, more demands for infrastructure, facilities and natural resources—generating pressure on the natural system;
- Lack of imposing innovative ways for building component level planning, regulation and development by development authority;
- Ineffective monitoring and integration among government organizations looking after various natural elements in the city.

These aspects can be addressed by adopting technological advancement by imposing smarter governance promoting sustainable environment by safeguarding from ill-effect of ongoing practices. Some solutions that can be considered as for practising way-forward are discussed in further sub-sections.

6.1 Concept of Integration of Agencies

It was identified that every agency is independently obtaining, recording, and monitoring on various parameters of their concern. However, a centralized database that is available for an integrated understanding of the scenario is entirely missing. The use of RDBMS based solutions through web-servers can be adopted by a leading organization, most suitably the Collector office.

6.2 Mobile and Web-Based Data Collection and Integration

Many citizens are using mobile phones and similar devices that are internet enabled. The authority may fetch big data of a variety and process the same post-screening operation for a specific purpose to obtain information. Such information can be integrated with the data server and be used for further visualization and understanding the instantaneous situations depicting for changes.

Remote sensing images captured by satellite-based electromagnetic wave sensors are available for free-of-cost with a spatial resolution of 30 m, and the Government of India can provide images with yet more excellent spatial resolution. A mechanism for obtaining time-series of such images on a regular interval and opting for analyzing these images will drastically help in detecting the change in the land use and land cover as well as many other parametric changes over a duration. Such operations can help in decision making and drafting suitable policies for land use structuring and zoning.

Use of sensor-enabled devices in operations of providing services can result in consumptive use of the electricity and other resources. Similarly, such devices can assist in the monitoring and reduction of waste and pollution to an extent. Solar powered streetlights with sensors can leverage the obtainment of information regarding movement of people and mass that will further allow for customizing service delivery, leading to optimization. Along with the installation of devices, the capacity building for humans involved will be among a significant challenge considering the massive scale of operations for a successful implementation.

6.3 Use of GNSS and GIS

GNSS is an abbreviated and popularizing term for Global Navigation Satellite System. In the mid of 1970s, the United States released navigation satellites in the spaces, known mainly as GPS—Global Positioning System that is operational till date with 24 satellites orbiting on six paths around the earth. It has enabled a citizen to navigate by instantaneous positioning on the surface of the earth using mobile phone devices. In addition to the GPS of USA, there are other constellations of satellites as Beidou of China, QZSS of Japan, GLONASS of Russia, NAVIC of India Galileo of Europe. These systems work on specific spectrum signature of satellite signals for the same functions—positioning, navigation and timing. Generally, the accuracy of stand-alone systems (single point positioning) is achieved to a few meters in a spherical manner. However, with employing solutions of Real-Time Kinematics (RTK) which are open source tools for serving through Differential positioning, can provide accuracy as high as that of a few centimetres.

The technology is available through interoperability of RTKLib software tools, and precise point positions are obtained in real-time, that too, at lower costs of capital investment and operations. The technology can be employed for various challenges discussed in preceding sub-sections. A few identified applications are discussed herewith:

- Location-based services for moving vehicles carrying solid waste collected from predefined spots—such real-time movement enable for getting the destination ready for arrival and further operational time optimization, leading to saving of fuel, energy, manpower, time and so on.
- Route identification and optimization based on waste collection volumes and setting up or rescheduling of waste collection frequency.
- The technology can also reveal the disturbances in the troposphere and ionosphere due to global actions. The effects of global phenomena on the local level can be established and put to interventions for the policy framework.
- The positioning in three dimensions can also be employed for monitoring towards catastrophic climatic events and for generating early warning systems. It can be instrumental in the events of tsunami and earthquake monitoring—the signals changes as it passes through the atmosphere and is affected by the quality of air, the presence of gases and so on.
- The technology can also enable for monitoring soil and vegetation moisture by indirect measurements.
- The differential GNSS can provide very accurate positioning for mass transportation vehicles thus resulting in savings of resources making the system utilization more efficient.
- Monitoring the level of water in river, lakes and sea for constant monitoring obtaining pattern of variations due to many reasons to be identified.

Henceforth, the loss made to the local and global environment is not measurable. The effects observed through the interrelationship of natural elements is still a matter of research and due for complete revelation. However, changes in the ongoing practices and adopting technology that allows for sustainable use of available resources will make operations smarter. Surat Municipal Corporation has led the forum of urban bodies in India for many activities and putting for such practices in routine will enable the citizens to prosper over a longer term.

References

1. S. N. Connections, Setting goals and developing specific, measurable, achievable, relevant, and time-bound objectives. Subst. Abus. Ment. Heal. Serv. Adm.
2. S. M. Corporation, History of Surat. Surat municipal corporation, 2011. [Online]. Available: https://www.suratmunicipal.gov.in/TheCity/History. Accessed 22 Feb 2017
3. SUDA (2014) SUDA Revised Development Plan.Pdf, Surat
4. C. Statistics (2011) City Mayors: Largest Indian cities. [Online]. Available: http://www.citymayors.com/gratis/indian_cities.html. Accessed 17 Jan 2017
5. World's fastest growing urban areas. City Mayors
6. Gumber A (2014) Annual survey of India's city-systems, 2nd edn. Janaagraha Centre for Citizenship and Democracy, Bangalore

7. The Free Encyclopedia Wikipedia (2018) List of cleanest cities in India. [Online]. Available: https://en.wikipedia.org/wiki/List_of_cleanest_cities_in_India. Accessed 24 Dec 2018
8. USGS-EROS (2018) LANDSAT 8 OLI/TIRS Collection 1—Path: 148 Row: 45 for Scene: LC08_L1TP_148045_20180117_20180205_01_T1. U.S. Geological Survey (USGS) Earth Resources Observation and Science (EROS) Center. [Online]. Available: https://lta.cr.usgs. gov/TM
9. R. C. D. Plan (2013) Revised City Development Plan (2008–2013). Surat
10. Surat Municipal Corporation (2016) TP Details: Surat Municipal Corporation. Surat Municipal Corporation. [Online]. Available: https://www.suratmunicipal.gov.in/Departments/TownPlanningTPDetails. Accessed 17 Jan 2017
11. Provisional Population Totals Urban Agglomerations and Cities. Census India 2011
12. District Census Handbook—Surat. Census of India, p 40
13. Surat Municipal Corporation (2011) Population of India, Gujarat, Surat District and SMC [1901–2001]: Surat Municipal Corporation. [Online]. Available: https://www.suratmunicipal. gov.in/TheCity/City/Stml9. Accessed 27 Dec 2018
14. Surat Municipal Corporation (2015) Details of city limit extension & population: Surat Municipal Corporation. [Online]. Available: https://www.suratmunicipal.gov.in/TheCity/City/Stml1. Accessed 27 Dec 2018
15. USGS-EROS (2005) LANDSAT TM Collection 1—Path: 148 Row: 45 for Scene: LT05_L1TP_148045_20110130_20161010_01_T1. U.S. Geological Survey (USGS) Earth Resources Observation and Science (EROS) Center. [Online]. Available: https://lta.cr.usgs. gov/TM
16. Patel PP, Bhatt BV (2015) Land use change detection in Surat using geospatial techniques. In: EIOCD 2015, pp 1–8
17. Climate-data.org, Climate Surat: temperature, climate graph, climate table for Surat—Climate-Data.org. [Online]. Available: https://en.climate-data.org/location/959693/. Accessed 24 Feb 2017
18. Bhasker B, Sharma ND (2015) Scope of modeling for urban land-use leading to climate change. Int J Adv Res Eng Sci Manag 1–7
19. Dhiman R, VishnuRadhan R, Eldho TI, Inamdar A (2018) Flood risk and adaptation in Indian coastal cities: recent scenarios. Appl Water Sci 9(1):5(1–16)
20. Patel D, Dholakia M (2010) Identifying probable submergence area of Surat city using digital elevation model and geographical information system. World Appl Sci J 9(4):461–466
21. Surat Municipal Corporation (2018) Surat—air quality monitoring. [Online]. Available: https:// www.suratmunicipal.gov.in/Home/AirQualityInfo. Accessed 28 Dec 2018
22. Surat Municipal Corporation (2013) City report Surat solid waste management project under JNNURM Jawaharlal Nehru National Urban Renewal Mission July 2013. Surat
23. Surat Municipal Corporation (2016) Hydraulics Department of SMC. [Online]. Available: https://www.suratmunicipal.gov.in/Departments/HydraulicHome. Accessed 27 Dec 2018
24. Surat Municipal Corporation (2018) Drainage: Surat Municipal Corporation. [Online]. Available: https://www.suratmunicipal.gov.in/Departments/DrainageIntroduction. Accessed 27 Dec 2018
25. Surat Municipal Corporation (2016) Approaches: Surat Municipal Corporation. [Online]. Available: https://www.suratmunicipal.gov.in/Departments/SolidWasteManagementApproaches. Accessed 27 Dec 2018
26. Surat Municipal Corporation (2018) Solid waste management statistics: Surat Municipal Corporation. [Online]. Available: https://www.suratmunicipal.gov.in/Departments/SolidWasteManagementStatistics. Accessed 27 Dec 2018
27. Census of India 2011 (2015) Surat District Population Census 2011, Gujarat literacy sex ratio and density. Census Population 2015 Data. [Online]. Available: https://www.census2011.co. in/census/district/206-surat.html. Accessed 13 May 2018
28. Gujarat Ecology Commission (2018) Objectives|About Us|Gujarat Ecology Commission. [Online]. Available: https://www.gec.gujarat.gov.in/showpage.aspx?contentid=8&lang= English. Accessed 28 Dec 2018

29. Gujarat Ecology Commission, gec_annual_report_2015-16.pdf, Gandhinagar
30. Gujarat Forest Department (2018) Gujarat Forest Statistics 2017–18 Principal Chief Conservator of Forests & Head of the Forest Force Gujarat State, Gandhinagar, Gandhinagar

Part VIII
Yokohama

Local Government and Technological Innovation: Lessons from a Case Study of "Yokohama Smart City Project"

Aki Suwa

Abstract Currently, most climate policies focus on the international and national level efforts. The international and national levels may set a strategic orientation, but the real effect of the strategies would be made through local actions. Cities and municipalities are the core of actions to cope with global and local environmental problems. In response to these problems, the cities are increasingly taking a strategic approach to climate changes to implementing overarching and systemic changes, by redesigning and reconfiguring the infrastructure networks through which energy is produced and consumed. Primarily, innovative technology, including those associated with smart city and smart-grid, force the local government to reconsider the different levels of technical capacities between them and industries. This study focuses on Yokohama city as a case to illustrate an ambitious energy technology innovation programme at the city-wide scale. It can fill the research gap of elaborating the Asian example of public-private sector cooperation, whether and how the allocate resources and expertise to deliver and experiment smart city as a green tech innovation.

Keywords Smart city · Local government · Technological innovation

1 Introduction

1.1 Concept of Sustainable Cities and Smart Cities

Today, nearly half of the world's population live in cities. With ever growing consumption of energy, carbon emission is typical by-products of urbanization. With the reflection on such the environmental impacts of the cities, the broader notion of "eco-city" and the related concepts and practices of 'sustainable urbanism,' have emerged. Sustainable urbanism typically engages with various aspects of environmental, economic, and social sustainability concerning the urban context, where the

A. Suwa (✉)
Faculty of Contemporary Society, Kyoto Women's University, 35 Imagumano
Kitahiyoshi-cho, 605-8501 Higashiyama, Kyoto, Japan
e-mail: suwa@kyoto-wu.ac.jp

© Springer Nature Singapore Pte Ltd. 2020
T. M. Vinod Kumar (ed.), *Smart Environment for Smart Cities*, Advances in 21st Century
Human Settlements, https://doi.org/10.1007/978-981-13-6822-6_10

"eco-city" would be best understood as an umbrella term that covers various notions of, and approaches to the sustainable urbanism [1].

At the same time, an increasingly stronger emphasis was placed on technology that may provide solutions for urban sustainability challenges. Smart city is an emerging concept to address sustainability of city in terms of controlling resources, energy and wastes in efficient ways. Any firm definition of smart city is yet to be established, but much debate has initiated: Initially, Information and Communication Technology (ICT) was considered as the main pillar of smart city, where cities and dwellers are interconnected as in "wired city" [2]. Recently, the concept evolved to cope with the local and global demand by employing available resources (people, energy, land, talent, environment, geographic location) in a way that maximises overall welfare and sustainable growth [3].

On this context, the European Union "Smart Cities and Communities Initiative" supports cities and regions in taking ambitious measures for 40% reduction of GHGs emissions through sustainable use and production of energy. The targets are to be achieved through approaches on energy efficiency, low carbon technologies, use of ICT on energy supply and demand, and building and transport management [4]. Thus, the use of information technology is one of the *many* key components of "Smart City" in the European understanding, in contrast to the Japanese context where the concept is understood more specifically from technological perspective.

Japanese academics, for example, sees the definition of smart city as "a city of high efficiency with integrated infrastructure by leveraging information technology to support urban life" [5]. The Japanese government also sees the smart city concept has strong association with information technologies, and its combined use with distributed energy sources, including energy management systems (EMSs) and energy storage devices for optimisation of available energies [6]. NEDO, a Japanese government agency for technological innovation, also presents "smart" community as a social boundary in which energy system and transportation system are merged together, where efficiency of material flow optimized through information and communication technologies [7].

There might be the cases where dominance of technological priority in Japan is the reflection of how the public sector sees the notion of "smart" and the degree of strength they associate it with ICT infrastructure, and the degree in which the private sector is involved in the economic benefits might arise through the public investment. In any case, the smart city and its integration with ICT require not just technological change but also need to make the changes to be embedded in the social context [8]. It poses a question how the social and technical aspects interact for the development, adoption and implementation of the technology, as will be explained in the later section. Having the recognition in mind, this paper focuses on hard and soft aspects of smart city and how a city energy management can and should incubate the relevant technological innovation.

1.2 Local Government as a Platform for Innovation

In response to the technological imperative to the smart city development, the cities are progressively taking a strategic approach to implementing overarching and systemic changes, by redesigning and reconfiguring the infrastructure networks through which energy is produced and consumed [9, 10]. These municipalities and cities are thus important determinants of effective climate technology adaptation, it would be valuable exercise to identify their role in local technological innovation.

There is also an increasing expectation for technological innovation to give solutions for the climate and sustainability problems. The governments of both developed and developing nations are placing urgency to the research and development (R&D) and implementation of the green technologies. These countries often offer incentives and speedy patent application procedure, to encourage firms and entrepreneurs to ensure innovation investment [11].

In addition to these incentives, there is a trend to support area-wide development. Cities, inter-alia, have been recognised as a platform for green technology experiments, designed to respond to the imperatives of mitigating and adapting to climate change [12]. Cities have been providing arenas to test and implement experimental concepts, not alone dating back to the historical examples of the garden city, compact city and other titles attached to. They are responding to the climate challenge, by formulating and implementing urban policies. A broad range of regional to urban spatial processes, including redesign of urban infrastructure and control of new and existing building environmental design and features, are part of such policies [13, 14].

With the advancement of green technology innovation, cities are becoming conductive in experimenting urban design futures [15]. As the technological imperative found the urban infrastructure platform as "testing field" for knowledge and technology validation, cities have become urban laboratories of knowledge production that will make them as "mechanisms to mobilise place to generate economic wealth and stimulate more resilient urban conditions, both through the creation of new landscapes and the retrofitting of existing ones" [16]. The urban laboratories potentially provide an answer to cities for low-carbon transition, by delivering innovative technology dissemination.

These perspectives for urban climate change experiments were classified into (1) policy laboratories to government experiments, (2) niche innovation and transformation, and (3) living laboratories and the design of urban future, in order to classify the experiment focuses [12]. Though overlaps in several aspects, these categories provide a basis to understand the recent increases of cities initiatives for undertaking climate experiments, using their urban infrastructure. In contrast to the laboratories in natural science, urban laboratory has social setting, but still reflects a semi-permeable border zone, with emphasis similarly placed on the role of places in facilitating knowledge production [16]. Urban laboratories are spaces of experimentation to test the relevance of a technology to the real life of people.

1.3 Smart City Governance

On the other hand, smart city concept, in itself, has emerged in the 1990s, when the focus was on the significance of Information and Communication Technology (ICT) upon modern infrastructures within cities. The initial argument was placed by the Californian Institute for Smart Communities on how cities and communities could be designed to implement information technology [17], though later the idea of then smart city was criticized with its strong emphasis on technical aspect. Smart city then became the core of urban labelling phenomenon, with a variety of indexes and visions attached to [18].

Also, the complication was exasperated partly because "smart city" agenda arises exactly at a time when, faced with global challenges of climate change, city governments and businesses realize their mutual opportunities for "green-tech" innovation. It gradually became a public-private partnership agenda, then predominantly the IT business opportunity. Yet, this development leads to the question of intellectual capacity and ownership of smart technology. Naturally, smart technologies and its intellectual properties are dominated by the private sector, rather than in universities and other academic institutions, especially in Japan, where industries hold comparative strength in applied sciences. Not to mention, how techno-scientific capacity can be accumulated in *local government* is also yet unknown, as they usually remain disconnected to the updated technological innovation.

Simultaneously, as touched upon in the previous section, crucial for any interactive technology is the field for testing: without the experimental demonstration, the incubated technology may fail to get the validity for its practical implementation, as the debate of "Give me lab and reduce carbon footprint" suggests [16]. One of the important components of smart-city concept, for example, is the use of ICT for smart-grid. The smart grid refers to a system in which energy transmission and distribution promote the stability of the electric power supply by utilising information and communication technology. It incorporates end-user information into energy management system, to enable demand-response energy supply [7].

The introduction of the smart-grid system is expected to play a significant role in promote expansion of renewable energy use by delivering reliable energy supply through advanced integrated mechanism, including demand-side control to accommodate unstable renewable power [19]. In order to optimise the potential of the smart-grid, it is necessary to ensure networks that enable two-way communication between the demand and supply endpoints. Importantly, smart-grid is able to detect the energy demand from consumers by collecting data, regarding when and how they use power. Through integration between supply and demand sides by establishment of smart-interface, smart-grid system would also become more resilient to power outages (ibid.).

Lack of the technical validity and the associated uncertainty on social implication of the smart grid may become an obstacle for the technology's diffusion. As well-known argument of Valley of Death, the term frequently used in the marketing and innovation discourses, meaning that taking an idea from basic concept to commercially available product often demands a complex and difficult expedition, involving a corridor phase for testing the market viability [20]. On this understanding, it becomes practical for industries to seek the testing field for the relevant technologies.

A question arises as to what role the local government should play to promote technological innovation, especially those need to be embedded as a city-wide infrastructure, and what is the efficient governance arrangement between local government and businesses to address shared responsibility for the providing the innovative infrastructure. This question is increasingly relevant, as the pace and size of technological development unprecedentedly fast and vast.

Many initiatives to promote changes fail or do not achieve the expected results because the localities do not have the technological capabilities to support new technologies. By understanding how cities learn to build those capabilities and what factors that influence the governance process, we may be able to promote technological changes more effectively. This study chooses Yokohama city as a case, because it has recently initiated an ambitious energy technology innovation programme at the city-wide scale. It can fill the research gap, by elaborating the Japanese example, of public-private sector cooperation, through the glass of whether and how the allocate resources and expertise to deliver and experiment smart city as a green tech innovation.

2 The Case Study of Technological Innovation Approach

2.1 Japanese Cities and Sustainability Initiative: Overview

Japan, the island nation in East Asia, is a fairly sizable archipelago of 6852 islands, although the four largest islands (Honshu, Hokkaido, Kyushu and Shikoku) account for 97% of its population. The last set of official figures pertaining to Japan's population were released at the time of the 2010 census and the final statistics showed there were 128 million people, the tenth largest country in the world. Japan is currently experiencing an unpreceded level of de-population. The downturn will continue and accelerated, as forecasted in Fig. 1, simultaneously aging becoming another demographic problem. Yokohama, the case study city this chapter has its foci, is perhaps less susceptible, with its second highest population level next to Tokyo, but it is yet not entirely immune to the de-population issue.

Many local governments are trying to revitalize their cities through a various programme, including, e.g. providing incentives for bearing children and inviting dwellers from bigger cities to settle into their domains. Sustainability and environmental issues are also taken as inspirational framework for regenerating local cities across the country, as it can generate attractive of the cities with sustainability benefits. In terms of sustainable urban development in general, there are a number of initiatives with different priorities, scales and denominations. These programs are administered by multiple government agencies, as well as private and

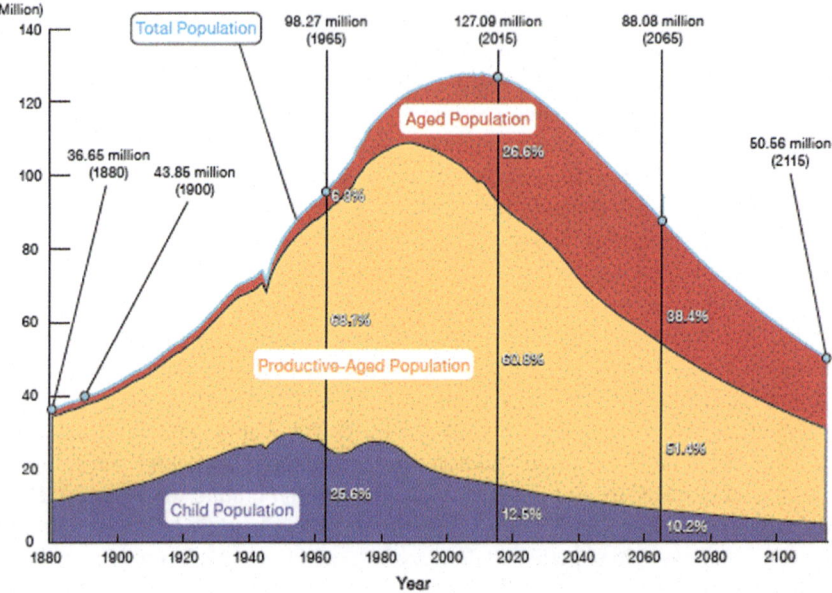

Fig. 1 Population trends in Japan from 1880 to 2100 [21]

non-governmental organizations. It is thus highly challenging to identify all the different schemes currently being implemented across the country. Among the clearest ones there is the "Eco-Model City" initiative developed by the Ministry of the Environment Japan and a "Future Environmental City" initiative of the Cabinet Office of Japan [22]. These initiatives are mainly focusing on environmental policies and actions of cities, whereas The Ministry of Economy, Trade and Industry later started to support local governments from the technological innovation perspective. Among these initiatives is the recent attempts of constructing "smart communities", to convert certain regeneration sites into "smart cities", by maximising optimum energy and resource efficiency. Initially, only Kiyakyushu and Chiyoda-ward of Tokyo Metropolitan government [23], were known for such initiative, but the situation was changed when the METI launched the "Next Generation Energy and Social System Demonstration Initiative", as will be explained in Sect. 2.3, but before going into the details of the initiative, Yokohama profile is to be provided in the following section [24].

2.2 Yokohama City

Yokohama city is the capital city of Kanagawa Prefecture located south of Tokyo in the Kanto region of the main Honshu Island, along the coastline of Japan's Pacific Ocean. With the area of 437.4 km^2, it is Japan's second largest city following Tokyo, alongside a population of about 3.65 million. Commercial districts are clustered around the Yokohama Station area, with industrial facilities concentrate around the bay area, while residential districts are designated inland.

In historical term, Yokohama's port was the first to be opened to foreign trade in 1859, while Japan maintained a policy of self-isolation. Since then it is the key place for foreign trade. In 1872, Japan's first railway was developed to link with Tokyo. Although large part of Yokohama was destroyed by the Great Kanto earthquake in 1923, followed by the Allied bombings during World War II, Yokohama city managed to recover and became one of the populated cities in Japan. Numbers of foreign enterprises have established their branches in Yokohama in order to have access to the international trading port. Yokohama city also adapted to the change by reconstructing its urban structure. For instance, the reclamation of the coast line to develop the commercial district called "MinatoMirai (meaning Port for Future)" is one of the largest scale urban re-generation projects in Japan, with its high-rise buildings on the site of a redeveloped bay area. The MinatoMirai District also embraces residential areas, covering about 2500 ha, developed since the 1970s until now [24].

Recently, Yokohama aims to brands itself as the city of environmental innovation, through city-level mitigation and adaptation for climate change action [23]. In 2006, it adopted "Yokohama Global Warming Measures Regional Promotion Policy", that sets the city-wide 6% GHGs reduction target by 2010. In 2008, Yokohama city has selected as one of the Environmental Model Cities, an initiative of the Japanese government. As an "Environmental Model City", Yokohama sets "Yokohama Climate Change Action Policy (CO-DO30)". These plans and policies are succeeded by Yokohama Global Warming Measures Implementing Programme in 2011 which established the mid-to-long term GHG reduction targets by 2020 (25%) and 2050 (80%) below the 1990 level [25]. With this background, it is important to introduce more renewable energies within the city, and the Yokohama government set the targets of reducing approx. 300,000 t-CO_2 by newly introduced PV for residents and industries, with another approx. 50,000 t-CO_2 through other sources of renewables (e.g. wind, mini-hydro and bio-diesel) (ibid.).

Yokohama city was also awarded the status of "Future Environmental City" under the similar governmental scheme. With the strong environmental presence and governmental credibility, the Yokohama city was nominated in 2010, under an initiative of Ministry of Economy, Trade and Industry (METI) for incubating prototype smart cities in Japan. Yokohama aims to install energy management systems under the Smart city initiative, by e.g. turning the MinatoMirai area as a showcase of environmental technology [14].

2.3 Yokohama Smart City Project

In 2010, with the growing interests into a smart city, using the technological grid control for energy management, METI has nominated four cities in Japan, including Yokohama city in Kanagawa, Keihanna city in Kyoto, Toyota city in Aichi and

Kitakyusyu city in Fukuoka as demonstration region to practically design, operate and test experimental smart-grid technologies. These four cities are selected among other candidates, because of their accumulated reputation over years in the environmental and technological capacities, e.g. Keihanna is strategically designed to accommodate information technology industry and institutions, whereas Toyota is known for international technology axis, Kitakyushu and Yokohama have been recognised as environmentally active, with previously awarded the Eco-cities by the MOEJ, as mentioned.

The METI scheme, coined as "Next Generation Energy and Social System Demonstration Initiative", was in principle a competitive grant to be allocated to bidding local authorities. In order to be successful candidates, local authorities have to demonstrate strong partnership with industrial stakeholders, to validate technological and financial credibility for smart technology project management.

The idea of making Yokohama as a "home" of smart technologies, though the METI framework, was originally conceived by Yokohama government office personnel. Mr. Nakajima, who was then the chief officer of Yokohama City Global Warming Policy Department, learned the smart grid technology and its implication to city development through his visit Hachinohe-city, Aomori Prefecture in Northern Japan, where the government funded project on micro-grid was carried out during 2003–2007.

During the conceptual phase, the idea of making Yokohama a successor of the micro-grid cities as in Hachinohe, remains a brain child of the officer, circulated within the relatively closed network of people who he knew. Though the idea already included a grand vision of developing relevant technologies using the city infrastructure as a platform, and to potentially export the technologies to worldwide market, it took some time before it really took off in the form of an official project of Yokohama City Government.

During the same conceptual phase, before the Yokohama city decided to apply to the METI funding, a similar project was envisaged by the different division of Yokohama city (City Planning Department) in cooperation with the Yokohama National University and the United Nations University Institute of Advanced Studies (UNU-IAS), a research institute based then in Yokohama. Their proposed project was also to furnish smart technologies within the main areas within the city to optimising energy balance though electricity and heat management systems, with a potential funding from Ministry of Education (MEXT). In the end their application did not succeed in obtaining the fund form MEXT, but some elements, including the local energy management through smart grid experiment, were similar to the Yokohama application to METI funding.

Upon the successful application to METI fund (about 74 billion yen for 5 years), the city officials accelerated their work to obtain the consensus and participation of potential business and industrial partners to the project. The process was not necessarily smooth, as one official confessed that Panasonic originally did not show much interest to be part of the project, whereas Toshiba demonstrated its willingness to participate. Also, the officials are required to consult to METI for their supervision to the overall project, while making legal arrangement with participating business entities, for example as to the confidentiality agreements on the technical specification.

Under the METI's Next Generation Energy and Social System Demonstration Initiative, Yokohama city started Yokohama Smart City Project (YSCP), as one of the largest scale smart-city experiments in Japan. YSCP became no longer a private concept, but a shared value and tasks allocated among the city officials of the Yokohama City Government, which is collaborating with 34 institutions, including Nissan, Panasonic, Toshiba, Tokyo Gas, Tokyo Electric Power and Accenture, to implement a various kind of technical and social systems related to smart technology. This experiment has a significant implication for future international marketing, as the Yokohama city plans to export the smart city model as an infrastructure package to developing world [26].

3   Achievement of YSCP

This section demonstrates the technical progress of Yokohama project in contributing to the development of smart city, paying a particular attention into the question as to what kinds of technological achievements cities can achieve in terms of smart city development, in cooperation with the industries.

3.1   Demand Response Demonstration

The Yokohama project is made up with numbers of initiatives, taken led by the prominent Japanese business entities. Figure 2 lists the main participating corporations and institutions.

These institutions voluntarily forming groups that focus on main 6 technological pillars: Community Energy Management System (CEMS), Building Energy Management System (BEMS), Factory Supervisory Control and Data Acquisition (FEMS), Home Energy Management System (HEMS), Supervisory Control and Data Acquisition (SCADA) and Electric Vehicle Energy Management System (EV-EMS).

The overall project is made up with a number of segmented experiments. The segmented tasks are carried out in urban, residential and industrial sites. The key initiative in the project is community energy management system (CEMS), which manages output fluctuation from renewables by integrating stationary battery with Home Energy Management Systems (HEMS) and Building Energy Management Systems (BEMS). HEMS controls combined patterns of multiple renewable electricity generation through e.g. photo voltaic (PV) and energy saving equipment at residential houses. Similarly, Building Energy Management System (BEMS) is designed to control systems at commercial districts.

Since the community power output fluctuation can be managed through HEMS and BEMS, Community Energy Management System (CEMS that integrates HEMS

	Participants
Manufacturers	Toshiba, Hitachi, Meidensha, NEC, Sony Energy Device, Sharp, Panasonic, Nissan, ORIX Automobile
System engineering	Accenture, NTT Facilities, NTT Docomo, Nikki Information System
Estate / Construction industry	Daikyo, Mitsui Real Estate Residential, Mitsubishi Real Estate, Daikyo, MM42 SPC, Taisei, Shimizu
Electricity / energy Company	TEPCO, JX Nikko-Nisseki Energy, Tokyo Gas, Sumitomo Electric Industry
Consultant / Academic institutions	Nikki, ORIX, Tokyo Institute of Technology, Kyoto University

Fig. 2 Main participants to YSCP [27]

and BEMS) was expected to become compatible with the regional electricity network protocol. With regards to economic incentives, CEMS would enable visualization of the energy use in the community (and the associated CO_2 emission). The advantage of the initiative is that it can achieve efficient energy management by managing both demand side and stationary energy storage. CEMS provides operational control to enhance locally produced and consumed energy to be used within the community in the sustainable manner, though there is no official quantitative GHGs reduction target specified to the YSCP.

A set of targets were established for the whole YSCP, as shown in Fig. 3. These targets have been achieved by the end of 2014, well before 2015 that was the official due date.

One of the most significant components of the project is demand response (DR) for HEMS domain. Demand response usually provides an opportunity for consumers to play a significant role in the operation of the electric grid, by reducing or shifting their electricity usage during peak periods in response to time-based rates or other forms of financial incentives. Automated demand response, or Auto-DR, is a platform and program to automate the energy curtailment process to initiate an "automatically" pre-programmed DR strategy. YSCP became one of the first large-scale platforms to test automated demand response (ADR) as well as the conventional DR in Japan and worldwide [28].

Years	2010	2011	2012	2013	2014
Photo Voltaic (target:27MW)	6.8MW	19.1MW	31.0MW	36.9MW	
Home Energy Management System (target:4000 houses)	66	995	2640	4230 houses	
Electric vehicle (target: 2000 vehicles)	427	1104	1859	2294 vehicles	
Central Energy Management System		Planning & development		Demand prediction, Demand response, Testing interface	
Supervisory Control And Data Acquisition				Short-period demand-supply adjustment	
Building Energy Management System				Demand management, Testing interface, Power to heat optimum control	
Home Energy Management System				Demand management, inter and intra-households energy management	
Electric Vehicle				Power to Vehicle energy management	

Fig. 3 Technical targets and progress management schedule [27]

CEMS as standing for Community Energy Management, was developed to predict community energy demand. The difficulty of CEMS usually lies where the energy demand from multiple and bulk consumers is unpredictable, but the developed CEMS system has enabled very high accuracy rate of the community energy demand prediction.

For BEMS, demand response was tested and proved to be effective in significantly reducing peak electricity demand. Electricity storage and power-to-heat optimum operation proved to be feasible in applying into the BEMS domain. The integrated use of power storage and renewable energy were also turned to be useful in FEMS aspect.

Fig. 4 Daikyo, one of the participants to YSCP, developed a condominium with home energy management system in the heart of Yokohama [29]

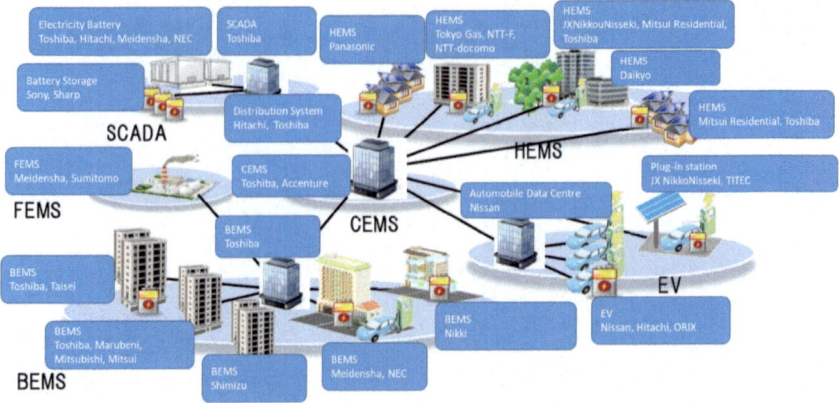

Fig. 5 Overall participants to the YSCP initiative (Yokohama City Government and 34 institutions, to carry out 15 projects) [27]

PV-generated power is effectively stored into electric vehicles (EV), pushing the self-consumption rate of PV power on site. The stored power is used during the period when there is less PV production.

SCADA, the cutting-edge technology for electricity management, was also tested in the field to demonstrate its feasibility of controlling multiple power storage devices. The multiple storage, acting as if one, to level gap between power demand and supply though a specially designed interface (Fig. 4).

Figure 5 illustrates the individual achievement for these technical pillars.

3.2   Project Governance Framework

Based on an interview with Yokohama City Government, the role of the local government to promote technological innovation was explored. The governance structure was particularly elaborated to understand arrangement between local government and businesses to address shared responsibility for the providing the innovative infrastructure at a city-scale. The following information was given at the interview with the personnel at City of Yokohama government in 2012.

As seen, the City of Yokohama decided to apply to the METI fund and won, with Toyota, Kita-Kyushu or Keihanna cities. After selected, Ministry of Economy, Trade and Industry (METI) exercised its power in final decision making for the details of the applicants' projects. METI, e.g. ensures that the projects in smart cities do not overlap with each other. Each selected city is supposed to undertake different kinds of smart city projects, meaning that in some cases, the selected cities gave up certain project components if METI asked them not to do.[1]

A project was set-up by the Yokohama City Government, with the initial support of a private IT company (Accenture). Yokohama city made agreements with "project leader" companies, and those companies made their own proposal to contribute to the overall projects. The leader companies organise a participant's association to establish self-governance mechanism. The participants association hold regular meetings every month. In order to supervise the industrial self-governance, the association create a body for an auditing process. The associations have strong influence over the whole project, e.g. most of the associations need to agree on what new projects to be included in YSCP. Once the project was defined out of the scope of YSCP protocol, then the projects are rejected, or given less importance (with less budget) by the association.

The participating industries play pivotal roles in knowledge development, experimentation, resource mobilization, legitimation, market formation or political advocacy processes in the project. The leader companies have the power to lead other 34 companies through the vertical integration, to function as a management body of YSCP. In terms of CEMS, Tokyo Electronic Power Company (TEPCO) have the critical power at the final stage of the decision-making process.

There are interactions among key actors involved in the project with horizontal and vertical nexuses. The association provide the horizontal dialogue among the leader companies. Besides the vertical relationship demonstrated by the Yokohama city which oversees the association, the 4 main projects (CEMS, HEMS, BEMS, EV) exhibit vertical integration of participants. For instance, the association member Toshiba is the leader of the HEMS projects, over other members, such as Panasonic and Tokyo Gas.

For the development of knowledge relevant to the YSCP operations and technologies, richness of technical experts in the Yokohama local government seems to be an asset. There are many people from private companies, as well as from national government, in the Yokohama City Government. Thus, the local government has the mixture of people with different background. This leads to knowledge sharing and networking among the City government administrators. Yokohama City Government is also known for higher penetration of "elite" university graduates, that partly explains relative advantage of taking up innovative activities, than other local governments in Japan. Also, externally, there are various seminars for knowledge sharing, including "HEMS club" organized by Toshiba, of which purpose is for the citizen to understand the technical outline of HEMS and its role in the overall projects.

Key actors have expected benefits (e.g., increased capacities, prestige, revenues etc.) from collaboration with certain actors and participating in the project. Apart from the obvious financial merit of getting access to the ample METI fund, the participating companies are interested in participating the project as they perceive it

---

[1]The Yokohama City Government, for example, originally planned to have a bigger component to test heat management, but they were advised by METI not to include the heat content into the Master Plan for YSCP (the heat element later became as a project component).

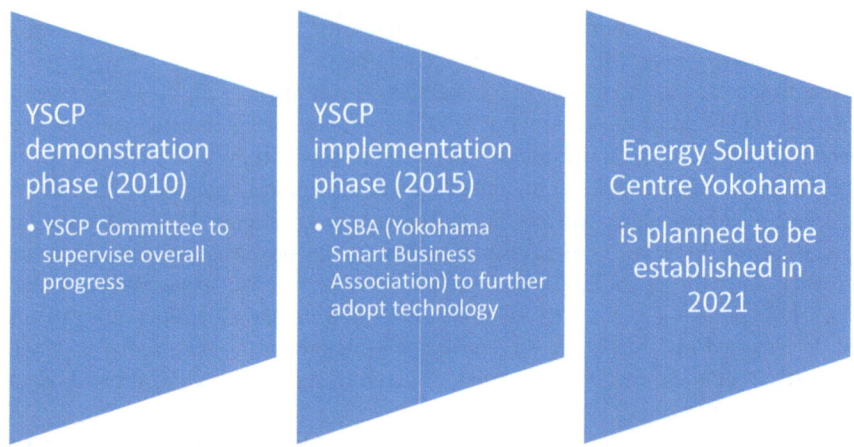

Fig. 6 Progress so far and the planned further development of Yokohama Smart City Project (YSCP)

is a good testing arena for the new technology. City of Yokohama does not specify where, but the participating industries have a fair amount of freedom to choose where exactly a project would be carried out (e.g. BEMS projects in a shopping malls and supermarkets).

Besides, a various factor outside of the governance fostered or hindered positive outcomes of the project. Since the Great Eastern Japanese Earthquake on March 11 in 2011, things have changed largely, especially the Yokohama residents' perception changed towards their energy use: Since the critical event of 3.11, public interests increased significantly, to the extent that energy gadget such as smart-meters and power storage battery gain more attention. Also, the introduction of Feed-in Tariff (FIT) in 2012 triggered residents' interest on electricity generation and how it can be managed. These factors seem to push response rates of residents higher though DR.

In 2015, after the completion of the YSCP, the Yokohama Smart Business Association (YSBA) was established to continue and further develop the inherited project achievement. YSBA is to ensure the technological components been tested so far to be adopted and widely used in the City government boundary, while maintaining the industrial and institutional network stemmed through YSCP experience (Fig. 6).

In this way, a business consortium, created under the auspices of the Yokohama City Government, to further develop intra and/or inter industrial network. Since the smart grid technology requires technological integration among different business expertise, this industrial cooperative network seen in the Yokohama would provide interesting and important implications on how to support business confederation that would make the real impact on the smart city infrastructure testing and emergence.

4 Discussion and Conclusion

The organization involved in the project, including the Yokohama City Government and the business stakeholders, learned the valuable experience and capability through different kinds of mechanisms. 'Learning' here will be understood as the various processes by which cities or organizations within a city build up and accumulate their technological and innovation capabilities. This research paid particularly attention in knowing how local government learn and build technological capabilities and shared governance structure to technological innovation, and what role the local government should play to promote technological innovation, especially those need to be embedded as a city-wide infrastructure.

Based on the Yokohama experience, we understood that the capacity of the business shall be optimised to bring innovative technology developed and implemented at cities. It turned out that when cities and the partner businesses seek to deepen their innovative capabilities rapidly, they will need to make partnership structure to efficiently share capacity and responsibility. The public administrators, often remain as a regulator to business operation, can play an inviting role to accommodate the enterprises, by facilitating actions that contribute to the technological change. It is similarly essential to identify what is the efficient governance arrangement between local government and businesses to address shared responsibility for the providing the innovative infrastructure.

Though the YSCP case, it tuned out the ability of the Yokohama City Government is particularly high to "integrate" the different project, simultaneously "distributing" technical responsibility to the selected participants. First, Yokohama City Government, using its geographical advantage of being close to the Tokyo and its vicinity where the Japanese leading industries to be based in, successfully invited a number of the active participants. These participants have high technical capacity themselves, but it seems Yokohama City Government exercised proactive a role to place well-coordinated governance structure to integrate the potentially conflicting business interests.

As the second step, the Yokohama City Government delegate main managerial responsibility to the selected participants, e.g. TEPCO, Toshiba and Nissan, to function as the leader of individual projects. This arrangement is important to ensure the high technical level of the whole initiative: It is often the case that the level of technical specification is restricted by the limited technological understanding of the supervising local authority personnel, as they are not usually educated as high-end engineers for smart energy applications. By distributing managerial responsibility, the participating industries were able to promote technological experiments effectively, under the professional supervision.

This paper initially posed a question as to what role the local government should play to promote technological innovation, especially those need to be embedded as a city-wide infrastructure, and what is the efficient governance arrangement between local government and businesses to address shared responsibility for the providing the innovative infrastructure. Thus, coming back to these questions, the answer is yes, and the YSCP well proved the point. As elaborated in this chapter, the initiation of an ambitious energy technology innovation programme at the city-wide scale, could bring interesting opportunity for the industries to test the new technology in the real-world context. High level of managerial skills for coordinating (including delegating) technical expertise is essential, while having essential motivation to understand and disseminate the technology in question at the city domain and beyond.

This chapter thus tried to fill the research gap of elaborating the Japanese example of public-private sector cooperation, whether and how to allocate resources and expertise to deliver and experiment smart city as a green tech innovation. It is understood how a learning mechanism is specifically connected to the construction of an innovative capability, how it contributed to technological change and consequently to the building of a specific innovative capability within and outside of city administration.

Acknowledgements The authors are grateful to United Nations University Institute of Advanced Studies, especially Dr. Csaba Putzai, to assist research, and Dr. Haruka Okada from University of Australia for her help on this work. Dr. Andrew Blok from University of Copenhagen, last but not least, is highly appreciated for his professional suggestions and comments.

References

1. Joss S (2015) Eco-cities and sustainable urbanism. In: Wright James D (ed) International encyclopedia of the social & behavioral sciences, vol 6, 2nd edn. Oxford Elsevier, pp 829–837
2. Dutton WH, Blumer JG, Kraemer KL (1987) Wired cities: shaping the future of communications. Annenberg School of Communications, Washington, DC
3. Caragliu A, Bo CD, Nijkamp P (2011) Smart cities in Europe. J Urban Technol 18(0048):65–82
4. European Commission (2009) A technology roadmap for the communication on investing in the development of law carbon technologies (SET-Plan). Available from http://setis.ec.europa.eu/
5. Fukuchi M (2011) Status of domestic/overseas smart cities. Knowl Creat Integr 19(5):6–19 (in Japanese)
6. ANRE (Agency for Natural Resources and Energy) (2014) ANRE's initiatives for establishing smart communities policy planning division, http://www.meti.go.jp/english/policy/energy_environment/smart_community/pdf/201402smartcomunity.pdf
7. NEDO (New Energy and Industrial Technology Department Organization) (2010) Renewable energy technology Hodpaper, http://www.nedo.go.jp/content/100107278.pdf
8. Mah DN, Van der Vleuten MJ, Ip JCM, Hills PR (2012) Governing the transition of socio-technical systems: a case study of the development of smart grids in Korea. Energy Policy 45:133–141
9. Bulkeley H (2010) Cities and the governing of climate change. Annu Rev Environ Resour 35:229–253
10. Hodson M, Marvin S (2009) 'Urban ecological security': a new urban paradigm? Int J Urban Reg Res 33(1):193–215
11. OECD (2012) Green technology and innovation: OECD Science, Technology and Industry Outlook 2012. https://doi.org/10.1787/sti_outlook-2012-en
12. Bulkeley H, Broto C (2013) Government by experiment? Global cities and the governing of climate change. Trans Inst Br Geogr 38(3):361–375
13. Saavedra C, Budd WW (2009) Climate change and environmental planning: working to build community resilience and adaptive capacity in Washington State, USA. Habitat Int 33:33

14. Balaban O, Puppim de Oliveira JA (2014) Understanding the links between urban regeneration and climate-friendly urban development: lessons from two case studies in Japan. Local Environ 19(8):868–890
15. Jonas A, Ward K (2007) Introduction to a debate on city-regions: new geographies of governance, democracy and social reproduction. Int J Urban Reg Res. https://doi.org/10.1111/j.1468-2427.2007.00711.x
16. Evans J, Karvonen A (2014) 'Give me a laboratory and I will lower your carbon footprint!'—urban laboratories and the governance of low-carbon futures. Int J Urban Reg Res 38(2):413–430
17. Alawadhi S, Aldama-Nalda A, Chourabi H, Gil-Garcia RJ, Leung S, Mellouli S, Nam T, Pardo TA, Scholl HJ, Walker S (2012) Building understanding of smart city initiatives. In: Scholl HJ, Janssen M, Wimmer MA, Moe CE, Flak LS (eds) Electronic government. EGOV 2012. Lecture Notes in Computer Science, vol 7443. Springer, Berlin. https://doi.org/10.1007/978-3-642-33489-4_4
18. Albino V, Berardi U, Dangelico RM (2015) Smart cities: definitions, dimensions, performance, and initiatives. J Urban Technol 22(1):3–21
19. Ling APA, Kokichi S, Masao M (2012) The Japanese smart grid initiatives, investments and collaborations. Int J Adv Comput Sci Appl 3(7):1–11
20. Markham SK, Ward SJ, Aiman-Smith L, Kingon AI (2010) The valley of death as context for role theory in product innovation. J Prod Innov Manag 27(3):402–417
21. National Institute of Population and Social Security Research (2010) Population Trends in Japan, http://www.ipss.go.jp/pr-ad/e/eng/03.html
22. Cabinet Office of Japan, Environmental Future City Initiative, http://www.kantei.go.jp/jp/singi/tiiki/kankyo/
23. Onishi T, Kobayashi H (2011) Low-carbon cities: the future of urban planning. Gakugei-Shuppan, Kyoto
24. Yokohama Convention & Visitors Bureau (2013) Yokohama Visitors' Guide, viewed 28 May 2013, http://www.welcome.city.yokohama.jp/eng/travel/about/
25. Yokohama City Government (2011) Global warming measures implementation plan, http://www.city.yokohama.lg.jp/ondan/plan/h2603jikkou/pdf/h2603honpen.pdf
26. Ministry of Economy, Trade and Industry (METI) (2010) Yokohama Smart City Project Master Plan, http://www.meti.go.jp/committee/summary/0004633/masterplan001.pdf
27. Yokohama City Government (2015) Yokohama Smart City Project (YSCP) General Meeting Final Report, http://www.city.yokohama.lg.jp/ondan/yscp/image/siryou3.pdf
28. Yoda T, Tanaka M, Itoh K (2017) Smart grid economics. Yuhikaku, Japan (in Japanese)
29. Port RE (2015) Daikyo developed a condominium in Kamio-oka, Yokohama, https://www.re-port.net/article/news/0000041808/

Part IX
Nairobi

Responsive Infrastructure and Service Provision Initiatives Framing Smart Environment Attainment in Nairobi

Romanus O. Opiyo, Silas M. Muketha, Wilfred O. Omollo
and Dennis Mwaniki

Abstract This chapter will seek to document and profile various initiatives in infrastructure development and related services which are deemed smart. The chapter will focus on transport (mobility), water and solid waste management in Nairobi which are seen as either catalysts or potential in attainment of smart environment in Nairobi. Initiatives such as application of Information and Communication Technologies (ICTs) through Kenyan mobile, web and SMS platform, wants to address mobility, water access and waste management challenges in Nairobi. These three sectors are also considered key and basic to all *Nairobians* regardless of their social status, hence their understanding of how they are provided and accessed is key in understanding how smart approaches in their provision and use can have positive effects in meeting the elusive smart environment. The chapter discusses various initiatives in terms of infrastructure provided, services associated with those infrastructure and application of digital technology and how these are likely to support attainment of smart environment.

Keywords Nairobi · Smart environment · Smart mobility · Water · Solid waste management

R. O. Opiyo (✉)
Department of Urban and Regional Planning (DURP), Centre for Urban Research and Innovations (CURI), University of Nairobi (UoN), Nairobi 30197-00100, Kenya
e-mail: romanop2000@gmail.com; ropiyo@uonbi.ac.ke

S. M. Muketha
Department of Urban and Regional Planning (DURP), University of Nairobi (UoN), Nairobi 30197-00100, Kenya
e-mail: smmuketha@gmail.com

W. O. Omollo
Department of Planning and Development, Kisii University, Kisii 408-40200, Kenya
e-mail: wochieng@gmail.com

D. Mwaniki
Urban Planning & GIS in GORA Corp., Nairobi 50497-00100, Kenya
e-mail: dmwaniki@gora4people.org

© Springer Nature Singapore Pte Ltd. 2020
T. M. Vinod Kumar (ed.), *Smart Environment for Smart Cities*, Advances in 21st Century Human Settlements, https://doi.org/10.1007/978-981-13-6822-6_11

1 Overview

The goal of both the global and local commitments such as the Sustainable development Goals and Kenya's vision 2030 is to have sustainable cities that are not only functional and responsive to citizens needs but which also provide opportunities for all—including access to basic services such as water, transport and waste management—which are all geared towards making these cities environmentally clean. Integration of technology into different development sectors results in smart initiatives and offers unique opportunities for sustainable development in different countries. Smart infrastructure provides the foundation for all of the key themes related to a smart city, including smart people, smart mobility, smart water, smart waste management, smart governance and smart digital layers [1]. Attainment of smart environment status and negative pressures associated with urbanization are motivating cities in Africa (including Nairobi) to embrace Information and Communication Technologies (ICTs), which are the main drivers of smart solutions, and which enable such cities to deliver smart infrastructure and services that support attainment of smart environments. The immediate need for cities in developing countries is to provide responsive urban infrastructure to meet the increasing pace of urbanization. In the process of meeting infrastructure demands, smart infrastructure applications provide a way for such cities to achieve leapfrogging in technology [2].

This chapter documents and profiles various initiatives addressing infrastructure and service provision in Nairobi across three sectors—mobility (transport), water and solid waste management. The focus on these three sectors is due to their important role for the city's existence, as well as their ability to contribute significantly to carbon emissions, which is a big threat to attainment of smart living environments. The chapter also highlights some of the smart and ICT driven initiatives and strategies that are being applied in Nairobi, or which can be adopted to enhance efficiency in each of these sectors, towards attainment of sustainable and smart environments.

The enviable smart environment to be attained in this chapter is related to Cook and Das [3] understanding of the concept. They defined 'smart environment' as one that is able to acquire and apply knowledge about an environment and adopt to its inhabitants in order to improve their experience in that environment. They further note that intelligent automation can reduce the amount of interaction required by the inhabitants, as well as reducing utility consumption and other potential wastages [3]. It is out of this concern that this chapter is focusing on various initiatives in the transport, water and waste management sectors In line with technology application in these sectors frame's Nairobi inhabitants experience about attainment of smart environment.

2 Context of Smart Infrastructure and Services in Relation to Attainment of Smart Environment in Nairobi

Nairobi is the capital city of Kenya and an important economic, transport and ICT hub in Africa. Nairobi city was first established as a transportation centre in 1899 because it offered a suitable stopping place between Mombasa and Kisumu cities. Nairobi also had adequate water supply from the nearby Nairobi River and Mbagathi River [4]. The historical association of Nairobi with transport and water is critical in understanding how the city's transport and water infrastructure and associated services are provided and in particular how adoption of smart technology is shaping the provision of these basic services including solid waste management, as this is also one of the major service provision challenges facing Nairobi and other cities in Africa which are seen as hubs of innovations and modernization.

According to the 2009 population census report, Nairobi had a population of 3,138,369 and it is estimated to cover 694.90 km^2 with a population density of 4,516.29 persons/km^2. The Institute of Economic Affairs (2011) indicates that, Nairobi employed 25% of Kenyans and 43% of the Country's urban workers and it generated over 45% of the GDP by the year 2006 [5]. Nairobi's growing population size is both an opportunity and a threat to attainment of smart environment status. The Nairobi population is rich in terms of skilled labour which is critical for accelerated innovations associated with smart ideas in provision of infrastructure and services which eventually will lead to attainment of smart environment. The challenge associated with this growing population is the rapidly growing demand for mobility, water and waste management infrastructure and services which are increasingly becoming a threat to attainment of smart environment in Nairobi. Each of these elements are discussed in depth in the subsequent sections of this chapter.

3 Smart Mobility and Nairobi's Environmental Status

Smart mobility systems include mass transit systems as well as individual mobility systems that feature bicycle sharing, ride sharing (or carpooling), and vehicle sharing and, more recently, on-demand transportation. The availability of, and reliability of such systems in cities is a great asset in accessing other goods and services, but also plays a major role in promoting environmental sustainability due to their associated low levels of energy and space consumption, reduced traffic congestion as well as reduced pollution. This section focuses on the available mobility infrastructure options in Nairobi and how they can contribute to attainment of smart environments. Focus is put in three kinds of transport sectors—road, railway and air transport infrastructure systems.

3.1 Nairobi Mobility Infrastructure

a. Nairobi City Roads and Associated Infrastructure

According to the Kenya Roads Board (KRB) roads inventory and conditions survey conducted in 2009, Nairobi has a total road length of 14,719 km, 53% (7,730 km) of which are paved [6]. Based on the reports assessment of the roads status, 45% of Nairobi's roads are in either fair or good condition, while 55% are in poor condition (Table 1). The poor condition of majority of the city's roads is a major cause for environmental concern, particularly because these can lead either to driving in low speed using heavy gear, or driving in dusty environments, contributing to particulate high emissions of Particulate Matter (PM2.5).

The Nairobi City County Integrated Development draft plan 2018–2022 (CIDP) notes that the City has 300 km of non-motorized transport (NMT) facilities coverage, 39,000 street lights, 52 surveillance cameras, 22 traffic lights management systems (signalized junctions), 198 bridges, 41 overpasses (flyovers), 15 bus terminus and 96 city county traffic marshals, all of which are aimed at management of traffic and protecting and improving mobility through NMT [7]. The CIDP however still describes the current status of provided facilities as inadequate in terms of coverage to meet current and future demands as envisaged in the Vision 2030 [7]. Provision of NMT infrastructure supports green mobility modes such as walking and cycling, which account for nearly half of the city's modal share (Table 2). NMT infrastructure and modes promote environmentally friendly transport as they contribute to very limited amounts of air pollution, greenhouse gas emissions and have minimal noise pollution. Expanding provision of such NMT support infrastructure in Nairobi as envisaged in addressing the missing links identified in Nairobi Transport Masterplan in 2006 (NUTRANS) is likely to contribute significantly in attainment of smart environment in Nairobi city.

b. Nairobi Railway and Associated Infrastructure

Nairobi City has a railway network of 75 km and a total of 15 functional railway stations which are: Embakasi, Makadara, and Nairobi main terminal, Dandora, Githurai, Kahawa, Kibera, Dagoretti, JKIA, Syokimau, Makadara and Imara Daima railway stations. The expansion of Nairobi platform will help to improve public transportation in Nairobi for socioeconomic development [10].

According to the Kenya Railways Corporation (KRC), the Nairobi Commuter Rail Service (NCRS) development is formulated as part of Nairobi Metropolitan Transport Master Plan with the aim of integrating rail transport with other modes of transport, more so road and air transport. The project will be developed in three (3) phases [11] (http://krc.co.ke/nairobi-commuter-rails/).

Phase 1: This will be developed within the existing railway corridors to provide commuter rail services between Nairobi Railway Station and the following destinations: Ruiru, Syokimau, Jomo Kenyatta International Airport, Kikuyu, and Embakasi Village. This development includes the building of 26 new modern passen-

Table 1 Road length and conditions in Nairobi City

Paved road condition	Length in Km	Percentage of paved	Unpaved road condition	Length in Km	Percentage of paved	Cumulative condition	Cumulative (km)	Cumulative percentage
Excellent	13	0.2%	Excellent	12	0.2%	Excellent	25	0.2%
Fair	2,753	35.6%	Fair	1,593	22.8%	Fair	4,346	29.2%
Good	1,819	23.5%	Good	527	7.5%	Good	2,346	15.5%
Poor	3,145	40.7%	Poor	4,857	69.5%	Poor	8,002	55.1%
Total	7,730	100.0	Total	6,989	100.0	Total	14,719	100.0

Source Kenya Roads Board, Roads Inventory Data [6]

Table 2 Modal share in Nairobi

Reference	Public transport (%)	Walking (%)	Cycling (%)	Private car (%)	Train (%)	Institutional bus (%)	Others (%)
NUTRANS 2005 [8]	32.7	47.1	1.2	15.3	0.4	3.2	0.2
JICA 2006 [9]	36.0	47.0	–	16.5	0.4		

ger handling stations at the existing railway stations and at new locations. Syokimau, Makadara and Imara Daima railway station are complete and operational.

Phase 2: This development will also be within the existing railway corridor and will extend commuter rail services to: Thika, Limuru, and Lukenya.

Phase 3: This development will be within new railway corridors and will target the outlying satellite towns such as: Ongata Rongai, Kiserian, Ngong, Kiambu, Ruai and Kangemi.

While the commuter rail system has great potential for providing smart mobility solution to Nairobi residents and visitors, the main challenge has been the slow pace of expansion of the rail infrastructure as well as limited modernization of the stations in an integrated manner. This has made railway transport unattractive to various people using other modes of transport, with particular preference for small-capacity bus transport options which are a threat to smart environment. It is however noted that expansion of the rail lines and modernization of the stations is still one of the most feasible strategies of the city realizing its dream of Mass Rapid Transit (MRT) which has a great potential of operationalizing High Occupancy Vehicles (HOVs), which is a contributor to attainment of smart environment.

c. Airport and Associated Infrastructure

Nairobi city hosts 3 airports; Jomo Kenyatta International Airport, Wilson Airport and Eastleigh Airport. Jomo Kenyatta International Airport (JKIA) is the biggest Airport in East and Central Africa, and is the focal point for major aviation activity in the region. Its importance as an aviation Centre makes it the pacesetter for other airports in the region. JKIA, located 18 km to the East of Nairobi City centre, is served by 49 scheduled airlines. JKIA has direct flight connections to Europe, USA, the Middle East, Far East, and many parts of Africa. JKIA has five cargo facilities with a capacity to handle 200,000 tonnes of cargo annually, and an animal holding facility which occupies 4,318.95 square feet. The Airport has a runway measuring 4,117 m long and 45 m wide on 4,472.2 ha of land [10]. Wilson Airport is the second airport in the County. It has two runways one that is 1,463 m long and 24 m wide while the other is 1,558 m by 24 m with displaced threshold giving a landing distance of 1,350 m [10].

One of the notable challenges facing access and mobility to Nairobi's main airport (JKIA) is poor integration of transport modes. Proposals to link the airport with commuter train services and Bus Rapid Transit (BRT) will help in addressing this major transport missing link. The recently proposed installation of a solar powered cargo facility owned by Swissport Kenya limited and in terminals A, B, C and D of JKIA is a major footstep in enhancing environmental friendly footprints in the transport sector. This will enhance use of renewable energy, which is environmental friendly which is fully supported by Kenya's Energy Regulatory Commission (ERC).

3.2　Nairobi Mobility Services

Nairobi experiences heavy traffic congestion in most roads especially during the morning and evening peak hours. This can be attributed to the increasing number of private vehicles in the city. Between 2004 and 2013, private cars in the city of Nairobi increased from 147,387 to 253,298 cars, representing a 70% increase. In 2014, it was estimated that 46,000 vehicles passed through the city to other destinations on a daily basis; and that 84,000 vehicles moved into the city while 86,000 moved out of the city on daily basis [12]. The rapid growth in the number of private cars is purely attributed to lack of reliable public transport, hence a sizeable portion of the residents opt to invest in owning a private car for trips which could easily be undertaken by public transport. These high numbers of low occupancy vehicles contribute greatly to air pollution through emission of greenhouse gases, which is more significant during peak hours.

Nairobi city, just like many other African cities lacks real public transport services that are purely operated by Government. In Nairobi, apart from Nairobi Commuter Train Services and the a few buses operated by the National Youth Service (NYS), most of the residents rely on privately owned and managed public transport services (para-transit services) commonly known as *matatus*. Currently Nairobi city is lacking public scheduled buses, Bus Rapid Transit (BRT) and Light Rail Transit (LRT) but there are proposals introduce them in the very near future, as provided through establishment of Nairobi Metropolitan Area Transport Authority (NAMATA), which was created through the presidential order, under the legal notice 18 of February 2017.

Motorcycle taxi commonly known as *bodaboda* is also an emerging feature of Public Service Vehicles in Nairobi. Other forms of mobility in the city include the use of the three wheeled auto rickshaw commonly known as *Tuktuk* and use of traditional taxis.

Over the past decade or so, several smart mobility services have emerged in the city of Nairobi, with a cumulative effect of eased movement for commuters. Some of these services include e-hailing taxi services, traffic notification applications, cashless travel services among others as discussed below. Each service targets a unique market in the large pool of commuters, and has varying levels of convenience and cost implication on the user.

(a)　Application of E-Hailing Taxi Services

ICT is an integral part of smart transportation [13]. Current exploitation of infor-mation technologies especially in application of mobile Apps in booking transport services including e-hailing taxi services (such as Taxify and Uber) and car-pooling is changing mobility patterns and options especially for the middle and high income earners in Nairobi who are able to pay for the services and associated conveniences. With internet usage in Kenya projected at over 67%, these ICT capabilities are primary inputs to smart mobility [14]. Such e-hailing services are likely to increase road safety and reduction of private cars moving one individual on the road, hence increased passenger capacities (car-pooling) which will further contribute to reduction in pollution and ultimately lead to attainment of smart environment. Such services are being extended to motorcycles and even online booking of long distance travel buses which reduces unnecessary movements, with a cumulative positive effect to the condition of the environment.

Other E-hailing related applications such as Ubabi Vanpooling society, which is a startup that offers a vanpooling solution for people who go to work in locations close to each other and those who leave work at relatively the same time, can vanpool to and from work. The Society was formed in response to a need to reduce traffic congestion in the City of Nairobi through encouraging private vehicle owners to leave their vehicles at home [15].

Another smart mobility technology is exhibited by Autotruck E. A Limited, which is a green technology company that fabricates and locally assembles electric powered three-wheeler light duty cabs and handcarts. It is set to reduce harmful emissions into the environment while offering taxi services and light transportation in African cities. The Ubabi vanpooling society and Autotruck E. A. Ltd are all part of 6 startups hosted by University of Nairobi C4dlab with the aim of achieving Transformative Urban Mobility Initiative (TUMI). From the orientation and focus of these two startups, the end result is to attain smart environment through various mobility technology and services.

(b)   Digital Matatu

Digital *matatu* is based on data collection using GPS-enabled cell phone applications, which are used to generate valuable information that allows for the development of way-finding tools, both digital and analog, for transit users in the *matatu* (para-transit) sector. The application of the digital matatu is important for reducing movements, trips and time spent by passengers in locating various terminus facilities and bus stops, and in accessing and using of matatus within the city of Nairobi which lead to promotion of smart mobility which contributes positively to attainment of smart environment.

(c)   Cashless Transport Fare system

This system was introduced to both regulate pricing for public transport services in Nairobi, and to also increase efficiency and convenience for commuters, who would load bus fare to special cards and use these to pay for individual trips. The cash-less fare was as a result of the then newly formed National Transport and Safety Authority (NTSA) published regulations instructing all Matatus to introduce and implement cashless fare systems as part of reform agenda for all Public Service Vehicles in Kenya in 2013 [16]. Beyond this, the cashless fare system was a good way of reducing paper money transaction and demand for such which in itself is a contributory factor to clean environment which indeed is a step towards attainment of smart environment.

The service was given special attention due to the choice by Matatu Owners to implement the next generation cross-platform for all Matatus. However, although the system had huge potential for improving services and creating order to both the passengers and the PSV operators and owners, it did not go far as its implementation by the matatu owners was only meant to conform to the requirements of the NTSA regulations.

(d) Proposed Mass Rapid Transit

There is a proposal by World Bank and other agencies to have a Mass Rapid Transit system in Nairobi to deal with growing mobility demand and to address the congestion the City is facing as informed by the demand forecasts in terms of peak hour peak direction traffic (PHPDT). Some of the proposals currently being discussed include 5 Bus Rapid Transit (BRT) corridors and a Light Rail Transit (LRT). The Nairobi Metropolitan Area Transport Authority (NAMATA) was formed purposely to provide for an integrated and sustainable public transport system within the Nairobi Metropolitan which covers 5 counties namely; Nairobi, Kiambu, Machakos, Kajiado and Murang'a.

The relevance of the mass rapid transit system in smart environment is that such systems support efforts geared towards lower energy consumption in terms of increase in fuel use efficiency and reduced greenhouse gas emissions per passenger per kilometer which is associated with high occupancy vehicles. This therefore shows the potentiality of attaining smart environment when this route is followed and embraced.

Other mobility issues related to land use are well articulated in the Nairobi Integrated Urban Development Master Plan (NIUPLAN) which is the first deliberate attempt to integrate land use in transportation planning in Nairobi. It also emphasizes on modal integration which are important contributors to efficient and sustainable mobility which are all key contributors to attainment of smart environment.

4 Smart Water Infrastructure and Services in Nairobi

This section begins with a description of the background to water and infrastructure and services provision. It then describes the current water infrastructure and service situation in Nairobi. This culminates in the analysis of the current water supply and demand nexus in the city. The section further describes various policy and legal interventions before looking at various WIS policies in the National Development plans in Kenya. Challenges facing water service provision in Nairobi have been described before immersing into smart water supply, use and management. The section goes on to outline the application of smart water technologies in Nairobi. Finally, the section ends with a conclusion and recommendation of achieving smart water infrastructure and services in Nairobi.

4.1 Background to Water Infrastructure and Service Provision

Various efforts to water infrastructure and service provision have advanced the supply, demand and market equilibrium as well as the provider and enabler approaches. The provider approach includes massive water infrastructure and service provisions by governments. The enabler approach, on the other hand, is geared towards mobilization of the private sector. The Kenya government from independence in 1963 adopted the provider approach to the provision of Water Infrastructure and Services However, it could not on its own meet the water infrastructure and services needs of a rapidly growing population. The shift in focus gradually led to commercialization and privatization of basic services including water [17].

Smart water infrastructure and services are necessary to avoid service disruptions and bottlenecks, and to also support a range of activities. However, it requires very high capital investment which is way beyond the means of urban residents. This gap makes the provider approach the viable option. Deliberate smart water infrastructure and services in an environment of other underlying factors is prerequisite to sustainability. The impacts of smart water infrastructure and services is the positive contribution that it can make to improve the other dimensions such as social, economic and environmental sustainability. Without embracing Smart water infrastructure and services may cause loss of opportunities to advance socially, economically, technologically or culturally [18].

The current initiatives towards water service provision in Nairobi are mainly twofold: development of new water projects, and enactment of policies and laws on better use and management of water resources. The first initiative includes several activities to increase water availability to the city, so as to bridge an ever increasing demand-supply gap. The ever-rapidly rising city population due to immigration and natural births in the city makes planning for water systems a herculean task. The city's water projects have exacerbated serious social, economic, environmental and even political ramifications to the counties. From the current supply of 444,500 cubic metres of water per day to the city, it means 40% which amounts to 178,200 cubic metres per day is lost. Accounting for the losses would have avoided further development of new schemes. Unfortunately, the losses continue to increase in absolute terms as the water demands increases.

The other initiative includes enactment of policy and laws to enable better use and management of water resources. International policies including the Sustainable Development goals (SDGs) identify access to water as a basic human right, while the World Health Organization (WHO) standards stipulate minimum standards for clean water. This is echoed in the Constitution of Kenya, Chap. 6 on the Bills of Right which states that "every citizen has a right to clean water" (Kenya, 2010). However, policy, legal and institutional initiatives have not comprehensively resolved the water problem in the city. The Water Act was enacted in 2002 to provide a framework to address the challenges of management of water supply in the country. The Act provided in its framework, various institutions at the national, regional and local

Table 3 Main water sources in Nairobi

S/no	Water facility	Details
1.	Thika Dam	Located in Ndakaini (Thika District). Completed in 1994 Utilises water from Thika river but is also linked to chania river through a 4 km tunnel which serves Ngethu Storage capacity is 70 million m^3
2.	Ngethu water works	Located in Gatundu North. Ngethu Phase 1: Commissioned in 1974, nominal design output is 61,000 m^3/day Ngethu Phase 2: Commissioned in 1984. Nominal design output is 157,000 m^3/day Ngethu Phase 3: Commissioned in 1997. Nominal design output of 222,000 m^3/day. Ngethu treatment plant is designed to produce a rated nominal output of 440,000 m^3/day of potable water (production)
3.	Sasumua Dam	Located in Njambini, Nyandarua. First stage completed in 1955 and second in 1968 Dam on Sasumua River but receives Kiburu and Chania waters Storage capacity is 15.9 Million m^3. Pipeline to Kabete is 60 km. Design yield is 59,000 m^3/day Current yield is 59,533 m^3/day
4.	Ruiru Dam	Located in Gthunguri, Kiambu Completed in 1950 Dam is on Ruiru River Storage Capacity is 2.90 million m^3. Pipeline to Kabete is 25 km Stores raw water with yield of 21,600 m^3/day The current yield is 19,343 m^3/day
5.	Kikuyu springs	Springs near Magana Flowers in Kikuyu. Has three springs Treatment is by chlorination only The yield is 4,000–5,000 m^3/day Pipeline to Nairobi is 10 km

Source NCWSC [51]

levels. Water Service Providers (WSPs) have the responsibility of providing water and sewerage services at the local level. This is the basis of creating the Nairobi City Water and Sewerage Company (NCWSC) (Table 3). On the other hand, Water Services Boards were created by the Act at the regional level to develop new water and sewerage infrastructure. This is the case of Athi Water Services Board (AWSB) which serves Nairobi and the neighbouring County of Kiambu. Despite the water sector reforms, the city of Nairobi City has over the years been experiencing, and still experiences serious shortage of water supplies. Even with the enactment of Water Act 2016 in line with the Constitution, the problem of water in the city of Nairobi still remains unresolved. Implementation of innovative technological solutions should be considered as a core component in the introduction of smart water services in Nairobi.

Table 4 Total current water supply to Nairobi with 444,500 cubic meters per day

Source	m³/day
Thika (Ndakaini Reservoir)	360,000
Sasamua reservoir	60,000
Ruiru reservoir	20,000
Kikuyu springs	4,500
Total supply	444,500

Source NCWSC [51]

4.2 Water Infrastructure and Service Situation in Nairobi

From Kenya's independence in 1963 until the 1970s, the main source of water for Nairobi was the Kabete project. The first and second Nairobi water supply projects (WS I and WS II) developed the new Chania Scheme to improve and expand the distribution of treated water. The third Nairobi water supply engineering project which was commissioned in 1985 realized the Thika dam. The dam has a storage capacity of 70 million cubic meters. With the two water projects, the gross water availability increased from 165 to 200 litres per capita in the period from 1976 to 1995 [19]. The supply generally matched the demand though the population had increased threefold during the same period. The fourth project which is ongoing is the Northern Collector Project (NCP). This new project aims to serve the city of Nairobi and 13 satellite towns up to the year 2035. The NCP aims at tapping about 40% of flood water to ease some of its adverse effect to the environment. NCP will provide an additional 140 million litres of water to the city and its environs. The distribution network for the city of Nairobi receives treated water from Kabete, Kyuna, Kiambu and Gigiri reservoirs. The distribution area is segmented into 13 zones based on the reservoir supplying the water to the zone. Pipes are densely installed in the Western area of Nairobi City and sparsely installed in the Eastern area [19]. Figure 2 presents a diagrammatic view of the water sources and production capacities for Nairobi.

4.3 Current Water Supply and Demand Nexus in Nairobi

According to the water resources management authority (WRMA 2010), Nairobi City County has a water deficit of approximately 200,000 cubic meters per day. The current estimated water demand for Nairobi is 650,000 m³/day compared to the production of 482,940 m³/day [20]. According to Athi water services Board (AWSB), the city depends on supply from five water sources, with the bulk of water supply coming from Thika, Sasmua and Ruiru Dams. The other source is the Kikuyu Springs. In 2014, these projects supplied the city of Nairobi (Table 4).

Table 5 Water demand forecast

Year	Water demand (m³/day)	Water supply (m³/day)	Gap (m³/day)
2000	363,400		
2005	450,200		
2010	557,700	444,500	113,200
2020	806,600	444,500	362,100

Source NCWSC [51]

From Table 5, it is clear the water demand is rapidly rising against a fixed supply which has led to water supply rationing in the city. An Environmental Impact Assessment carried out by AWSB in 2009 revealed that 50% of Nairobi's population living in informal settlements, have water consumption of about 34, 500 m³/day. This amounts to only 8% of NCWSCs total daily supply to Nairobi City. The report by AWSB (2009) also indicates that about 75% of the city residents get water from pushcart vendors and kiosk resellers. According to the report, the vendor's prices ranged between KES 5 and KES 10 per 20-liter plastic jerry can. This amounted to between KES 250/m³ and KES 500/m³, with an average of KES 375/m³ which is 7 times more than the official regulated rate of KES 53/m³ for domestic consumers.

4.4 *Legal Intervention*

The Kenya government embarked on improving access to safe and clean water right from attainment of independence. The main theme initially was "water for all by the year 2000", however, the government missed the target and only an estimated one million out of a total population of 31 million had access to clean water in the year 2000 [21]. To overcome the problems in the previous water Act 372, the government moved towards commercialization of WIS services in line with the structural adjustment programmes through enactment of water Act 2002. The Act provided for the conservation, control, appointment and use of water resources in Kenya. The Water Act 2002 favoured commercialization which required that local authorities form autonomous water and sewerage companies with independent boards of directors to provide water services. The revenue received from water sales was expected to be reinvested in WIS service improvement. However, in line with the Kenya Constitution of 2010, the Water Act 2002 was repealed and replaced with Water Act 2016 which was aligned with the constitution.

4.5 Challenges Facing Water Service Provision in Nairobi

The City Momentum Index (CMI), developed by [22], rated Nairobi as number ten in the globe and as the most dynamic city in Africa. While innovation and technology were some of the determinants used to rank the cities, the technologies have not been holistically embraced in the water sector. Despite the ranking, Nairobi is facing serious water supply, use and management challenges more so involving water mismanagement arising from old and dilapidated infrastructure, poor billing systems, and illegal connections. According to Republic of Kenya [23] the main challenges facing water infrastructure and service provision are; control of unaccounted for water (UfW), Inefficiencies in old and dilapidated distribution network and demands for water & Sewer rehabilitation and expansion. Other challenges include old billing and customer management system that have inherent functional problems that hamper efficiency in operations and manual operational processes. Other challenges cited by NCWSC are obstruction and encroachment on to way leaves for water and sewer lines. For example, there are structures built on top of water and sewer lines. Also cited are problems of Illegal water and sewerage connections, destruction of sewer manholes to access sewerage for farming and staff over-establishment of over 500 employees. Political interferences, Massive Rural—Urban migration and Massive unplanned developments have also been cited.

4.6 Smart Water Supply, Use and Management

Shahanas and Sivakumar [24] Describes Smart water as a combination of intelligent water infrastructure with data analytics in a way that leads to actionable information [24]. Choi et al. [25] Defines a smart water city as a city that uses integrated water resource management together with smart ICT technologies. Integration can promote coordinated development and management of water, land and related resources. As a result, it can maximize economic and social welfare in an equitable manner without compromising the sustainability of vital ecosystems. According to Britton et al. [18], the architectural components of an ISWM includes stakeholder integrated information system (SIIS), decision support system (DSS), and ICT infrastructure and control.

Smart Water Management is a water management strategy capable of integrating and managing the entire process of the water cycle from analysis of current situations to purification, distribution, as well as using and recycling of water resources scientifically and systematically [25]. Ensuring adequate provision of water to city users is a major challenge, especially in regions experiencing a sustained high influx of urban residents and high per capita consumption. As clean and safe water is fundamental to any socioeconomic development, city leaders need to adapt smart water management and their reliance on technologies to support sustainable development and cater for rapidly growing water needs [26].

Smart Water Infrastructure has the potential to contribute significantly towards improved service delivery and efficiency of water services providers; reducing costs and water losses, streamlining operation and maintenance, and improving data and asset management in Water Service Providers, allowing for information-based decision making [27]. Smart Water Service entails improving water efficiency for the benefit of the city and its residents [26]. This is through using data and technologies as enablers, to increase efficiency of irrigation networks and reuse of water, allow for deferred investment in water production plants and optimization in pipe renewal, as well as preserve water resources by enhancing leakage management and enabling demand response programs (Veolia, n.d.).

Smart Water Service if applied in Nairobi would help curb the current unaccounted-for water crisis by managing leakages and reservation of water sources. Some of the smart water initiatives that have been implemented in Nairobi include smart water metering, and integration of various technologies for enhanced water systems as discussed below.

(a) Smart Water Metering (AMR/AMI)

Smart metering is a component of the smart water management that allows a utility to obtain meter readings on demand (daily, hourly or more frequently) without the need of manual meter readers to transmit information. Smart Metering has three components:

Automated Meter Reading (AMR) is a technology that automatically collects consumption data from a water meter or energy metering device [28]. This data is used for billing purposes; to Analyses usage and manage consumption, and to identify or resolve technical problems. AMR systems promote more cognizant water usage by revealing exactly how a site is using resources, and where reductions can be made to improve efficiency and lower costs. AMR continually gathers data and can provide this information on a real-time basis. Usage data can be viewed at any time, and once collected, is immediately stored in a repository of historical consumption information for comparative or analysis purposes.

Automated Metering Infrastructure (AMI) a fixed network system, with smart meters providing two-way communications between the water meter and the utility [27]. AMI improves the efficiency of water utilities and eliminates the costs of routine meter reading. Combined with GIS meter data management, AMI increases the accuracy and precision of the meter reading and reduces re-reads. The result is an accurate and timely readings that are ready for billing, with an identification of failed and failing meters before actual billing, improving the utility's cash flow.

Automated billing system involves taking the meter number and generating a bill for that meter by use of the database information collected from the meter reading [27]. When installed in Nairobi it would help in achieving SWM by reducing cases of billing errors thus making the system efficient.

According to Arniella [27] smart meters can benefit the water utility, the environment, and the utility's customers by: Lowering the cost of meter reading by eliminating manual meter reading; Enhancing employee safety by reducing the number

Fig. 1 Automated meter.
Source KAPS [29]

of personnel on the road; Reducing billing errors and disputes; Monitoring the water system in a timely manner; Enabling flexible reading schedules, reducing delays in billing of commercial accounts; Providing useful data for balancing customer demand; Enabling possible dynamic pricing; Benefiting the environment by reducing pollution from vehicles driven by meter readers; Assessing Non-Revenue Water in real time or short intervals; Facilitating the data to establish the night water consumption patterns, analyzing the minimum night flows (MNF), and offering a more detailed feedback on water use patterns; Enabling customers to adjust their habits to lower water bills; Providing real-time billing information, reducing estimated readings and re-billing costs; Reducing customer complaint calls and increasing customer satisfaction; Improving the monitoring of potential meter tampering and water theft; and Detecting water line leaks sooner, so they can be repaired faster.

While smart meters have many benefits, they also present challenges to water utilities, customers, and the environment. Installation of these require Front-end capital investment; Long-term financial commitment to the new metering technology and related software; Ensuring the security of metering data and preventing cyber-attacks; Transitioning to new technology and processes with proper training; Managing public reaction and customer acceptance of the new meters; Managing and storing vast quantities of metering data and Disposing of the old meters.

Kenya Airports Parking Services (KAPS), Smart Water Metering System (SWMS) is the first of its kind in Kenya. According to Kenya Airports Parking Services (KAPS) [29] (Fig. 1), a Smart meter design for a SWMS is basically made of three building blocks:

Advanced/automated metering infrastructure(AMI)
Automated Meter reading(AMR) devices
Automated Meter Management(AMM) platform.

KAPS brings out the concept of automated meter management platform. This platform brings together the benefits of simplified water reading with the ease of managing data. KAPS has used a customer-oriented simple interface where; the water utility company can perform all technical measurement settings and customize functionalities of the system to suit their requirements. Finally, the customer can be able to log in and view analytical data of their water consumption [29].

(b) Internet of Things (IoT) for Smart Water Infrastructure and Services in Nairobi

The Internet of Things (IoT) refers to the use of intelligently connected devices and systems to leverage data gathered by embedded sensors and actuators in machines and other physical objects. [30] Points out that the definitions of the IoT vary depending on the context, the effects and the views of the person giving the definition. However, most of the definitions related to this vision have much in common, such as, the ubiquitous nature of connectivity, the global identification of everything, the ability of each thing to send and receive data across the Internet or the private network they are connected into [30].

The IoT application provides an efficient control and monitoring approach for water utility in order to reduce the current water loss. IoT uses internet and web-based applications that can help water utility operators improve water management systems. The IoT could prove to be one of the most important methods for developing more utility-proper systems and for making the consumption of water resources more efficient. Water demand in Nairobi has continuously increased over time arising from factors like population growth, spatial growth of the city or even lifestyle changes.

The solutions to the water problem in Nairobi are twofold. First, is supply-oriented solution that identifies and exploits new water sources. This has been the practice in Nairobi with new sources being more than 50 km from the city. If this trend continues the next sources will be from Mt. Kenya about 150 km away or even Lake Victoria 400 km. This of course has its own socio economic and environmental challenges. Secondly, is the demand-oriented solution that would consist smart exploitation, use and management of scarce water resource [31]. Reporting for the business daily, and further observed by the CEO of NCWSC, automated metering devices, are to be installed for the top 10,000 water users. NCWSC aims to connect 90,000 of its quarter of a million customers to the smart meters in five years, hoping to rid its system of illegal connections and inaccurate billings. However, Connecting 90,000 of a possible 250,000 water consumers against a population of 4 million is an indication that the problem is far from being solved. The missing link is the about 60% residents in informal settlements who have no connections and therefore do not pay for water. Moreover, procurement and installation of such meters require embracing a holistic approach that integrates different aspects of smart WIS.

There are two important characteristics associated with recent trends in water supply management; Functional integration and geographical distribution. These elements must be connected and automated in a proper way so that the entire system can be operational and functional. This would best done through innovations like the IoT. IoT can be used to enable real time water quality monitoring system.

Water quality monitoring is defined as the collection of information about the physical, chemical and biological characteristics of water at set locations and at regular intervals in order to provide data which may be used to define current conditions, establish trends, etc. [31]. Water quality is affected by both point and non-point sources of pollution, which include sewage discharge, discharge from industries, runoff from agricultural fields and urban run-off. Other sources of water contamination include floods and droughts and due to lack of awareness and education among users. The need for user involvement in maintaining water quality and looking at other aspects like hygiene, environment sanitation, storage and disposal are critical elements to maintain the quality of water resources. Main objectives of online water quality monitoring include measurement of critical water quality parameters such as microbial, physical and chemical properties, to identify deviations in parameters and provide early warning identification of hazards. Also, the monitoring system provides real time analysis of data collected and suggest suitable remedial measures [32].

(c) GIS application for Smart Water Infrastructure and Services in Nairobi

Geographic information systems (GIS) are an organized collection of computer hardware, software, and geographic data, supported by trained personnel to efficiently capture, store, update, manipulate, analyze, and display all forms of geographically referenced information. GIS operates primarily on spatial data but also include events and specific data [27]. One of the most applications of GIS is network analysis where the geographical phenomenon is analyzed in context of network such as streets, water network pipe, telephone and electric line and so on.

GIS technology can be used to achieve optimum management of water in Nairobi. Mapping of water resources is therefore increasingly important for effective water utility planning, use and management. Access to sufficient information of urban water distribution network is very difficult majorly because the establishment of urban water network is underground. Geo-referencing of the underground utilities and saving them in GIS geodatabases is the best way of locating them. Priorities in water management start with basic information; data is collected, imported, saved and managed in GIS [33]. Secondly, the water distribution networks are modeled and analyzed using GIS network analysis tool. Supported with comprehensive data, GIS is a very powerful technology for analyzing water utilities in critical conditions.

Water quality assessment with aid of GIS and IoT, are powerful tools for establishing relationships on water quality between impacts due to natural as well as human activities and their effect. GIS can be applied to analyze selected indicator parameters of existing water quality data. The analyses results show that some of indicator parameters of the water sources and distribution are spatially or regionally variable while others are randomly distributed [34]. Interpretations and assessments of these results then assists in getting information with respect to water quality conditions and status of sources and distributions [33]. This information may be used for implementing control measures and improvement to be in place.

(d) ICT Application to Smart Water Infrastructure and Services in Nairobi

As Kenya gears itself towards meeting Vision 2030 goals, scarce water resources in Nairobi puts pressure on the capital [35]. Finlay et al., goes on to point out that the application of ICTs in water management would need to employ both mobile technologies used at the local level, and the roll-out of broadband infrastructure which together would hold strong potential for water management. According to

Ndaw [36], potential benefits of integrating ICTs in water and sanitation projects as has been achieved in Nairobi are as follows:

Reduces the duration and costs of monitoring and inventory activities. Accurate data and information management systems are a precursor for sound management and decision support systems.

Improves efficiency gains of water service providers. ICTs can enable shortened response time, reduce travel distance and maintenance costs, optimize operations (production costs, energy efficiency etc.) and improve quality of service.

Improves collection rates of water service providers through ICT based-payment systems. Some of the most common ICTs adopted by utilities are e-payment systems which offer payment facilitation and increased reliability in billing and payment recovery, reduced administrative and payment transaction costs, and improved revenue collection. The Kiamumbi Water Trust (KWT) in Kenya established an M-PESA payment system in December 2010, enabling 550 households to settle their monthly water bills via mobile phone.

Ensure better services to the poor. Mobile phones, especially, are particularly well placed to serve the development needs of the poorest and most vulnerable populations. In Kenya, "Jisomee Mita" is an application that enables water consumers to use a mobile phone to query and receive current water bills, at a frequency of their convenience.

Strengthen citizen voice and accountability framework. ICTs can be used to promote public participation and create a system of transparency and accountability. "MajiVoice", a platform for communication between citizens and utilities, was successfully tested in Nairobi and enabled complaints rose from 400 to over 4,000 per month and 94% of submitted complaints closed from 46% in initial months. Other initiatives include Maji SMS and Maji Data.

Short Message Service (SMS) notifications and access to internet-based water bill are tools that allow water users and system managers to understand current water systems conditions and make informed forecasts. Other ICT initiatives include applications such as billing System, Meter Reading System (MRS), Customer complaints Management system (e.g. MajiVoice), Financial Management system, Procurement system, Dam Monitoring System, customer information management, automated communication with customers, water quality management and being developed is the Laboratory Information Management system (LIMS) for water quality [37].

5 Smart Waste Management

An increase in global population coupled with economic growth has resulted to an increase in the amount of wastes generated each year. In 2011, amount of wastes discharged globally was estimated at 10.4 billion tons, which is projected to increase to 14.8 billion tons by 2051. However, the amount generated by developing countries accounts for 56% of the world's total waste. Some of the problems associated with wastes generated from these countries include poor collection and low transportation capacity, open dumping, improper treatment of hazardous wastes [38]. The Sustainable Development Goal 11 (SDG 11) is on making cities inclusive, safe, resilient and sustainable. To attain this goal, one of the targets set is to reduce the adverse per capita environmental impact of cities by paying special attention to air quality and municipal and other waste management by the year 2030 [39]. This evidently sets the agenda for pragmatic, innovative and sustainable Solid Waste Management (SWM) systems. This section explores the potential for applying smart technologies in SWM in Nairobi.

5.1 Situational Analysis in Nairobi

Solid waste generation in Kenya has been increasing significantly, attributable to rapid urbanization. The current amount generated is about 4 million tones/year, and is projected to double by 2030. However, the rise in waste generation has not been accompanied by an equivalent increase in the capacity of county governments in dealing with the challenge [40]. As regards legal framework, issues pertaining to SWM are addressed under various legislations. Article 42 of the Constitution of Kenya (2010) [41] confers every Kenyan with a clean and healthy environment. Article 2 of the Fourth Schedule of the Constitution further gives the County Governments the responsibility for SWM. Regulation two (2) of the Environmental Management and Coordination (Waste Management) Regulations (2006) [42], additionally requires any person whose activities generate waste, to collect, segregate and dispose or cause to be disposed of such waste in a sustainable manner. Concerning physical planning, Part Three (Form PPA1) of the Physical Planning Act (Cap 286) [43] states deposit of solid wastes on land constitutes development and as such need to be controlled. Owing to this, the Act requires developers to state the method of solid waste disposal upon application for development permission. In 2015, the National Environment Management Authority (NEMA) of Kenya developed the National Solid Waste Management Strategy (NSWMS) [44]. The strategy seeks to establish a common platform for action between stakeholders to systematically improve SWM in Kenya. The main guiding principle is "zero waste principle" where wastes are seen as a resource that can be harnessed to create wealth, employment and reduce pollution of the environment. NSWMS is, however, not particular on any new and innovative smart strategy that may be perused towards sustainably addressing the prevailing problem of SWM in Kenya.

Despite the existing legal framework, to date, Nairobi City is still contending with the challenge effective SWM. For instance, a study done by UN Habitat [8], found that 30–40% of waste generated in Nairobi is not collected and that only 50 percent of the population is served. According to the NSWMS [44], although Nairobi City generates approximately 2,400 tons of waste per day, 65% remains uncollected. This could however be more owing to the ever rising population induced analogous increasing industrial activities from many sectors of the city. Currently, the end disposal of Nairobi's waste is open dumping at a site located at Dandora, in the

Fig. 2 Garbage trucks drive to the Dandora dumpsite in Nairobi. The dumpsite was declared a health hazard for the neighbouring residents in 2001, but chemical, hospital, industrial, agricultural and domestic waste are still dumped here and left unprocessed. *Source* Koech [48]. Adapted from Nairobi City Water and Sewerage Company Limited-Strategic Plan 2014/15—2018/19

Eastland's section of the city, located 7.5 km southeast of the Central Business District (CBD), covering an area of 26.6 hectares. Although it is several kilometers away from the CBD, the rapid growth of the city's population has resulted in settlements encroaching upon the dumpsite.

Areas adjacent to the Dandora dumpsite (Dandora, Korogocho and Kariobangi estates) experience several problems caused by various activities carried out at the dumpsite. These problems need to be addressed as a matter of urgency to save the deteriorating situation. As observed [45], the situation is giving the county government officials sleepless nights as they seek ways to tackle the garbage collection problem. Most city residents have tales of heaps of uncollected garbage lying in their neighbourhoods. The challenge is no different on some of the streets in the city centre, where piles of garbage remains uncollected for weeks, limiting use of the streets and emitting a foul smell.

5.1.1 Previous Policy and Strategy Interventions

There have been several attempts in the past to address the problem of SWM in Nairobi. From 1996 to 1998, the Government of Japan carried out a study on SWM in Nairobi City through its technical assistance programme. This study included the formulation of a Master Plan composed of a collection and transportation plan, a waste reduction and recycling plan, and a final disposal plan. The report recommended for institutional and legal restructuring plan, a private sector involvement and financial improvement plan, a waste collection system improvement plan, and a construction plan of a new final disposal site were proposed [46]. In 2010, a study entitled, "Preparatory Survey for Integrated Solid Waste Management in Nairobi City in the Republic of Kenya" was also undertaken by Japan International Cooperation Agency (JICA) in conjunction with the City Council of Nairobi. The study sought to review the current situation of SWM in Nairobi City and revise the Master Plan prepared in 1998. Among the recommendations included construction of new sanitary landfill site and closure of Dandora dumpsite; setting up of the Preparatory Unit (PU) in the Department of Environment (DoE) for the SWM Public Corporation (SWMPC); drafting of By-law for the establishment of SWMPC; and preparation for the introduction of the step-wise franchise system [38]. The recommendations were however largely not implemented.

Based on these earlier proposals, and in an attempt to address the earlier limitations, a preparation of a report, "Integrated Solid Waste Management Plan for the City of Nairobi, Kenya 2010–2020", was commissioned by UNEP and developed by a team constituting representatives from, among others, the City Council of Nairobi, and the University of Cape Town and University of Nairobi. Among the recommendations made were reducing waste quantities by introducing policies and instruments that regulate wasteful behaviour; extending resource recovery, both in terms of materials and energy, through source separation as an essential component of sustainable waste management and building environmentally sound infrastructure and systems for safe disposal of residual waste, replacing current disposal sites which need to be rehabilitated [47]. In the same year (2010), another study entitled, "Solid Waste Management in Nairobi: A Situation Analysis", was also undertaken by the United Nations Environment Programme (UNEP) on behalf of the City Council of Nairobi [47]. The study aimed at preparing technical document for the Integrated Solid Waste Management (ISWM) to explain in more detail the thinking, rationale, calculations, modelling and assumptions made in the development of the Specific ISWM Actions. Among the intervention areas proposed in the report included reducing waste generation at source; getting general waste collection and safe disposal right; zoning of waste collection; and waste separation at source with incentives. Figure 2 illustrates a key challenge facing SWM in Nairobi.

Further, in 2015, the County Government of Nairobi Assembly enacted the Nairobi City County Solid Waste Management Act, one of its kind in Kenya, in an attempt to address the challenge [49]. Its objective is to make provision for the management of solid waste in the county and for related matters. The Act under Section 17 makes it the duty of every occupier or owner or agent of a house, or other premise to clean or cause to be cleaned ten metres radius around his or her house or other premises or any area otherwise in his or her control, but which shall not include a main road or street. It moreover under section 18 commits the county government to provide appropriate waste containers for the disposal of solid waste in the public streets and other public places. Regardless of this noble initiative that was as well operationalized through comprehensive regulations, the problem of SWM still escalates in the city.

The foregoing review demonstrates that although several studies have been conducted on how the problem of SWM in the City of Nairobi could be sustainably addressed, most of them are limited to general administrative and functional areas such as waste reduction; collection; enforcement and financing. A scarcity in strategy and literature therefore exists on how the contemporary technological advances may be used to provide other alternative strategies in SWM.

5.2 Towards Smart Waste Management in Nairobi City

a. Adoption of Remote Sensing and GIS

GIS and Remote Sensing are indispensable, efficient and low cost tools in the study of environmental and developmental issues in recent times [50] Advances in remote sensing and image analysis techniques have made it possible to accurately and cost effectively study and measure variables of an extensive area of land in real time. This applies to SMW. Making use of a combination of remote sensing and field mapping, it is possible to implement spatial and temporal modeling to undertake the waste generation and disposal patterns in Nairobi in support of effective SWM strategies. This has a potential for contributing towards economic and environmental savings through the reduction of travel time, distance, fuel consumption and pollutants emissions. In particular, the techniques can be integrated to attain the following aspects of SWM, which are a basis for more complex analysis of patterns, and formulation of workable strategies:

(a) Mapping the location and spatio-temporal extent of unplanned solid waste disposal sites;
(b) Quantifying the amount of solid waste generated;
(c) Based on generation rates, predicting future amount of wastes likely generated; and
(d) Determining the optimal or least cost path/route towards the disposal sites.

Currently, the County Government of Nairobi lacks a database and information indicating the spatial location of solid waste dumping sites (legal or illegal) in the city's neighbourhoods. The use of GIS technologies thus provides the city with an opportunity for effective solid waste management. For example, using these smart mapping technologies offers the city authority a system to identify the most suitable locations for solid for waste collection sites within neighbourhoods, in consequence promoting efficiency in service delivery and budgeting. In addition, once mapping has been undertaken, it is possible to quantify the amount of solid waste generated from each site as well as to determine the optimal or least cost path/route to the disposal sites.

b. Adoption of ICTs in Monitoring Waste Generation

In this chapter, it is conjectured that the challenges facing SWM in Nairobi may be timely responded to through a further adoption of "smart waste management" concept. This is because one of the outstanding problems that epitomizes several neighborhoods of the city is that in cases where public bins are provided, there is usually no monitoring mechanism put in place by the County Government to determine how well these receptacles are used. The ideal requirement is that as soon as receptacles are filled, they should be emptied to avoid disposal of wastes on the ground. Generally, monitoring of solid waste generation, their collection from designated public receptacles and transporting them to disposal sites is usually a costly and time consuming undertaking. As such, the County Government invests a lot of resources in both collection and transportation of solid wastes.

It is thus suggested that with the advanced ICT development and applications in Kenya, the opportunity should be embraced as one of the options for providing an effective SWM in Nairobi, especially at the storage point where residents should have access to public solid waste collection bins. The objective is to monitor the extent at which the bins are used, thereby avoiding the time consuming process involved in physically monitoring if they are full or collecting wastes from bins which are not full, thereby escalating transportation costs.

A proposal for "smart waste management" application which can be readily integrated with android enabled smart phones is recommended. This entail investing in *intelligent* bins strategically placed in neighborhoods and fitted with ultrasound sensor nodes, programmable Interface Controllers, linked with Global System for Mobile Communications and GPS (Global Positioning System) to notify the server database on the coordinate and status of each bin. The database will maintain the details of each bin such as location, capacity, temperature and humidity. Since the bins have sensors, when users deposit wastes into them, their threshold levels will be continuously monitored at the Control Centre through a Graphical User Interface. In this case, as each bin nears their maximum holding capacity, the signal transmitted raises alarm on the need to empty them in order to prevent overflow, a major determinant for unsustainable open dumping around the bins leading to environmental degradation. Additionally, since the spatial location of each bin can be determined through GPS, an optimal least cost transportation path can be determined making it easier to locate a bin that is due for emptying. Moreover, through an android based mobile application, residents may query the system to determine the direction to the nearest bin within the neighborhood, including their status.

Monitoring the capacity of bins through of sensors presents an opportunity for leveraging on a more efficient SWM in Nairobi as it provides a system that monitors the status of the garbage bin and provides information to the concerned authorities to plan and manage the collection intervals from the bins. As observed in Nairobi City, one of the reasons why sold wastes overflows from bins is because they are not emptied at planned interval's.

6 Conclusion

The discussions in this chapter indicate that Nairobi faces huge environmental challenges from the existing infrastructure services that range from unsustainable mobility systems and unbalanced demand for water and goods increasing the environmental footprint, to challenges of solid waste management. A diversity of ICT-driven innovations which constitute smart systems are available at the global and local levels that can significantly address the prevailing challenges, while also creating more opportunities for rapid growth towards environmental sustainability.

Though globalization and modernization are the key drivers of smart infrastructure and associated services, they are yet to be fully appreciated in relation to how they can enhance attainment of smart environment, hence are purely seen as initiatives geared towards improving access and quality of life of citizens living and working in Nairobi. This also has implications related to the digital divide, where some services are associated such as smart taxis are associated with costs which is likely not to benefit the urban poor who at times are forced to walk long distances to access services due to lack of transport fees associated with alternative means of transport or forced.

This chapter finally concludes that the role of smart infrastructure and related services in attaining smart environment in Nairobi will be heavily dependent on the deliberate effort by various level of governments to link the two through policies, financial budgets and plans and proper management and governance of such projects which should also be based on affirmative in order to take the interest of the urban poor and other vulnerable groups such as those living with disability, children and the elderly among others. This will lead to embracing of the infrastructure and related services by many whose cumulative effect will be positive in terms of attainment of smart environment in Nairobi.

References

1. Mwaniki D (2017) Smart city foundation, the core pillar for smart economic development in Nairobi. In: Kumar TV (ed) Smart economy in smart cities. Springer, New Delhi, p 1094
2. Deloitte (2014) Africa is ready to leapfrog the competition through smart cities technology. Available at http://www2.deloitte.com/content/dam/Deloitte/za/Documents/publicsector/ZA_SmartCities_12052014.pdf
3. Cook DJ, Das SK (2006) Designing smart environments: a paradigm based on learning and prediction. https://eecs.wsu.edu/~cook/pubs/wmsn06.pdf
4. Owuor SO, Mbatia T (2008) Post independence development of Nairobi City, Kenya. Paper presented at Workshop on African Cities, Dakar in Senegal. 22–23 September 2008. https://profiles.uonbi.ac.ke/samowuor/files/2008_dakar_workshop.pdf
5. Institute of Economic Affairs (IEA) (2011). Nairobi city scenarios
6. Kenya Roads Board (KRB), Roads inventory data
7. Nairobi City County (2018) Nairobi City County Integrated Development Plan (2018–2022)
8. NUTRANS 2005: Nairobi Urban Transport, 2005, The World Bank
9. JICA (2006) The study on master plan for urban transport in the Nairobi metropolitan area

10. Nairobi County Government (2018) Draft Nairobi County integrated development plan (2017–2022)
11. http://krc.co.ke/nairobi-commuter-rails/
12. JICA (2014) JICA strategy paper on solid waste management. Retrieved from http://gwweb.jica.go.jp/km/FSubject1801.nsf/ff4eb182720efa0f49256bc20018fd25/1710a8711dffc41b492579d4002d25af/$FILE/JICA_Strategy%20Paper%20on%20Solid%20Waste%20Management_2014Edtion(Eng).pdf
13. Mitullah WV, Opiyo RO (2016) Mainstreaming non-motorised transport (NMT) in policy & planning in nairobi: institutional issues and opportunities. J Sustain Mob (JSM) 3(1)
14. Opiyo RO, Mwau BC, Mwaniki DM, Mwang'a KM (2017) Attaining E-democracy through digital platforms in Kenya. In: Vinod K (ed) E-democracy for smart cities, advances in 21st century human settlements. Springer Nature Singapore Pte Ltd, pp 441–460. http://www.springer.com/gp/book/9789811040344
15. https://ubabivanpooling.com
16. Mwaniki C (2014) Matatus sign to M-Pesa ahead of cash fare ban. Business Daily Africa. Nation Media Group. Retrieved 16 Dec 2017
17. Hancke GP, Hancke GP Jr (2012) The role of advanced sensing in smart cities. Sensors 13(1):393–425
18. Britton TC, Stewart RA, O'Halloran KR (2013) Smart metering: enabler for rapid and effective post meter leakage identification and water loss management. J Clean Prod 54:166–176
19. Nairobi City County (2014) The project on integrated urban development master plan for the city of Nairobi in the Republic of Kenya Final Report-Part I: current conditions. Technical Support from Japan International Cooperation Agency (JICA)
20. Water Resources Management Authority (2010) Environmental impact assessment report
21. Wambua S (2004) Water privatization in Kenya. Global Issue Papers, No. 8
22. Jones Lang LaSalle (JLL) (2017) City momentum index 2017 edition
23. Republic of Kenya (2002) Water act 2002: laws of Kenya. Government Printer, Nairobi
24. Shahanas KM, Sivakumar PB (2016) Framework for a smart water management system in the context of smart city initiatives in India. Procedia Comput Sci 92:142–147
25. Choi GW, Chong KY, Kim SJ, Ryu TS (2016) SWMI: new paradigm of water resources management for SDGs. Smart Water 1(1):1–12
26. Veolia Approach to Sustainable and Smart City-smart water (n.d.) Retrieved from mea.mkg.smarthub@veolia.com
27. Arniella EFP (2017) Evaluation of smart water infrastructure technologies. Inter-American Development Bank
28. United States Environmental Protection Agency (2017) The United States water infrastructure
29. Kenya Airports Parking Services (2015). Innovative Water Management System. Retrieved from http://www.kaps.co.ke/kaps-innovative-water-management-system/
30. Turcu C, Turcu C, Gaitan V (n.d.) Water utility monitoring and control. IOT Approach
31. Mutegi M. pmutegi@ke.nationmedia.com. Business Daily
32. Geetha S, Gouthami S (2016) Internet of things enabled real time water
33. Hatzopoulos JN (2002) Geographic information systems (GIS) in water management. In: Proceedings of the 3rd international forum integrated water management: the key to sustainable water resources
34. Chapman D (1996) Water quality assessments. In: Water quality assessments—sediments and water in environmental monitoring
35. Finlay, Adera (2012) (n.d.) Developing country experiences and emerging research priorities
36. Ndaw MF (2015) (n.d.) Unlocking the potential of information communications technology to improve water and sanitation services
37. Otuke (2016) (n.d.) Role of information communication technologies in water management
38. JICA (2010) Preparatory survey for integrated solid waste management in Nairobi City in the Republic of Kenya. Retrieved from http://open_jicareport.jica.go.jp/pdf/12005443.pdf
39. UNDP (2017) Sustainable development goals. Retrieved from http://www.undp.org/content/undp/en/home/sustainable-development-goals.html

40. Haregu TN, Ziraba AK, Mberu B (2016) Integration of solid waste management policies in Kenya: analysis of coherence, gaps and overlaps. Afr Population Stud 30(3):2876–2885. Retrieved from http://aps.journals.ac.za/pub/article/view/889/701
41. Republic of Kenya (2010) Constitution. Nairobi, Government Printer
42. Republic of Kenya (2006) Environmental management and coordination (waste management) regulations. Retrieved from https://www.nema.go.ke/images/Docs/Regulations/Waste%20Management%20Regulations-1.pdf
43. Republic of Kenya (1996) Physical planning act. Government Printer, Nairobi
44. NEMA (2015) National solid waste management strategy. Retrieved from https://www.nema.go.ke/images/Docs/Media%20centre/Publication/National%20Solid%20Waste%20Management%20Strategy%20.pdf
45. Mwololo M (2016) Managing solid waste remains a nightmare for Nairobi County. Retrieved from https://www.nation.co.ke/lifestyle/dn2/Nairobi-City-sinking-in-trash/957860-3188156-12j42w6/index.html
46. Oduor R (2017) We need to do better in terms of waste management in Nairobi County. Retrieved from https://www.potentash.com/2017/03/03/need-better-terms-waste-management-nairobi-county/
47. UNEP (2010) Integrated solid waste management plan for the City of Nairobi. Retrieved from http://www.unep.or.jp/ietc/GPWM/data/T3/IS_6_4_Nairobi_ISWMplan_draft1_19Feb.pdf
48. Koech G (2016) Counties to be ranked on how well they manage solid waste. Star Newspaper. Retrieved from https://www.the-star.co.ke/news/2016/09/24/counties-to-be-ranked-on-how-well-they-manage-solid-waste_c1425646
49. County Government of Nairobi (2015) Solid waste management act. Government Printer, Nairobi
50. Mohammedshum AA, Gebresilassie MA, Rulinda CM, Kahsay GH, Tesfay MS (2014) Application of geographic information system and remote sensing in effective solid waste disposal sites selection in Wukro Town, Tigray, Ethiopia. Int Archiv Photogrammetry Remote Sens Spat. Inf Sci 50(2):115–118
51. NCWSC (2014) Strategic Plan, 2014/2015-2018/2019. Retrieved from https://www.google.com/search?client=firefox-b-d&q=nairobi+sewerage+company+strategic+plan#

Part X
Dubai

Smart Dubai: Sensing Dubai Smart City for Smart Environment Management

Ummer Sahib

Abstract Every city has a heart and soul of its own and may claim same rights as a human being. In early 2017 the court in the northern Indian state of Uttarakhand ordered that the Ganges and its main tributary, the Yamuna, be accorded the status of living human entities [1]. The decision, which was welcomed by environmentalists, means that polluting or damaging the rivers will be legally equivalent to harming a person. The judges cited the example of the Whanganui River, revered by the indigenous Māori people, which was declared a living entity with full legal rights by the New Zealand government. The judges said the Ganges and Yamuna rivers and their tributaries would be "legal and living entities having the status of a legal person with all corresponding rights, duties and liabilities". So, if we are to recognize the entire city to be a living entity, just like a smart human being, a smart city would also require constant monitoring of its health. Just like a human being using a smart band or a smart watch to sense his/her vital parameters such as temperature, blood pressure, glucose, heart rate, lipid profile, Vitamins etc. the smart city may also require several thousand sensors installed in its arteries and touch points to sense its vital organs. Just like a smart band transmitting vital statistics of a human being to a cloud, the measurement of sensors from the smart city touch points also requires to be transmitted to the cloud. Just like the human vital statistics are subjected to triggering health alerts in the event of spotting an anomaly, the smart city sensor data will also have to be subjected to rigorous analysis to observe if an anomaly of ill health to the smart city is spotted in any of its key node and alerts are triggered to the city guardians for immediate treatment. Furthermore, the process of diagnosis, prognosis, treatment and monitoring of human health becomes applicable for a smart city too. This chapter is divided into 4 sections, essential for smart city Practitioners drawn from smart Dubai's use cases. Part 1 examines the technology and application of Internet of Things (IoT) for establishing a smart environment. This part will focus on the environmental parameters that can be sensed in a smart city? What kinds of sensors are required? What data do they generate and how do they get transmitted to the cloud? Part 2 delves into IoT Platform where the IoT data is stored, managed

U. Sahib (✉)
Informap Technology Center LLC, Office HC2, Tiger Tower 1,
Al Tawun St, Al Tawun Area., PO Box 38098, Sharjah, United Arab Emirates
e-mail: us@informap.ae

© Springer Nature Singapore Pte Ltd. 2020
T. M. Vinod Kumar (ed.), *Smart Environment for Smart Cities*, Advances in 21st Century Human Settlements, https://doi.org/10.1007/978-981-13-6822-6_12

437

and secured along with the survey of IoT platform providers? Part 3 illustrates smart environment use cases of IoT and the parameters that can be sensed in a smart city. Part 4 gives a detailed account of the initiatives taken by Smart Dubai for smart environment management that can be adapted or replicated in other smart cities around the world.

Keywords Smart dubai · IoT · IoT sensors · IoT platform · IoT use cases · Solar park · Smart environment · IoT challenges · DEWA · Dubai municipality · Solar energy · RTA

1 Introduction

Fast-forward to 2021, a smart city dweller checks the air quality on the dashboard of their smart phone app before they start their morning walk or making a trip to a particular area in the city. Along with traffic jam colour and carbon footprint on Google map, there may be yet another layer showing green, yellow, orange and violet illustrating the live city zonal air quality at macro and micro level. Hazard Alerts will be flashed for people of particular age or vulnerable age group asking them not to go outdoors. Those who are driving are able to ask their digital assistant about the live air quality status, just like the live weather broadcast. The overall impact is the city dweller becoming conscious of the environment that they live and the awareness that the city where they live has to be cared for. This to a certain degree may motivate the city dweller to reduce their carbon footprint and to take more participative and proactive approach in exercising their food and transport choices, reducing the usage of plastic and participate in help improving the city environment.

The dashboard on the city administration monitoring center would be rather different. It will display the vast network of citywide sensors streaming live data to the city's cloud platform and the various levels of alerts and respective actions to be taken to manage the environment. In other words, Smart city administration has to spend more of their time on diagnosing the causes and sources of environmental dis-ease, Prognosis, treatment and continuous monitoring with the sole aim of making the city environment cleaner and healthy.

The City's IoT platform control center on the other hand will be looking after the working of the sensors, maintenance, battery, their data communication, data storage, back up and the overall data security. All this is completely hidden from the end-user. As Marsk Weiser said in Scientific American 1991, "The most profound technologies are those that disappear. They weave themselves into the fabric of everyday life until they are indistinguishable from it".

1.1 Smart City Versioning

Welcome to the smart City 3.0, driven by city residents, where IoT, Robotics, Autonomous vehicles, Big data, AI and geo analytics will rapidly change the way we plan, manage and live in a smart city.

The smart city version denotes the stage of smart city in its evolution [2]. Smart City 1.0 is largely technology driven with the sole aim of tapping ICT for digitalization and improving connectivity between the silos or making them transparent. Smart City 2.0 is technology enabled, city-led, in which cities use technology to help determine the future of their city and to improve the quality of life of its residents. Smart City 3.0 is beyond technology or city driven. It is citizen driven through co-creation, through social inclusion, through e-democracy, enterprise creation and building social capital. The main challenge that a city administration faces is that their city is a bit different and the non-replicability of many of the smart city practices that are successful elsewhere. If the ultimate of aim of the smart city is to make a city of happy residents, then citizen participation and e-democracy becomes the central pillar to achieve it.

1.2 Electronic Skin of the Earth

In 1999 article [3], Neil Gross writes; "The skin is an uncanny piece of engineering. It processes immense amounts of data on temperature, pressure, humidity, and texture. It registers movement in the air, gauges the size of objects by the distance between points of contact, alerts us to danger, and prepares us for pleasure. But the skin does more than register superficial events, it's a controller. It sends signals to regulate blood flow, activate sweat glands, alert immune cells to marauding invaders, and block ultraviolet light. Even when skin dies, it is utilitarian: Dead cells accumulate in layers to prevent unwanted penetration.

In the next century, planet earth will don an electronic skin. It will use the Internet as a scaffold to support and transmit its sensations. This skin is already being stitched together. It consists of millions of embedded electronic measuring devices: thermostats, pressure gauges, pollution detectors, cameras, microphones, glucose sensors, EKGs, electroencephalographs. These will probe and monitor cities and endangered species, the atmosphere, our ships, highways and fleets of trucks, our conversations, our bodies–even our dreams."

The above two paragraphs well captures the essence of what this chapter is set to unveil. The recognition of the city as a living being with its own heart and soul and the idea of city's electronic skin form the context of our discussion for the next session on IoT.

1.3   Smart Environment Insights

The Environmental Insights explorer [4] is an online tool that Google has launched along with Global Covenant of Mayors for Climate and Energy [5] to help visualize the carbon foot print of a city at building level. The map illustrates the carbon footprint computed from the emission data collected for all of its buildings, all the car trips, bus and subway rides, and other transportation used by city residents [6]. The ultimate purpose of this initiative is to monitor how the 9000 enrolled cities committed to cut emissions in line with the goals of the Paris Agreement fares with other cities in its sustainability efforts.

The Environmental Insights Explorer uses aggregated data from Google to derive city specific data, including distance driven by mode, the volume and type of buildings, and solar production. It then applies regional assumptions from the Climate Action for Urban Sustainability (CURB) tool to estimate the mix of vehicle and fuel types and the energy consumption of buildings. Finally, the standardized greenhouse gas (GHG) emissions factors is applied per type of vehicle, type of fuel or electricity generation.

Although only a few cities are listed now, it highlights the immense use in such an online tool that allows collecting emission data in a pre-defined standard, store in the open platform and to carry out various analysis. It is of particular help for the City administration who are challenged for funds to use Environmental insights as a common framework. Planners can carry out impact assessment of carbon foot print on various plan options to the extent of computing emissions of a planned road or the net impact of an increase in housing FSI. Using this tool, City Administration can make decisions through collaboration with many other departments and agencies and resident participation. This inclusive approach enables the cities to achieve citizen's happiness, which is the ultimate aim of a smart city (Fig. 1).

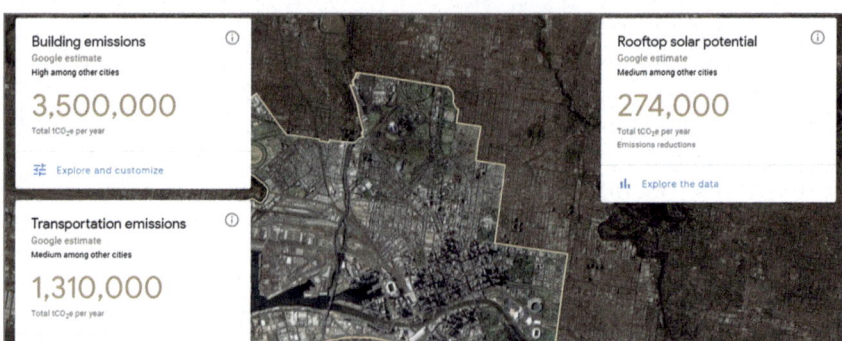

Fig. 1 Google's environment insight: emissions for Melbourne City, Australia. *Source* https://insights.sustainability.google/places/ChIJv_FYgkNd1moRpxLuRXZURFs

2   Role of IoT in Industrial Revolution 4.0

The Internet of Things (IoT) is and will fuel Industrial Revolution 4.0 as the primary data collector in real-time and near real-time as well as a responsive action driver through edge computing. While most of the IoT data are one dimensional, coupling it with geo-location and operational data will enable rich visualization and geoanalytics. The ultimate purpose of IoT in a smart city is to churn sensor data to knowledge and gain wisdom to help make policy decisions (Fig. 2).

IoT remains the most hyped technology acronym today with stunning figures. Gartner [7] expects to see 20 billion internet connected things by 2020 and growing annually to an eleven Trillion Dollar industry by 2025. As such, IoT cannot be avoided or ignored. As Ayne Rand said: "You can avoid reality, but you cannot avoid the consequences of avoiding reality".

2.1   Definition of Connected Things

Internet has been for people by people to share text, images, voice and videos. With the UN declaring Internet access as a human right (UN Report March 2011), being connected is a right as human being. IoT connects things to the internet with the ability to sense, communicate, touch, control and to share their experience with other things. Just like how we human beings interact with other people with our 5 senses; see, hear, taste, touch and smell; things also get to interact and collaborate with other things. For instance; The AC thermostat and lights turn off sensing there is no one in the room, garage door opens and closes for authorized vehicles using Automatic Number Plate Reader (ANPR), sprinkler turns on when the soil moisture gets dry and closes upon reaching the right moisture.

Fig. 2   The knowledge pyramid DIKW. *Source* Wikimedia Commons, by Longlivetheux, CC-BY-SA

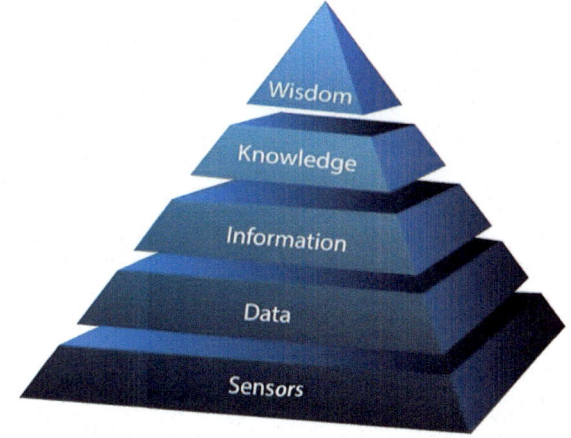

There are different acronyms used to define IoT and examining them is essential to capture their respective perspectives.

The term "Internet of Things" (IoT) [8] was first used in 1999 by British technology pioneer Kevin Ashton to describe a system in which objects in the physical world could be connected to the Internet by sensors. Ashton coined the term to illustrate the power of connecting Radio-Frequency Identification (RFID) used in corporate supply chains to the Internet in order to count and track goods without the need for human intervention. Today, the Internet of Things has become a popular term for describing scenarios in which Internet connectivity and computing capability extend to a variety of objects, devices, sensors, and everyday items.

Gartner's definition captures what is IoT and its purpose from the business perspective?

> The Internet of Things (IoT) is a network of dedicated physical objects (things) that contain embedded technology to communicate and sense or interact with their internal states or the external environment. The connecting of assets, processes and personnel enables the capture of data and events from which a company can learn behavior and usage, react with preventive action, or augment or transform business processes. The IoT is a foundational capability for the creation of a digital business.

There are 2 other acronyms used in the IoT domain; IoE and IoS. The Internet of Everything (IoE) [9] brings together people, process, data, and things to make networked connections more relevant and valuable than ever before. Turning information into actions that create new capabilities, richer experiences and unprecedented economic opportunity for businesses, individuals, and countries. IoE treats people themselves as sensors or devices that people carry such as their mobile phones and wearable devices as Sensors. The U.K.-based technology and development organization called The Technology Partnership however asserts that IoT technology should be called Internet of Sensors (IoS).

High capital asset companies like GE refer the IoT to the "Industrial Internet". Two more concepts are in the horizon called, Building Internet of Things (BIoT) and Industrial Internet of Things (IIoT).

So, there are different acronyms and variants of IoT as defined above that one should be familiar with. Notwithstanding, in this chapter, we will stick with the most commonly used acronym, IoT.

To summarize; Three characteristics of IoT are to be particularly reiterated here, apart from being web enabled;

Ability to sense, communicate each other and react in the event of sensing an exception or danger.

An IoT device can be programmed to "Think" to the extent it is required to "think" and respond as programmed. This makes the IoT emulate some of the human sensory qualities or response system.

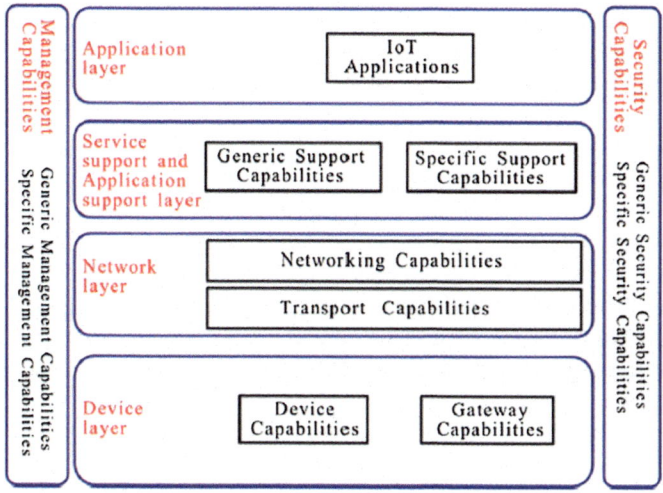

Fig. 3 IoT reference model by Stephen Miles. *Source* ITU-T Y.2060

2.2 Overview of IoT Reference Model

IoT reference model contains 4 layers plus the security and management capabilities that are required across all 4 layers. This section illustrates the various layers and capabilities that are required for IoT to function as identified by ITU (International Telecommunication Union) [10] (Fig. 3).

a. Device Layer: The devices have sensory capabilities that can communicate in three different means. Device can simply bounce their data to another peer device or communicate directly to the cloud or via a gateway. Similarly, device can also receive commands, if the network permits. Devices also can sleep to save energy and wake up only when it is time to push data or when an exception occurs.
   A gateway is required if multiple devices have to use a single device for communication; e.g., ZigBee and Bluetooth technology protocols uses a WIFI or GSM (3G/4G) enabled gateway for communication. Long-term evolution networks (LTE), Ethernet or digital subscriber lines (DSL) can also be used by the gateway for communication. WIFI also provides the MAC address through which indoor position is possible.

b. Network Layer: This layer provide relevant control functions of network connectivity, such as access and transport resource control functions, mobility management or authentication, authorization and accounting (AAA). The Transport capabilities focus on providing connectivity for the transport of IoT service and application specific data, as well as the transport of IoT-related control and management information.

c. Service support and application support layer: This layer consists of Generic and specific support capabilities. While the generic are common capabilities that can be used by different IoT applications, such as data processing or data storage, the specific support caters to the requirements of diversified applications.

d. The application layer supports end user applications that may include analysis and visualization.

e. Management capabilities: This covers the overall management of the system such as configuration, accounting, performance, security, Fault detection and maintenance. Capabilities such as device management (remote device activation and de-activation), battery, diagnostics, firmware and software updating, device working status management, local network topology management, traffic and congestion management are essential. The Specific management capabilities are closely coupled with application-specific requirements.

f. Security capabilities: Generic security capabilities are independent of applications. At the application layer level, they include; authorization, authentication, application data confidentiality, integrity protection, privacy protection, security audit and anti-virus. At the network layer, they include authorization, authentication, signaling data confidentiality and signaling integrity protection. At the device layer; authentication, authorization, device integrity validation, access control, data confidentiality and integrity protection. Specific security capabilities may be required for application such as mobile payment gateway.

2.3 IoT Ecosystem

Since IoT is currently evolving, there is ambiguity and overlaps. Nevertheless, an understanding of the players, their roles and operating models will help understand what to source and from whom.

The providers can be classified into 4; namely Device, Network (carrier), platform and applications as illustrated in Fig. 4. Similarly, the operator models also vary depending on the markets that they operate and the size of the project that they undertake. In certain cases, the carrier or the network operator provides the device, connectivity and platform thereby the application provider just focus on building the application as required by the end user. This may be suitable for small countries or cities. However, in most cases the application provider provides both device and applications.

Operationally, the IoT ecosystem has 5 components as described below:

a. Sensors and Devices for Data Collection: Sensors come as a standalone single hardware unit that measures a single parameter such as Temperature or may have multiple built-in sensors. In some cases, an IoT device has multiple sensors embedded internally or has open ports to connect wired sensors. The sensor units push their data at regular interval or when an exception occurs, as it is programmed. There are 3 means of data push; One way is to simply push data

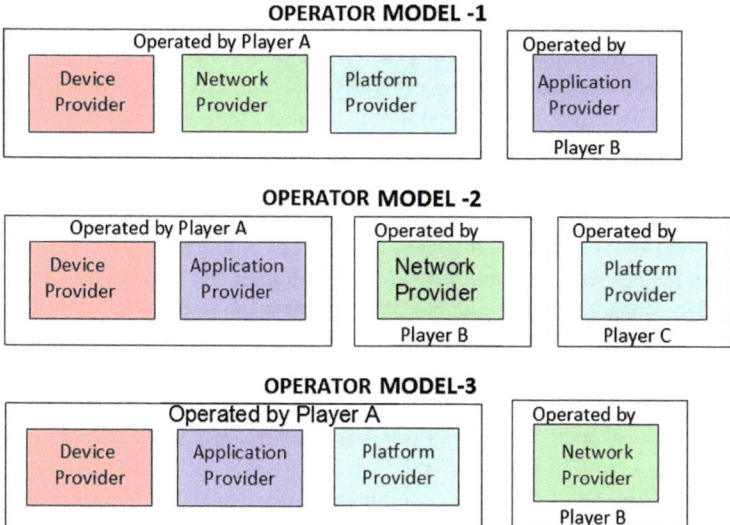

Fig. 4 IoT ecosystem operators—business models

to another peer device through Bluetooth/Zigbee, eventually getting pushed to a gateway. Second way is; the sensor unit itself pushes data to the cloud and third way; is to push through a gateway.

b. Gateway is an essential component in the IoT system. Gateways facilitate data flow securely between IoT edge devices and the cloud. Many sensors/devices can "talk" to a gateway and they act as bridges between sensors/devices and the cloud. Since the gateway is connected to a Wi-Fi/Ethernet/GSM and has a power source, the sensors need to communicate over shorter distances only, thus boosting sensor's battery life. The gateways can also communicate with sensors/devices over varying protocols and then translate that data into a standard protocol such as MQTT to be sent to the cloud. Gateways can pre-process (edge/fog computing) and filter the data being generated by sensors/devices to decrease transmission, processing and storage requirements. Gateways can also reduce latency (data transmitting time) in critical applications by performing processing on the gateway itself rather than in the cloud. Using a gateway is also secure as it reduces the exposure of sensor's direct connection to internet, thereby reducing the chances of sensors getting hacked.

c. Connectivity: IoT requires two kind of connectivity. Inter-sensor/device connectivity and internet connectivity. The inter-sensor connectivity can be achieved through short range connectivity technologies such as NFC, zigbee, Bluetooth or RFID. In this scenario, sensors connect using BLE range of 50–100 m to form a peer to peer mesh network of tiny low cost sensor nodes. Gateway assigned to these sensors receives data and transfer to the cloud using either GSM, Wi-Fi, Ethernet, cellular, satellite or low-power wide-area networks (LPWAN) LORA or Sigfox. In certain cases, sensors or devices may directly use LPWAN to transmit data directly to the cloud, thus skipping the need for a gateway.

Choosing the right connectivity option depends on the specific IoT use case and the tradeoff between power consumption, coverage range and data bandwidth. Though the perfect connectivity option would be low power consumption, long range and high bandwidth, it does not exist. The cellular and satellite connectivity consumes higher power and provides High Range and High Bandwidth. Since cellular connectivity is available only where there is GSM coverage, satellite connectivity would be required in places where GSM is not available. Wi-Fi, Bluetooth and Ethernet consume lower power and cover Low Range but allow higher bandwidth. LPWAN connectivity such as Sigfox and LORA covers wide range and consumes very low power, but provides only low payload (data bandwidth). Recently, Low Power Wide Area (LPWAN) cellular technologies such as LTE-M and NB-IoT offers a combination of lower cost, broader coverage and better battery life with globally available and secure cellular networks.

d. Data Processing: Once the IoT data reaches the cloud platform, a number of processing is carried out by the software to identify the sender, time stamping and pushing raw data to client server. The client server could carry out processing as simple as detecting a temperature exception or complex such as motion sensor to detect intrusion and trigger programmed actions.

e. Application for User Interface: This component allows viewing of data and to configure the type of action. It could be an alert to the user by means of an email or an SMS. The application may also allow users to have a dashboard to check in near real-time, generate report and program an automatic action to be carried out if a certain condition is met. This could be turning on the A/C if temperature exceeds or set off an alarm or alert police control if an intrusion is detected.

2.4 IoT Value Settings

IoT can be used in several settings starting from humans, home, factory and cities. McKinsey Global Institute [11] identifies the following nine "settings" where IoT creates value (See Fig. 5).

A "settings" lens helps capture all sources of value; we identify nine settings where IoT creates value

Setting		Description	Examples
	Human	Devices attached to or inside the human body	Devices (wearables and ingestibles) to monitor and maintain human health and wellness; disease management, increased fitness, higher productivity
	Home	Buildings where people live	Home controllers and security systems
	Retail environments	Spaces where consumers engage in commerce	Stores, banks, restaurants, arenas—anywhere consumers consider and buy; self-checkout, in-store offers, inventory optimization
	Offices	Spaces where knowledge workers work	Energy management and security in office buildings; improved productivity, including for mobile employees
	Factories	Standardized production environments	Places with repetitive work routines, including hospitals and farms; operating efficiencies, optimizing equipment use and inventory
	Worksites	Custom production environments	Mining, oil and gas, construction; operating efficiencies, predictive maintenance, health and safety
	Vehicles	Systems inside moving vehicles	Vehicles including cars, trucks, ships, aircraft, and trains; condition-based maintenance, usage-based design, pre-sales analytics
	Cities	Urban environments	Public spaces and infrastructure in urban settings; adaptive traffic control, smart meters, environmental monitoring, resource management
	Outside	Between urban environments (and outside other settings)	Outside uses include railroad tracks, autonomous vehicles (outside urban locations), and flight navigation; real-time routing, connected navigation, shipment tracking

SOURCE: McKinsey Global Institute analysis

Fig. 5 IoT value settings. *Source* Mckinsey Global Institute Analysis

2.5 IoT Devices: Sensors for Smart Environment

This section gives an overview of what can be sensed in a smart city in general and sensors that are required for smart environment monitoring in particular. There are 2 components in an IoT device (edge); sensors and actuators that enable interaction with the physical world. The illustration [12] below depicts its respective function. While sensor probes a particular phenomenon and transmits to the cloud as data input, the actuators receives a command or instruction from the application or AI system or programmed in the edge itself, to carry out a particular action in the event of detecting an exception to manipulate the physical environment such as turning on a sprinkler in the park at a particular time for a certain duration (Fig. 6).

Sensor to **Actuator** Flow

Fig. 6 Sensor to actuator flow. *Source* https://bridgera.com/iot-system-sensors-actuators)

It is the sensors and the actuators that provide a digital nervous response system. Location is sensed using the GPS chip, Camera lens act as the eyes, microphone lends the ears, sensor probes can feel light, temperature, pressure and sensor probes can smell too!

The environmental parameters that can be sensed through IoT devices are illustrated in the following infographic [13] (Fig. 7).

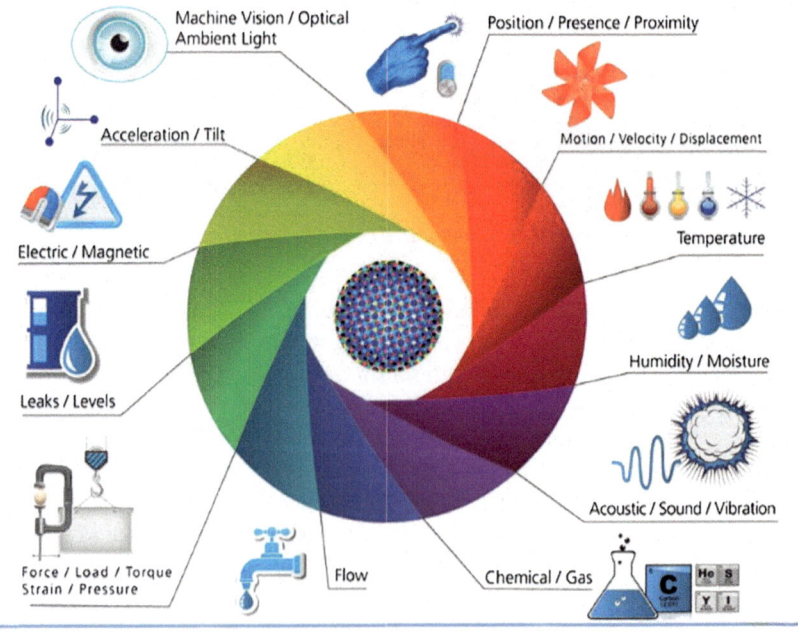

Fig. 7 Environmental parameters that can be sensed. *Source* https://www.postscapes.com/ [13]

Sensors collect data from the environment or object under measurement. The data includes device identification, timestamp, location, and sensed variable's readings. As these devices produce an avalanche of data, it will be important to choose which data is useful and which can be ignored. Depending on the data type, volume and velocity, decision has to be made if the data should be transmitted to a dedicated in-house server or to the cloud. Some data should be processed immediately, i.e., time-sensitive data, threat detection, immediate crash statistics, abrupt shutdowns, etc. either in the edge or in the cloud and immediate response system should be activated. Otherwise, data that gets stored for deep process and analysis should be pushed directly to the cloud to avoid network clutter.

Finding a good quality sensor that communicates with the available communication protocols to the IoT platform is itself a daunting task. Therefore, a list of available COTS (Commercially off the shelf) environmental sensor providers are compiled below as a helpful reference.

Ref	Sensor provider	Description/comment	Website
1.	Libelium Comunicaciones Distribuidas S. L. Spain	Spanish company specialized in providing smart environment solutions. Sensor and gateways support for most of the communication protocols	https://www.the-iot-marketplace.com/solutions/smart-environment
2.	SAYME	Spanish company providing sensors for waste collection and street lighting	http://www.sayme.es/
3.	Evvos, Luxembourg	Provides a weather station that can measure 12 weather variables including: air temperature, relative humidity, vapour pressure, barometric pressure, wind speed, gust and direction, solar radiation, precipitation, lightning strike counter and distance	https://www.evvos.com/
4.	Spectrum Technologies, Illinois, USA	Provides sensors for plant soil and moisture measurements	https://www.specmeters.com/soil-and-water/

(continued)

(continued)

Ref	Sensor provider	Description/comment	Website
5.	SENSONEO j. s. a. Bratislava, Slovakia	Provides unique ultrasonic smart Sensors that monitor waste in real-time using IoT resulting in overall waste collection cost reduction by at least 30% and carbon emission reduction up to 60% in cities	https://sensoneo.com/smart-waste-management-for-cities/
6.	Sensit Inc., Redlands, CA, USA	Specialized sensor provider for soil erosion monitoring	http://www.sensit.com/
7.	Sensile Technologies Switzerland	Provides remote monitoring of meters and tank levels for the oil and gas industry	https://www.sensile.com/solutions/gas-tank/

3 Iot Platform

As discussed above, the IoT sensors (e.g. a moisture sensor) collect data from the environment or perform actions in the environment (e.g. watering the lawn in the public park). The connectivity allows transmitting data and receives commands directly or through gateway. An IoT platform, as the name suggests, forms the support software that connects everything in an IoT system. The platform is responsible for analyzing the data it is collecting from the sensors and making decisions (e.g. knowing from moisture data that it just rained and then telling the irrigation system not to turn on today). The platform also provides an interface for users to interact with the IoT system through a web-based dashboard that shows moisture trends and allows officials to manually turn irrigation systems on or off. An IoT platform facilitates communication, data flow, device management and the functionality of applications. With varying kinds of IoT hardware and the different connectivity options, there needs to be ways of making everything work together. An IoT platform does exactly that as shown below [14] (Fig. 8).

To summarize, IoT Platform help:

- Connect hardware, such as sensors and devices
- Handle different hardware and software communication protocols
- Provide security and authentication for devices and users
- Collect, visualize, and analyze data the sensors and devices gather
- Integrate all of the above with other web services

City administration may opt for COTS (Commercially off the shelf) IoT platform instead of the expensive and time consuming process of hiring software developers or depend on the IT department to build everything in-house. Using a COTS IoT

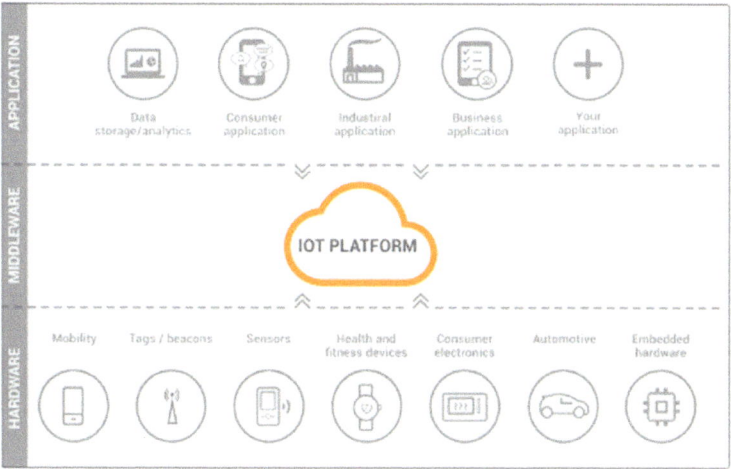

Fig. 8 IoT platform middleware. *Source* https://www.kaaproject.org/what-is-iot/

platform may be cost-effective and at the same time allows implementing the system far more quickly. There are over 500 IoT platforms and selecting the right platform may be a very formidable task.

3.1 IoT Platform Functionality

In order to choose the right IoT platform, City administration must ask "What IoT Platform functionality are required?" to support the use cases considered. To help with this question, following summary [14] is presented in the four typical layers of the IoT stack.

The Things are connected to the platform either directly or through gateways where the data is received and processed. The app and analytic stack handles the analysis, visualization and reporting (Fig. 9).

3.2 IoT Platform Selection

The selection process can only start once the catalogue of use cases and the desired applications are complete. Based on the functionality required, the selection of a COTS IoT platform provider shall be subjected to the following selection criteria.

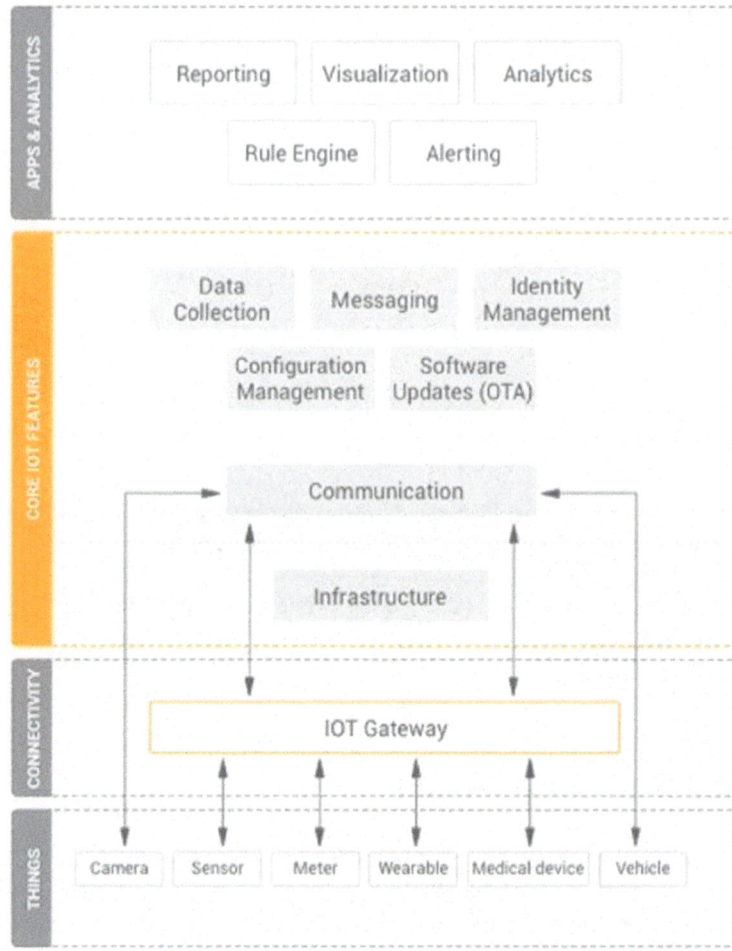

Fig. 9 IoT platform functionality—technology stack. *Source* https://www.kaaproject.org/what-is-iot/

Ref	Selection parameter	Description of what to look for
1.	Track record of the IoT platform vendor and reliability	Credibility, how long in business and customers. Minimum 4+ years is usually is essential
2.	Ecosystem and protocol support	Connectivity protocols that are supported. Multiple protocols are essential. Event processing and interfacing/integration
3.	Device and gateway support	How much time does it take for the hardware onboarding/provision/management systems

(continued)

(continued)

Ref	Selection parameter	Description of what to look for
4.	Interoperability	Programming languages, application enablement
5.	Platform scalability	Ability to support far more IoT devices than the initial project. Scalability, customizability, ease of use, code control, integration with 3rd party software, deployment options
6.	Security	Authentication, certification and encryption. Need to have security by design in an end-to-end way. GDPR compliance on privacy for use cases involving personal data and identifiers

3.3 Survey of IoT COTS Platform Providers

The well known IoT commercially off the shelf platform providers [15] are listed as follows in the order of popularity to help evaluate their width and depth of service offering.

Ref	Platform provider	Description of IoT platform
1	Amazon	Amazon Web Services (AWS) IoT platform: makes a lot easier for developers to connect sensors for multiple applications ranging from automobiles to turbines to smart home light bulbs
2	Microsoft	Microsoft Azure: the Azure IoT suite comes with Azure Stream Analytics to process massive amounts of information in real-time
3	Google	Google Cloud: with the ability to handle the vast amount of data using Cloud IoT Core. Advanced analytics owing to Google's Big Query and Cloud Data Studio
4	PTC (Parametric Technology Corporation)	ThingWorx: a platform for the rapid development of enterprise applications designed for smart, connected sensors, devices, and products. It enables innovators to rapidly create and deploy game-changing applications, solutions and experiences for today's smart, connected world
5	IBM	IBM Watson: backed by IBM's hybrid cloud PaaS (platform as a service) development platform, the Bluemix, Watson IoT enables developers to easily deploy IoT applications
6	Samsung electronics	Artik: is known for providing complete security to products which often gets neglected. It aims to provide a fast and open platform that is responsible for developing and managing products
7	Cisco	Cisco IoT Cloud: platform is for mobile operators offering, also partnered with national farmers' federation in Australia to provide internet of things in agriculture solutions

(continued)

(continued)

Ref	Platform provider	Description of IoT platform
8	Hewlett Packard	Universal of Things: offers its clients scalability by providing solutions to the vast majority of their problems. The platform offers support that can be deployed over the cloud or locally. HPE Universal of Things Platform has been successfully used in smart cities and automobile industry
9	Thunder	Salesforce: is focused entirely on customer engagement. Provides business users with much a much more comprehensive and integrated perspective on customers, without requiring technical expertise or the services of a data analyst
10	Bsquare	Datav: is a complete Internet of Things (IoT) software solution allowing business and industrial concerns to use data generated by myriad connected devices in order to drive better business outcomes
11	Siemens	Mindsphere: Open IoT operating system that connects products, plants, systems, and machines, to harness the wealth of data generated by the Internet of Things (IoT) with advanced analytics
12	Ayla network	Ayla: platform provides a complete solution to securely connect any device to any application while offering all of the tools necessary to provision and manage connected products at scale
13	Bosch	Bosch IoT suite: Bosch cloud offers its customers complete safety and reliability while storing the data on its secure servers. Using platform as a Service (PaaS) Bosch offer its service economically
14	Altair engineering	Carriots: is a PaaS platform. It also widely used for Machine to Machine development. Several network operators have chosen Carriots platform
15	Oracle	Oracle integrated cloud: Offers big data analysis for real-time IoT data, device virtualization, endpoint management and high speed messaging. Users can receive notifications directly on their devices
16	General Electric	Predix: offering a PaaS platform to provide better decision making insights in real time, GE's Predix is made for mainstream sectors like healthcare, transportation, aviation and energy. It helps in development of IoT applications that can process real-time operational data aiding better decision making
17	Apache	MBED IoT device: platform provides an open source service. It includes an operating system, cloud services and developer tools that make setting up of commercial products and its functioning easy
18	LTI	Mosaic: Specializes in cloud services and end-to-end services for various industries. Mosaic uses Rapid Software as a Service (SaaS) platform to enable the digital transformation of businesses
19	Mocana corporation	Mocana: The platform is aimed to provide security to industrial IoT devices and industrial clouds

Microsoft Azure IoT Suite is the most favoured platform [16] with 29% developers favouring it, following by Google Cloud IT second (24%), Amazon AWS IoT third (19%), Cisco IoT service (13%), IBM Watson IoT (10%) and Thingworks PTC (7%) following it. While AWS is more popular among smaller organizations, Google's

Cloud IoT is being adopted more by employees in larger companies. There may be other smaller platforms that are worth considering for certain use cases.

3.4 Survey of Open Source IoT Platform Providers

Organizations that have in-house IT department and the required human resources to develop and maintain the IoT platform, using open source should be considered seriously before opting for COTS platform.

The table below is a web review of some [17] open source IoT platforms:

1	Kaa Enterprise	Kaa IoT is one the most efficient and rich open source IoT cloud platform. Supports unlimited number of connected devices with cross device interoperability. Real time device monitoring with the possibility of remote device provisioning and configuration
2	MACHINNA.io	This IoT platform provides a web enabled, modular and extensible JavaScript and C++ runtime environment for developing IoT gateway applications. It also supports a wide variety of sensors and connection technologies including Tinkerforge, bricklets, Xbee and many others including accelerometers
3	ZETTA	Zetta is a server oriented platform that has been built around Node.js, REST and a flow based reactive programming development philosophy linked with the Siren hypermedia APIs. They are connected with cloud services after being abstracted as REST APIs. These cloud services include visualization tools and support for machine analytics tool like splunk. It creates a gero-distributed network by connecting end points such as Linux and Arduino hacker boards with platforms such as Heroku
4	GE PREDIX	GE's platform as a service software for industrial IoT is based on the concept of cloud foundry. It adds asset management, device security and real time, predictive analytics that also supports heterogeneous data acquisition, access and storage. GE predix was developed by GE for its own operations and consequently has become one of the most successful of the enterprise IoT platforms and with the recent partnering of GE and HPE, the future looks even better
5	ThingSpeak	ThingSpeak is another IoT platform that lets analysis and visualization of the data in MATLAB and eliminates the need to buy a license for the same. It helps collect and store sensor data in private channels while giving the freedom to share them in public channels. It works with Arduino, particle photon and electron and many more applications. It is used mostly for sensor logging, location tracking and alerts and analysis. It also has a worldwide community which is quite helpful in itself
6	6th to 10th Open source Platforms	DeviceHive, DSA (Distributed Services Architecture), Eclipse, Open connectivity foundation and OpenHAB are the remaining 5 open source platforms that could be considered

3.5   IoT Security Vulnerability and Mitigation

IoT devices interact with the physical world much different than our computers, tablets, or smartphones as they do not run on Windows or Mac OS that have built-in security. IoT devices are connected to the internet as points of entry without a security firewall. Here lies the root cause of the security breach! IoT Security is a prime concern, considering the estimated 20.4 billion devices projected to communicate between them and connected to the internet by 2020. This is around 2.6 devices per person globally and a hockey stick growth is predicted by 2025.

As Donald Rumsfeld, former US secretary of Defense, said in his memoir Known and Unknown "There are **known knowns**. These are things we know that we know. There are **known unknowns**. That is to say, there are things that we know we don't know. But there are also **unknown unknowns**. There are things we don't know we don't know." IoT's security vulnerability can also be compared in the same way.

The major security concerns that we certainly know consists of data privacy, device passwords, devices interacting unintentionally, user safety of devices due to poor make and quality posing cyber security risk. Other security risks reported are; spoofing of smart electricity meters to under report energy consumption, data transmission loss, data storage safety, software security, use of non-standard devices and lack of standardization. But, there are several unknown unknowns that will unfold as we start using.

The known measures to mitigate include; avoid counterfeit/fake products, proactive maintenance, identify spoofing, ability to auto-recover in case of accidental crash during data transmission, protect itself from malicious programs that could be externally inserted, authentication framework that would permit data transfer only between authenticated devices, authorization of access control limited to just one or two responsible officers, prevention of access to personal data, contingency plan in the event of data breach or what could be done with a compromised device.

The National Institute of Standards and Technology (NIST) released a draft publication in September 2018; NIST Internal Report (NISTIR) 8228 [18], Considerations for Managing Internet of Things (IoT) Cyber security and Privacy Risks. This report identifies three higher level serious security vulnerabilities of IoT devices:

a. **Can Cause Physical Damage**: IoT devices have the ability to make changes to physical systems such as HVAC, elevators, sprinkler systems, and others that, if commandeered, could pose physical damage or safety risks.
b. **Difficulty to Access**: Since the firmware or OS are proprietary, IoT devices cannot be remotely accessed, managed, or monitored in the same way as conventional IT devices. It becomes difficult or impossible for IT personnel to monitor and patch each device, as it may involve doing tasks manually for large numbers of IoT devices. As a result, IoT devices are deployed with the default credentials. Hackers use this vulnerability to launch SSH password brute force attack or recruit IoT devices as botnets to use in distributed denial-of-service (DDoS) attacks.

c. **Varied Security Policies**: The availability, efficiency, and effectiveness of cyber security and privacy capabilities are often different for IoT devices than conventional IT devices. This means organizations may have to select, implement and manage additional controls, as well as determine how to respond to risk when sufficient controls for mitigating risk are not available.

NIST recommends following three high-level risk mitigation goals:

a. **Protect Device Security**: Prevent any IoT device from being used to conduct attacks; including participating in distributed denial of service (DDoS) attacks against other organizations and eavesdropping on network traffic or compromising other devices on the same network segment. Apply strict password changing policies and use separate third party security firewalls to manage IoT security.
b. **Protect Data Security**: Protect the confidentiality, integrity, and/or availability of data (including personally identifiable information [PII]) collected by, stored on, processed by, or transmitted to or from the IoT device. This applies to any IoT device with one or more data capabilities.
c. **Protect Individuals' Privacy**: Protect PII (Personally Identifiable Information) on IoT devices that process PII or directly impact individuals.

NIST report further recommends three mitigations that organizations must take throughout the IoT device lifecycle. They are; Understand risk consideration and mitigation challenges, adjust organizational policies and processes to address those challenges and implement updated risk mitigation practices for IoT devices.

3.6 IoT Implementation Success

IoT experts recommend a number of do's and don'ts for the success of IoT implementation. Some of the important ones are captured below:

	Do's for IoT implementation	Don'ts for IoT implementation
1.	Secure senior business sponsorship	Confuse data with insights
2.	Focus on business requirements	Build a solution for all your needs
3.	Start small, think big. Minimum Viable Product MVP	Over-pivot on the technology

(continued)

(continued)

	Do's for IoT implementation	Don'ts for IoT implementation
4.	Build Multi-disciplinary teams	Focus on future requirements
5.	Engage the right experience	Develop your own IoT infrastructure
6.	Keep security central in all you do	See middleware as the IoT solution
7.	Use Agile development process	Think connectivity is not a problem
8.	Design for flexibility and change	Underestimate the importance of quality sensors
9.	Take an ecosystem approach to IoT	Forget that IoT is devices and software
10.	Allocate budget for maintenance	Overlook security

4 Smart Environment Use Cases of IoT

There are 2 distinct IoT waves through which sensing is carried out; the first wave uses embedded devices and sensors and the second wave uses people as sensors through the devices that they are carrying.

The most popular smart environment use cases [19] that use embedded IoT devices and sensors are compiled in the table below.

5 Smart Dubai IoT Strategy

Smart Dubai launched its IoT Strategy [20] in October 2017 with a mission to build the world's most advanced IoT ecosystem in Dubai. This is to be implemented in 4 phases by 2020 in collaboration with Dubai Electronic Security Centre (DESC). Phase 1 is Coordination; to set IoT policies across government departments. Phase 2 will address integration and Convergence, to harmonize the efforts towards implementing the IoT Strategy. Phase 3 will focus on optimization of management framework for IoT ecosystem and Phase 4 will move to final "Blockchain Journey", to establish a Self-regulating IoT ecosystem based on Blockchain and the "economy of things" (Fig. 10).

Dubai IoT Strategy is comprehensively designed to leverage a citywide partner network, to secure and promote Dubai's digital wealth and deliver efficiency benefits and peace of mind to all city residents, visitors, business owners and public officials. Smart Dubai has introduced the framework and implementation guidelines for Public Key Infrastructure (PKI) and Digital Certificates that will help secure a trusted IoT ecosystem. The PKI will define roles, policies, and procedures to create, manage, distribute, use, store and revoke digital certificates and manage public encryption. The Digital certificate is guaranteed by the Dubai Government to allow exchange of information securely over the internet using the PKI. These two key initiatives

	Embedded sensor use cases	Usage, impact and benefits
1.	Smart signals https://miovision.com/traffic-link/	To monitor and manage traffic signals remotely. Makes connected city by supporting vehicle to intersection (V2I) technology, while enabling transit priority and emergency preemption. By optical real-time counting and connecting intersections and forming a grid that talks each other, a busy traffic intersection can talk to other signals and give more time. Miovision knows the work route of the subscriber and can tell the 3 intersections on the route that are congested and advise other intersections that are not yet congested
2.	Smart intersections https://gridsmart.com/	The bell-shaped smart camera "sees" the center of the intersection. The system actuates the intersection, provides video feed for situational awareness, and collects and records data on everything happening at the intersection, counts, speed, vehicle classification, and many others
3.	Smart parking https://parksmart.org/	Uses a network of parking sensors and video cameras to obtain real-time parking spaces availability. Users can check the availability of free parking through a mobile app in their desired slot. Smart parking remembers the car where it is parked and then illuminates the poles that help lead the way and turns off illumination to 10% once person passes through
4.	Fleet management	Uses vehicle telematics to manage mobile resources. Public transport users are able to see their vehicle in near real-time using mobile app. Public transport and utility fleet communicate when it is time for scheduled maintenance or replacement
5.	Smart street lighting http://www.lighting.philips.com/	Motion sensors are used to turn on light and control luminosity upon detecting a user. Ability to monitor how much energy street lights is using and which ones need repair. The poles also are fitted with sensors to collect air quality throughout the city
6.	Smart analytics for transportation http://citilogik.com/	Uses pedestrian, vehicle, and rail journey data to get a comprehensive view of a city transport system, including travel patterns, peak, off-peak, seasonal, and historic patterns allowing cities to make informed decisions for future transportation projects and improvements
7.	Smart transit options https://transitscreen.com/	Help commuters to get a real-time comparison of train times to bus arrivals and make better choices. This saves time and help make cities healthier, sustainable, and more accessible for everyone

(continued)

(continued)

	Embedded sensor use cases	Usage, impact and benefits
8.	Secured workforce and residents connectivity/Wi-Fi https://veniam.com/	Wi-Fi hotspots in vehicles and controlled spaces, such as ports and container terminals ensure that all mobile workers and assets are securely connected. Connectivity helps secure citizens
9.	Energy management http://c3iot.com/	Energy Management enables large enterprises, government agencies, and utilities to achieve data-driven, cost effective energy management. Benchmark building performance, track sustainability goals, and implement environmental stewardship. Authorities are able to monitor energy generation and distribution infrastructure for optimizing both proactive and reactive maintenance schedule. To monitor electricity consumption, compare time series or location specific. On consumer side, inculcate consciousness on the energy that they use and for what, thereby strive to reduce carbon footprint
10.	Water control/management, smart irrigation https://www.altairsmartworks.com/smartcities	Humidity and soil sensors deployed near irrigation sprinklers in parks and other public areas can be used to measure moisture content of the ground. When combined with data from weather forecast, it can be used to control sprinkler systems and dispense water only when humidity level falls below a pre-determined threshold and there is no forecast of rain in near future
11.	Trash/garbage collection, smart waste management	Sensor deployed in garbage containers throughout the city can inform how full the container is. Depending on data from the sensors and their location, trash collection routes can be optimized to only clear the containers that are above a pre-determined level
12.	Water flow monitoring	Water flow measurement sensors can be used to capture and transmit real time data about water flow. Unexpected water flow issues could indicate leaks; providing early warning alerts to public works teams in the city
13.	Fire detection and fire prevention	Sensors can be used to monitor hazardous conditions that typically results in fire. Emergency response teams can be alerted so timely help can be provided
14.	Flood detection	Using sensors in low lying areas to detect possible flooding conditions will allow cities to complement existing water gauges and successfully monitor and respond to hazardous conditions
15.	Air/particulate quality	Air quality sensors deployed across the city (e.g. park, large buildings, light poles or other points of interest) provides much better and accurate air quality readings than current method of macro readings. This data can be used to alert vulnerable citizen population

(continued)

(continued)

	Embedded sensor use cases	Usage, impact and benefits
16.	Water quality monitoring	Using potable water monitoring solutions, cities are able to monitor quality of drinking water and alert in case of contamination
17.	Snow level monitoring	Deploying sensors to monitor snow build up on road surfaces, sidewalks, etc. will provide public works team within a city to reduce risk associated with slippery conditions to both pedestrians and motorists
18.	Smart pole	The smart pole provides efficient LED based smart lighting and Wi-Fi hotspot solutions. The LED saves electricity consumption by reducing to 10% and illuminates the path for usage. The pole can have a camera, microphone that can listen, and particle sensors that can sense particulate matter like dust or pollen. The network of smart poles can give Insight into the pollution levels and correlate to diseases. For instance; Increase in pollen counts can result in the increase in the number of patients in the hospitals with breathing difficulties. This smart pole network can be used as a warning system to those who are suffering from certain illness or allergies
19.	Urban noise level monitoring	This sensor module monitors urban noise pollution and allows comparing noise levels with the number of bars and restaurants in an area
20.	Ozone level monitoring	Ozone sensors are used for monitoring ground-level ozone; known to irritate the nose and lungs
21.	Gas leak monitor https://temboo.com/iot-applications	Undetected gas leaks can be deadly and costly. An application that monitors a gas line for leaks and closes the remote control of a gas valve to prevent further leakage
22.	Overhead water tank monitor https://temboo.com/iot-applications/water-management	Water resource management is becoming increasingly important. IoT sensors can sense the water levels in a tank, alerts when water levels are too low, and allows remotely refilling from a reserve
23	Smart fire equipment monitoring	Monitor appropriate storage of fire equipment to ensure compliance with fire safety regulations and receive alerts when attention is required
25	Soil quality monitor	Nurturing crops on a large-scale farm can require time consuming testing of soil quality. IoT can be deployed for automated recurring watering schedule or sense moisture level to allow remote control of sprinklers, ensuring that crops receive the right amount of water without waste
26	Energy management in cold chain industry and chiller vans	IoT sensors can regulate the ideal temperature thereby reducing the energy consumption, reduce food wastage and ensuring food is kept fit for consumption

(continued)

(continued)

	Embedded sensor use cases	Usage, impact and benefits
	People carried devices as sensors	Usage, impact and benefits
21	Smart phone as sensors for urban mobility. Mobility as a Service (MaaS) platform https://moovit.com/	Moovit turns bus and train riders into near-real-time sensors to aggregate the world's largest repository of transit data, from over 200 million users in 80 countries across 2500 cities. It Provides a sustainable and integrated mobility platform and help improve both service and rider experience. Moovit notify other moovit subscribers about a delay in the train if one is living close by the train and advises to take a bus instead and further tells which bus number to take
22	Smart phone sensors to capture road pothole http://www.streetbump.org/	Volunteers use the Street Bump mobile app to collect road condition data while they drive. The smart phone app fixed on the dashboard uses GPS and Accelerometer data to pinpoint the location of potholes that need repair. 3G accelerometer can detect every time the car jumps up and down. When the smart phones accelerometer detects a motion in y axis, it records the GPS location of where that bump occurred and transmits to the road department who can fix it when it is smaller at a reduce the cost. Street Bump helps residents keep the Streets Smooth
23	Tourist destination crowd management using smart phone app https://www.vizalytics.com/	Vizalytics partnered with the city of Copenhagen to combine IoT and Analytics capabilities to provide powerful insights into passenger liner tourist behavior using the Copenhagen card. The app tracks the tourists to find their travel direction, mode of travel, stoppage duration, visited attractions, unvisited attractions and near real-time crowd density at any point of time. Notifications can be sent to encourage or discourage people from going into a particular attraction as it is crowded and divert them to visit attractions that are not crowded
23	Seismic warning system	Make use of people who can alert utilities to earthquakes and automatically shutoff gas distribution networks in affected areas or acoustic sensors that can detect gunshots and alert emergency response by blinking nearby streetlights

Fig. 10 Smart Dubai IoT implementation road map. *Source* https://iot.xische.com/

will facilitate the trusted sharing and exchange of valuable data through the IoT with confidence. The IoT will connect to Dubai Pulse platform, which is the digital backbone of the smart city project.

To guide the implementation of a comprehensive, citywide IoT encompassing both the public and private sector and benefitting business and individuals, Smart Dubai pursued a methodical approach to plan, deliver and manage the city's Internet of Things across 6 strategic domains namely; governance, management, acceleration, deployment, monetization and security. UAE is voted 17 out of 105 as the best prepared country in the cyber security ranking by UN's ITU.

5.1 Smart Dubai Platform and Dubai Data

In 2016, Smart Dubai launched the Smart Dubai Platform (SDP) in partnership with Strategic Partner Du [21], the telecom provider. The SDP is conceived as the central operating system for the city, supplying access to city services and data for all individuals, private and public sector entities. Dubai data is a complementary initiative that aims to achieve a seamless, efficient, impactful and safe data governance and data sharing at the city level contributing to Dubai's smart transformation.

The UNECE and ITU chose Smart Dubai Platform in its flipbook "Towards the Sustainable Development Goals" [22] in the series of case studies as a successful measure into the global spotlight for adoption. As such, the SDP offers a very good model for other cities to learn and adapt to their own city. The major highlights of SDP and Dubai Data are as summarized below:

a. To build the most comprehensive city platform, Smart Dubai evaluated 3 alternatives; Build it themselves, Purchase off the shelf and Public Private Partnership. After a rigorous selection exercise, PPP was opted due to several advantages it posed. Smart Dubai's public mission was combined and integrated with the predominantly commercial mission of the private sector partner Du. PPP has enabled utilization of private capital to supplement public sector investment, thus freeing up additional capital for Smart Dubai to invest in other city projects.

b. SDP united all layers of Dubai's ICT architecture; IoT management and data aggregation, geo-location data support, connected infrastructure, data orchestration, Digital IDs and payments, IoT and data management, dashboards and analytics and to provide Platform as a Service (PaaS).

c. The social impact of SDP will help fulfill Smart Dubai's vision of making Dubai as the happiest city in the world. The direct benefits to residents are for example; A mother can check when vaccination is due for a child, father gets alerts about excess energy consumption, tourists utilize public transport efficiently, commuters enjoy reduced traffic jams and online services and an urban planner or developer achieving operational cost-savings into data-driven research and development about Dubai through the data available in SDP.

d. The environmental impact of SDP are far reaching; collect city wide environmental data such as air quality, noise, pollution, energy, water, land and sea; DubaiNow online services will reduce significant number of trips which in turn reduces GHG (Green House Gas) emissions and will also reduce the stress on transport infrastructure; enablers to achieve city resources efficiencies and related consumption reduction tools will positively impact the environment.

e. SDP is scalable and will help Smart Dubai to carry out sustainable city initiatives such as unifying standards, collaborating with data providers, incentivizing participation and tracking the impact that can be transferred to other cities.

f. Dubai Data initiative established a radically different operating model for its data. It laid clear governance frameworks to ensure data is managed as a strategic asset, use of open standards to ensure that data sets can easily be leveraged by various city constituents, nurturing the development of a flourishing 'data market place' in which all stakeholders can use Dubai Data to create new sorts of social and economic value.

Through Dubai Pulse [23] residents can access/export data, gain visual data insights and all stakeholders can use DevZone as a complete deployment environment in the cloud under PaaS/IaaS (Infrastructure as a Service) model.

6 Smart Environment Initatives of Smart Dubai

Before carrying out any action, it is important for a smart city to articulate its clear vision and the actions to achieve that vision together with the key performance indicators for each action. Dubai leadership articulated its vision to "become the happiest city on earth". In 2016, Dubai chose to be the first pilot partner to implement the ITU-standardized Key Performance Indicators (KPIs) for Smart Sustainable Cities [24]. In the first year of its ITU-Dubai pilot program, other cities such Valencia, Singapore, Buenos Aires, and Montevideo signed up for similar projects.

The ITU and UNECE's definition for smart, sustainable city is: "A smart sustainable city is an innovative city that uses information and communication technologies (ICTs) and other means to improve quality of life, efficiency of urban operation and services, and competitiveness, while ensuring that it meets the needs of present and future generations with respect to economic, social, environmental as well as cultural aspects".

Following table summarizes the Environmental indicators as per the ITU-T International Standards:

Parameter	KPI indicator	KPI definition
Air quality	Application of ICT based monitoring system for particles and toxic substances (air quality)	Proportion of city area covered by outdoor ICT based monitoring system for particles and toxic substances (air quality) monitoring
Air quality	Air pollution intensity	Level of particles and toxic substances is based on five aspects: ground-level ozone, particulate matter, carbon monoxide, sulphur dioxide, and nitrogen dioxide, measured by the Air Quality Index (AQI)
CO_2 emissions	GHG emissions (Green House Gas) vapour, carbon dioxide, methane, nitrous oxide and ozone	Amount of GHG emissions per capita tonnes CO_2 eq/inh
Energy	Electricity use for street lighting	Electricity used for street lighting per km

(continued)

(continued)

Parameter	KPI indicator	KPI definition
Energy	Energy saving in households	Energy saving in households compared to a baseline. The baseline may be either a previous measurement or a reference value
Water, soil and noise	Application of city water monitoring through ICT	Proportion of the city water resources (rivers, lakes etc.) monitored by ICT with respect to availability
Water, soil and noise	Quality of city water resources	Quality of water resources (rivers, lakes etc.). Pollution of water resources including acidity, organic, floatables, algae, chemical substances and bacteria, etc.
Water, soil and noise	Recycling of waste	Proportion of waste recycled compared to total collected waste
Water, soil and Noise	Green areas surface	Proportion of municipal territory allocated as publicly accessible green areas

Of the six strategic objectives established by Smart Dubai [25] on its trajectory to transform Dubai into a smartest city in 2021, the clean Environment Enabled by Cutting-Edge ICT Innovations aims to:

a. Leverage ICT to ensure sustainability and quality of the Emirate's resources (water, air, energy, and land) for residents and visitors.
b. Deploy leading-edge, ICT powered demand and supply side strategies to improve resource efficiency and conserve consumption, and
c. Digitally transform utilities, manufacturing, transportation, and waste treatment sectors to reduce the Emirate's Carbon footprint for a cleaner, healthier environment.

Following sections capture the initiatives carried out by Smart Dubai strategic partners:

6.1 Green Initiative

Dubai endeavors to transform itself into a sustainable and green economy and to become the world's cleanest city in 2050 [26]. The ambitious goal is to reduce carbon footprint and achieve 75% renewable energy target by 2050. DEWA (Dubai Electricity and Water Authority) has brought down the solar tariffs to nearly US$ 3 cents for the 800 MW third phase of MBR solar park; the target is to bring it down to US 1 cent per KW. 1500 MW of projects are under execution in Dubai and this capacity will expand to 5000 MW by 2030 at a single site. The MBR Park will offset 1.4 million tons of carbon dioxide emissions.

What is driving this renewable solar revolution is the falling prices of photovoltaic cells and technology used to harness solar power efficiently. With the West and some European countries ending subsidies and introducing policies that are counter-productive to the growth of the industry, the UAE is sticking with solar and advancing with confidence. The speed of implementation of projects sets the City apart. What's commendable is people's participation where by Dubai Electricity and Water Authority's involve citizens and residents in the production of clean energy through Shams Dubai. The scheme encourages building owners to install solar panels to produce electricity.

Government also launched a number Green mobility projects that called on government entities to purchase 10% green vehicles and to use Euro 6 engine specific buses to cut down CO_2 emissions. RTA (Roads and Transport Authority) also plans to convert 50% of existing taxis to hybrid by 2021. Around 50 electric car taxis have been added including Tesla to support green mobility and plans are in place to add another 150 vehicles to reach 200 by 2019.

Dubai will also organize a competition for universities to design buildings based on solar energy in 2018 and another one in 2020. By 2030, Dubai aims to reduce water and electricity demand by 30%

The Ultimate objective is to have Green economy, green mobility and green city.

6.2 Dubai Carbon

Established in 2011 [27] by the Dubai Supreme Council of Energy (DSCE) and the United Nations Development Program (UNDP), Dubai Carbon runs a number of programs and projects to cater for the transition to a low-carbon and green economy. Programs such as the Carbon Ambassadors, Dubai Green Economy Partnership (Dubai GEP), Green Jobs and the Green deal facilitate capacity building and training for utility sector employees. The projects of Dubai Carbon Centre of Excellence (DCCE) focus on reducing food waste, generating bio-fuel from left-over food, rooftop solar panel design, energy saving lamps, and improve energy efficiency of existing exhaust fans. DCCE also created a set of 60 energy efficiency tips for residents and developed national greenhouse gas inventory systems and standards for energy attribute tracking systems. InSinkErator, part of Emerson has invented food waste disposers for home and commercial use that can power a 10 W LED light bulb for 17 hours with 1 lb of left-over food [28]. Thus Putting food waste directly into a digester can generate fuel and divert food waste from going to landfills.

6.3 Dubai Autonomous Transportation Strategy

The Dubai Autonomous Transportation Strategy [29] aims to transform 12% of city trips through autonomous driverless systems by 2021 and 25% of the total transportation in Dubai to autonomous mode by 2030. It is expected to bring AED 22 billion in annual economic revenues in several sectors by reducing transportation costs, carbon emissions and accidents, raising the productivity of individuals as well as saving hundreds of millions of hours wasted in conventional transportation.

The strategy will help cut transportation costs by 44%, resulting in savings of up to AED 900 million a year, reduce environmental pollution by 12% saving AED 1.5 billion, increase public transport efficiency ratio by 20% saving AED 18 billion, increase in individual productivity hours yearly by 13% Saving AED 396 million, reduce traffic accidents and losses by 12%, equivalent to savings of AED 2 billion annually and also reduce the spaces allocated for parking.

The strategy includes the launch of 'Dubai World Autonomous Transportation Challenge' as a global RFP to encourage the world's most innovative international companies, academic institutions and centers of research and development to test the latest advances in this technology by providing transportation solutions and scenarios that are realistic and tailored for the streets of Dubai. The participating companies will compete in order to apply the concept of autonomous transportation in Dubai within two main tracks. The first track named 'Last Mile Transportation' will transfer passengers from metro stations to a group of nearby destinations to increase the attractiveness of public transportation and the second track will use autonomous buses within specific neighborhoods to move people between specific destinations such as shops, schools, clinics and service centers.

According to studies by the World Economic Forum, 58% of the populations of major cities and up to 70% in Dubai prefer using autonomous transportation citing potential for increased productivity and elimination of the need for searching parking spaces.

Dubai also started constructing cycling tracks to provide suitable and attractive options for residents and tourists to encourage the sport of biking [30]. In 2018, Dubai has over 250 km of dedicated cycling tracks and the RTA plans to reach 560 km by 2022. These cycle lanes are intended to be used as environment-friendly means of mobility or practicing sport. Government also is promoting bike-sharing providers who offer bikes throughout the city. Dubai aims to be among the most cycling-friendly cities in the world ahead of Expo 2020.

6.4 Paperless Dubai

In 2021, Dubai government aims to go completely paper-free [31], eliminating more than 1 billion pieces of paper used for government transactions every year saving time, resources and the environment. As a fully paperless government, 100% of internal and customer transactions will be digitized from 2021. This means government will no longer issue or ask for paper documents across all of its operations.

The paperless strategy would apply technology as well as legal framework to address digital procedures to enable paper-free transactions. This could save enough money to feed 4 million children, prevent 130,000 trees from being cut down and save 40 h of productivity to give people more time to spend doing what they love.

As part of the paperless strategy and to motivate residents use online or mobile apps for government services [31], Dubai launched the "week without service centers" from October 21 to 25, 2018. During the entire week, 40 government entities offered around 1100 services only online and all customer service centers remain closed. The unified government services smart app "Dubai Now" app provided 50 smart services from 22 government entities enabling residents to carry out transactions on utilities, transport, driving, visas, housing, security, education, business and employment.

UAE TRA also has introduced SmartPass that allows app and online users to access UAE Government services using a single account. Users just need to authenticate once to access multiple on-line Government e-services. SmartPass provides identity verification and authorization and the user need to memorize only the SmartPass username and password to access all services. The UAEPASS provides a single digital identity that allows the user to access services for both local and federal government entities in addition to other service providers. The solution introduces mobile based authentication to users who can validate their identity using a smart phone. It also allows users to digitally sign and validate documents to minimize visits to service centers. The UAEPASS focus is on paperless society thus reducing carbon emissions from the use of paper and print accessories.

6.5   Clean Energy and Sustainable Living

Dubai Clean Energy Strategy 2050 [32] aims to diversify the energy mix so that clean energy will generate 7% of Dubai's total power output by 2020, 25% by 2030 and 75% by 2050. The strategy consists of five main pillars: Infrastructure, Legislation, Funding, Building capacities and skills, and an Environmentally-friendly energy mix.

Under the infrastructure pillar, Mohammed bin Rashid Al Maktoum Solar Park, the largest single-site Concentrated Solar Power (CSP) park in the world with a total investment of AED 50 billion ($13.6 Billion) is being built with a planned capacity to produce 5,000 MW to serve 800,000 homes by 2030. 13 MW Phase 1 was commissioned in 2013 and 200 MW Phase 2 in 2017. The 800 MW phase 3 and 700 MW Phase 4 will start from 2017 to be commissioned by 2020. It will also include 260 m high, world's tallest solar tower and world's largest solar thermal storage capacity. The MBR Park alone will reduce 6.5 million tons of carbon emissions annually.

The infrastructure pillar also includes a comprehensive innovation center that focuses on renewable energy, producing electricity using solar power and smart grids and water networks. It also includes the establishment of a new free zone called Dubai

Green Zone, dedicated to attracting R&D centers and emerging companies in clean energy.

The second pillar focuses on the establishment of a legislative structure supporting clean energy policies through the Shams Dubai. The third pillar of funding strategy is to establish a Dubai Green Fund valued AED 100 billion (USD27.23 Billion) for investment in R&D on clean energy and its applications. The fourth pillar aims to develop the capabilities of employees through global training programs in clean energy, in cooperation with international organizations and institutes, as well as international companies and R&D centers. The fifth pillar aims to create an environmental-friendly energy mix with solar energy generating 25%, nuclear power 7%, clean coal 7%, and gas 61% by 2030. The mix will gradually increase the use of clean energy sources to 75% by 2050.

Dubai Electricity and Water Authority (DEWA) have launched a number of smart environment initiatives that involves resident participation under the "Green Dubai" [33] program. These initiatives aim to transform Dubai into the smartest city in the world and enhance the quality of life. In 2011, the Dubai Supreme Energy Council formulated the Dubai Integrated Energy Strategy 2030, which requires renewable electricity sources to contribute to 15% of Dubai power needs. Around 80% of the total (596,000) numbers of mechanical and electro-mechanical meters in Dubai are replaced with smart meter and Dubai is on track to become smart water meters city by the end of 2019.

Dubai has embarked on massive plans to reduce reliance on conventional energy sources and increase renewable energy. DEWA's initiatives for a smart green Dubai are summarized below:

a. **Shams Dubai**: Aims to connect solar energy to buildings as a part of the Distributed Renewable Resources Generation program. It encourages household and building owners to install PV panels to generate electricity, and connect them to DEWA's grid. The electricity is used on site and the surplus is exported to DEWA's network. An inclusive approach of bringing the Customers, Consultants, Contractors, suppliers and manufacturers on a single digital platform. The SHAMS DUBAI CALCULATOR Web app allows marking the roof area on a map and computes monthly and annual figures of KWh of power, Kg of Carbon Emissions avoided, number of Trees Grown and km of Car Usage. Till December 2017, DEWA has connected 557 buildings in Dubai with a total capacity of 24.3 MW and is working towards doubling that number by 2030.

b. **My Sustainable Living**: To help residents make their home the most electricity and water efficient, this program lets customers to check, compare and monitor electricity and water consumption through the app. The consumption graph allows seeing one's own consumption and compare against other homes in the locality. Through the ideal home initiative, DEWA encourages customers to compete among each other in adopting conservation and best sustainable practices.

c. **High Water Usage Alert**: Smart water meters customers are notified of any high water usage if their last 48 h water consumption is higher than their average daily water consumption during the past three months. The alert indicates the possibility of internal water leakage within consumer premises, such as a broken pipe in order for the consumer to undertake preventive maintenance. Alert notification is triggered once in 15 days for 3 months and if the usage is not reduced, High Water Usage level is treated as normal water consumption. SMS or E-mail alerts are triggered on the registered number and through DEWA smart application. Previously it took 40 days, before resident were notified of the leakages. Of 800,000 customers, 100 leakages per month are reported. Around 20,000 water leakage cases and 4,700 faults were detected through this initiative. The Moro $10\times$ initiative and digital DEWA help Dubai residents to go for smart homes in which electronic devices in their home are connected to one cloud.

d. **Electric Vehicles (EV)**: Dubai ranks among the highest car oriented cities with 550 vehicles per 1,000 residents and a relatively high travel index of 1.84 in 2016, resulting in high CO_2 emissions. In line with the E-sayyara campaign launched by the Dubai Supreme council of Energy (DSCE), there is a very strong campaign to promote eco-friendly vehicles and encourage residents to buy e-cars. Around 55 new models of EV's are planned to be introduced in 2019 with several incentives to promote green mobility. As of 2018, there are 4000 e-cars on Dubai roads and the target is 270,000 e-cars by 2020. Similarly, Dubai Taxi Corporation's experiment proves that Hybrid cars can save 30% in fuel consumption and reduce 30% carbon dioxide emissions, thereby opting 50% hybrid taxis' by 2021.

e. **EV Green Charger**: Under the Electric Vehicle Charging Stations initiative, DEWA has installed 200 eV Green Charger stations for charging electric vehicles across Dubai. This initiative contributes towards DEWA's efforts to encourage the use of environmentally friendly electric vehicles in the emirate in order to reduce carbon emissions and support sustainable modes of transport in Dubai. Most of the charging stations are stationed in government offices, airports, petrol stations, shopping malls, commercial offices, clinics and hospitals, residential complexes and establishments. 3 types of green chargers are provided; Wall Box Charger, Public Charger, Fast Charger.

The Fast Charger (43 kW AC with Type 2 Socket, 50 kW DC ChadeMO and Combo CCS Sockets) provides an 80% charge within 20 to 45 min depending on the type of car and battery capacity. Most of these are installed at petrol stations. Public Charger (2×22 kW AC, with double Type 2 Socket) provides full charge in 2 to 4 h, depending on the type of car and battery capacity.

Wall-Box (22 kW AC, with single Type 2 Socket) provides a full charge in 2 to 4 h, depending on the type of car and battery capacity.

Green Charging facility requires the consumer to have a Green Charger card. Prior to September 2017, the cost of charging electric cars at public electric vehicle charging stations was 29Fils ($0.079) per kW, which is a great saving compared to fuel-powered cars. From September 2017 till December 2019, DEWA will provide free charging in public electric vehicle charging stations for electric vehicle owners registered in the Green Charger initiative. This initiative is to encourage the public to use electric vehicles in Dubai and to contribute to the protection of the environment. The Roads and Transport Authority (RTA) also provides incentives for electric vehicles, including free assigned parking, exemption from

RTA electric vehicle registration and renewal fees, exemption from Salik tag fee, and arranging a special sticker for number plates.

In addition to the fixed charging station, the Roads and Transport Authority also has deployed 13 Mobile Charging Stations for Electric Vehicles to take care of emergency charging needs. The charging vehicle can travel to recharge the customer's vehicle on demand to help reach the nearest electric recharging station. The idea is to boost public confidence in the use of electric vehicles as a reliable alternative to fuel-powered vehicles.

The ultimate aim of the Green Charger program is to reduce CO_2 emissions by 16% by 2021 in line with Dubai's Carbon Abatement Strategy.

f. **3D Printed Self Cooling Homes**: To achieve smart Dubai's plan for 3D printing 25% of Dubai homes, a home with 3D printed walls using eco-friendly geopolymer cement that cuts carbon emissions by 10% was exhibited in the 2017 Future cities show [34]. The low CO_2 concrete is fire resistant, salt and acid resistant, durable and due to thermal insulation, the need for AC in buildings will be 60% less during hot summer. Considering Dubai having 25,000 ongoing construction projects that consume 10 million tons of cement per year, an equal amount of CO_2 is emitted. If 25% of these buildings can be 3D printed, there will be a huge reduction in CO_2 emissions. 3D printing will also save construction costs, time and manpower and construction site will be clean and safe. If aligned with solar panels on house roof tops as part of DEWA's shams project and use transparent films in future windows to generate solar energy, this will be a big breakthrough.

6.6 Smart Environment Initiatives in Dubai Silicon Oasis (DSO)

DSO is the test bed for many of the greenfield smart city use cases in Dubai. Following are the initiatives undertaken by DSO towards smart environment.

a. **Green Roofs and Green Walls** [35]: DSO has 14% green area of the total 7.2 km^2 that it covers. The idea of green roofs was implemented starting from DSO HQ buildings covering 2000 m^2 and increasing to 2800 m^2. The green walls coverage has around 350 m^2. These initiatives help in creating urban ecological habitats, reduction of thermal loading to buildings thus lowering heat and cooling costs. It also help lower CO_2 emissions, reduction of heat island effect due to lesser reflected area, storm water attenuation via rain water harvesting system and recharge local hydrological cycle. Additionally, the foliage helps humidification through evapotranspiration as air purification plants are efficient filters of pollution, help reduce noise and have positive urban psychology uplifting effect on those who see it.

b. **Smart Parking** [36]: The IoT project has been carried out by Du for road surface parking using LoRaWAN communication to ease the search of free parking slots. The parking sensor consumes low power lasting 3 years, withstand high temperature, smaller, accurate, reliable and have faster detection. The sensors are inexpensive and easy to install. Data collected from the parking sensor nodes are transmitted using LoRaWAN base station to a cloud platform where the data is visualized, analyzed, and converted into smart city services.

c. **Smart Street Light poles**[37]: The smart street lights are programmed to provide a normal visibility of 25% until triggered to full power by approaching vehicles and pedestrians. This reduces the energy costs by 35%, maintenance costs by 42%, reduces carbon impact and prolongs the life of electric bulbs. The smart poles works on solar and are equipped with a CCTV Camera, Digital screen for ad signs, SOS communication button, WIFI connection, IoT gateway, CO_2 monitor, wind direction, Humidity cum Temp display and reports maintenance schedule.

d. **Smart Irrigation**: This innovative system currently waters more than 3000 palm trees and 70,000 m^2 of landscaped area covering half of all the green areas. This system has reduced irrigation water consumption levels by 30–40% and operational costs by 55%. DSO aims to implement this system across all green areas covering 950,000 m^2 by 2020.

e. **Smart Waste Management:** Introduced in 2014, IoT level sensors are installed on 130 garbage bins to measure the level of waste. The trucks need to visit only when bins are full, thus reducing the trips to pick up waste. This technology reduced the operations costs by almost 65% and helped lower CO_2 emissions.

f. **Smart lighting:** Replacement of traditional with LED lights, reducing power consumption by 23%. The bulbs lower carbon emissions and smart infrastructure alerts when lights need maintenance.

g. **Smart Bench:** This 100% solar powered smart bench is placed in DSO's Villa park. Equipped with sensors, photovoltaic modules and charging facility, the smart bench produces energy for consumption and can withstand Temperature from −45 to +60 degree Celsius and Humidity 1–100% shutting down if there is heavy rain. It also offers USB, cable, wireless charging and System sensor analyses every device inside the bench.

6.7 Smart Environment Initiatives in Dubai Municipality (DM)

Dubai Municipality launched its corporate policy [38] in 2016 on climate change in line with the emirate's vision for green sustainable development and to support in transforming Dubai into a low-carbon green economy. The policy outlines the main climate change issues targeting both the reduction of emissions and adaptation to the impact of climate change with regard to all operations and activities of municipality services. Other DM initiatives include Car Free Day initiative, Blue Carbon Project, Coastal Protection Project as well as projects concerned with the extraction of methane gas from the landfills and waste-to-energy conversion.

In June 2017, Dubai Municipality announced [39] its 2017–2021 Air Quality Strategy with an ambitious Dh500 million(USD136 million) plan to develop advanced monitoring systems and 'Smart Air Quality Stations' including Mobile monitoring stations to improve the air quality. The Environment Department responsible for the project analyzed data from past several years to create over 300 digital maps pinpointing the locations in which pollution levels increase, their sources and how they spread taking into account various climactic factors such as variations in temperature, humidity, speed and direction of winds. This forms a solid foundation for air quality monitoring. The Continuous monitoring of green house gas (GHG), ambient noise, electromagnetic radiation, air quality and water quality aims for a smarter environment.

a. **DM SAT-1 Environmental Monitoring Nano satellite**: The Mohammed bin Rashid Space Centre (MBRSC) and the Dubai Municipality(DM) signed an agreement in Dec 2016 to design and build the region's first environmental monitoring Nano metric satellites, named DM SAT1 [40]. Equipped with a multispectral remote sensing camera, DM1's main goal is to monitor air quality, pollution and greenhouse gas emissions (water vapor, carbon dioxide, methane, nitrous oxide and ozone) levels in the atmosphere. Weighing 15 kg and orbiting at an altitude of 600 km with a revisit of 14 times a day enabling it to observe the same location at least once every 5 days. Expected to be launched in Q2 2019, DM-SAT1 will help fulfill the commitment to reduce green gas emissions to the lowest possible levels by 2025.

b. **Smart Air Quality Stations**: Dubai Municipality has installed 13 smart stations [41] to monitor the air and water quality in different locations of the emirate covering various urban uses in the industrial, commercial and residential areas as well as on the major roads of the emirate. These smart stations are meant to monitor and control any emissions of air pollutants resulting from various sectors such as energy production, transport sector and the industrial sector. The target is to achieve 90% clean air quality in 2021 from 88% in 2015. Data from the smart stations have been used to identify hot spots that have the greatest vulnerability of contaminants. It is then used to frame policy and legal framework to penalize violators. DM is also able to communicate air quality to its residents through its online dashboards in order to increase environmental awareness and its adverse effect on public health and how residents can participate in reporting to help reduce the pollutants. Dubai environment portal allows residents to check near-realtime air quality in all 13 stations.

c. **Mobile Air Quality Monitoring Truck**: In Feb 2018, DM rolled out the German made monitoring vehicle [42] to collect and send real-time data to complement the 13 permanent stations (Fig. 11).
   This first environmentally-friendly mobile air quality monitoring station is a state-of-the-art truck equipped with artificial intelligence systems and tools. It will move around the city and provide real time data on air pollution to officials so that preventive and corrective action can be taken immediately. Industries and other firms found to pollute air can be identified onsite and ordered to take

Fig. 11 Dubai municipality smart air quality monitoring truck. Photo taken in GITEX 2018

remedial measures. This Euro 5 standards truck operated by the energy generated by the solar panels located on top of the station is able to monitor seven elements and compounds through a 100-metre spiral pipe, which will take samples of the polluting elements from the chimneys, including nitrogen oxides and hydrogen chloride. It has also sensors for monitoring 75 components and compounds of toxic pollutants and odours as well as the monitoring of levels of radioactivity, noise and meteorological data. The station is also equipped with 20 sensors and can monitor greenhouse gases causing global warming.

d. **Odour Management System**: The odour Management System aims to sense odour emissions efficiently and to quantify its impact outside the sewage plants.

e. **Smart Irrigation Management**: The smart irrigation system is implemented to manage all irrigation operations centrally. The system can be controlled from the office and it utilizes prevailing weather conditions, current and historic evapotranspiration data, soil moisture levels and other relevant factors to adjust water application to meet the estimated needs of green areas while minimizing watering excesses. The system uses sensors to provide Monitoring and control of the irrigation network, reporting and controlling upstream and downstream water levels, releases of water from tanks and gate valves operation, On-off, pressure, velocity and flow rates. This smart city project has helped reduce irrigation water usage by 30–40%.

f. **Air and Noise Sensor Monitoring Network**: The purpose of this smart city environment initiative is to build an extensive multi-sensor urban measurement network to specifically focus on noise and air pollution caused by road traffic. This initiative places IoT in various locations in the city and monitors noise and air pollution instantly and ingests the data into Smart city platform for analysis.

Cost effective sensors are used to implement this sensor network. The potentially lower data quality sensors are overcome by adding intelligence to the network at various levels.

g. **Dust Storms Prediction**: The UAE Ministry of Climate Change and Environment (MoCCAE) announced in October 2018 that it has set up a satellite system to monitor and forecast the air quality or dust storms 3 days in advance [43]. Since the prevalence of dust increases the cases of respiratory diseases such as asthma. This forecasting system can prevent sensitive groups from going outdoors.

h. **Smart Paint**: Dubai Municipality plans to apply smart paint across its public parks, bridges and tunnels to reduce air pollution [44]. Tests are carried out on 3000 m^2 area of park, which is equal to planting 3000 trees. Known to reduce air pollution by 20%, the smart Paint is made of materials that absorb carbon dioxide from the air. It can be applied in different colors thus serving the purpose of beautifying the city while enhancing air quality and preserving public health. The smart paint will be introduced in crowded areas across Dubai to help reduce carbon emissions of vehicles. The paint works with CristalACTiV photo catalytic technology where its titanium dioxide (TiO_2) transforms water vapor into hydroxyl and peroxyl free radicals at the surface. These free radicals break down nitrogen oxides (NO_x) coming from emission of vehicles once it comes into contact with the paint's surface. The harmful NO_x is then converted to nitric acid that is rapidly neutralized by alkaline calcium carbonate particle in the paint.

i. **Smart and Sustainable Waste Management**: Dubai produced a daily domestic waste of 9781 Tons and an annual waste of 3.57 Million tons in 2017. With a target of reducing waste reaching landfill by 75% in 2021, DM is carrying out a number of initiatives for waste recycling [45]. Some of the notable ones are: **Smart Sustainability Oasis recycling centers** located near public parks and municipality centers that allow residents to deposit 18 types of recyclable materials. There are 23 of them [46] operational in Dubai which are solar powered, self-efficient recycling center with built in sensors and CCTV cameras that are directly connected to headquarters.

Second project is installing **150 Bigbelly waste containers** that use solar power for 100% of its energy needs making it carbon neutral. Bigbelly is fitted with a compaction capability to hold six to eight times more waste than the average street bin. The volume sensors installed in the bin triggers compaction when the waste reaches a certain level. Each of the Bigbelly stations is geotagged allowing the headquarters to monitor the efficiency of the bins based on their location. The sensor also notifies headquarters to empty when the bins are almost full, allowing for logistics efficiency.

The third initiative is; **Nafith Smart Gate System** which is a fully automated entry management system at Dubai Municipality landfill sites. RFID plus Automatic Number Plate Recognition (ANPR) and integrated software are utilized to control entry of vehicles at sites, gather weight information and automatic credit deduction.

The fourth initiative is; the installation of **underground waste compactors** to address the waste disposal needs in busy and heavily populated public areas. This system has compacting mechanism to add storage, eliminates odors, protects the machinery from vandalism and provides a better aesthetic look in the area.

The Fifth initiative is a campaign called "**My City My Environment**" through which DM encourages resident participation in recycling. Through this, 2 bins are provided for segregating recyclables from general waste for residents. A number of private corporates and residential communities participate in embracing the green idea. The "green truck" distributes a bin to each subscriber and collects the recycles weekly to supply as raw material to factories. Some communities such as "Emirates living" provide all homes with a free outdoor recycling bin. "Take my junk" collects unwanted things from the door step.

j. **Solid Waste-to-Energy Plant**: In line with the national agenda to reduce waste landfills and to protect the environment from methane gas emitted by landfills, DM announced the Middle East's largest solid waste-to-energy plant in 2016 [47]. This waste incineration project is the first of the four projects to produce 7% of Dubai's total energy from clean energy sources by 2020. Built at a cost of USD544·5 M in Warsan District2, the plant will be operational in Q2 2020. The plant can process 2,000 metric tons of municipal solid waste per day to produce 60 MW of power.

k. **Dubai Coastal Zone Monitoring and Forecasting Program:** The Coastal Zone & Waterways Management Section (CWMS) of DM Environment Department has deployed smart coastal zone monitoring portal. Its near-realtime open web portal [48] provides a dashboard that shows live Beach conditions, wave forecast, interactive map, beach Cameras and latest sea conditions. There are 28 live cameras lined along Dubai city beaches with a 5 min refresh interval. The interactive ocean map view shows wind direction and speed, waves, currents, water temperature, salinity, sea surface height, inundation, bathymetry and areas of warning. Users can click on any of the 20 + monitoring stations on the map and view all the measured parameters and forecasts.

l. **Water Bottle Refill Monitoring**: To prevent using 5 gallon water refills more than the permitted 33 times, DM has enforced drinking water bottling companies to affix a laser printed QR code as a unique mark on the water bottle [49]. The smart sticker placed on the lid of each bottle, each time it is packaged in the production line, is equipped with digital technologies and secure features similar to those used to protect bank notes from forgery. The QR code on the lid holds data on factory name, date and time of production. Customers and inspectors can scan the QR code using the mobile app **watersmartrace** to check the number of times the bottle was refilled to know safety of the water for drinking.

m. **Drones Environment Inspectors**: DM has employed aerial drones to nab industrial polluters [50] who are harming air quality across the emirate, resulting in a 40% decline of Environmental industrial violations. The cement factories are asked to install air and odor pollution control units and monitoring units that are linked with DM's air quality database to check if they are exceeding permissible environmental limits.

n. **Environmentally-Friendly Vacuum Cleaners**: To remove sand and waste accumulated on the streets, DM has deployed 15 vacuum cleaners [51]. The equipment is 100% electrical, silent, freedom of movement, ability to maneuver widely, smooth starting and rapid spread to achieve the desired goal. They can collect a large quantity of waste as they have a high capacity of about 240 L. Additional features include energy saving night lights and warning panels installed on the front and rear. It easily works for 16 h non-stop with high flexibility.

o. **Smart Park**: Dubai Municipality is in the process of converting all public parks and facilities into smart parks [52]. Mamzar Park that adds a number of creative innovations which stretches the imagination of what can be done to make the park smart. Smart paint to purify the air, solar powered waste recycling bin, sensors to detect waste to empty, smart benches with Wi-Fi charger, smart oasis that turns humid air into drinking water and cools in summer, smart drones for rescue, smart band to track and trace children, virtual reality education for children on plants and an app to avail various facilities.

p. **Green Building**: Dubai is aggressively promoting Green building initiatives. Dubai Municipality, DEWA and other zonal agencies promotes the use of green building materials and have instituted graded certifications [53]. The Emirates Green Building Council (EGBC) is the national agency for green key certification, a standard of excellence in the field of environmental sustainability and sustainable operation in the tourism industry. Major commercial developers have secured green key certification for its hotels and resorts. Dubai Municipality has published green building guidelines in its website and has established Dubai Green Lab equipped for testing and certification of green materials and equipment.

q. **Recycling Used Cooking Oil to Biofuel**: The 14,000 eateries in Dubai discarded around 303,000 L of cooking oil per day in 2017 and in 2020 this will rise to 19,000 eateries and 378,000 L [54]. Through public-private partnership with Dubai Municipality, 60% of the discarded cooking oil is collected and processed in a recycling plant that converts 70% of the oil waste to clean water used for irrigation, 20% to fertilizer for farming and remaining 10% into environment-friendly biodiesel.

6.8 Smart Environment Initiatives from RTA

Dubai Roads and Transport Authority (RTA) is a strategic partner of Smart Dubai. RTA has launched a several smart environment initiatives. Some of the important initiatives are covered below:

a. **Smart Traffic Signals**: Dubai Roads and Transport Authority(RTA) have replaced 400 traffic lights into connected signals using 3G and wireless systems [55] with a seamless connection to the traffic control center. These wireless connected signals have high usability, efficiency and can be easily maintained.

It eliminates the lag in the traffic signal timing and is cost-efficient compared to the wired signals that required an intensive infrastructure in terms of cables and telephone lines to run the service nearby each signal.

b. **Smart Pedestrian Signals**: RTA has also installed 15 smart signals [56] in some of the busy pedestrian crossings in Dubai and 10 more are planned for 2019. The smart signal is triggered by sensors connected to a ground optical system synchronized with the operation of the signal. It reads pedestrians traffic on the pavement before crossing and the pedestrian path while crossing. It automatically adjusts the remaining signal time according to that reading to ensure a safe and smooth crossing of higher number of pedestrians. It is of particular benefit for pedestrians such as senior citizens, people of determination, and those with luggage or stroller who require higher crossing time.

c. **Robot Cleaner**: Dubai Roads and Transport Authority (RTA) has deployed robots [57] beginning 2019 to clean the floors of Metro stations. The Robot is 375 kg, 4 feet tall with a tank capacity of 90 L. It is environment-friendly and will reduce water usage through the built-in water purification system. Water consumption is reduced by up to 76% and the onboard chemical dosing system can save on cleaning solutions by up to 70%. It has spinning brushes to reach tight corners that can be used for mopping, sterilization and vacuuming. Working at a speed of 2 km per hour, the robot can run up to 4 h covering 1260 m^2 of floor area.

6.9 Other IoT and Smart Environment Applications

There are a number of other smart environment initiatives carried using ICT technologies that contribute to a smarter environment. They are as follows:

a. **World's Largest Vertical Farm**: Dubai Emirates Airlines has announced [58] to build a 147 m, 130,000-square foot, $40 million facility in Al Maktoum International Airport at Dubai World Central to produce 2,700 kg of vegetables, equivalent of 900 acres of farmland. Vertical farming has significantly smaller carbon footprint than traditional farming and uses 99% less water than outdoor fields with only 0.003% of the space. There are other vertical farms such as Badia farm which is the first vertical indoor farm that produce pesticide free vegetables. Other urban farming initiatives such as Aquaponics/Hydroponic farming, urban rooftop garden, Bio dome Greenhouses etc. are taking roots.

b. **Green Hotel Initiative**: Dubai sustainable tourism in a drive to help 700+ hotels save energy and reduce waste issued "12 Steps Towards Sustainability User Guide" [59] that illustrates energy conservation, water conservation, waste management and best corporate practices. The notable points are to use energy-efficient appliances, equipment, and lighting; install water-saving systems for showers, sinks and gardens, Reduce consumption of chemicals and hazardous waste and place recycling containers in each floor.

c. **Air Quality Mobile App**: The Environment Agency (EAD) in Abu Dhabi has launched a mobile app called Plume report [60] that offer live air quality data to residents and alert about air quality before they venture into outdoor activities. Due to the desert environment, seasonal sandstorms can trigger respiratory distress for sensitive groups such as children, elderly and those who are ill. 20 stationary and 2 mobile stations that collect data every minute is transmitted in near real-time for the Plume app users.

d. **Sustainable City**: Dubai based private developer; Diamond developers have developed a 46-hectare gated community that has 500 villas that has roof top solar panels, greenery and biodomes. The 10,000 m^2 innovation center set to be completed by mid-2019 will produce 40% surplus solar energy to export [61]. The sustainable city also has 42% reduction in power and 30% in water consumption and has helped the community avoiding 6500 tons of CO_2 emissions per year. The Sustainable city's total emission in 2017 is 8.76 tonne CO_2e (Carbon dioxide equivalent) against Dubai's 22.5 tons of CO_2e annually.

e. **Smart Environment-Friendly Road Paving**: The UAE Ministry of Infrastructure development's experiment using recycled materials including rubber tyres and organic materials in paving roads [62] found a 50% cost reduction and increase road's lifetime by 15 years. The waste cut rubber tyres, broken glass and concrete are mixed with the recycled asphalt to boost road's durability.

f. **AI Lab to Check Air Quality**: The UAE Ministry of Climate Change and Environment opened a lab that uses AI to monitor and predicts air quality across the UAE 3 days in advance using data received from 41 monitoring stations [63]. Residents can check instant readings and calculated forecasts of the air quality index (AQI) for the whole UAE through an App.

g. **Public Participation in Recycling**: In order to spread environment awareness and encourage public on recycling Reverse Vending Machines (RVM) are placed in high footfall locations [64] by Sharjah's environmental management company Bee'ah. By depositing a recyclable into the RVM, a receipt is printed with unique code that can be scanned through an app for a lucky draw. This incentive based approach help accelerate from a linear model of consumption to a resource recovery model for a sustainable future.

h. **Student Innovations**: In the "Think Science 2016", students of UAE University won for the idea to create paper from sand [64] by binding sand and polymer pellets. Inspired by the use of environment friendly rock stock, that is biodegradable requiring little energy and no water, they substituted sand that is locally available. Similarly, in the Expo2020 theme "what works forum" [65], students presented the idea to trap electrical charge from the pressure of moving cars. The energy thus trapped on high traffic volume roads can be transferred to power the street lights.

i. **Pneumatic Waste Collection Systems (PWCS)**: The global leader in the vacuum waste collection, the Swedish company Envac, has implemented the longest running PWCS systems [66] in Jumeirah Beach Residence (JBR), Dubai. Two separate systems have been in operation since 2007 and collect a combined total of 35 tons of solid waste each day from 36 residential towers, four hotels and over

100 restaurants. Installed at JBR as a retrofit almost one year after construction commenced, the Envac system became essential due to the lack of truck access to the buildings in JBR. The system created additional storage space for the residents with the removal of the need for conventional garbage rooms. The PWCS reduced the number of hours waste collection trucks spend at JBR each day by 90% and eliminated the storage of waste bins on the side of the road awaiting collection.

j. **Air Cognizer App**: This app needs a special mention here due to its disruption of Air Quality measurement using a smart phone app instead of using any sensors. Developed by a team of Machine Learning enthusiasts and Developers interning at IIT Delhi India as a part of Celestini Project India 2018 [67], the app computes Air Quality Index (AQI) from a photograph taken on the smart phone app with sky. The app combines data from the photo as well as data pulled from nearest meteorological station to compute the AQI. This may be good enough for cities that are highly polluted such as Delhi.

k. **Tesla's Waste Disposal Machine BlackHOLE**: Deployed in Ladakh to help solve the tourist waste problem, this machine [68] uses super plasma heat decomposition technology to turn non-biodegradable waste into ceramic ash that can be used for building roads and houses without any requirement for fuel and power. It can process up to 1 ton of garbage a day and can be operated by any layperson. This seems to be a viable solution to reduce the waste that goes into the landfill.

l. **City Air Management CyAM**: This software platform from Siemens [69] displays real-time air quality derived from sensors across a city and predicts values for the upcoming three to five days. These air-quality forecasts are computed with the aid of algorithms that tap into an artificial neural network and draw on historical and current data on air quality as well as weather and traffic patterns. The advantage of this system is its ability to recommend a selection of actions chosen from a set of 17 measures that can be implemented at short notice in order to improve air quality. For e.g. establishing low-emission zones, reducing speed limits and offering local public transportation services at no charge for a limited period. The system can simulate effectiveness of a particular measure.

7   IoT Implementation Challenges

Government agencies in smart cities are overwhelmed by the rapid changes witnessed in the IoT industry. The city administration is confronted with a number of challenges starting from the mandate, new infrastructure cost and maintenance, existing infrastructure and legacy systems, data accuracy, stakeholder cooperation, data security, flexibility to reprogram, multiple vendor management, service provisioning, Interoperability, Legal constraints and standardization.

Lack of sufficient resources and dedicated staff knowledgeable about the technology may be a very big challenge. So, most of the agencies either wait for the technology to mature or look for a champion who have implemented so that they can avoid the pitfalls. IoT as a technology itself is evolving and there are a number of challenges one must be aware in order to take care of them. Following section describes some of the common challenges articulated by industry leaders in their interviews and panel discussions:

a. **Loosely Coupled Solution**: IoT consists of a number of different components that need to operate in tandem for the whole system to work. This requires considerable time and efforts to set up and maintain the whole system. If one component is faulty, the whole system fails.

b. **Fragmentation of the IoT Market**: It is a challenge to manage the whole value chain. The IoT ecosystem consists of hardware device providers, radio communication manufacturers, data service providers, platform providers, cloud service providers and system integrators. To identify the right partner for each component and to have them present locally is indeed a challenge.

c. **Hardware Standards and Interoperability**: There are many communication protocols available in different countries. Accordingly, the hardware devices availability and their pricing differ. The same is the case with the cloud services and the applications developed in any IoT project. So, it becomes a challenge to look for a sensor connected to a certain cloud platform through a particular communication protocol.

d. **Wireless Technology**: Every year there is a new radio technology emerging and the users wait for the winner wireless technology to emerge and consolidate. This probably makes investment in the IoT project very uncertain. Since the infrastructure cannot be replaced quickly in the public sector, there is a delay in the project. This fragmentation seems to be continuing as each technology has its own pros and cons. As a result, it is necessary to have a dual communication in an IoT device due to the uncertainty and to avoid vendor lock-in in the future.

e. **Marketing Message**: There is a lot of hype built in the market and the reality is that most of them are work in progress and one or two years ahead of its availability. The hype ends with nice prototypes to make PoC's. Also, many vendors who release IoT products do not have certifications.

f. **Data Accuracy From IoT Devices**: The IoT sensors do not replace the conventional testing laboratories. These sensors should not be treated as calibrated high accurate equipment. On the contrary the near real-time data from IoT must be used as an indication to carry out detailed laboratory examination.

g. **Data Capturing Nodes**: Public agencies often go for a larger network of sensor nodes without really considering the quality and accuracy of the data that the sensors transmit. It is better to collect high quality accurate data from just the right nodes rather than massive amount of poor quality low accuracy data from several sensors. Quality and not quantity that matters. The sensor nodes require regular re-calibration and maintenance in order to maintain accuracy. Funds should be budgeted accordingly for this activity.

h. **Security**: Security is a challenge due to the loosely coupled solution. Security is required at all levels; privacy/confidentiality, authentication and data integrity. However, it is not recommended to use all security layers for every single application as it can be very time consuming and unnecessary. First make something that works and is usable, then secure it, but be conscious of the process. Higher security will be more complex and costly to develop and maintain. Around 80% of organizations that use IoT have experienced IoT related security breach. As IoT grows, organizations need to take steps to protect their networks and devices. Without gaining visibility of IoT activities, organizations are highly vulnerable to attack.

i. **Lack of Skilled Workforce**: Lack of a workforce prepared to program 20 billion devices that will be connected to the Internet by 2020. A Hybrid Engineer with cross-pollination of very different skill sets that need to work together are required in order to build IoT solutions that ultimately solve the customer's problems. It is projected that the market will need at least 4.5 million developers globally by 2020. Technology Universities need to take the lead on this and should impart necessary skills to guarantee the future of the IoT.

j. **IoT Maintenance Funding**: Many of the public sector IoT implementation dysfunction due to the lack of maintenance. Since IoT maintenance requires around 30 to 40% of the capital investment, allocation of the maintenance budget is important for a sustainable operation.

k. **Success Measure**: IoT solution must start by identifying the "pain points" and "pleasure points" together with the success measures. KPI's must be identified and the method of measurement must be defined with clear checklist. Constant monitoring of the KPI with the flexibility to adapt and if necessary, change has to be instituted.

l. **Securing Public Data**: It is extremely important to ensure privacy of the people by keeping it anonymous. At the same time, public has to be informed clearly what data is being collected, how is it secured, what is going to be done with the data, how is it going to be used and most of all how will they benefit.

8   Summary of Findings

As cities face increasing pressure on urban environment, tapping IoT technologies is vital for its sustainability and improving the quality of life of its residents and visitors. This chapter highlights smart city version 3.0 where citizen participation and co-creation are the essential ingredients for a happy city. Following points capture the essence of this chapter with a focus on the approaches taken by Smart Dubai.

a. **Clear Vision and Leadership**: The success story of Smart Dubai can be attributed to its visionary leadership and the support that it receives from the government. The institutionalization of the smart city with a transformation road map, backed by a clear vision to become the "Happiest city on earth" by 2021 may be THE single most motivator. Although each city requires a different approach and a different strategy, articulating how city residents will benefit through digitalization is important to solicit their participation.

b. **Mandate and Measurement**: Following the vision, clear mandate and higher committees to monitor progress of the initiatives on regular basis with policies, regulations and legal framework is essential. Smart Dubai in this case followed a well-coordinated approach of bringing together all of its government departments and authorities under one umbrella thus breaking silos in offering smart services. It also took an inclusive approach of bringing the private sector also into the fold. Implementation of continuous measurement criteria to measure the effectiveness of the service through happiness meters and through various other indicators allows making necessary changes if needed.

c. **Road Map**: A clear road map has to be set that needs to be published so that all the stakeholders including the residents are aware of the outcome of each initiative and how it will directly benefit them. Smart Dubai has a well-structured road map [70] website that gives a dynamic catalogue of all current and planned initiatives and services from government and private sector entities towards accomplishing the mission of Smart Dubai.

d. **Recognition**: Human beings respond to incentives! Creative Innovation to improve government services and recognizing excellence is the key to motivate each government department to become smart. Every city in the world has to learn from Smart Dubai on how its leadership has instituted prizes and awards that have not left anyone who need to be recognized. Almost every week there is a prize or an award ceremony that recognizes the innovators, winners of hackathons and students with a smart idea. The residents are well informed on smart initiatives by the media with a number of incentives to use smart services.

e. **Strategic Objectives to Use IoT**: Tapping IoT should be a means to an end. The focus should be to clearly understand what the final target is and then move towards identifying the steps on how to achieve it. Determine where IoT is required and how it is going to be used has to be well defined and aligned to international standards such as WHO. Smart Dubai has identified "Smart Environment" as one of its six strategic objectives as defined below [71]:

   i. Leverage ICT to ensure sustainability and quality of the Emirate's resources (water, air, energy, and land) for residents and visitors.

   ii. Deploy leading-edge, ICT powered demand and supply side strategies to improve resource efficiency and conserve consumption.

   iii. Digitally transform utilities, manufacturing, transportation, and waste treatment sectors to reduce the Emirate's Carbon footprint for a cleaner and healthier environment.

f. **IoT Strategy**: A citywide IoT strategy that takes care of the IoT ecosystem for the whole city including public and private stakeholders is essential to ensure the security and optimization of the resources. Smart Dubai launched its IoT strategy in 2017 that will coordinate the IoT policies across government departments and to establish a self-regulating IoT ecosystem.

g. **IoT Platform**: Selecting an agile IoT Platform that can absorb required changes needs a versatile solution provider. IoT is still emerging and developing very fast, but not yet fully matured. As such, vendor lock of IoT sensors, devices, data carriers and other enablers must be avoided. Often it is difficult and expensive to switch vendors and to the extent possible, it is better stick to multiple data transmission protocols. A unified city platform that will act as a central operating system for the city providing the dashboard to access city services for the residents seems to be good model. The Smart Dubai IoT Platform (SDP) set up with a private strategic partner has been chosen by UNEC and ITU for other cities to learn and adapt to their own city.

h. **IoT Security**: Designing and establishing a security strategy with well-defined security policies is critical element of IoT implementation. The entire IoT ecosystem is vulnerable for attacks and therefore adherence to security certification and using certified devices is a must. Smart Dubai's establishment of Dubai Electronic Security Centre (DESC) emphasizes the importance given to this aspect. Aggregation of data from all IOT connected devices using secure network protocols across the city and single dashboard access for organizations and residents reassures sharing privacy data for secure transactions.

i. **Open Data and Visualization**: Data is becoming the fuel of the 4th industrial revolution. Since IoT generates a huge volume of data, big data storage, analysis and visualization provides the end benefit for city residents. The big data need to be organized, managed and analyzed through data mining and data analytics. Hosting the application in more than one data centers rather than doing it for a single location may guarantee uninterrupted service and reduced latency in response times in case of any disasters. As the application will grow from proof of concept to deployment, scalability becomes imperative and has to be factored in hardware, data storage and funds for maintenance is also essential. Smart Dubai implemented open data and data visualization platform call Dubai Pulse through which residents can gain visual insights.

j. **Green Initiatives**: Dubai has implemented a number of initiatives to reduce carbon foot print and achieve 75% renewable energy target by 2050. Dubai experience shows that multitude of projects such as Dubai's solar park, electric vehicles, autonomous vehicles and green vehicle initiatives are required to achieve the goal of green and sustainable city. Dubai's paperless strategy aims to carry out all transactions without paper by 2021. Promoting residents to use mobile apps, smart pass and UAE pass for digital signature are aimed to achieve a paperless society.

k. **Clean Energy Strategy**: Dubai's diversification of energy mix and an aggressive target to produce clean energy in a phased manner from 7% by 2020 and increase to 75% by 2050 ensures sustainable living. Smart cities should consider setting long term goal in this respect and formulate capacity building and lay the necessary infrastructure to fulfill the goal progressively.

l. **Conservation of Water and Electricity**: Residents end benefit to reduce utility bills through installation of smart meters and to receive alerts on detection of leakage is conservation actions carried out by DEWA. At the same time, residents are also incentivized to install solar panels to generate for self-use and become an energy supplier. Similarly, Dubai's promotion of electric vehicles and charging station are smart initiatives to reduce energy consumption.

m. **Smart Environment Use Case POC**: Dubai Silicon Oasis (DSO) has been chosen as a District to test all the use cases such as green roofs and walls, smart parking, smart street light poles, smart irrigation, smart waste management and smart benches before roll out. Many smart cities have tried the same concept and are a good practice to emulate. Being a new development, DSO offers a green field test bed. Notwithstanding, the same idea can be applied by choosing a brown field district to test the IoT implementation before rolling out to cover the entire city.

n. **Smart Environmental Monitoring**: IoT can be effectively used to improve air quality, detect water leakages, reduce energy bill, reduce carbon footprint, control water quality, alert high UV index, respond to floods, prevent forest fires etc. Dubai municipality has implemented several smart IoT projects to monitor air pollution, water pollution, sound, odor, solid waste management, coastal monitoring and conversion of solid waste to energy to reduce landfills thereby reducing greenhouse emissions. Introducing a Mobile air quality monitoring to collect real-time data using a number of IoT sensors is a smart practice that can be replicated. Mobile monitoring can also be carried out using IoT sensors mounted on any vehicle or on city buses to monitor air pollution. Dubai RTA (Roads and Transport Authority) has also carried out a number of initiatives using IoT for traffic light, pedestrian crossings and deploying electric and hybrid vehicles for taxis and public transport to reduce carbon emissions.

9   Conclusions

Applying IoT technologies for Urban Environment monitoring is a powerful option to collect required datasets for diagnosing the root cause of the problem and to find alternative treatments. Smart Dubai experience proves that by applying IoT technologies, carbon emissions can be cut by 10–15%, water consumption can be lowered by 20–30%, and solid waste per capita can be reduced by 10–20%.

Building automation systems can reduce Greenhouse gas emissions by 3–6% and dynamic pricing of utilities can help shift load to off-peak periods and IoT sensors can prevent water leakages. Green mobility through e-hailing, micro transit for last mile, fuel-efficient fleet, intelligent traffic signals, congestion pricing can also help cut traffic emissions.

Using IoT sensors and sharing air quality through apps can help reduce negative health effects along with the measures to identify polluting industries and to tackle them from the source, as Dubai has done. Beijing was able to reduce air pollution by 20% through monitoring sources of pollution and traffic regulation. Smarter solid waste management has the benefit of collection cost reduction, recycling and reduce landfills.

A low-pollution and low-emission environment coupled with clean resources will ensure a sustainable development path for a smart city. A combination of initiatives, big bang and smaller ones are required to effectively reduce carbon emissions and to make the city more sustainable. Adapting an inclusive approach through public and private sector partnerships and active public participation can achieve the end goal of creating a clean and healthy city.

References

1. https://www.theguardian.com/world/2017/mar/21/ganges-and-yamuna-rivers-granted-same-legal-rights-as-human-beings
2. https://safesmart.city/en/smart-cities-3-0/
3. https://www.bloomberg.com/news/articles/1999-08-29/14-the-earth-will-don-an-electronic-skin
4. https://insights.sustainability.google/
5. https://www.globalcovenantofmayors.org/
6. https://www.fastcompany.com/90233731/a-new-use-for-google-maps-calculating-a-citys-carbon-footprint
7. https://www.gartner.com/imagesrv/books/iot/iotEbook_digital.pdf
8. https://www.internetsociety.org/iot. The internet of things: an overview, Oct 2015
9. https://newsroom.cisco.com/ioe
10. ITU-T Y.2060, Telecommunication Standardization Sector, Of Itu (06/2012), Series Y. Global information infrastructure, internet protocol aspects and next-generation networks. Next generation networks—frameworks and functional architecture models. overview of the internet of things
11. http://www.mckinsey.com/insights/business_technologythe_internet_of_things_the_value_of_digitizing_the_physical_world
12. https://bridgera.com/iot-system-sensors-actuators/
13. https://www.postscapes.com/what-exactly-is-the-internet-of-things-infographic/
14. https://www.kaaproject.org/what-is-iot/
15. https://internetofthingswiki.com/top-20-iot-platforms/634/;   https://www.khaleejtimes.com/editorials-columns/uae-leads-solar-energy-revolution
16. https://www.forbes.com/sites/louiscolumbus/2018/06/06/10-charts-that-will-challenge-your-perspective-of-iots-growth/#241e1cc83ecc
17. https://www.techtic.com/blog/top-10-open-source-iot-frameworks
18. https://csrc.nist.gov/publications/detail/nistir/8228/draft
19. https://aws.amazon.com/smart-cities/
20. https://iot.xische.com/
21. https://www.dayofdubai.com/news/smart-dubai-announces-smart-dubai-platform-strategic-partner-du
22. https://www.itu.int/dms_pub/itu-t/opb/tut/T-TUT-SMARTCITY-2017-PDF-E.pdf
23. https://www.dubaipulse.gov.ae/
24. http://www.itu.int/en/ITU-T/ssc/. Implementing ITU-T international standards to shape smart sustainable cities—the case of Dubai
25. https://2021.smartdubai.ae/
26. https://www.khaleejtimes.com/dubais-leap-into-green-economy
27. http://dcce.ae/project-experience/
28. https://www.menaherald.com/en/economy/energy/insinkerator-contributes-uae-state-energy-report-third-consecutive-year
29. https://www.dubaifuture.gov.ae/mohammed-bin-rashid-approves-dubai-autonomous-transportation-strategy/
30. https://www.rta.ae/wps/portal/rta/ae/home/about-rta/cycling-track
31. https://www.smartdubai.ae/initiatives/paperless

32. https://www.dewa.gov.ae:renewable-energy/publications/solarpark
33. https://www.dewa.gov.ae/en/customer/innovation
34. https://www.pressreader.com/uae/khaleej-times/20170403/281642485016918
35. https://www.dsoa.ae/en/dubai-smart-city/
36. www.libelium.com/resources/white-papers/
37. https://www.dsoa.ae/en/news/dubai-silicon-oasis-installs-first-smart-street-lights-in-the-uae-in-collaboration-with-du/
38. http://emirates-business.ae/dubai-municipality-rolls-out-first-corporate-climate-change-policy/
39. https://www.khaleejtimes.com/nation/dubai/dubai-has-grand-plan-for-cleaner-air
40. https://mbrsc.ae/en/page/dm-sat-1-nanosatellite
41. http://www.dubaiairenvironment.dm.gov.ae/home/index
42. https://gulfnews.com/news/uae/environment/mobile-air-quality-monitoring-station-to-track-down-violators-in-dubai-1.2168334
43. https://www.khaleejtimes.com/news/weather/Know-about-UAEs-dust-storms-three-days-in-advance
44. https://www.khaleejtimes.com/nation/dubai/soon-Smart-Paint-to-reduce-air-/pollution-across-dubai-1
45. http://en.envirocitiesmag.com/articles/generating_economic_development_through_integrated_waste_management/smart_and_sustainable_waste_management_in_dubai.phpt
46. https://lovindubai.com/feature/recycling-in-dubai
47. http://www.constructionweekonline.com/article-39667-dubai-to-construct-544m-waste-to-energy-plant/
48. http://www.dubaicoast.ae/en-gb
49. https://www.khaleejtimes.com/nation/dubai/new-smart-system-to-track-drinking-water-bottles-in-dubai
50. https://gulfnews.com/news/uae/environment/drone-inspections-help-cut-pollution-by-half-1.2263928
51. https://gulfnews.com/going-out/society/new-environmentally-friendly-vacuum-cleaners-on-dubai-roads-1.2190317
52. https://gulfnews.com/uae/environment/dubais-al-mamzar-beach-park-goes-smart-1.60338562
53. https://www.dewa.gov.ae/en/consultants-and-contractors/policies-and-regulations/circulars-and-forms/green-building
54. https://gulfnews.com/uae/environment/dubai-recycles-50000-gallons-of-cooking-oil-everyday-1.2009473
55. https://www.emirates247.com/news/emirates/rta-completes-final-phase-of-connecting-light-signals-with-traffic-control-center-2017-01-17-1.646553
56. https://rta.ae/wps/portal/rta/ae/home/news-and-media/ArchivedNews/ArchivedNewsDetails/archived+news+details/exhibiting+smart+pedestrian+signal+in+gitex+2017
57. https://www.khaleejtimes.com/technology/robots-to-keep-dubai-metro-stations-looking-clean
58. https://www.thenational.ae/uae/environment/dubai-government-agrees-on-deal-to-start-up-12-vertical-farms-in-the-city-1.748247
59. https://dst.dubaitourism.ae/
60. https://www.khaleejtimes.com/technology/air-quality-app-to-help-you-plan-outdoor-trips-in-uae
61. https://www.khaleejtimes.com/nation/dubai/new-sustainable-theatre-to-open-in-dubai
62. https://www.khaleejtimes.com/news/general/used-rubber-tyres-organic-materials-pave-uae-roads
63. https://www.msn.com/en-ae/news/uae/uae-ministry-opens-ai-lab-to-help-residents-check-air-quality/ar-BBPrN7m
64. https://gulfnews.com/uae/environment/beeah-offers-chance-to-win-by-recycling-1.1963473
65. https://www.thenational.ae/uae/uaeu-students-seek-to-make-paper-from-sand-1.662584

66. http://www.envac.ae/projects/jumeirah_beach_reference
67. https://sites.google.com/view/aircognizer/air-cognizer/technology
68. https://india.smartcitiescouncil.com/article/teslas-unique-technology-help-solve-waste-problem-ladakh
69. https://www.siemens.com/press/en/pressrelease/?press=/en/pressrelease/2018/corporate/pr2018070230coen.htm
70. http://roadmap.smartdubai.ae/
71. https://2021.smartdubai.ae/smart-environment
72. https://www.postscapes.com/iot-gateways/
73. https://www.iotforall.com/what-is-an-iot-platform/
74. https://www.i-scoop.eu/internet-of-things-guide/iot-platform-market-2017-2025/#Selecting_an_IoT_platform_business-related_and_functional_criteria
75. http://www.openiot.eu/
76. https://www.controleng.com/single-article/six-iot-implementation-challenges-and-solutions/39a725cebb69aeb73f65dfe6992be2b9.html
77. http://www.saviantconsulting.com/blog/iot-implementation-challenges-enterprises.aspx
78. https://www.itproportal.com/features/three-iot-implementation-challenges-and-how-to-overcome-them/

International Collaborative Research: "Smart Environment for Smart Cities" and Conclusions of Cities Case Studies

T. M. Vinod Kumar

Abstract This chapter has two parts. In the first part, the organizational details of the international research collaborative project "Smart Environment for Smart Cities" is discussed. In the second part are presented in consultation with the team leaders of the city study, their general conclusions of the study.

Keywords Organization study · Results

1 Smart Environment for Smart Cities

This "Smart Environment for Smart Cities", is an outcome of collaborative research aimed at developing the Design and Practice of Smart Environmental Resources Management for Smart Cities. The Smart Environment Resources are common pro-prieties where an active role of Government and People are required and hence its management is a joint and synchronous effort of E-Democracy, E-Governance, ICT and IOT system in a 24 h 7-day framework on any environment resource in any smart cities. The environment has no political or spatial boundaries and so it is a shared blessing/wealth and all countries need to be equally responsible in how we manage this common wealth. This commonwealth can become common misery of people if not properly managed.

The smart environmental resources management is a practice that uses information and communication technologies, Internet of Things, Internet of Governance (E-Governance) and Internet of People (E-Democracy) along with conventional resource management tools to realise the coordinated, effective and efficient management, development, and conservation that improves ecological and economic welfare in an equitable manner without compromising the sustainability of development ecosystems and stakeholders. This book will present 15 city case studies from 6 countries

T. M. Vinod Kumar (✉)
School of Planning and Architecture, New Delhi, India
e-mail: tmvinod@gmail.com

Besant Nivas, Jayanthi Road, P.O. Kolathara, Kozhikode, Kerala 673655, India

© Springer Nature Singapore Pte Ltd. 2020   491
T. M. Vinod Kumar (ed.), *Smart Environment for Smart Cities*, Advances in 21st Century
Human Settlements, https://doi.org/10.1007/978-981-13-6822-6_13

centred on one or many environmental resources each in study city. The Smart Environment energises, reorganises and transform the legacy environment to a smart environment embracing highly responsive ICT and IOT, creating a sustainable and harmonious environment, responsibly sharing this effort, by smart communities.

These 15 city studies conducted in this book start with a background chapter. The state of the art of eco system studies have a history of progress to create a world view which often is alarming but environmental action towards climate change, air pollution, and alleviation of ill health's due to the environment has not progressed as well as environment knowledge creation. Both aspects are equally important, and one cannot lag other. Our ancestors have kept the environment in a better shape than us which explains our own existence today. Otherwise we would have been lost among many species that is no more living. There was a strong cultural system practiced by households and communities among our ancestors based on a way of life of many religions which maintained a robust environment in a sustainable manner.

Smart Environment for Smart Cites can be conceived and developed emphasizing the sum of integration of internet of enabling ICT technologies for the smart environment, embracing a carbon neutral environment monitored in real time with a transportation system that performs intelligent and smart mobility based on real-time information drawn from big data using electric energy for movement. The authors of this book have adopted only one simple and dependable way of looking at Smart Environment and smart environment governance through smart activation of Smart People. Smart People need not embrace Capitalism since they have found that without much capital, not owning a brick and motor store, or a mall, they can be part of Amazon or eBay seller which is the biggest market place in the world sharing the ICT-enabled marketing and logistic system at low marginal cost and price. In the same way, they can also be a part of the largest taxi services like Uber with only one taxi at his disposal while Uber does not own even one taxi. They can be part of Air B and B without owning a hotel defeating the central concept of Capitalism. Since Uber just provides location-based computer and smart phone platform for the taxi service it does not have a huge and very expensive (with very limited benefit) bureaucracy that dictates movements unlike Socialism and Communism. Since real-time information and big data guide all these economic activities, socialism and its consequent ills of proliferating and expensive and mostly corrupt bureaucracy for tax payers with no value addition to the urban economy are no use for the Smart city economic development and smart environment.

During the colonial period, everything about life was centred around the bureaucrats of the colonialist with a Stockholm syndrome. Even when many countries became free from colonialism this dependency on bureaucracy was evident as if bureaucrat regulator can give leadership to create a smart environment which by the very job of the bureaucrat cannot. At best, he can implement policies of the elected government enshrined in legislations in close cooperation with people with a history of more failures than success. This does not mean any regulation or no law and order in Smart Environment. Like tax compliance by electronic filing of Goods and Service Tax, income and wealth tax, Smart E-Governance is there to replace age old and dysfunctional bureaucracy enshrined in a brick and motor building in a prominent place in the city, a legacy of the sixteenth century in many colonial countries. We consider Smart People and smart community and their E-democracy as all powerful and capable building blocks of Smart Cities for Smart Environment replacing capitalism and socialism as well as the largely dysfunctional bureaucracy of socialism at one go. For Smart People, there should be opportunities in Smart Cities for continuing education and skilling for Smart People to make them smarter today compared to yesterday in a rapidly urbanising world with new skills. This calls for the closer relationship of community with academia which does not exist today. Smart People can be everyone in a city irrespective of their wealth, educational qualification and social background and, therefore, an inclusive concept since all of them have a constructive role to play in Smart Cities to create the smart environment. They can be below the poverty level or above, which does not matter but all of them should have the wish to be Smart People and can be part of never-ending learning mode to be smart. Smart People through their E-Democracy and E-Governance plan, design and govern the Smart Environment system in Smart Cities. Smart People are the creators, governors, regulators, managers and maintainers of Smart City Environment. The required Smart Cities technologies which are ICT-enabled can easily be designed by Smart People's creativity in collaboration with academia and business community, and prototypes are made in Fabrication Laboratory (Fab Lab) if located in Smart Cities for later adoption and mass production and use. Being the creators of the Smart metropolitan city technologies, Smart People can maintain, repair, innovate and evolve the existing technologies to more cost-effective and functionally superior, next generation technologies which can be shared profitably with other smart cities for creating a smart environment the international goal. No one is running away from Smart People's creation and redevelopment of next generation Smart city technologies like some international firms.

2 The International Collaborative Research Projects on Smart Cities

This book is the sixth in a series Professor T. M. Vinod Kumar conceived, coordinated, implemented and edited about articulating the various roles of Smart People in Smart Cities.

The first book entitled "Geographic Information System for Smart Cities" [1] was aimed at creating a comprehensive self-awareness of city functioning every second and every day in real time which is the foundation of Smart city. Geospatial technologies, sensors and analytics can be used to reach the awareness and use it in real time for various types of use by Smart People. How it can be used for a variety of urban issues commonly observed globally is what that book is all about.

These Smart People thereby progress towards their self-directed goals, such as they demand Smart Living, Smart Environment and Smart Economic Development. They aspire to the highest level of quality of life in Smart metropolitan or megacity

city environment which they can very well afford and can expand many folds the economic development opportunities to satisfy higher income and employment needs to sustain Smart People. No smart person in a city is an island or elite, but they share a common destiny and common urban space, urban realm, and social and physical infrastructure. Government as the regulator is required that none of the Smart People is denied of all city provides for irrespective of their income level and social status or they are above or below the poverty level. Hierarchy of government exists in a city, but their governance needs to be for a Smart city that is fully aware of itself every second and as against Government who comes to know about the issue when a case is filed in the court which takes many decades to get a final judgement. The existing governance systems are obsolete being a product of sixteenth century or earlier designed for colonial rulers, built on the model of East India Company's administration in India or elsewhere which cannot be used or Smart city economic development and for the smart environment creation. However, those who aspire to live in Smart Cities are in the twenty-first century and no more part of an exploitative empire under the iron hand of a colonial administrator. Therefore, the twenty-first century Smart Cities require Smart city e-governance system that was the subject matter of the second book entitled "E-Governance for Smart Cities" [2]. This book is all about E-Governance in Smart Cities in action. It is divided into three parts, State of the Art Surveys, Domain Studies and Tools and Issue of E-Governance in Smart Cities.

The third book in this series is, "Smart Economy in Smart Cities" [3]. This book explores possibilities for rapid change in the income level and employment opportunities of those Smart People below or above the poverty level in a Smart city, and to make the NDP growth rate to a desired higher level consistently for the next many decades. Then, the current trend of urban local economic development is required to be converted to Smart city Economic Development. A, 10% NDP growth rate envisaged for the next three decades in India and many other countries can only be realized through Smart city Economic Development. Smart Cities and the related conceptualization boast of the Smart Economy but not much has been systematically researched or documented about it so far. This calls for a study of many cities across the world to document what constitutes a Smart Economy. There are two groups of cities being studied in this book. Some of them have been designated as Smart Cities by learned societies, but others are not but aspire to be Smart Cities. These call for different approaches to research design and studies. It was seen from case studies both these cases in different countries emphasize different approaches, establishing that there are no cook book solutions. The cities being studied in this book are spread across several major continents and regions, including North America, Europe, Africa, Indian subcontinent and East Asia. They are Ottawa in Canada; Stuttgart in Germany; Bologna in Italy; Dakar in Senegal; Lagos in Nigeria; Nairobi in Kenya; Cape Town in South Africa; New Delhi, Varanasi, Vijayawada, and Kozhikode in India; Hong Kong in China, Cape Town, Dakar, Nairobi and Lagos in Africa.

The fourth book in this series is "E-Democracy for Smart Cities" [4]. The world over, participatory democracy is worshipped and preached but what is practised is representative democracy at the city level and beyond. It is believed that in meta cities, megacities and metropolitan cities, an only representative democracy with elected representatives will work. However, democracy practised in small cities like Athens in Greece, and Licchavi in India in ancient times and many parts of the world documents face-to-face democracy in practice. In these cities, everyone in a city sat together and jointly decided on all aspects of the city during war or peace time. Citizens not only participated in decision making but acted together as one government and even as administrator for a task and as a regulator. There were no permanent administrators then. With the advent of ICTs in Smart Cities of the twenty-first century, it is possible to go back to the face-to-face democracy that, by any measure, is much superior to representative democracy. The fourth book is all about E-Democracy in Smart Cities in action. It is divided into three parts, State of the Art Surveys, Domain Studies and Tools and Issue of E-Democracy in Smart Cities.

The fifth book in the series is "Smart Metropolitan Regional Development: Economic and Spatial Design Strategies" [5]. Here metropolis also includes meta cities with 20 million and above population, mega cities with 10 million and above population and metro cities with one million and above. Here these cities however large these cities may be, need to be converted to smart metropolis using the specific design of economic and spatial strategies and not by purchasing smart technologies alone. The city studies for the "Smart Metropolitan Regional Development" result in many insights on many smart spatial and economic strategies using the Internet of Things, Internet of Democracy and Internet of Governance oriented to the specific issue of a town and its potential; taking into consideration that the Smart metropolitan city is an integrated six systems in which Smart Economy is an integral part. It can relate to Smart Mobility, Smart Environment or Smart Living. Based on the elaboration of Smart metropolitan city System, if one must develop any metropolitan-, region, then a country-and city specific economic and spatial design strategies for a Smart metropolitan city, must be designed based on a local ecological and cultural system of the city and not a type universal design. Location-specific and culturally acceptable economic and spatial strategies can be locally evolved, governed and managed. This is the only way local culture will find expression in the Smart metropolitan city using specific economic and spatial strategies by utilizing local, creative talents of smart people in many institutions in Smart Cities. There are 16 cities being studied in this project namely Pittsburgh in USA, Stuttgart in Germany and Naples in Italy, Dakar in Senegal, Conakry in Guinea, Abuja in Nigeria, Johannesburg in South Africa and Nairobi in Kenya, Ahmedabad-Gandhi Nagar, Bangalore, Chandigarh, Jaipur, Kozhikode, New Delhi, Surat in India and Hong Kong and greater Pearl River Delta Region from China.

The sixth book in this series is this book that will be published in 2019 by Springer-Nature is entitled "Smart Environment for Smart Cities" as a product of international collaborative research. This book is aimed at developing the Design and Protocol and Practice of Smart Environmental Resources Management for Smart Cities. Environment Resources are common proprieties where an active role of Government and People are required and hence its management is a joint and synchronous effort of E-Democracy, E-Governance and IOT system in a 24 h 7-day framework on any environment resource in any smart cities.

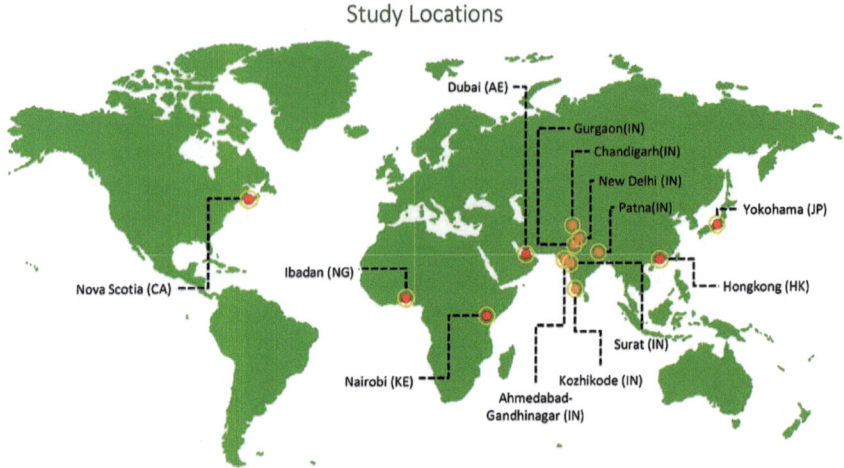

Fig. 1 Participating study areas for "Smart Environment for Smart Cities"

The smart environmental resources management is a practice that uses information and communication technologies, Internet of Things, Internet of Governance (E-Governance) and Internet of People (E-Democracy) along with conventional resource management tools to realise the coordinated, effective and efficient management, development, and conservation that improves ecological and economic welfare in an equitable manner without compromising the sustainability of development ecosystems and stakeholders.

Figure 1 presents the location map of the studies conducted in this book.

This book will present many city case studies (Hong Kong in China, Ahmedabad, Gandhi Nagar, Chandigarh, Kozhikode, New Delhi, Patna, Surat, Gurgaon from India, Yokohama in Japan, Nairobi in Kenya, Ibadan in Nigeria, and Dubai in UAE), that is centred on one or all environmental resource each in a city.

3   The Sixth Book on Smart Environment for Smart Cities

The editor and coordinator of the book series T. M. Vinod Kumar and many authors who participated in the earlier five books felt that there is a gap in knowledge about Smart Environment in a smart city. Funding for such collaborative research project was another issue. Universities and research centres dominated in collaborating these six-smart metropolitan city research projects. We also found that along with Universities, some not-for-profit national and international networks and institutions, city governments and regional governments in certain countries also came forward to participate in this collaborative research programme. The editor and coordinator of the project again felt that this international project shall not seek any external funding other than the internal resource mobilization from within the participating universities.

4 Design of the Collaborative Research Programme

Research Collaborations worked out is purely voluntary and without any financial support that binds a project together. Since collaborators are universities, Government, research institutions, professional networks and not-for-profit associations from six countries, complete independence for pursuing the research was there, free of the baggage of ideologies of granting organisation. They need not accept existing smart cities policies of study cities in their research. Coordinator Editor of the project has no financial or administrative control over any institution participating in the project since he was not in receipt of any grants and did not distribute it. Typologies of the institutions involved in this international project are given in Fig. 2. All these autonomous institutions are guided by the highest standard of scholarship and timely completion of research and publication.

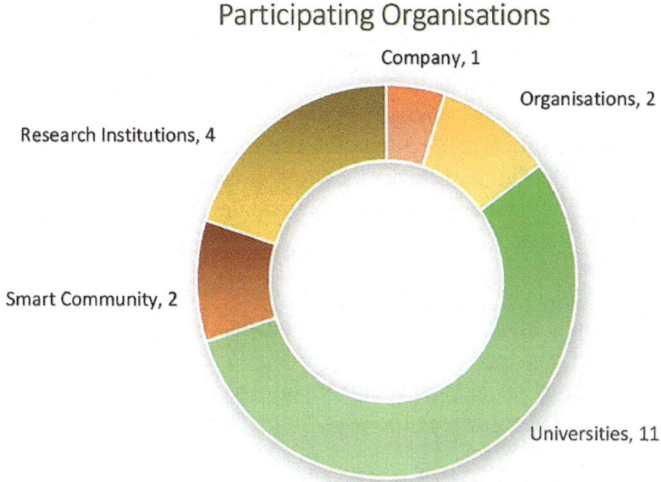

Fig. 2 Typologies of research institutions participating in this book

5 Research Questions on Smart Metropolitan Regional Development

The kind of collaboration in this international research project requires that all participating institutions shall formulate their own research questions and research the methodology which is of use to the country where these study cities are located. Depending upon the type of city some of which are leading Smart Cities, and some are not Smart Cities, the approaches must differ?

However, the paucity of empirical evidence on the Smart Environment for Smart Cities opens a new area of research: What strategy intervention brings about Smart Environment? This is the central focus of the book.

Do cities awaken social, cultural development and ecological (environmental) management through smart city development? This question lies at the heart of the proposed international collaborative research programme, and unpacking it gives us four interrelated research questions, as follows:

I. What constitutes a Smart Environment? This will need identification of the key ingredients and their role in making Smart Environmental and Spatial changes in Smart Cities.
II. What changes the Smart Environment brings to social development, cultural preservation, heritage conservation and ecological management? This calls for understanding the inter-linkages between Smart Environment in Smart Cities on one hand, with social development, cultural preservation, heritage conservation and ecological management on the other.
III. How and what processes facilitate the changes to the smart environment? Do cities bring to social development, cultural preservation, heritage conservation and ecological management? These may include:

a. innovation–diffusion (by ICTs and other modes),
b. spatial planning,
c. sectoral planning (including economic, social development, cultural preservation and ecological management),
d. heritage conservation and management plan and
e. institutional and governance processes, among others.

IV. How and what changes can be brought to improve the processes to achieve improved/optimal results? These changes related to the various processes as mentioned in Research Question iii.

A deeper understanding of changes in the social, cultural and ecological system of the Smart city with the advent of Smart Environment and Smart People for smart city development is the focus of study. This research programme and the institutions selected for this purpose as academic collaborators are an effort to address this research gap.

6 Scope of Research

The following outlines the areas that may be covered when conducting research under the "Smart Environment" programme. This is an indication only, and it is left to the team to decide what is appropriate.

I. A time series study of changes in the urban environment and identifying distinct features of evolving to the Smart environment.

II. Study of theories of environment interventions at the smart city level and modelling for study city.

III. The concept of accessibility in the Master Plan and its changes with respect to the increasing use of ICT in Smart Cities for the smart environment.

IV. Changing the role of the hierarchy of service areas or watersheds in a Smart environment in a Smart city as influenced by the increasing use of ICT.

V. Evolving structure of metropolitan urban agglomeration and changes required in a Smart city.

VI. Evolving structure of cities in urban agglomeration and changes required in view of the increase in the use of ICT.

VII. Change of spatial standards in a Smart metropolitan city.

VIII. Changes required in zonal policies and plans.

IX. Study of Town and Country Planning legislation and suggest changes as per the special requirement of Smart environment in smart cities.

X. Change of role of community-based organizations (for example, Residential Associations in India) in a Smart city with an increase in the use of ICT.

XI. Change of role of Ward Committee in a Smart city with an increase in the use of ICT.

XII. Change of role of the Municipal Council in a Smart city with an increase in the use of ICT.

XIII. Change of role of the Planning system in a Smart city with an increase in the use of ICT.

Note: The scope of research can be further elaborated by the collaborating institutions but need not be uniform for all study cities. Each department of university participating in this research programme shall incorporate relevant Smart Environment Development features appropriate to the goals of each department. The coordinator of this project does not intend to dictate the direction of the research and have a diverse group of collaborating universities, and they should orient their study strictly based on academic goals of their department.

7 Study Cities

The study city will be selected as a study area by each of the collaborating universities independently, which will be the place the one-year and two-semester combined effort to conduct this research. Universities participating in this programme adopted different types of collaboration. Some universities used, their doctorate and post-doctorate students, while others used students at masters and first professional degree level. A post-doctoral student in the department can work on a narrow subject area in the study as individual work. While graduate and undergraduate students can work on design solutions for the Smart Environment, and Research institutions can charter their own strategic areas of research.

8 Project Details

One City will be selected as a study area by each of the collaborating universities independently, which will be the place the one-year and two-semester combined effort to conduct this research. Universities participating in this programme adopted different types of collaboration.

The project details of the study city are given in Table 1.

9 Way of Working the Programme

9.1 Integrating Smart Environment Research with Academic Programmes

This international collaborative research programme, with the participation of 6 countries and 15 study cities as tabulated above was conducted by many diverse university departments, research institutions and others as shown in the graph and table.

9.2 Role of Students

This international collaborative research programme is essentially meant for students since they are the future and being part of an internal academic programme of the university. We consider they are the main actor and shall be given an important role in this programme. Perhaps many that age group will live in the Smart Cities than their older faculty. Under the direction of faculty new concepts were introduced in the studio and empirical studies were conducted around these concepts.

Table 1 Project details

Project Details
Authors, Countries and Institutions

S No.	Study City	Country	Authors	Institutions
1	Patna	India	Veena Aggarwal	The Energy and Environment Resources Institute, New Delhi,
			Bibhu Prasad Nayak	Tata Institute of Social Sciences, Hyderabad
2	Dubai	UAE	Ummer Sahib	Informap Technology Centre, Dubai
3	Ahmedabad-Gandhinagar	India	Jignesh Bhatt	Dharmsinh Desai University, Nadiad
			Omkar Jani	Gujarat Energy Research and Management Institute, Gandhinagar
			Chetan Bhatt	Government Engineering College, Sector-28, Gandhinagar
4	Nova Scotia	Canada	Barry Gander	CATA, i-CANADA, i-Valley, Canada
			Bill Hutchison	i-CANADA, Canada
			James Boxall	Department of Earth Sciences, Dalhousie University, Canada
			John Reid	Canadian Advanced Technology Alliance, Canada
			Patrick Maher	Community Studies & Outdoor Leadership University Teaching Chair, Community Engaged Teaching & Scholarship, Canada
			Robert Maher	Senior Research Scientist in Applied Geomatics Research, Canada
			Terry Dalton	i-valley, Canada
			Tony Walters	i-valley, Canada
5	Kozhikode	India	T.M. Vinod Kumar	School of Planning and Architecture, New Delhi
			Mohammed Firoz	National Institute of Technology, Calicut
			Bimal Puthuvayi	
			Praveen Sankaran	
			P.S. Hari Kumar	Centre for Water Resources Development and Management, Kozhikode
6	G Nagar	India	Asfa Siddiqui	Indian Institute of Remote Sensing, Dehra Dunn
			Pramod Kumar	
7	Ibadan	Nigeria	Femi Aiyegbajeje	University of Ibadan, Ibadan, Nigeria
			Femi Olokesusi	University of Lagos, Lagos, Nigeria
8	Chandigarh	India	Prabh Bedi	School of Planning and Architecture, New Delhi
			Mahavir	
			Neha Tripathi	
9	Nairobi	Kenya	Romanus Opiyo	Department of Urban and Regional Planning, University of Nairobi, Kenya.
			Silas Muketha	
			Dennis Mwaniki	UN Habitat, Nairobi, Kenya
			Wilfred Ochieng	Planning and Development Department, Kisii University, Kisii, Kenya
10	New Delhi	India	Shovan K. Saha	School of Architecture and Planning, Sharda University, Gautam Budhh Nagar, India
			Achintya Kumar Sengupta	
			Mahendra Sethi	
			Sewa Ram	School of Planning and Architecture, New Delhi
			Saumya Saxena	National Institute of Urban Affairs, New Delhi, India
11	Hong Kong	China	Sujata Govada	Institute for Sustainable Urbanisation, UDP International, Chinese University of Hong Kong
			Edwin CHAN	Research Institute for Sustainable Development, Hong Kong Polytechnic University, Hong Kong
			Jin-Guang Teng	Research Institute for Sustainable Development, Hong Kong Polytechnic University, Hong Kong
			Anqi WANG	Department of Building and Real Estate, Hong Kong Polytechnic University, Hong Kong
			Timothy Rodgers	Institute for Sustainable Urbanisation, UDP International, Hong Kong
12	Gurgaon	India	Asmita Bharadvaj	Ansal University, Gurugram, India
13	New Delhi	India	Ashok Kumar	School of Planning and Architecture, New Delhi
14	Yokohama	Japan	Aki Suwa	Kyoto Women's University
15	Chandigarh	India	Kshama Gupta	Indian Institute of Remote Sensing
			Mahavir	School of Planning and Architecture- New Delhi
			Kshama Puntambekar	School of Planning and Architecture Bhopal
			Pramod Kumar	Indian Institute of Remote Sensing
16	Patna	India	Veena Aggarwal	The Energy and Environment Resources Institute, New Delhi
			Bibhu Prasad Nayak	Tata Institute of Social Sciences, Hyderabad
17	Surat	India	Bhasker Vijaykumar Bhatt	Bhaikaka Centre for Human Settlements (APIED), Vallabh Vidyanagar, Anand
			Shashikant Kumar	
			Neeraj D. Sharma	Sitarambhai Naranji Patel Institute of Technology & Research Centre, Umrakh, Bardoli

9.3 Role of Faculty

The faculty is the designer of the program within the framework of existing curricula in design studios and theory courses of each participating university.

   I. The project duration is one academic year or two semesters.
  II. They guide and monitor student work as usual as part of the academic programme.
 III. They monitor students' input to the monthly progress report.
  IV. They rewrite the output of the project for a book to be published by an international Publisher giving due credit to their work.

9.4 Co-design and Co-production of Knowledge

This international collaborative research programme is founded on the principles of co-design and co-production of knowledge. In today's interconnected world, such collaboration is physically and intellectually possible—thanks to the Internet and ICTs. The collaborative aspect of the research programme will be actualized in the form of:

   I. Co-design the programme with the partner academic institutions.
  II. Co-production of knowledge through an interactive process of sharing, reviewing and finalizing research findings.
 III. Within each partnering institution, co-design and co-production of knowledge can be implemented through design studio/laboratory work between faculty and students.

9.5 Research Output

The key output of the "Smart environment for smart cities" research programme will be a book edited by the coordinator Professor T. M. Vinod Kumar, to be published by Springer-Nature, an internationally reputed publisher in 2019.

10   Bulletin

During the conduct of research about 12 months through in 2018, two Bulletin has been used to communicate with the international teams of researchers. These Bulletin highlights study city profiles selected by the various study teams independently and introduces to the research network the research methodologies adopted, and the profiles of authors of the research output for the book, "Smart Environment for Smart Cities". The Bulletin is jointly edited by a Bulletin team among authors, Jignesh Bhatt, Aki Suwa and Bimal. P.

The list of Bulletin produced with respective case studies is shown in Table 2.

Table 2 List of Bulletins

No.	Country	City	Bulletin number/date
1	UAE	Dubai	31 March 2018
2	Hongkong	China	31 March 2018
3	Ibadan	Nigeria	31 March 2018
4	Nairobi	Kenya	31 March 2018
5	Nova scotia	Canada	31 March 2018
6	Yokohama	Japan	31 March 2018
7	Ahmedabad-Gandhinagar	India	30 June 2018
8 and 9	Chandigarh 1 and 2	India	30 June 2018
10	Gandhinagar	India	30 June 2018
11	Gurugram	India	30 June 2018
12	Kozhikode	India	30 June 2018
13 and 14	New Delhi 1 and 2	India	30 June 2018
15	Jaipur	India	30 June 2018
16	Surat	India	30 June 2018
17	Patna	India	30 June 2018

A cover page of the Bulletin 3 and one sample page of the Bulletin is given below

Bulletin presenting ongoing research of groups participating in International Collaborative Research Project to be published as a book in the year 2019 by Springer-Nature

For private circulation only

Editorial Team
JIGNESH BHATT, AKI SUWA & BIMAL P.

Smart Environment for Smart Cities

2/2, Year 2018, 30th of June

Presented in this Bulletin

AHMEDABAD-GANDHINAGAR ~ CHANDIGARH ~ GANDHINAGAR ~ GURGAON ~ KOZHIKODE ~ NEW DELHI ~ PATNA

By Authors: Jignesh Bhatt, Omkar Jani, Chetan Bhatt, Prabh Bedi, Mahavir, Neha Tripathi, Kshama Gupta, Mahavir, Kshama Puntambekar, Pramod kumar, Asfa Siddiqui, Pramod Kumar, Asmita Bharadvaj, T.M. Vinod Kumar, Mohammed Firoz, Bimal P., P.S. Harikumar, Praveen Sankaran, Shovan K. Saha, Achintya Kumar Sengupta, Mahendra Sethi, Sewa Ram, Saumya Saxena, Ashok Kumar, Veena Aggarwal, Bibhu Prasad Nayak, Bhasker V. Bhatt, Shashikant Kumar, Neeraj D. Sharma

A page of Bulletin is as shown below.

SMART ENVIRONMENT: THE UNIFYING SPARK FOR A SMART COMMUNITY
CANADA
Barry Gander, Bill Hutchison, James Boxall, John Reid, Patrick Maher, Robert Maher, Terry Dalton, Tony Walters

HOW SMART ENVIRONMENT PRACTICES CAN ACCELERATE THE OVERALL DEVELOPMENT OF A "SMART COMMUNITY", FOCUSING ON NOVA SCOTIA – THE WORLD'S LARGEST 'SMART COMMUNITY'

Theme expansion

- Nova Scotia is a region covering 55,000 sq. km.'s – seven times larger than the world's largest city (New York).
- It is a well-defined area, bounded by the ocean on three sides, with a rich cross-cultural history that, over centuries, has melded the peoples together into a self-referencing entity...a community.
- With the advent of the unifying power of ICT and Internet-enabled communications, the province is uniting into a holistic Smart Region or Smart Community -- effectively, a unified city with diverse occupations and landscapes, and one central identity and focus.
- The study of and smart use of the environment is one of the strongest drivers for ecological and cultural evolution in Nova Scotia

- This gives readers the benefits of:
  - o Analysis and study of the biosphere to determine the effects of Smart Environment practices;
  - o Lessons that readers can identify with, expanding the number of cities and audiences for the book;
  - o A sharp focus on coordination and its role in successful cross-sector Smart holism;
  - o Tools that can be brought into play for Smart practices, increasing the effective solutions that readers can identify with; and
  - o A great variety of population densities and needs, leading to requirements for approaches that are multi-faceted and varied.
- The framework of the book is captured in the comparison of the Smart Community with the Smart Environment:

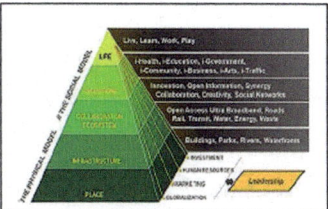
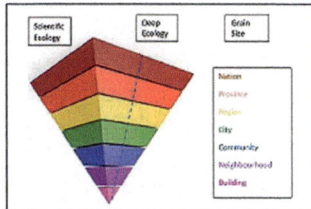

11   Summary of Conclusions and City Case Studies

Team leaders of authors were requested to prepare a summary of conclusions of their city studies and the Editor compiled it and is given below.

11.1   Smart Environment for Smart Cities

Many environmental issues that affect all of us had accelerated in the last decade. An overview view of environmental issues of the universe is presented as a backdrop for research studies of this book in the First Chapter in a very brief manner. These are our global environmental challenges. The chapter studies these major global environmental issues and shows how it is being solved. Then, the smart environment is defined followed by the definition of smart cities. City society through their legal framework of constitution brings about the creativity of design of an environment and spatial strategy to take these cities to the next level facing all local challenges through their official Master Plans. Our Cities are continuously being exposed to environmental changes and require immediate and intermediate range strategies to face it which is not there in a 20-year Master Plan of cities. These periodic environmental strategies of cities call for differing approaches to intervene in the emerging global situation. To reshape these emerging environmental challenges, there is a need to design a smart environment in smart cities. It is the local culture based on religion and faith which made us interact with the environment in a responsible way in the past centauries. This is discussed in this chapter with the Vedic and Buddhist approach presented as examples.

Main conclusions brought about in Chap. 1 is summarised below.

1. Smart Environment has been defined in this book. "Smart Environment is a knowledge based environment that develops extra ordinary capabilities to be self-aware, how it functions 24 h and 7 days a week and communicate, selectively, in real time knowledge to citizen end users for a satisfactory way of life with easy public delivery of services, comfortable mobility, conserve energy, environment and other natural resources, and create energetic face to face communities and a vibrant urban economy even at a time there are National economic downturns" This is similar to the definition of a smart city used in this book.

2. Just like Smart City system, it is a six-component integrated system such as Smart People, Smart Mobility, Smart Economy, Smart Governance, Smart Environment, and Smart Living; Smart Environment can also be considered as a six-component system namely for example for water element of environment, Smart water community, Smart Water Mobility, Smart Water Economy, Smart Water Governance, Smart Water Environment and Smart Water Centred Living. The same type of smart environment system can be designed for say Smart Forest.

3. Three types of design need to be generated for the smart environment. They are ICT and IOT system specific to the local environment needs, E-Environment Democracy and Environment Governance. All these three designs shall have linkages with six Smart Environment System.

4. IoTs are a network of technologies which can monitor the status of physical objects, capture meaningful data, and communicate that data over a wireless network to a computer in the cloud for software to analyse in real time and help determine action steps. Typically, each data transmission from a device is small, but the number of transmissions can be frequent. Each sensor will monitor temperature, pressure, rainfall or water quality and a natural part of the environment such as an area of ground to be measured for moisture or chemical content (pollution).

5. As many IoT nodes are spread across the study area, the setup requires a robust communication technology that covers a wide geographical area and can handle a huge amount of data traffic. For the unconstrained IoT nodes we use the traditional LAN, MAN, and WAN communication technologies such as WIFI, optic fibre, Ethernet, and broadband.

6. In a smart environment system, there are design components and protocol component. While design can be generic, the protocol can be specific to environment issues.

7. The environment is the eco system. The Scientific study of the eco system has progressed considerably around the world which may be called scientific ecology. We have now a basket of descriptive models of the eco system, diagnostic models and predictive models which can be calibrated using mathematical tools. We have analytic tools to determine the environmental impacts. In other words, the knowledge base of the environment is available but requires more development within the university system of research to answer emerging environment issues which is ever changing. These models have remained within the confines of universities but have not come down to the field for applications where it is required to sort out environmental issues on a day to day basis. All means available knowledge should be used to share these environment knowledge bases based on the most up-to-date data.

8. The environment or eco system become smart when environment knowledge base, or ecosystem knowledge base is available at the place and to people it is required for acting. This calls for developing analytics for a common man based on these models available in cloud computing system accessed by apps in the smart phone and can be used to sort out emerging issues with or without human intervention using automata. This can only be executed at household and community level faced with identical problems addressed in the apps. The job of creating apps for common men based on the latest environment information and analytical models is the unfinished or never attempted jobs of universities or business. The other unfinished or never attempted job of the university and Non-Governmental Agencies is to strengthen community and households to be smart with minimal but up to date knowledge base of the eco system and

analytic apps that enable the household/community to sort out the issues using apps in smart phones.

9. The domain of knowledge base required to manage the eco system issue at one location cannot be found in one department of studies in these universities alone which make this operation of linking university, developing ICT and IOT infrastructure and smart phone apps for the field more complex. The best remedy is community concentrate first on the most pressing environmental issues and then add the next one to the list subsequently. For example, if a community has water related and sanitation related issues with no interconnections with each other and water issue is the most pressing issue, then start with water issue and a water community who subsequently embrace other issues.

10. There is a continuous training requirement for the community to make them capable of intervening scientifically and based on most current information to face the environment issues which can be an important task of the local university or business. Here the university/business train people with differing education level to face the issues to tackle the problems using mobile computing.

11. Environmental issues and required timely intervention require an appropriate ICT and IOT infrastructure to be developed in the study area for effective interventions again by the university/business which they are capable of. The technical management of these systems to tackle environment issues can also be designed by the local university/business. Internet based technical protocol for managing the ICT and IOT system can also be developed by the local university. Alternatively, this can be a livelihood of young university graduates doing related business.

12. In addition to ICT and IOT design and deployment for issue-based environment management at community and household level, there is a need to develop E-Environment Democracy Applications and E-Environmental Governance Applications as per specific needs.

13. E-Environment Democracy is the rightful role of citizens to participate in the environment management of smart cities as per the constitutions of the country using electronic means such as high-speed fibre optic internet network, mobile computing as well as Internet of Things that can transform the community to smart community with dominant role of citizen, and technology in smart cities. Majority of environment resources such as the air we breathe, the water we use, the public domain urban space and property we use every day are all common property resources and E-Democracy is the only way it can be used efficiently, responsibly and rationally.

14. The issues E-Environment Democracy faces are spatial environment issues which can only be solved by the domain eco system knowledge base. It requires spatial analytics using Geographic Information System based well-developed suites of Spatial Decision Support Systems mobile applications to intervene effectively. There is not much development taking place in this area and smart cities can form smart cities association and jointly develop such applications with local universities and share freely.

15. E-Environment Democracy ideally should follow Deep Ecology principles stated in this chapter integrated with scientific ecology to create a smart environment. This calls for a change of view on environment emanated from the first industrial revolution that advocates that the environment resources need to be conquered to meet the never-ending greed resulted in colonialism and slave traffic and greed-based living. These approaches should perish and replace with thoughts discussed in this chapter based on a Vedic view of the environment and Buddhist view of the environment or any other religion being practiced.

16. E-Democracy requires directive principles which are stated in this chapter.

17. Most of the E-Governance applications are not based on water or other natural resources management using existing legislations. They are mostly centred around office management of clerical operations such as issuing birth and death certificate or caste certificate or land parcel GIS. The absence of E-Environment mobile applications to monitor and stop point source pollution or non-point source pollution is not there and the existing implementation of environment governance is far from satisfactory.

18. E-Environment Governance and E-Environment Democracy can only be implemented by the smart community including households since they know the issues and they want a remedy. A registration under Indian Societies act or similar in other countries makes this community a legal entity in India which can access Commercial Bank loans and are accountable to Government. These smart community shall be armed with technology, domain knowledge base, fully operational E-Democracy, E-Governance and continuing education set up to be effective.

19. A smart city should not be considered as a technological artefact. It is an ecological and cultural system. The ecological system defines the environment issues and can be diverse at every location within the city and may differ from other cities. A cultural system that interacts with the environment arises out of the way of life which is influenced by religious practices.

20. The manifestation of the cultural system in a city is determined by people who live there with their differing religions and caste compositions in a country that follows secular constitutions like India. These religions influence their followers to adopt a distinct world view of the environment as per their scriptures which are discussed in this chapter for two sample religion the Hinduism and Buddhism. They also have environment ethics on how to intervene with the environment. The Spatial Decision Support System Mobile applications should consider these facts while formulating applications as a robust value system based on religious belief.

21. Smart Community is the creator and sustained of the smart environment. Their cultural system is the key to community action which cannot be dictated from outside. The environmental view of the religion they practice is that one which directs their action.

22. The Environment view of Vedas gives how Veda view the environment and establish a protocol for environmental intervention by people and same with Buddhism. This is discussed in this chapter.

23. The way many schools of Buddhism such as Mahayana, Theravada, Chan and Zen Buddhism and Buddhist Tantric practices view and intervene with nature in a different way reinforces the cultural system. This is discussed in this chapter. Buddhism consider all elements of the environment as equal and even nonliving. Humans are not superior to these elements. The conditioning of mind and practice of mindfulness is the basis of intervention with the environment as Buddhist practice.

24. Religious and caste leader has a role in shaping the intervention of local community with nature and creating a smart environment as much as the practitioners of scientific ecology.

25. Most commonly used e-tools such as e-discussion, e-initiative, e-petition, e-consultation, e-feedback, e-complaints, e-polls, e-voting, e-campaigning, e-budgeting, e-meetings, e-democracy games, e-award and web casting may be used for e-democracy.

11.2 Smart Environment for Smart and Sustainable Hong Kong

Smart Environment implements smart resource management for the environment, in which ecology and biodiversity should play a vital role in providing a stimulating milieu for people to live, work and enjoy in the dense urban environment. Plentiful open spaces provide a place for social interactions and leisure activities as well as facilitate an inclusive and cohesive society that will provide physical, psychological and social health benefits for everyone. A smart environment should have an easily accessible natural environment, promote Green Buildings and sustainable neighbourhoods, implement energy saving techniques, utilize sustainable materials, and manage waste, water and electricity usage efficiently. The Government should also advocate for environmental protection and take up the responsibility in educating the public on the dangers of unsustainable living and introducing policies and regulations to safeguard the natural environment from excessive development or detrimental emissions, such as pollution control and management. The public needs to be more educated to become proactive and smart about safeguarding the environment and the planet and influencing others by their responsible behaviour. As the government takes the lead with strong leadership, other stakeholders including the private sector, NGOs, institutes and community are encouraged to change their social behaviour, hence saving energy. Currently, Hong Kong has still a long way to go especially compared to other developed countries such as Singapore. The provision of open space is lower than other international cities. There is also a lack of awareness in recycling and energy saving amongst Hong Kong residents, which would need further public education and enforcement through regulations.

In recent years, Hong Kong has been pursuing a holistic vision of building a low-carbon and sustainable city. At the highest level, Hong Kong is working towards

Smart Environment through the development of territorial policies including Hong Kong 2030+, Climate Action Plan 2030+, Smart City Blueprint and Harbourfront Protection. As part of the Greater Bay Area (GBA), Hong Kong is collaborating with China cities nearby to work closely to improve the overall environment. Locally, there are also other government policies and environmental initiatives tackling air quality, noise pollution, water quality and resources as well as energy saving. "A Clean Air Plan for Hong Kong" proposed a series of air quality management strategies in place to reduce vehicle emission through stricter vehicle emission standards, port regulations and cleaner energy sources. The government also imposed different controls to reduce noise production, encouraged the use of technology and designs for sound absorption, as well as considered careful city planning of major roads and residential to minimise the impact of noise pollution. At the same time, Hong Kong has made a continuous effort on sewage treatment and clean-up work to enhance the water quality of Victoria Harbour. Hong Kong is successful in water resource management in adaptive re-use of reclaimed water for road cleaning, irrigation and toilet flushing, thereby reducing the need for fresh water. Besides water resource management, Hong Kong also promotes resource management by raising awareness and become more conscious of waste management for both consumers and manufacturers. The government not only advocates for proper recycling and less use of plastic bags with levies and public outreach, but also requiring electronic suppliers to provide free removal service to collect electronic waste for recycling use. In addition, Hong Kong has set up a Food Wise Campaign to avoid and reduce food waste through working with NGOs, educational institutes and other government departments to increase awareness. Lastly, the government led by example when it comes to energy saving through their Zero Carbon Building and T-Park, which demonstrated how to construct low-carbon sustainable buildings and converting waste into energy. These policies and initiatives energise and encourage the community and businesses to work together towards a more sustainable living and working practices.

Apart from policies targeting specific environmental issues, there are Smart Environment projects that provide more all-round strategies to achieve all aspects of Smart Environment. Projects such as Energising Kowloon East, Green Deck, Walk DVRC and Smart Cultural Triangle Precinct (SCTP) are some of the examples where different stakeholders attempt to achieve Smart Environment. Energising Kowloon East is a government initiative to transform Kowloon East into a new business district. The conceptual masterplan aims to create a green, walkable and vibrant environment, which demonstrated strategies with Smart Environment principles. The proposed Kwun Tong Promenade and revitalization of green spaces would provide a recreational space using spaces which were originally cargo area or open spaces underneath flyovers. Moreover, it would contribute to creating a pedestrian-friendly linkage to the waterfront. The transformation of Kowloon East also provides the opportunity to adopt green buildings and thereby improving the sustainability of the area with higher energy efficiency and lower carbon emission. On the other hand, Green Deck is a project proposed by the Hong Kong Polytechnic University (PolyU) to improve the pedestrian connection at Hung Hom Station; a major public transport hub in Hong Kong. PolyU proposed an upper deck above the vehicular way to utilise the vertical space to create a pedestrian space connecting by the existing footbridges. The Green Deck would provide additional space to the already congested pedestrian footbridges, minimise the air and noise pollution coming from vehicles on ground level, as well as creating an urban oasis in the dense area. Although from a different perspective, both Energising Kowloon East and the Green Deck project aims to provide additional open space within Hong Kong's compact environment. The Walk DVRC initiative, however, tries to redefine the use of road spaces and reclaim them as pedestrian and open space. The initiative recognises Hong Kong's built environment should put the city in an advantage to create a pedestrian friendly environment with its highly developed public transport system. Similarly, the proposed SCTP aims to create a walkable environment connecting the heritage areas to create a unique cultural precinct. SCTP also make use of the preservation of the historic buildings as open spaces of the area. The involvement of institutes and community advocate groups is a positive sign for Hong Kong. Although these projects are only in the conceptual phase, they have illustrated that the thinking and mind-set of the Smart Environment are conscious within the community and there are enthusiasts willing to take forward through grassroots initiatives.

Hong Kong still has a long way to go in order to achieve a smart and sustainable environment. Currently the society lacks smart thinking mind-set and awareness in environmental protection. Public education is essential to raise awareness and consciousness about the importance of the environment and encourage the public to be more proactive. Moreover, long term environmental policies are fundamental. The current policies are non-statutory and piecemeal, which lacks an over-arching plan with coordination. Future policies shall be well-organised and coordinated with a forward-looking and sustainable perspective in the long run. In conclusion, Hong Kong has the potential to achieve a smart environment. Making use of existing assets such as the harbour and good transport link with the support from the community and the general public, Hong Kong is expected to transform into a smart city with the smart environment.

11.3 *Automation Based Smart Environment Resources Management in the Smart Building of the Smart City*

With objectives to provide high quality life and excellent living-working environment, smart cities have been observed globally coming up by urban renewal and infrastructure development. Energy requirements of such smart cities have been satisfied by smart grids and smart buildings are the basic building blocks. Therefore, smart monitoring of environment is quite vital and crucial.

The third chapter of this book keeps its major emphasis not only on automatic monitoring of critical environmental parameters but also suggest technological approaches along with guidelines and recommendations along with interesting case study examples. The main focus of this chapter has been to discuss the importance of smart buildings for smart cities and present various inputs that bring that smartness, with building environment monitoring in particular.

11.4 Smart Open Spaces for Smart Chandigarh

Chandigarh, 'The City Beautiful' was planned as the capital city of undivided Punjab. It is located near the foothills of the Shivalik range of the Himalayas in Northwest India and is independent India's first planned. In 1966 Punjab state was reorganized into Punjab, Haryana and Himachal Pradesh. Chandigarh, administratively came under the central government but continued being the capital of Punjab and Haryana with a total area of 114 km^2.

Chandigarh was planned in 1954 for a population of half a million. The population of the city in 2011 was 1.06 million [6]. The Master Plan visualized a population of 1.5 lacks in phase one, later to be expanded to accommodate a total of 5 lakhs in the second phase. An important aspect of the plan of Chandigarh was that it has a periphery area extending 16 km around the city. Planning along the lines of Garden City concept, Le Corbusier envisioned the periphery to be a large green space whose agriculture would support the city and the city in turn would generate employment for the local people. As the food bowl of the city, the periphery would meet its daily requirements as well as that of construction of the city by providing local construction material. The vision of its planner, Le Corbusier to retain the agricultural character of the periphery was strengthened by the enactment of the New Punjab Capital (Periphery) Control Act in 1952 [7].

Combining modernity with ecology, eight km long linear-park, known as the Leisure Valley, runs through the city from its north eastern tip to its south-western end. The Rose Garden, Bougainvillea Garden, Shanti Kunj, Fitness Track, Topiary Park, Terrace Garden, Hibiscus Garden, Garden of Fragrance, Garden of Annuals, Garden of Herbs and Shrubs, Champa Park, and Botanical Garden form a part of the green belt in the city [2].

Corbusier designed an integrated system of seven roads to ensure efficient traffic circulation referred to as the 7 Vs. The city's vertical roads run northeast/southwest known in local language as 'path' and the horizontal roads run northwest/southeast known as 'marg'. They intersect at right angles, forming a grid of network for movement. This arrangement of road-use leads to a hierarchy of movement, which also ensures that the residential areas segregated from the noise and pollution of traffic.

The work place of the city, the industrial area comprises 2.35 Km^2, is located in the extreme southeastern side of the city near the railway line. The prevalent wind direction in the area is north west to the south east in the winter and reverse in the summers. The area has been allocated and planned for non-polluting and light industry and is directly connected to the civic centre by a V-3 road. A wide buffer of fruit trees has been planted to screen off the industrial area from the rest of the city. Tree plantation and landscaping has been an integral part of the city's Plan.

There emerged gaps between what was visualized and what came up as the Chandigarh grew. Violation to The Punjab New Capital (Periphery) Control Act, 1952 started in 1962. Numerous violations to the Act have taken place, starting as early as 1962, when the cantonment was established, followed by the setting up of townships of Mohali and Panchkula in the adjoining states of Punjab and Haryana. These two satellite towns and Chandigarh collectively are known as the Tri-City.

In the last four decades, the towns of Panchkula and Mohali have grown around Chandigarh not only in terms of population but have spread in the area too turning it into a vast and economically vibrant region. These developments by the state government took place in the Chandigarh city's greenbelt, an area under green activities like agriculture and trees. The controlled area gradually became uncontrolled. Punjab and Haryana governments instead of protecting the green belt established satellite towns adjacent to Chandigarh to benefit from the economic opportunities provided by it. The Periphery Control Act which was notified in order to ensure the sustainability of the rural landscape around Chandigarh, due to economic and developmental pressures failed to do so. Increasing urbanization and sprawl impacted the environment of the region.

Even though the green cover within the city limits of Chandigarh has been maintained, the rapid urbanisation of the periphery specially since after 2008 has been impacting the core of the periphery area. There have been increased cases of water logging in the city, which can be attributed to rampant urbanisation in the periphery as there is a loss of recharge zones in the region. 1021 Sq. KMs of periphery area falling within Punjab has been notified as Greater Mohali metropolitan area. Subsequently, the Haryana government added approximately 6 Km^2 of the area to Panchkula Development Plan.

Lack of a coordinated vision for the region clubbed with a multiplicity of governments are primarily seen as the challenges in maintaining the agricultural character of the Periphery. Other than this there needs to be watershed level planning clubbed with afforestation so as to address the rising issues of water logging, not only in Chandigarh but in the Tri-City.

It must be noted that an integrated plan of the periphery area is an essential requirement. Punjab government has prepared a metropolitan plan for great Mohali region which encompasses all of the Periphery area falling under the state of Punjab. Haryana government has similar plans for developing five townships in the Periphery area. All this development in the forthcoming years will lead to Chandigarh becoming a very large and dense metropolitan spread over more than 1600 Km^2, all in violation of the New Punjab Capital (Periphery) Control Act.

The path of urbanisation in the region may not be reversible, however, measures can be taken by the authorities to ensure that the region is planned as a single entity maintaining the basic character and ethos of the Plan initially laid out by Le Corbusier.

To achieve the Green Space vision a four–point strategy has been proposed that is adopting a coordinated approach towards the planning of the Periphery and the Core; conserving and retaining the water and tree cover; adopting modern techniques of greening the region and basing the planning for the region on the principles of the watershed.

The challenges that are arising in the planned city of India can be addressed through smart tools which essentially include geospatial technologies encompassing remote sensing and GIS for the purpose of online monitoring of the green spaces. It is pertinent that not only the extent of the green spaces, both in the city limits of Tri City but also in the Periphery area but also the quality be monitored and upgraded from time to time. Close monitoring of these aspects will help the authorities in improving the green index and per capita ratio.

Some strategies for maintaining the green character of the region could be through adoption of smart laws that enforce 'Green on Buildings' by creating green walls and roofs. The impact of this can be monitored through sensors and geospatial technology models can be generated to monitor the heat island effect. This will help the authorities in taking corrective measures to reduce the heat island effect.

It must be noted that the city and its environs are to be considered as ecological space. The authors emphasize that Smart Green Spaces for Chandigarh and its Periphery has to be based on technology within the framework of sustainability. It is pertinent that through smart green spaces the city builds on its social capital. Planning and implementing for the smart green spaces will reduce the adverse impact of urbanisation and at the same time maximize the intangible benefits and build of the aspects of culture in turn boosting the economy of the area.

11.5   Smart Environment Through Smart Tools and Technologies for Urban Green Spaces

Chapter 5 demonstrates the potential of Geospatial Technologies and ICT technology for improved assessment, monitoring and management of Urban Green Spaces of Chandigarh city.

The study presents that remotely sensed data provides multi resolution satellite data sets at varying spatial and spectral resolutions in different wavelength regions which can be utilized to analyse the extent of UGS, type of UGS, the proximity of built-up to UGS and monitoring of vegetation health and stress, park cool island effect in urban areas. The analysis of the extent of green through RS data and multi-criteria analysis reveals that sectors in the early phase of Chandigarh's development have uniformly distributed UGS while sectors in later phases of development have less amount of UGS. Some of the sectors in the later phase of development has green cover nearly 10% which is due to the presence of urban villages in those sectors. This contrast in the spatial distribution of UGS emphasizes the need to address the lack of planning norms in urban villages which is becoming one of the major challenges in fast growing urban areas. The monitoring of vegetation health gains prominence in urban areas as urban vegetation is continuously under stress due to lack of moisture, restricted growth area and effect of atmospheric pollutants due to vehicular and industrial emissions. The availability of red edge band in remote sensing satellites such as Sentinel 2A, World view 2 and 3, RapidEye etc. and hyperspectral sensors provides unique opportunities for assessing the vegetation stress in urban vegetation as it is sensitive to stress induced changes in chlorophyll. The estimation of chlorophyll content in various parts of Chandigarh through Sentinel-2A data shows lower values of chlorophyll content in avenue tress and urban forest as compared to the neighborhood, central parks and plantations which indicates the stressed condition of vegetation in these areas. The avenue trees are stressed due to increasing traffic and resultant air pollution while urban forest faces a threat due to human interference and cutting of trees. Thermal remote sensing satellite data imaged in have been utilized in this study to assess the park cool island effect as one moves away from the UGS which presents the necessity of uniformly distributed green cover at the distance of 150–160 m for regulating the urban micro climate. The study also demonstrates the use of web based and mobile app tools for monitoring and management of UGS by integrating Geospatial technologies with ICT tools. The tool is designed to compute a functionality index for UGS which can be utilized for grading of UGS, to identify management issues and to encourage authorities or Resident Welfare associations responsible for management to provide an improved environment in UGS. The developed mobile app can be enhanced for the collection of geo-tagged tree inventory data, to identify dead and diseased trees for felling and to obtained geo-tagged information on new plantation. This tools can help citizens to participate in the monitoring and management of UGS. The study provides an insight into the utilization of geospatial technologies as innovative tools in the hand of planners for the study, assessment, monitoring and management of UGS in urban areas. It can be an effective and smart tool in preserving and monitoring green and open spaces in an urban area.

11.6 A Solar Intensive Approach for Smart Environment Planning

In this era of threat towards the non-replenishable source of energy, the need for a renewable source of energy production multiplies manifolds. The time has arrived when there is a grave need to understand consequences of using non-renewable sources of energy leading to climate change and inevitable for countries to relish their energy deposits to meet energy needs [8]. Renewable energy is that source of energy which encourages zero CO_2 emissions and is constantly replaceable in the human timescale [9]. The various sources of renewable energy such as sunlight, wind, rain, tides, geothermal heat and waves can be tapped efficiently using effective modern techniques capable of identifying the resource potential, both spatially and temporally. Techniques like Remote Sensing (RS) and Geographic Information Systems (GIS) can be helpful in mapping the resource zones and justify its exploitation amply in accordance to the site locations [10].

Considering the inescapable energy demand in urban areas, and to meet the other pillars supporting the city's infrastructure, the Smart Cities Mission was launched in June 2015. The 'Smart Cities—Mission Statement and Guidelines' document prepared by the Ministry of Urban Development, Government of India suggested 'assured electricity supply' as one of the core infrastructure elements in a Smart City. The guidelines also discussed 'Smart Solutions', clearly indicating 'Renewable Sources of Energy' as one of the solutions for 'Energy Management' [11].

Solar energy is one of the more promising sustainable energy sources due to its accessibility and its abundance in nature. Solar energy is clean and safe and hence its demand for Sustainable urban development is increasing manifolds. Over recent years, the technology that uses the PV (photovoltaic) effect for the production of electrical power has progressed immensely. Private investors and local authorities in the Government are showing a keen interest in capturing the solar potential. Solar Energy is said to provide 2850 times potential than the current global energy needs. Solar energy is intercepted by the earth's atmosphere at an annual average rate of about 1.3–1.4 kW/m^2 (Rogner 2000; Sorensen 2000). It is estimated that the maximum influx at the earth's surface is about 1 kW/m^2 if we take the earth's global radiation into consideration. At this rate, the ratio of potentially useable solar energy to current primary energy consumption is approximately 9,000–1 (Rogner 2000). Rooftop PV (RTPV) system leaders in the world were Japan, USA and Germany and recent growth is witnessed in countries like Italy, Australia and China [12].

The Ministry of New and Renewable Energy (MNRE) has created a Solar City Programme supporting 60 cities in India which are to be developed as Solar Cities. It is expected that the program may lead to reducing the conventional energy demand by 10%. It was proposed during the 11th Five Year Plan period. So far Solar City Master Plans have been prepared for nearly 48 cities including Agra, Gandhinagar, Rajkot, Surat, Thane, Shirdi, Nagpur, Aurangabad, Imphal, Chandigarh, Gurgaon, Faridabad, Bilaspur, Raipur, Agartala, Guwahati, Jorhat, Mysore, Shimla, Hamirpur, Jodhpur, Vijayawada, Ludhiana, Amritsar, Dehradun, Panaji and New Delhi (NDMC area). Eight cities are to be developed as "Model Solar Cities", the ministry said, adding Nagpur, Chandigarh, Gandhinagar and Mysore have so far been selected for this. [13, 14]. The Jawaharlal Nehru National Solar Mission (JNNSM) was one of the eight missions under the National Action Plan for Climate Change (NAPCC). The first phase of the mission envisaging 1300 MW capacity was targeted to be completed during Phase I of the solar Mission [15].

India is bestowed with abundant solar energy due to its convenient location near the equator. Due to the advantage of location, India receives solar energy equivalent to over 5000 trillion units per year and country has the potential to generate more than 1000 billion units annually from solar energy. This is primarily since India has around 250–300 sunny days in a year along with solar insolation of 4.5–6.5 kWh per sq. m. per day in the most part of the country. If the energy is harnessed effectively, it can help in reducing the energy deficit scenario and meet the electricity demand of the country without any carbon emission. Indian recent estimates on Solar Energy indicates that it has nearly 20–100 GW potential [16]. Many states and cities across the country have aspired to use their solar reserves like Karnataka, Tamil Nadu, West Bengal, Gandhinagar in Gujarat, Rajasthan, Andhra Pradesh, Chhattisgarh and Orissa [16].

There are few studies illustrating the potential of solar energy in the country. The studies indicate the abundance in regions like Gangetic Plains and Plateu, Western part of India with Gujarat plains and hill region as well as West Coast Plains and Thar region receiving annual global insolation of 5 KWh/m^2/day [17].

The solar potential of an area can be estimated using various techniques like:

- Using in situ measurements
- Data from dispersed weather stations measuring solar radiation
- Data from meteorological parameters aiding estimation of radiation
- Data using specially designing statistical and physical models using satellite derived parameters
- Data from satellite imagery using statistical and physical models.

Remote sensing data from various polar and geostationary satellites bearing high spatial resolution is being used in the research industry for estimation of solar insolation using various theoretical, empirical and statistical techniques. Satellites like METEOSAT, GMS-3, and Geostationary Operational Environmental Satellite (GOES) are being used for the same purpose. From 2000 to 2007, daily, direct and diffused radiation was estimated in six locations in India using statistical models using METEOSAT images with a spatial resolution of 5 km incorporating aerosol component prevalent in the atmosphere showing 12% RMSSE error [18].

Many software's were developed to be used for the estimation of solar rooftop potential of the rooftops of the city. Models like SolarFlux in Arc Info GIS module [18, 19], solar radiation algorithm in GIS Genasys [20] and model Solei using MS Windows linked with GIS IDRISI using simple empirical formulas were developed for the same purpose [21]. Later, advancements in models to be used for micro-scale applications were developed in Solar Analyst extension of ArcGIS based on upward-looking hemispherical viewshed model using digital elevation model (DEM) as input [21]. This model is considered suitable for detailed level studies. Also, GIS based model available in an open source platform called GRASS GIS was available in the form of r.sun. This model provided more flexibility than any other model to calculate all the three components of radiation viz. direct, diffuse and reflected with more precision. The model can estimate both clear sky and cloudy sky conditions incorporating certain important parameters like Aerosol Optical Depth, Linke Turbidity, Albedo, etc. in the form of raster maps [22, 23].

With the advancement in the field of using GIS based technology for renewable energy potential; especially solar potential; many web-based applications were made for the same. The most widely used by the European community is the PV-GIS. It is a web based solar radiation database, specially designed for the calculation of PV potential in Europe. This platform allows query for incident irradiation for different inclination angles for various months and time of the day at any location in Europe [24]. European Solar Radiation Atlas was also built by the same team of experts. A web based solar radiation atlas was also prepared as a joint initiative of Latvia, Lithuania, Poland and Estonia. The atlas is comprised of satellite based solar surface irradiance, direct irradiance and direct normalized irradiance maps. Baltic solar atlas is a great regional supplement to EUMETSAT climate atlas and AEMET solar atlas for Spain.

In the same league, understanding the need for solar resource data, NCEI, Department of Energy's National Renewable Energy Laboratory (NREL), the National Aeronautics and Space Administration, the Northeast Regional Climate Center, and several universities and companies collaborated to create the National Solar Radiation Database (NSRDB). It was updated in 2012 containing data from 1991 to 2010 for over 1500 stations in the United States. Similarly, in the Solar Energy Centre (SEC)-NREL collaborative project on "Solar Resource Assessment" under the Indo-US Energy Dialogue, solar maps were generated using satellite imagery-based measurements. The maps are available for the entire country at a 10 km spatial resolution for entire India. Solar maps containing Direct Normal Irradiance (DNI) and Global Horizontal Irradiance (GHI) from January 2002 to December 2008 initially. The maps are later updated extending the data up to 2014 using weather satellite METEOSAT measurements [24].

Promoting the need for rooftop solar power potential, a web-GIS tool was prepared for Chandigarh by The Energy Research Institute (TERI). For India it is first of its kind cloud based open source tool using web based GIS technology [25, 26].

This study is undertaken to assist urban planners to estimate the potential of solar rooftop energy to make cities smart. Climate change and global warming are the recent problem cities are facing due to pollution in the atmosphere and other relate issues. There is a grave need to use renewable and clean energy solar energy resource available in plenty. Solar rooftop energy is one of the best options because it uses the land as a resource neither you have to invest in a land resource to use solar energy, anyone can produce solar energy on their rooftop and if production is more in a city, it can be used for public use by connecting it to the grid. Here, the attempt of utilizing low budget and easily available stereo pair data for the city of concern (in this study, Gandhinagar) so as to assess the rooftop potential of the planned built fabric.

The study focuses on generating high quality Digital Surface Model (DSM) using stereo pair imagery of Gandhinagar and using a physical model approach for preparing a DSM model. The accuracy achieved was remarkable in terms of built up area with an RMSE of only 0.27 pixels using 10 GCPs collected by DGPS survey. The built fabric was extracted using manual digitization of the whole sector (2368 Buildings). Insolation over rooftops was calculated using Solar Analyst extension developed by Fu and Rich, 2000 in Arc GIS. The results clearly justify the energy recovery for the whole sector. Detailed analysis is done for 25 different land use buildings to know recovery of current consumption using solar rooftop potential. It is concluded that using a smaller area of the rooftop, we can easily recover current consumption for Residential building and partially recover for Commercial, Mixed and hospital building use.

In order to respond to the growing urbanization process and subsequent energy demand, methodological approaches which implement alternative urban models are required to support the indispensable change towards more energy efficient cities. Smart energy is one of the important pillars to make city smart and smarter in the days to come.

11.7 Smart Water Management for Smart Kozhikode Metropolitan Area

Smart Water is an important component of smart environment since life, sustenance, growth and death of the environment depends on water. Water Resources in a Metropolis is a key Environmental Resource that links to households, community, land and land use with far reaching impacts. Water in Metropolitan area is utilized for basic needs of human, animal and vegetation and urban non-agricultural activities. The concept of smart water is derived from the concept of smart cities having one to one relationship in their building blocks.

Smart Water Management is a very high responsive intelligent digital system operated by IOTs and ICTs, clouds, related computer-based models along with humans to identify water related issues and even automatically using artificial intelligence solving it in real time without human interventions. Water in the study area is studied in conjunction with the spatial distribution of community in a watershed and issues of households, and present and potential use pattern for households for community wellbeing and economic development is the basis of Water Resources Management in Kozhikode Metropolitan Area (KMA).

Efficient and equitable management of drinking water is one of the greatest challenges the humanity faces currently. Right for fare use of water is directly linked to the ecological context of human existence. Private ownership of water resources and its management had resulted in an adverse environment and livelihood results. However, the public ownership and socialistic management of water resources appear to be ideal but had proven to be inefficient and leads to corruption. This chapter identifies certain local community practices, which are practiced for centuries, and proven to be efficient and equitable. The chapter further looks, how such practices can be implemented to Kozhikode Metropolitan Region, to take care of its diverse issues related to water resource management. Smartness imparted to the region through the development of ICT framework and other components of the smart city may be utilised to enhance the water resource management. The research attempts to achieve the principles of water democracy [17] with the help of the IoT and ICT framework of the smart region.

The region was mapped extensively to take stock of its resources, and geographic peculiarities. The region was divided into several watersheds. Each of these watersheds was analysed to match its water requirements and resources. Various issues faced by the people in these watersheds were enumerated and four major issues namely scarcity, contamination, ground water depletion, and salinity were identified.

Several case studies were done to get state-of-the-art solutions to the problems faced by these communities. Various smart solutions were identified for these watersheds to forecast, detect and monitor the water resource utilisation. An IoT, ICT framework is proposed to integrate various solutions activities required for the management of the water resources. Smart water communities were proposed for the management of the water resources, which is envisaged as an agile body to constantly monitor the resources and its equitable usage. Three nodal institutions in the region were identified to help these communities to evolve solutions for their unique problems and impart training to them.

An SDSS framework is proposed to help these communities to function efficiently. The SDSS automates many of the managing tasks and accomplishes them mostly without human interference. The SDSS was designed by looking at various legal provisions and responsibilities of various governing bodies.

11.8 Visualising Environmental Impact of Smart New Delhi

Recognizing the cultural and developmental diversity of India, national policies and programmes have been consistently generic in nature, enabling every State government, to interpret them according to their respective uniqueness and evolve suitable detailed programmes and action plans for implementation. The Smart City Mission is no exception from this point of view. However, characterised by their inimitable activity patterns justaxposed with streaks of cosmopolitan, regional and local nuances, Indian metropolises exhibit their own culture and images. Like cities anywhere in the world, Indian cities are typically summed results of a number of tangible and intangible dynamic driving forces.

Consequent to 1991 economic liberalization, as India strives to graduate into a prosperous urban society, so did her capital city New Delhi, since its inception in 1931. In the beginning, New Delhi was akin to an alien entity keen to govern the subcontinent but standing at a little distance from Delhi. Presently it is one of the cities within Delhi, adsorbing and responding to the dynamics of local, regional, national and international forces. The pace at which New Delhi evolved was determined not only by the normative evolutionary forces but also drastic but lasting politico-cultural changes within and without South Asia. The fact that some of those changes are tangible and others are intangibly posed a tough and yet fascinating challenge for the concerned agencies and individuals to visualise the character of New Delhi in future with sufficient accuracy and clarity. Even so, visualizing the environmental impact of such changes within New Delhi and beyond is necessary for the interest of sustainability.

Unlike specific industrial or development projects, standard procedures or process for evaluating the environmental impact of development and growth of cities do not exist. For urbanizing and developing nations like India, it is critically important to achieving socio-economic development for an increasingly large proportion of the citizens at a minimum cost to the environment. The unavailability of a standard method of assessing the environmental cost of city development, this paper presents a concept to that end.

In this regard, while the idea of transforming New Delhi to a Smart City through the Smart City Mission is commendable, New Delhi has adapted itself to the changing aspirations of the citizens and the dynamics of national development since its inception. Nevertheless, there is a need to transform New Delhi into a smart city that the rest of the Indian metropolises would willingly look up to.

11.9 Amidst the Governance Challenges in Environmental Management and Sustainable Urbanization in Surat

The urbanization across the globe is increasing and taking the place of some of the other natural resources. Managing environment is seldom addressed as a complete responsibility with integration among various government organizations. Environmental management is a crucial element in any urban settlement. If not managed properly, it can result in several instead of a variety of short-term and long-term adversities. Urbanization is dynamic in nature with mostly, increasing populations generating diverse demands and creating pressure on natural resources available in the surrounding. With an urban spread, the Urban Local Bodies need to empower itself in terms of physical and intellectual capital to cater to the needs for managing and governing the utilization of resources and avoid exploitation.

In the case of Surat, history of which dates back to 300 B.C. [27], as well following the principle of human settlements. It is flourishing on the banks of River Tapi and Mindhola river. Both are perennial rivers although, the level of discharge carrying capacity and flow allows for Tapi to serve the requirements of water demanded. With a population of about 5 million persons settled in about 1.4 million houses, the municipal administration is taking well care to provide all comfort through infrastructure and public services.

It is the eighth largest city and ninth largest urban agglomeration [28] in the country of India. It became the fourth fastest growing city [31] by the year 2016. In the year 2013, it was awarded as The Best City by Annual Survey of India's City-Systems (ASICS), and in the year 2014, the city was awarded as Runners-up for 'Quality of life' and 'Quality of City-Systems' [29]. The city has reported making leading progress over several facets in terms of Smart cities movement initiated by the Government of India. The Ministry of Urban Development declared it to be the third Cleanliest City of India in the year 2010 [30]. The municipal administrative limits are extended to the extent of 326.515 km^2 within the urban agglomeration of 985 km^2 [31]. About two decades back, the city had a spatial spread over merely 8.18 km^2 [32]. The Census of India, 2011 revealed that the city was housing about 44.67 Lakh citizens [36] within the administrative boundary of Surat Municipal Corporation with the population of urban agglomeration of about 60 Lakh residing in an area of about 722 km^2.

The Surat Municipal Corporation (SMC) is very active in taking the initiative where human efforts are minimized and optimize resource utilization. Especially for the solid waste management, since the year 1995, post-outburst of an epidemic of plague following a flood in Tapi river, the city emerged among the top list of cleanest cities in India. The city is prone to monsoon floods when high-tide in the

Sea hinders the flow of city stormwater to the river. Since the year 2001, the SMC has managed municipal finance in a way so that there is no debt on the account and yet capital investments along with regular operation and maintenance of created assets are observed [33].

An analysis was conducted to identify the extent of the rise in the built-up spaces using the satellite images (downloaded from EarthExplorer of USGS) for the years of 2000, 2009 and 2018. The images used were for the months of January for specific years so as there are fewer cloud covers and consistent visibility. In about 18 years, it is found that the built-up spaces have increased to the extent of about 250 km^2.

The climate of the city is actively moderated by the Arabian Sea (to the West) and the Gulf of Cambay. The climate is classified to be tropical savannah having Summer to have a duration from March to June [34]. The monsoon begins with June-July, and average precipitation is about 1200 mm annually [35].

The region has a dense canal network to support the agricultural activities, and the system is dependent on the Tapi river water. The canals supply water from Ukai Dam and Kakrapar Weir located at about 85 and 60 km towards east from the city of Surat. The canal network was constructed in the duration of 1980s and are well maintained by the Surat Irrigation Circle, Government of Gujarat. Over a period, the cropping pattern has changed to more of cash crops and rotation is not stringently followed. The region is well connected by means of having linkages of NH-48 (Mumbai-Delhi), NH-53 (towards Kolkata), railway (Mumbai-Delhi and central India locations), a domestic airport (in the process of conversion to an international status), and ports of Magdalla and Hazira in the West. The proposal of a coastal highway of Gujarat, expressway of Vadodara to Mumbai, Dedicated rail freight corridor, ring roads and other roads are at various stages of different proposals.

The regional topography has a mild slope towards the west (the Arabian Sea), and the city of Surat has spread of about 18 m above the Mean Sea Level. The region is rich with water resources. Many natural drains and tanks are existing with a major perineal river Tapi flowing from the centre of the city.

A study was carried out using the satellite images made available by the Earth-Explorer, USGS. The NDVI values for the year 2000 ranges from −0.004353 to 0.246034 whereas for the year 2018, the value ranges from −0.000231 to 0.178733. The barren lands are increasing in the study area. Although, since the year 2000, thick vegetation or forests are not identified in the study region and the land seems to be covered with sparse vegetation only.

The city is booming with industrial establishments. It is among the prime reasons for the prosperity in the region and considerable migration of people from across the nation in search of opportunities. The city itself has more than 50,000 working units as small to medium scale enterprises.

Based on the discussions in the chapter, the SMC capable of catering to the needs for supplying treated water as per the norms set by the central government. An underground network of sewerage enables for safe and separate collection of sewage for treatment and disposal to nature harmlessly. Stringent adherence for treated effluent parameters is observed so that minimal disturbance to natural streams and ecosystems thereof is observed. The solid waste management by the department of SMC is taking an approach of integration with various activities in the city. It is integrated with operations related to water supply, sewage, health and so on [36]. The city generates about 4000 MT of solid waste every day at an average of about 700 gm per person [37]. The waste is collected by means of the door-to-door garbage collection system, container lifting, hotel-kitchen waste management and night scrapping-brushing activities. It is brought at transfer stations from collection points. The waste is transferred to the treatment site by means of the secondary transfer system. A scientific landfill activity is performed at the disposal site located at Khajod (Southern part of the city) since the year 2006. There is a strict observance of no open burning of solid waste in the city. Biomedical waste is collected by NGO and treated separately by means of shredding and incineration in an autoclave, observing safe disposal [38].

Several organizations are playing a role in managing various natural resources and the urban environment. The chapter discussed the governance paradigm related to these organizations and challenges faced and opportunities by employing smarter alternatives for making a sustainable environment. The loss made to the local and global environment is not measurable while attaining the development status for Surat. The effects observed through the interrelationship of natural elements is still a matter of research and due for complete revelation. However, changes in the ongoing practices and adopting technology that allows for sustainable use of available resources will make operations smarter. Ideas for integrating different organizations bringing on a shared database platform and visualizing a whole effect on the environment is among significant actions anticipated. The organizations need to develop and adopt the GNSS-based applications to have information about various actions in real-time. The use of remote sensing and GIS-based systems need to be incorporated extensively. Surat Municipal Corporation has led the forum of urban bodies in India for many activities and putting for such practices in routine will enable the citizens to prosper over a longer term along with other parastatal and government organizations.

11.10 Local Government and Technological Innovation: Lessons from a Case Study of Yokohama Smart City Project

Currently, most climate policies focus on the international and national level efforts. The international and national levels may set a strategic orientation, but the real effect of the strategies would be made through local actions. Cities and municipalities are the core of actions to cope with global and local environmental problems. In response to these problems, the cities are increasingly taking a strategic approach to climate changes to implementing overarching and systemic changes, by redesigning and reconfiguring the infrastructure networks through which energy is produced and consumed. The municipalities and cities are important determinants of effective climate technology adaptation, it would be valuable exercise to identify their role in local technological innovation.

Primarily, innovative technology, including those associated with smart city and smart-grid, force the local government to reconsider the different levels of technical capacities between them and industries. Smart city, in itself, has emerged in the 1990s, when the focus was on the significance of ICT on modern infrastructures within cities. The initial argument was placed by the Californian Institute for Smart Communities on how cities and communities could be designed to implement information technology, though later the idea of the then smart city was criticized with its strong emphasis on the technical aspect. The smart city then became the core of urban labelling phenomenon, with a variety of indexes and visions attached to.

Also, the complication was exasperated partly because "smart city" agenda arises exactly at a time when, faced with global challenges of climate change, city governments and businesses realize their mutual opportunities for 'green-tech innovation. It gradually became a public-private partnership agenda, than predominantly the IT business opportunity. Yet, this development leads to the question of intellectual capacity and ownership of smart technology. Naturally, smart technologies and its intellectual properties are dominated by the private sector. In contrast, how techno-scientific capacity can be accumulated in local government is yet unknown, as they usually remain disconnected to the updated technological innovation.

On the other hand, crucial for any interactive technology is the field for testing: without the experimental demonstration, the incubated technology may fail to get the validity for its practical implementation, as the debate of "Give me a lab and reduce carbon footprint" suggests. Lack of technical validity and the associated uncertainty on social implication may become an obstacle for the technology's diffusion. As the well-known argument of Valley of Death, in between the stages of research and development, there is a corridor phase for commercial viability. On this understanding, it becomes practical for industries to seek the testing field.

A question arises as to what role the local government should play to promote technological innovation, especially those need to be embedded as a city-wide infrastructure, and what is the efficient governance arrangement between local government and businesses to address shared responsibility for the providing the innovative infrastructure. This question is increasingly relevant, as the pace and size of technological development unprecedentedly fast and vast.

This paper initially posed a question as to what role the local government should play to promote technological innovation, especially those need to be embedded as a city-wide infrastructure, and what is the efficient governance arrangement between local government and businesses to address shared responsibility for the providing the innovative infrastructure. This chapter focuses on the Yokohama City Government and the business stakeholders, earned valuable experience and capability through different kinds of mechanisms. Learning' here will be understood as the various processes by which cities or organizations within a city build up and accumulate their technological and innovation capabilities. This research paid particular attention in knowing how local government learn and build technological capabilities and shared governance structure to technological innovation, and what role the local government should play to promote technological innovation, especially those need to be embedded as a city-wide infrastructure.

Based on the Yokohama experience, we understood that the capacity of the business shall be optimised to bring innovative technology developed and implemented at cities. It turned out that when cities and the partner businesses seek to deepen their innovative capabilities rapidly, they will need to make partnership structure to efficiently share capacity and responsibility. The public administrators, often remain as a regulator to business operation, can play an inviting role to accommodate the enterprises, by facilitating actions that contribute to the technological change. It is similarly essential to identify what is the efficient governance arrangement between local government and businesses to address shared responsibility for providing the innovative infrastructure.

Though the YSCP case, it tuned out the ability of the Yokohama City Government is particularly high to "integrate" the different project, simultaneously "distributing" technical responsibility to the selected participants. First, Yokohama City Government, using its geographical advantage of being close to the Tokyo and its vicinity where the Japanese leading industries to be based in, successfully invited a number of the active participants. These participants have high technical capacity themselves, but it seems Yokohama City Government exercised proactive a role to place well-coordinated governance structure to integrate the potentially conflicting business interests.

Thus, the YSCP well proved the point. As elaborated in this chapter, the initiation of an ambitious energy technology innovation programme at the city-wide scale, could bring interesting opportunity for the industries to test the new technology in the real-world context. High level of managerial skills for coordinating (including delegating) technical expertise is essential, while having essential motivation to understand and disseminate the technology in question at the city domain and beyond.

11.11   Responsive Infrastructure and Service Provision Initiatives Framing Smart Environment Attainment in Nairobi

In conclusion the ongoing initiatives and potentials of planned investments in mobility, water and waste management sectors in Nairobi is a pathway to the attainment of the smart environment. Globalization through innovation and modernization is noted as one of the most significant drivers to these initiatives which are smart as seen by how they respond to the felt needs of various users.

The chapter concludes that even though as Nairobi is making progress in improving the three discussed infrastructure and related services courtesy of the above mentioned smart initiatives, the city governors and stakeholders should take into consideration the following factors;

a. Informality—With the majority (over 50%) of Nairobi city residents living in informal settlements, this cerate's challenge in terms of deriving a reliable database of estimating the demand for such services, and estimating their consequent contribution to the much envisaged smart environment since documentation of such settlements development and expansion is undocumented.

b. Management and governance—Infrastructure projects are very lucrative and they attract a lot of resources. The problem associated with such projects is on how these projects are managed including the inadequate transfer of need relevant technical skills from the foreign expatriates to local people on how such projects should be operated, hence leading to an overreliance on foreign expertise which is expensive and unsustainable. Another related concern is governance in terms of lack of progressive political succession and commitment to several on-going initiatives, the new crop of political leaders may abandon good initiatives just for political expediency, this shows that some of such initiatives may not get sufficient backing and traction for implementation, expansion or management, they might even lack budgetary allocation.

c. Cost and locational Factors—significant portion of Nairobi residents live in informal settlements and a significant portion of them may not be in a position to enjoy some of the benefits associated with the discussed technology be it e-hailing taxi services due to cost involved, smart water management system since most of them do not have portable water, connection to integrated waste management system due to location and access challenges associated with where they live payment paid is still to fully.

d. Initiative drivers and motives- Some of these initiatives are driven by donors whose motives are not well defined, as some of them generally lack clear local context, for example the current debate on the proposed Nairobi Bus rapid Transit (BRT) as part of the proposed solution to Nairobi city congestion. As a result of that the paratransit owners and operators feel that they have not been consulted and they are likely to lose out on investments and employment opportunities. This affects the ownership and subsequent sustainability associated with such initiatives which will be done just to please the donors instead of linking such initiatives with the bigger picture where the smart environment is likely to be one of the positive achievements associated with such infrastructure and services.

e. Technology Infrastructure Expansion and reach- Just like other African cities, smart initiatives also have their limitation associated with technology accessibility. Technology and smart initiatives have been known to create both digital dividend and the digital divide. Those who tend to benefit are those who are able to afford and have requisite skills required to apply and enjoy the benefits of the technology, this means that the urban poor who are part and parcel part of the city may not fully enjoy the fruits of smartness associated with such infrastructure and related services, which may hinder the progress made in attaining of smart environment in Nairobi.

Despite all these concerns, Nairobi's hope of attaining smart environment is premised on the unstoppable pivot role of globalization which is real and which is driven with innovation, which is driven with ICT which is changing how key infrastructure and related services such as mobility, water and waste management will be provided. These will require addressing some of the above-mentioned factors hindering the provision of responsive infrastructure in an innovative manner which will have a positive cumulative effect on the environment which is deemed to be smart.

11.12 Sensing Dubai Smart City for Smart Environment Management

As cities face increasing pressure on the urban environment, tapping IoT technologies is vital for its sustainability and improving the quality of life of its residents and visitors. This chapter highlights smart city version 3.0 where citizen participation and co-creation are the essential ingredients for a happy city.

The success story of Smart Dubai can be attributed to its visionary leadership and the support that it receives from the government. The institutionalization of the smart city with a transformation road map backed by a clear vision to become the "Happiest city on earth" by 2021 is THE single most motivator. Although each city requires a different approach and a different strategy, articulating how city residents will benefit through digitalization is important to solicit their participation.

Tapping IoT should be a means to an end. The focus should be to clearly understand what the final target is and then move towards identifying the steps on how to achieve it. Determine where IoT is required and how it is going to be used has to be well defined and aligned to international standards such as WHO. Smart Dubai has identified "Smart Environment" as one of its six strategic objectives and laid a city-wide IoT strategy that takes care of the IoT ecosystem for the whole city including public and private stakeholders. The Smart Dubai IoT Platform (SDP), set up with a private strategic partner, acts as a central operating system for the city providing residents access to city services through smart phone apps and real-time dashboards.

Dubai has implemented a number of initiatives to reduce carbon foot print and achieve 75% renewable energy target by 2050. Dubai solar park, electric vehicles, autonomous vehicles, green vehicles and paperless transactions contribute to this vision. Residents benefit on reduced utility bills through the installation of smart meters and receive alerts on detection of leakage. The residents are also incentivized to install solar panels to generate for self-use and become an energy supplier.

Dubai applied IoT for the smart environment in its green field city Dubai Silicon Oasis (DSO). Use cases such as green roofs and walls, smart parking, smart street light poles, smart irrigation, smart waste management and smart benches are quite successful. Dubai Municipality's use of a truck for Mobile monitoring and Dubai RTA (Roads and Transport Authority)'s initiatives using IoT for a traffic light, pedestrian crossings and deploying electric and hybrid vehicles for taxis and public transport reduces carbon emissions.

Applying IoT technologies for Urban Environment monitoring is a powerful option to collect the required datasets for diagnosing the root cause of the problem and to find alternative treatments. Smart Dubai experience proves that by applying IoT technologies, carbon emissions can be cut by 10–15%, water consumption can be lowered by 20–30%, and solid waste per capita can be reduced by 10–20%.

References

1. Vinod Kumar TM and Associates (Editor) 2014 "Geographic Information Systems for Smart Cities" Copal Publishing Group, New Delhi
2. Vinod Kumar TM (ed) (2015) E Governance for Smart Cities. Springer, Singapore
3. Vinod Kumar TM (ed) (2016) Smart economy in smart cities international collaborative research: Ottawa, St Louis, Stuttgart, Bologna, Cape Town, Nairobi, Dakar, Lagos, New Delhi, Varanasi, Vijayawada, Kozhikode, Hong Kong. Springer, Singapore
4. Vinod Kumar TM (ed) (2017) E-Democracy for Smart Cities. Springer, Singapore
5. Vinod Kumar TM (ed) (2018) Smart metropolitan regional development: economic and spatial design strategies. Springer, Singapore
6. Government of India (2011) District Census Handbook. Chandigarh, Directorate of Census Operations
7. Chandigarh Administration 2015, Chandigarh Master Plan 2031
8. Warren C, Lumsden C, O'Dowd S, Birnie R (2005) Green on green: public perceptions of wind power in Scotland and Ireland. J Environ Plann Manag p 853–875
9. María del Rosario Iglesias (2013) Application of remote sensing-GIS to renewable energy resource assessment
10. Ramachandra T (2007) Solar energy potential assessment using GIS. Energy Educ Sci Technol pp 101–114
11. G. o. I. Ministry of Urban Development (2015) Smart cities—mission statement and guidelines, New Delhi
12. McHenry M (2012) Are small-scale grid-connected photovoltaic systems a cost-effective policy for lowering electricity. Energy Policy 45:64–72
13. Kandt A (2012) Indian solar cities programme: an overview of major activities and accomplishments, Denver, Colorado
14. MNRE (2015) Status note on solar cities as on 15.09.2015. Ministry of New and Renewable Energy
15. Gambhir A, Dixit S, Toro V, Singh V (2012) Solar rooftop PV in India
16. Gupta J (2011) Interviewee, the untapped gigantic potential of solar rooftop projects in India. [Interview]
17. Ramachandra T, Jain R, Krishnadas G (2011) Hotspots of solar potential in India. Renew Sust Energy Rev pp 3178–3186
18. Dubayah R, Rich P (1995) Topographic solar radiation models for GIS. Int J Geogr Inf Syst vol. 9
19. Hetrick W, Rich P, Barnes F, Weiss S (1993) GIS-based solar radiation flux models. Am Soc Photogr Remote Sensing, GIS, Photogr Model 3:132–143
20. Kumar L, Skidmore A, Knowles A (1997) Modelling topographic variation in solar radiation in a GIS environment. Int J Geogr Inf Sci 11(5):475–497
21. Hofierka J, Suri M (2002) The solar radiation model for open source GIS: implementation and applications. In: Proceedings of the Open source GIS - GRASS users conference, Trento, italy, 2002

22. Fu P, Rich P (2000) The solar analyst 1.0 user manual. Helios Environ Model Inst
23. Hofierka J (1997) Direct solar radiation modelling within an open GIS environment. In: Proceedings of the joint European GI conference, Vienna, 1997
24. Suri M, Huld TA, Dunlop ED (2005) PV-GIS: a web-based solar radiation database for the calculation of PV potential in Europe. Int J Sust Energy 24(2):55–67
25. NREL, NREL: National Renewable Energy Laboratory (2016). [Online] Available http://www.nrel.gov/international/ra_india.html. Accessed 12 Dec 2016
26. Datta A (2015) Development of Web-GIS tool for estimating the rooftop solar power potential for Indian solar cities
27. Corporation SM (2011) History of Surat, Surat Municipal Corporation [Online]. Available https://www.suratmunicipal.gov.in/TheCity/History. Accessed 22 Feb 2017
28. Statistics C (2017) City Mayors: largest Indian cities 2011. [Online] Available http://www.citymayors.com/gratis/indian_cities.html. Accessed 17 Jan 2017
29. World's fastest growing urban areas. City Mayors
30. Gumber A (2014) Annual survey of India's city-systems, 2nd edn. Janaagraha Centre for Citizenship and Democracy, Bangalore
31. The free encyclopedia Wikipedia (2018) List of cleanest cities in India. [Online] Available https://en.wikipedia.org/wiki/List_of_cleanest_cities_in_India. Accessed 24 Dec 2018
32. Plan RCD (2013) Revised city development plan (2008–2013), Surat
33. Surat Municipal Corporation (2016) TP details : Surat Municipal Corporation. Surat Municipal Corporation. [Online] Available https://www.suratmunicipal.gov.in/Departments/TownPlanningTPDetails. Accessed 17 Jan 2017
34. Provisional Population Totals Urban Agglomerations and Cities, Census India 2011
35. Climate-data.org (2017) Climate Surat: temperature, climate graph, climate table for Surat—Climate-Data.org. [Online] Available https://en.climate-data.org/location/959693/. Accessed 24 Feb 2017
36. Bhasker B, Sharma ND (2015) Scope of Modeling For Urban Land-Use Leading To Climate Change. Int J Adv Res Eng Sci Manag pp 1–7
37. Surat Municipal Corporation (2016) Approaches: Surat municipal corporation. [Online] Available https://www.suratmunicipal.gov.in/Departments/SolidWasteManagementApproaches. Accessed 27 Dec 2018
38. Surat Municipal Corporation (2018) Solid waste management statistics : Surat Municipal Corporation. [Online] Available https://www.suratmunicipal.gov.in/Departments/SolidWasteManagementStatistics. Accessed 27 Dec 2018

Printed by Printforce, the Netherlands